NAVAL WARFARE PUBLICATION

NAVAL SPECIAL WARFARE SEAL TACTICS

NWP 3-05.2

DEPARTMENT OF THE NAVY
OFFICE OF THE CHIEF OF NAVAL OPERATIONS

CLASSIFIED BY: Not Applicable	PRIMARY REVIEW AUTHORITY:
WARNING NOTICES: Not Applicable	SECONDARY REVIEW AUTHORITY:

ORIGINAL

DISTRIBUTION

NAVAL SPECIAL WARFARE PERSONNEL RECOVERY OPERATIONS

CNO (N312, N851)
USSOCOM (SOIO, SORR, SOOP, SOAL)
CINCLANTFLT (N02C)
CINCPACFLT (N3DC)
CINCUSNAVEUR (N9)
COMSECONDFLT
COMTHIRDFLT
COMFIFTHFLT
COMSIXTHFLT
COMSEVENTHFLT
NIWA
NAVWARCOL
NATDEFU
COMNAVSPECWARCOM
COMNAVSPECWARGRU ONE
COMNAVSPECWARGRU TWO
COMSPECBOATRON ONE
COMSPECBOATRON TWO
NAVSPECWARDEVGRU
NAVSPECWARCEN
NAVSPECWARCEN DET LITTLE CREEK
NAVSPECWARCEN DER KEYWEST
NAVSPECWARUNIT ONE
NAVSPECWARUNIT TWO
NAVSPECWARUNIT THREE
NAVSPECWARUNIT FOUR
NAVSPECWARUNIT TEN
SEAL TEAM ONE
SEAL TEAM TWO
SEAL TEAM THREE
SEAL TEAM FOUR
SEAL TEAM FIVE
SEAL TEAM EIGHT
SDV TEAM ONE
SDV TEAM TWO
SPECBOATUNIT TWELVE
SPECBOATUNIT TWENTY
SPECBOATUNIT TWENTY-TWO
PC-2
PC-3
PC-4
PC-7
PC-8
PC-9
PC-10
PC-11
PC-12
PC-13
PC-14

SEAL TACTICS

CONTENTS

LIST OF ILLUSTRATIONS

CHAPTER 2 SEAL TACTICS

RECORD OF CHANGES

Change No and Date of Change	Date of Entry	Page Count Verified By (Signature)

GLOSSARY

A

a-frame. An A-shaped bipod used in rope installation to create artificial height.

acclimatization. The adaptation to a new environment, especially to a higher elevation.

ambient light. The existing or encompassing light currently in the surrounding atmosphere.

apex. The top of an A-frame; the crossing point of the A-frame poles.

antiterrorism. Defensive measures used to reduce the vulnerability of individuals and property to terrorist acts, to include limited response and containment by local military forces. Abbreviated as AT. (Joint Publication (JP) 1-02)

area of operations. An operational area defined by the joint force commander for land and naval forces. Areas of operation do not typically encompass the entire operational area of the joint force commander, but should be large enough for component commanders to accomplish their missions and protect their forces. Abbreviated as AO. (JP 1-02)

arroyo. A natural watercourse or water-carved gully in a desert region.

assault. 1. The climax of an attack; closing with the enemy in hand-to-hand fighting. 2 In an amphibious operation, the period of time between the arrival of the major assault forces and the completion of the amphibious task force mission. 3. To make a short, violent, but well-ordered attack against a local objective, such as a gun emplacement, a fort, or a machine gun nest. 4. A phase of an airborne operation beginning with delivery by air of the assault echelon of the force into the objective area and extending through the attack of assault objectives and consolidation of the initial airhead.

assault climber. A trained military mountaineer who acts as a guide, advisor, and scout or is involved in special warfare in mountainous terrain.

assistant patrol leader. The designated second person in charge of a patrol under command of a patrol leader. Abbreviated as APL.

avalanche. A falling mass of snow, ice, or rocks.

B

balance climbing. Climbing without the use of artificial aids, the use of a rope, or other protection.

beach landing site. A geographic location selected for across-the-beach infiltration, exfiltration, or resupply operations. Abbreviated as BLS.

belay. To tie oneself to an anchor point; to provide braking safety to prevent a hazardous fall.

bergschrund. The large crack or crevasse that usually appears at the head of a glacier between the active and inactive ice, snow, or rock.

bight. A bend in a rope that does not cross itself.

brake hand. The hand used to wrap the rope around the body to increase the friction on the rope when belaying or rappelling.

burnoose. A one-piece hooded headgear worn by Arabs and Moors in the hot desert climate.

C

camming. The action of placing an irregularly shaped object, usually a chock, into a crack so that when force is applied it turns to fit more solidly.

camouflage. The use of natural or artificial material on personnel, objects, or tactical positions with the aim of confusing, misleading, or evading the enemy. (JP 1-02)

change step. A method of switching from one foot to the other on the same foothold without an intermediate step.

chimney. A large vertical crack in rock.

chock. A rock or other object that can be wedged or cammed in a crack to provide an anchor when climbing.

clandestine operation. An operation sponsored or conducted by governmental departments or agencies in such a way as to assure secrecy or concealment. A clandestine operation differs from covert operations in that emphasis is placed on concealment of the operation rather than on concealment of identity of the sponsor. In special operations, an activity may be both covert and clandestine and may focus equally on operational considerations and intelligence-related activities. (JP 1-02)

clean. To remove the pitons, chocks, natural protection, and runners from a pitch that has been climbed. Usually done by the last man in a party.

combat search and rescue. A specific task performed by rescue forces to effect the recovery of distressed personnel during wartime or military operations other than war (MOOTW). Doctrine states that the Services and United States Special Operations Command maintain forces responsible for performing combat search and rescue (CSAR) for their respective forces. There may be situations beyond the capabilities of conventional CSAR forces, however, when the specialized skills of Naval Special Warfare (NSW) may be required to recover isolated personnel (e.g., NSW forces conducting strike rescue of downed aviators in hostile a environment). Abbreviated as CSAR. (JP 1-02)

combined. Between two or more forces or agencies of two or more allies. (JP 1-02)

communications electronic operation instruction. Communications instructions containing frequencies, circuit assignments, key mats, and call signs. The instruction is promulgated by the Naval Special Warfare Task Group (NSWTG). Abbreviated as CEOI.

compromise. The known or suspected exposure of clandestine/covert personnel, installations, or other assets or of classified information or material, to an unauthorized person. (JP 1-02)

concealment. The protection from observation or surveillance. (JP 1-02)

cornice. An overhanging lip of snow formed by the action of the wind depositing snow on the lee side of a ridge crest.

counter-drug operations. Those active measures taken to detect, monitor, and counter the production, trafficking, and use of illegal drugs. Special operations forces (SOF) responsibilities include training host nation counter narcotics forces, gathering intelligence, and conducting specific direct action operations.

counter-guerrilla warfare. Operations and activities conducted by armed forces, paramilitary forces, or nonmilitary agencies against guerrillas. (JP 1-02)

counter-narcotics. Those activities taken by a government to defeat production, transport, distribution, and use of narcotics.

counterinsurgency. Those military, para-military, political, economic, psychological, and civic actions taken by a government to defeat insurgency. (JP 1-02)

counterpoise. Creating a state of balance in communications by setting up an opposing force, as done with antennas.

counterterrorism. Offensive measures taken to prevent, deter, and respond to terrorism. Abbreviated as CT. (JP 1-02)

cover. 1. The action by land, air, or sea forces to protect by offense, defense, or threat of either or both. 2. Those measures necessary to give protection to a person, plan, operation, formation, or installation from the enemy intelligence effort and leakage of information. 3. The act of maintaining a continuous receiver watch with transmitter calibrated and available, but not necessarily available for immediate use. 4. Shelter or protection, either natural or artificial (DOD). 5. Photographs or other recorded images that show a particular area of ground. 6. A code meaning, "Keep fighters between force/base and contact designated at distance stated from force/base." (JP 1-02)

covert operations. Operations that are so planned and executed as to conceal the identity of or permit plausible denial by the sponsor. They differ from clandestine operations in that emphasis is placed on concealment of identity of the sponsor rather than on concealment of the operation. (JP 1-02) In special operations, an activity may be both covert and clandestine.

crampon. A metal framework attached to a climbing boot that has points to provide secure footing on hard snow and ice.

crevasse. A split in the surface of a glacier caused by an irregularity in the surface over which the glacier is moving.

D

deadman. An anchor that is buried in the ground or snow. Some are constructed so that they are forced deeper as weight is applied to them.

deception. Those measures designed to mislead the enemy by manipulation, distortion, or falsification of evidence to induce him to react in a manner prejudicial to his interests. (JP 1-02)

deconflict. To prevent a hostile encounter between forces, usually meant between friendly forces in the same AO.

dehydration. The process of losing an abnormal amount of water or body fluids because of sweating or low fluid intake.

demolition. Destruction of structure, facilities, or material by use of fire, explosives, or other means.

desert. An arid barren tract incapable of supporting any considerable population without an artificial water supply.

dipole. A radio antenna consisting of two horizontal rods in line with each other with ends slightly separated.

direct action mission. Short-duration strikes and other small-scale offensive actions by SOF to seize, destroy, capture, recover, or inflict damage on designated personnel, facilities or material. In the conduct of these operations, SOF may employ raid, ambush, or direct assault tactics; emplace mines and other munitions; conduct standoff attacks by fire from air, ground, or maritime platforms; provide terminal guidance for precision-guided munitions; and conduct independent sabotage. Abbreviated as DA. (JP 1-02)

E

element. The Sea Air Land (SEAL) platoon is the largest operational element that will normally be employed to conduct a tactical mission. A platoon consists of sixteen SEALs and may be divided into two squads or four elements referred to as fire teams. For clarification purposes, the term "element" may be used to refer to the entire platoon or a single person such as a point element.

escarpment. A steep slope, usually a line of cliffs, extending for a long distance.

essential elements of information. The critical items of information regarding the enemy and the environment needed by the commander by a particular time to relate with other available information and intelligence in order to assist in reaching a logical decision. Abbreviated as EEI.

evasion and escape. The procedures and operations whereby military personnel and other selected individuals are enabled to emerge from an enemy-held or hostile area to areas under friendly control. Abbreviated as E&E.

evasion and recovery. The full spectrum of coordinated actions carried out by evaders, recovery forces, and operational recovery planners to effect the successful return of personnel isolated in hostile territory to friendly control. Abbreviated as E&R. (JP 1-02)

exfiltrate. The movement of personnel or units out of, or away from areas under enemy control.

extraction. The removal of personnel, patrols, or platoons from areas under enemy control by some type of platform.

F

fastrope. A means of inserting forces from a helicopter by sliding "firepole style" down a line. Usually only the hands are used for braking, but with heavy equipment, feet can be used as well.

firn. Old snow that has survived for more than one full year. The firn line on a glacier is the dividing line between the accumulation and ablation zones.

fixed rope. A rope fixed to primary and intermediate anchors to aid untrained or heavily laden men over difficult terrain. Usually emplaced by assault climbing teams.

flashflood. A sudden rush of water down a desert arroyo, often caused by distant heavy rains that collected and flowed.

foreign internal defense. Foreign internal defense (FID) is generally a joint and interagency activity. The primary FID mission for NSW forces is to train, advise, and otherwise assist friendly government military and paramilitary forces to protect their societies from subversion, lawlessness, and insurgency in support of theater and U.S. national objectives. FID activities normally are of long duration and require patient adherence to support of national policy. Abbreviated as FID.

forward operational base. In SO, a base usually located in friendly territory or afloat, established to extend command and control or communications, or to provide support for training and tactical operations. Facilities may be established for temporary or longer duration operations. They may include an airfield or an unimproved airstrip, anchorage, or a pier. The FOB may be the location of the SO component headquarters or smaller unit (an NSWTG or Naval Special Warfare Task Unit (NSWTU) in the case or NSW operations) which is controlled and/or supported by a main operational base. Abbreviated as FOB.

front pointing. A technique used in ice climbing in which the climber supports his weight on the horizontal front points of his crampons.

G

glacier. A perennial mass of ice and snow that slides or flows downhill under the pull of gravity.

glissade. Sliding down a snow or ice slope in a standing, squatting, or sitting position.

grenadier. The primary individual in a patrol who carries an M79 or M203 40mm grenade launcher attached to his rifle.

guerrilla force. A group of irregular, predominantly indigenous personnel, organized along military lines to conduct military and paramilitary operations in enemy-held, hostile, or denied territory. (The overt element of the resistance force.)

guerrilla warfare. Military and paramilitary combat operations conducted in enemy-held or hostile territory by irregular, predominantly indigenous forces. Guerrilla warfare may also be conducted in politically denied areas.

guide hand. The hand opposite the brake hand when belaying or rappelling,.

H

half hitch. A loop that runs around an object in such a manner as to lock on itself.

heat cramps. Painful and dangerous body cramps created by loss of fluids and sodium chloride.

heat exhaustion. A condition of weakness, nausea, dizziness, and profuse sweating, resulting from exertion in a hot environment.

herringbone. Walking straight up a slope with the feet placed to each side as in a "duck-walk", for greater traction.

high mobility multipurpose wheeled vehicle. Abbreviated as HMMWV, generally pronounced "HUM-V."

hummock. A small knoll.

hygiene. Conditions, or practice of cleanliness, conducive to good health.

hypothermia. The general lowering of the core body temperature due to heat loss and insufficient heat production.

I

ice axe. A tool with a head on one end and a spike on the other, which is used for cutting steps on snow or ice, as a point of balance on steep terrain, for probing crevasses, for belaying on snow, and as a general aid to climbing and safety on snow and ice.

ice fall. A jumbled mass of crevassed ice where a glacier slides over a steep irregularity on the sliding surface.

ice piton. A tubular piton hammered into the ice as a protection anchor. Usually used in pairs at right angles because of their directional nature.

ice screw. A piton with threads, which provides a secure anchor in ice.

indigenous. Born, produced, or growing naturally in an region or country. Indigenous personnel would be from the local area.

infiltration. The movement through or into an area or territory occupied by either friendly or enemy troops or organizations. The movement is made, either by small groups or by individuals, at extended or irregular intervals. When used in connection with the enemy, it infers that contact is avoided. (JP 1-02) In SO, infiltration is presumed to be the undetected movement of forces through or into a target area.

in extremis. A situation of such political or military significance and exceptional urgency, as determined by the National Command Authorities and/or a theater commander, that immediate action must be taken principally to avoid or minimize loss of life or degradation of political or military opportunities to respond.

infrared. Wavelengths of light just beyond the red end of the visible spectrum. Abbreviated as IR.

initial rally point. A well-defined point, easily distinguishable visually and/or electronically, used immediately after insertion as a starting point for the run to the target. Abbreviated as IRP.

insertion. The placing of personnel or a unit into an enemy's AO by pre-planned means, normally from a mobile platform of some type.

insurgency. An organized movement aimed at the overthrow of a constituted government through use of subversion and armed conflict.

intelligence. 1. A product resulting from the collection, processing, integration, analysis, evaluation, and interpretation of available information concerning foreign countries or areas. 2. Information and knowledge about an adversary obtained through observation, investigation, analysis, or understanding. (JP 1-02)

interoperability. The ability of forces to provide and accept services to/from other forces and operate effectively together.

J

joint. Alludes to activities, operations, organizations, etc., in which elements of two or more Military Departments participate.

K

katadyn pump. A small, lightweight, hand-pump filtering system capable of removing bacteria and impurities from water.

kermantle. Literally "core and sheath." A rope constructed of an outer woven sheath and an inner core of continuous filaments.

L

laid rope. A rope constructed of three bundles of continuous filaments twisted together.

landing zone. Any specified zone used for the landing of aircraft.

laser target designator. A device that emits a beam of laser energy used to mark a specific place or target. The ruby laser is visible to the human eye as red light. The neodymium: yttrium aluminum garnet (ND: YAG) laser is not visible. The ND:YAG laser is the most common military laser and is used by all the services for range finders and target designators. Abbreviated as LTD.

lay. The twist, usually a right-hand twist, in a laid rope.

leeward. On the lee side, opposite of the side from which the wind is blowing.

lip balm. Petroleum based ointment applied to lips and other areas to help prevent the skin from drying out by exposure.

loop. A bend in the rope in which the top crosses itself.

M

malleable iron. A soft, easily deformed iron used for military pitons with a relatively short useful life.

meal-ready-to-eat. Packaged military field rations that can be eaten hot or cold. Abbreviated as MRE.

military crest of a hill. An area on the forward slope of a hill or ridge from which maximum observation covering the slope down to the base of the hill or ridge can be obtained.

mirage. An optical effect in the desert, which may appear as a pool of water, caused by bending or reflection of light rays in heat.

moat. The large crack at the side of a glacier where the ice has separated from the rock.

mobile communications team. Mobile Communication Teams (MCT) are elements of Naval Special Warfare Groups (NSWG), organized, trained, and equipped to provide command, control, and communications needs to deployed NSW forces.

moraine. An accumulation of rock debris carried on and deposited by a glacier. Lateral moraines are located on either side of a glacier. Medial moraines are located at or near the middle of the glacier and are the result of lateral moraines from tributary glaciers that have merged with the main glacier. Terminal moraines are located at or near the termination of the glacier.

N

naval special warfare. A specific term describing a designated naval warfare specialty and covering operations generally accepted as being unconventional in nature and, in many cases, covert or clandestine in character. These operations include using specially trained forces assigned to conduct unconventional warfare, psychological operations, beach and coastal reconnaissance, operational deception operations, counterinsurgency operations, coastal and river interdiction and certain special tactical intelligence collection operations that are in addition to those intelligence functions normally required for planning and conducting special operations in a hostile environment. Abbreviated as NSW. (JCS Pub 1-02).

naval special warfare task element. A Naval Special Warfare Task Element (NSWTE) is a small NSW planning and operational unit, ashore or afloat, that conducts NSW operations to include SEAL, SDV, and SBU operations. Elements are assigned to a specific NSWTU ashore or afloat. The SEAL, SDV, or SBU Detachment Officer (03) or Chief Petty Officer (E7) in Charge can, and often does, perform the function of the Task Element leader. Abbreviated as NSWTE.

naval special warfare task force. A Naval Special Warfare Task Force (NSWTF) is the U.S. Special Operations Command Navy component assigned to the theater Special Operations Command's (SOC) Joint Special Operations Task Force (JSOTF). An NSWTF is tailored and task organized from Commander, Naval Special Warfare Command (CNSWC) resources. It serves as a regional battle staff ashore. Its mission is to provide command, control, and administrative services; coordinate mission and tasking assignments; secure and maintain logistics support; and provide secure communications, basing support, and other services. It will have one or more subordinate NSWTGs. Abbreviated as NSWTF.

naval special warfare task group/task unit. A provisional naval special warfare organization that plans, conducts, and supports special operations in support of fleet commanders and joint force special operations component commanders. Abbreviated as NSWTG/TU.

nuts. A common usage name for chocks.

O

oasis. A fertile place in the desert, due to the presence of water.

objective hazard. Hazards over which the climber has no direct control, such as rockfalls, avalanches, and corniced ridges.

objective rally point. The rally point designated by a patrol that is near the objective/target of the mission. Abbreviated as ORP.

open book. A crack with offset sides that form an angle usually greater than 70 degrees and less than 110 degrees.

operations security. A process of identifying critical information and subsequently analyzing friendly actions attendant to military operations and other activities to: 1. Identify those actions that can be observed by adversary intelligence systems. 2. Determine indicators hostile intelligence systems might obtain that could be interpreted or pieced together to derive critical information in time to be useful to adversaries. 3. Select and execute measures that eliminate or reduce to an acceptable level the vulnerabilities of friendly actions to adversary exploitation. Abbreviated as OPSEC. (JP 1-02)

overt operation. Operations planned and executed without attempting to conceal the operation or identity of the sponsor. (JP 1-02)

P

paramilitary forces. Forces or groups distinct from the regular armed forces of any country, but resembling them in organization, equipment, training, or mission. (JP 1-02)

party climbing. Trained climbers tied together to a climbing rope to increase their mutual safety and ability on difficult rock.

patrol leader. The individual designated in charge of a SEAL patrol organized to conduct a mission. Within the context of this manual, Patrol Leader, Platoon Commander (PC), and Officer in Charge (OIC) are interchangeable terms. This responsibility can be executed by both officer and enlisted personnel. Abbreviated as PL.

patrol leaders order. A mission brief encompassing situation, mission, administration and logistics, command and control, and communications and signals. Abbreviated as PLO.

pitch. A section of climbing between two belay points. The length of a pitch depends on the length of the climbing rope and on the proximity of suitable belay points.

piton. A small spike or tube, with an eye or ring, that is inserted into cracks in rock or into ice. It provides a snaplink attachment for the climbing rope to aid in protection and progression.

point man. The individual (or element) in a patrol that precedes all others while patrolling and searching for dangers. He is the eyes and ears of the front of a patrol. Abbreviated as PT.

propagation. Radio waves travelling through space spreading out over a greater area.

psychological operations. Planned operations to convey selected information and indicators to foreign audiences to influence their emotions, motives, objective reasoning, and ultimately the behavior of foreign governments, organizations, groups, and individuals. The purpose of psychological operations is to induce or reinforce foreign attitudes and behavior favorable to the originator's objectives. Abbreviated as PSYOPS. (JP 1-02)

pulk. Flat bottom sled pulled by one or two skiers and used for transporting equipment over snow-covered terrain.

R

rack. A selection of pitons, chocks, and runners carried by a climber for emplacement as protection while climbing.

radioman. The individual designated within a patrol to carry the radio and communicate as directed by the patrol leader. Abbreviated as RTO.

raid. An operation, usually small scale, involving a swift penetration of hostile territory to secure information, confuse the enemy, or to destroy his installations. It ends with a planned withdrawal upon completing the assigned mission. (JP 1-02)

rally point. Any position along a route of infiltration or exfiltration where the patrol can rendezvous. Abbreviated as RP.

rappel. Descent by means of a rope from a higher position to a lower position, such as from a helicopter or down a cliff.

rear security. The last individual or element in a patrol formation with responsibility for security of the patrol from behind. Abbreviated as RS.

reconnaissance. A mission undertaken to obtain, by visual observation or other detection methods, information about the activities and resources of an enemy or potential enemy; or to secure data concerning the meteorological, hydrographic, or geographic characteristics of a particular area. (JP 1-02)

rock nubbins. A projected rock feature, suitable for an anchor.

round turn. A single, complete wrap of a rope around an object, without crossing itself, providing 360 degrees contact.

rules of engagement. Directives issued by competent military authority which delineate the circumstances and limitations under which United States forces will initiate and/or continue combat engagement with other forces encountered. Abbreviated as ROE. (JP 1-02)

runner. A closed loop of rope or webbing used to lengthen the attachment of an anchor to the climbing rope or sling.

running end. The loose or working end of the rope.

S

sabotage. An act or acts with intent to injure, interfere with, or obstruct the national defense of a country by willfully injuring or destroying, or attempting to injure or destroy, any national defense or war material, premises, or utilities, to include human and natural resources. (JP 1-02)

salinity. The term denoting the degree of metallic salt dissolved in water such as in oceans or seas.

scree. Small unconsolidated rocks or gravel, fist size or smaller, located mostly below rock ridges and cliffs.

seal delivery vehicle team. SDV Teams are organized, trained, and equipped to operate and maintain the NSW inventory of combatant submersibles (SDVs and Advanced Seal Delivery Systems) and submarine deep dive systems called Dry Deck Shelters.

seal team. SEAL Teams are maritime SOF organized, trained, and equipped to plan, conduct, and support a variety of Navy and joint special operations, in all operational environments and levels of conflict. A SEAL Team has eight 16-man platoons. A SEAL platoon consists of two officers and 14 enlisted men. The platoon is subdivided into two 8-man squads, but can be further subdivided into 4-man elements called fire teams or 2-man operation pairs. The size and composition of the SEAL detachment can vary to meet mission requirements.

self-arrest. A method of stopping one's fall on snow or ice by using an ice axe.

slab. A relatively smooth portion of rock lying at an angle.

sling rope. A length of rope, ideally four meters in length, which is used to make runners, rappel seats, safety ropes, and other aids.

snaplink. A steel, aluminum, or alloy link, with a spring-loaded gate on one side, used to attach the climbing rope to pitons or slings. Also known as a carabiner.

snowbridge. A bridge of snow over a crevasse that is usually formed by wind action.

snow plume. The cloud of loose snow caused by high winds, usually on ridges.

special boat unit. Those US Navy forces organized, trained, and equipped to conduct or support naval special warfare, riverine warfare, coastal patrol and interdiction, and joint special operations with patrol boats or other combatant craft designed primarily for special operations support. Also called SBU.

special operations command, research, analysis, and threat evaluation system. Provides near real-time intelligence and imagery products to USSOCOM and its components. Abbreviated as SOCRATES.

special operations. Operations conducted by specially organized, trained, and equipped military and paramilitary forces to achieve military, political, economic, or psychological objectives by unconventional military means in hostile, denied, or politically sensitive areas. These operations are conducted during peacetime competition, conflict, and war, independently or in coordination with operations of conventional, non-special operations forces. Political-military considerations frequently shape special operations, requiring clandestine, covert, or low visibility techniques and oversight at the national level. Special operations differ from conventional operations in degree of physical and political risk, operational techniques, mode of employment, independence from friendly support, and dependence on detailed operational intelligence and indigenous assets. Abbreviated as SO or SPECOPS. (JP 1-02)

special operations forces. Those active and reserve component forces of the military services designated by the Secretary of Defense and specifically organized, trained, and equipped to conduct and support special operations. Abbreviated as SOF. (JP 1-02)

special reconnaissance operations. NSW Special Reconnaissance (SR) is a human intelligence function that places U.S.-controlled "eyes on target" in a hostile, denied, or politically sensitive maritime environment for reconnaissance and surveillance activities. On SR missions, NSW forces complement national and theater intelligence collection assets and systems by obtaining specific, well defined, and time-sensitive information of strategic or tactical value. The information collected can concern the capabilities, intentions, and activities of an actual or potential enemy, or it can be data concerning the meteorological, hydrographic, or geographic characteristics of a denied area. NSW SR may also be used to complement other collection methods that might be constrained by weather, terrain masking, or hostile countermeasures. NSW forces may conduct these missions unilaterally or in support of naval or other conventional operations. Abbreviate as SR.

spindrift. Small amounts of loose snow usually encountered during storms on steep faces in the form of spindrift avalanches. These are normally more annoying than dangerous.

spread eagle. A position in climbing when the climber's arms and/or legs are overextended. It is a difficult and usually dangerous position from which to recover.

stacking. Placing pitons or chocks in combination so that they will then fit an odd shaped crack.

standing part. The stationary or non-working end of the rope.

st. elmo's fire. The slight crackling and blue light on metal objects during lightning storms.

stirrup. A type of ladder, usually made out of a sling rope or webbing, used in direct aid climbing.

subject hazard. A hazard over which the climber can exercise direct control, such as selection of proper equipment.

subversion. Action designed to undermine the military, economic, psychological, political strength, or morale of a regime. (JP 1-02)

sun block. Oils or ointments which, when applied to the body, prevent penetration of ultraviolet rays and sunburns.

sunburn. Inflammation of the skin caused by overexposure to sunlight.

sun screen. A substance used in suntan preparations to protect the skin from excessive ultraviolet radiation.

sunstroke. A heatstroke caused by direct exposure to the sun, especially in very hot, arid climates.

T

talus. Accumulated rock debris that is fist size or larger, often much larger, as in boulder fields.

target analysis. An examination of potential targets to determine military importance, priority of attack, and weapons required to obtain a desired level of damage or casualties. Abbreviated as TA. (JP 1-02)

target information package. A compilation of essential information about a specific target. The package includes enemy order of battle (air, land, and sea), target description, vulnerabilities, lines of communication, general and localized photography, and other all-source intelligence information. Abbreviated as TIP.

terrorism. The calculated use of violence or threat of violence to instill fear; intended to coerce or to intimidate governments or societies in the pursuit of goals that are generally political, religious, or ideological. (JP 1-02)

thermal imaging sensor. An optical sensor that converts thermal energy (heat) to light. Abbreviated as TIS.

top rope. Any time a climber is belayed from above. This is a good procedure for the beginning climber.

topographic. Representing vertical and horizontal positions of terrain features in a measurable form.

topographic crest. Highest point of a hill, ridge, or mountain.

trafficability. Refers to the extent to which the terrain will permit continued movement of any type of traffic.

traverse. Ascending or descending diagonally, or moving horizontally instead of straight up or down.

tussock. A tuft or clump of grass.

U

unconventional warfare. A broad spectrum of military and paramilitary operations conducted in enemy-held, enemy-controlled, or politically sensitive territory. Unconventional warfare includes, but is not limited to, the interrelated fields of guerrilla warfare, evasion and escape, subversion, sabotage, and other operations of a low visibility, covert, or clandestine nature. These interrelated aspects of unconventional warfare may be prosecuted singly or collectively by predominantly indigenous personnel, usually supported and directed in varying degrees by an external source(s) during conditions of war or peace. Abbreviated as UW. (JP 1-02)

V

verglas. A thin coating of ice that forms on exposed objects by the freezing of a film of super-cooled moisture. Verglas is extremely hard to climb on because it is usually too thin to accommodate crampon points.

W

wadi. A shallow, usually sharply defined depression in a desert region caused by flooding during rainy seasons.

white-out. A condition in which either driving snow or an unbroken cloud cover on a snowfield makes it impossible to distinguish depth or horizon.

wind-chill. The combined effect of ambient temperature and wind. It greatly increases the effect of cold, and adds to the chances of frostbite or hypothermia.

wind slab. A layer of snow formed by wind-transported snow on the lee side of ridges. It is very dense and susceptible to dangerous avalanche conditions.

windward. The side from which the wind is blowing.

wrap. Placing a rope around an object.

LIST OF ACRONYMS AND ABBREVIATIONS

A

ABCCC. Airborne Command, Control, and Communications.

AFMIC. Armed Forces Medical Intelligence Center.

AMS. Acute Mountain Sickness.

AO. Area of Operations.

AOR. Area of Responsibility.

APL. Assistant Patrol Leader.

AT. Antiterrorism.

AW. Automatic Weapons Man.

B

BDU. Battle Dress Uniform.

BENT. Before Evening Nautical Twilight.

BLS. Beach Landing Site.

BMNT. Beginning of Morning Nautical Twilight.

C

C^2. Command and Control.

CARVER. Criticality, Accessibility, Recuperability, Vulnerability, Effect on Populace, and Recognizability.

CAS. Close Air Support.

CASHWORTH. **C**onserve energy, **A**lways test holds, **S**tand upright on flexed joints, **H**ands kept low, handholds should be waist to shoulder high, **W**atch your feet, **O**n three points of contact, **R**hythmic movement, **T**hink ahead, and **H**eels kept lower than toes and inboard.

CEOI. Communications Electronic Operation Instruction.

COLD. Keep clothing **C**lean, avoid **O**verheating, wear clothing loose and **L**ayered, and keep clothing **D**ry.

COMSEC. Communications Security.

CRRC. Combat Rubber Raiding Craft.

CSAR. Combat Search and Rescue.

CSST. Combat Service Support Team.

CT. Counterterrorism.

CW. Continuous Wave.

D

DA. Direct Action.

DF. Direction Finding.

DOD. Department of Defense.

DZ. Drop Zone.

E

E&E. Escape and Evasion.

E&R. Evasion and Recovery.

EEI. Essential Elements of Information.

ELINT. Electronic Intelligence

EOD. Explosive Ordnance Disposal.

EP. Extraction Point.

EPW. Enemy Prisoners of War.

EW. Electronic Warfare.

F

FAC. Forward Air Controller.

FID. Foreign Internal Defense.

FOB. Forward Operating Base.

FOUO. For Official Use Only.

FRP. Final Reference Point.

FT. Fire Team.

G

G or GN. Grenadier.

GPS. Global Positioning System.

GOTWA. Going – where you are going and what you are going to do.

Others – who you are taking with you.

Time – maximum amount of time you will be gone.

What if – what the person in charge will do, if you do not return on time.

Actions – actions to be taken by each element upon enemy contact.

H

HEDP. High Explosive Dual Purpose.

HF. High Frequency.

HM. Hospital Corpsman.

HN. Host Nation.

HUMINT. Human Intelligence.

I

IAD. Immediate Action Drill.

IR. Infrared.

IRP. Initial Rally Point.

L

LBE. Load Bearing Equipment.

LP. Listening Post.

LRPR. Long Range Patrol Rations.

LTD. Laser Target Designator.

LUP. Lay-up Point.

LZ. Landing Zone.

M

MASINT. Measure and Signature Intelligence.

MCT. Mobile Communications Team.

MEDEVAC. Medical Evacuation.

MEDTRAP. Medical Threat Risk Assessment Projection.

METT-T. Mission, Enemy, Terrain, Troops, Time Available.

MRE. Meal-Ready-To-Eat.

N

NAVSOF. Naval Special Operations Forces.

NBC. Nuclear, Biological, and Chemical.

NGFS. Naval Gunfire Support.

NOFORN. Not releasable to Foreign Nationals.

NSW. Naval Special Warfare.

NSWG. Naval Special Warfare Group.

NSWMPG. Naval Special Warfare Mission Planning Guide.

NSWTE. Naval Special Warfare Task Element.

NSWTF. Naval Special Warfare Task Force.

NSWTG. Naval Special Warfare Task Group.

NSWTU. Naval Special Warfare Task Unit.

NVEO. Night Vision Electro Optics.

NWP. Naval Warfare Publication.

O

OCOKA. **O**bservation and Fields of Fire, **C**over and Concealment, **O**bstacles, **K**ey Terrain, and **A**venues of Approach.

OD. Olive Drab.

OIC. Officer-in-Charge.

OP. Observation Point.

OPDEC. Operational Deception.

OPSEC. Operational Security.

ORP. Objective Rally Point.

OTC. Officer in Tactical Command.

OTB. Over-the-beach.

P

PC. Platoon Commander.

PFD. Personnel Flotation Device.

PL. Patrol Leader.

PLO. Patrol Leader's Order.

PLS. Personnel Locator System.

POW. Prisoner of War.

PT. Point Man.

R

R. Rifleman.

RCW. Ration, Cold Weather.

RIB. Rigid Hull Inflatable Boat.

ROE. Rules of Engagement.

RP. Rally Point.

RS. Rear Security.

RTO. Radio Telephone Operator.

S

SBU. Special Boat Unit.

SDV. SEAL Delivery Vehicle.

SEAL. Sea Air Land.

SLLS. Stop, Look, Listen, and Smell.

SO. Special Operations.

SOC. Special Operations Command.

SOCRATES. Special Operations Command Research Analysis and Threat Evaluation System.

SOF. Special Operations Forces.

SOP. Standard Operating Procedure.

SPECOPS. Special Operations.

SR. Special Reconnaissance.

SURVIVAL. **S**ize up the situation, **U**ndue hast makes waste, **R**emember where you are, **V**anquish fear and panic, **I**mprovise/Improve, **V**alue living, **A**ct like the natives, and **L**earn basic skills.

T

TA. Target Analysis.

TACMEMO. Tactical Memorandum.

TIP. Target Information Package.

TOT. Time on Target.

TTP. Tactics, Techniques, and Procedures.

TU. Task Unit.

U

UHF. Ultra High Frequency.

USSOCOM. United States Special Operations Command.

UW. Unconventional Warfare.

V

VHF. Very High Frequency.

W

WP. White Phosphorous.

X

XRP. Extraction Rally Point.

CHAPTER 1
Introduction

1.1 INTRODUCTION

1.1.1 Purpose and Scope. Naval Special Warfare (NSW) Sea Air Land (SEAL) teams are a maritime, multi-purpose, combat force designated by the Secretary of Defense to be task-organized, trained, and equipped to plan, conduct, and support a variety of special operations (SO) in all environments and levels of conflict. A clear understanding of NSW capabilities and limitations is essential to the planning, coordination, and execution of assigned missions. NSW forces are capable of operating independently and with limited support in hostile environments. However, the success of NSW operations depends upon many factors such as critical intelligence information, detailed mission planning, rehearsals, and strategic mobility. This Naval Warfare Publication (NWP) provides information for a better understanding of tactics, techniques, and procedures (TTP) of basic NSW operations.

1.1.2 Doctrine and Principles of War. Doctrine is a statement of the fundamental principles that guide the employment of military forces in support of national objectives. It is authoritative (rather than directive) in nature and requires judgment in its application. SEAL tactical doctrine is in consonance with the principles of war. These principles are discussed below. Their consideration will usually be involved in the selection and planning of SEAL TTP for specific mission execution.

1.1.2.1 Objective. Every military operation must be directed toward a clearly defined, decisive, and attainable objective. The military objective of a nation at war must be to apply whatever degree of force is necessary to attain the political purpose for which the war is being fought. When the political purpose is the total defeat of the adversary, the strategic military objective will most likely be the defeat of enemy armed forces and the destruction of their will to resist. Strategic, operational, and tactical objectives cannot be clearly identified and developed however, until the political purpose has been determined and defined by the National Command Authority (NCA). SO objectives may often be as political, economic, or psychological as they are military. SO objectives in war focus predominantly on enemy military vulnerabilities without direct force-on-force confrontation.

1.1.2.2 Offensive. Seize, retain, and exploit the initiative. The principle of the offensive suggests that offensive action, or maintenance of the initiative, is the most effective and decisive way to pursue and to attain a clearly defined, common goal. This is fundamentally true in the strategic, operational, and tactical senses. Offensive action, whatever form it takes, is the means by which the nation or a military force captures and holds the initiative, maintains freedom of action and achieves results. The side that retains the initiative through offensive action forces the foe to react rather than act. SO are inherently offensive.

1.1.2.3 Mass. Concentrate combat power at the decisive place and time. In the strategic context, this principle suggests that the nation should commit, or be prepared to commit, a predominance of national power to those regions or areas where the threat to vital security interests is greatest. Since every possible contingency or trouble spot cannot be anticipated, or planned for, it is absolutely essential for planners and forces to retain flexibility of thought and action. Special Operations Forces (SOF) are not employed to mass in the conventional sense. Acceptance of attrition or force-on-force battle is not applicable for SOF. SOF must concentrate their combat power covertly, indirectly, and at decisive times and places. In SO, concentration of force relies on the quality and focus of tactics, timing, and weaponry.

1.1.2.4 Economy of Force. Allocate minimum essential combat power to secondary efforts. As a reciprocal of the principle of mass, economy of force in the strategic dimension suggests that, in the absence of unlimited resources, a nation may have to accept some risks in areas where vital national interests are not immediately at stake. If the NCA must focus predominant power toward a clearly defined primary threat, it cannot allow attainment of that objective to be compromised by diversions to areas of lower priority. This involves risks, requires astute strategic planning and judgment by political and military leaders, and again places a premium on the need for flexibility of thought and action. SOF may be employed strategically as an economy of force measure to allow the concentration of other forces elsewhere. This may be particularly effective when SOF are employed in conjunction with indigenous forces to create a "force multiplier" effect, or when SO are conducted for the purpose of deception.

1.1.2.5 Maneuver. Place the enemy in a disadvantageous position through the flexible application of combat power. In the strategic sense, this principle has three interrelated dimensions. The first of these involves the need for flexibility in thought, plans, and operations. Such flexibility enhances the ability to react rapidly to unforeseen circumstances. The second dimension involves strategic mobility, which is especially critical in order to react promptly and to concentrate and project power on the primary objective. The final strategic dimension involves maneuverability within the theater so as to focus maximum strength against the enemy's weakest point and thereby gain the strategic advantage. The object of maneuver is to concentrate or to disperse forces in a manner designed to place the enemy at a disadvantage, thus achieving results that would otherwise be more costly in personnel and material. SOF do not maneuver against an enemy in the classical sense. In SO, maneuver implies the ability to infiltrate and exfiltrate denied areas to exploit enemy vulnerabilities. When employed, maneuver implies the ability to adjust the plan in order to concentrate and strike the enemy where and when he is most vulnerable and to disperse to avoid his strengths.

1.1.2.6 Unity of Command. For every objective, ensure unity of effort under one responsible commander. Unity of command means directing and coordinating the action of all forces toward a common goal or objective. Coordination may be achieved by cooperation; it is, however, best achieved by vesting a single commander with the requisite authority to direct and coordinate all force employed in pursuit of a common goal. It is axiomatic that the employment of military forces in a manner that develops their full combat power requires unity of command. To achieve unity of effort, SOF organizes with clean, uncluttered chains of command.

1.1.2.7 Security. Never permit the enemy to acquire an unexpected advantage. Security enhances freedom of action by reducing friendly vulnerability to hostile acts, influence, and surprise. Security measures, however, should not be allowed to interfere with flexibility of thought and action, since rigidity and dogmatism increase vulnerability to enemy surprise. Therefore, detailed staff planning and thorough knowledge and understanding of enemy strategy, tactics, and doctrine can improve security and reduce vulnerability to surprise. Security is paramount to SO. Planning is often compartmented and planning staffs are kept small. Within a compartmented activity, all must share information. Intelligence, cover, and deception are all integrated throughout the planning and execution of SO to enhance security and achieve surprise.

1.1.2.8 Surprise. Strike the enemy at a time or place, or in a manner, for which he is unprepared. Concealing one's own capabilities and intentions creates the opportunity to strike the enemy while he is unaware or unprepared. Surprise is difficult to achieve, yet very important to the force for it can decisively affect the outcome of battles. With surprise, a force can obtain success disproportionate to the effort expended. Factors contributing to surprise include speed, employment of unexpected tactics and methods of operations, effective intelligence, deception operations, and operations security (OPSEC). The achievement of surprise must be a primary capability of SOF. SO require bold, imaginative, and audacious actions, but these must be tempered with patience and forethought.

1.1.2.9 Simplicity. Prepare clear, concise plans and orders. Guidance, plans, and orders should be as simple and direct as the attainment of the objective will allow. Political and military objectives and operations must be presented in clear, concise, and understandable terms; simple and direct plans and orders cannot compensate for ambiguous and cloudy objectives. In military application, this principle promotes strategic flexibility by encouraging broad strategic guidance rather than detailed and involved instruction. Although SOF may often use sophisticated and unorthodox methods and equipment, the plans and procedures that drive their employment must be simple and direct in order to facilitate understanding, withstand the stress of operational environments, and be adaptable to changing situations.

1.1.3 Tactics, Techniques, and Procedures

1.1.3.1 Tactics. The term tactics refers to the employment of units in combat, or to the ordered arrangement and maneuver of units or their elements in relation to each other or the enemy, in order to use their full potential. With respect to doctrine, tactics are the theoretical methods by which doctrinal principles are achieved. For example, to achieve the principle of surprise, SEALs may employ various deception, electronic warfare (EW), and infiltration tactics.

1.1.3.2 Techniques. Techniques provide the detail of tactics, and refer to the basic methods of using people and equipment to carry out a tactical task. For example, to exercise the principle of maneuver, SEALs may employ the tactic of infiltrating by night, using high altitude low opening parachuting, combat swimming, or other such specialized techniques.

1.1.3.3 Procedures. Procedures are the lowest level of detail and are used to standardize or to make routine the performance of critical or recurring activities. They are often promulgated in the interest of interoperability or safety. For example, there are specific procedures for the conduct of high altitude low opening parachuting.

1.1.4 Primary Users. NSW Task Groups/Task Units (NSWTG/TU), SEAL patrol leaders (PL), assistant PLs (APL), and patrol personnel are the primary users of this NWP.

1.1.5 Manual Organization. This manual should be used in conjunction with the NSW Mission Planning Guide (NSWMPG) that presents the typical mission planning cycle and outlines the use and benefit of the phase diagramming system and contingency planning checklist. Chapter 1 is an introduction to principles applicable to the conduct of SO. Chapter 2 is divided into sections, addressing TTP for SEAL operations in general. The other chapters are focused on TTP and considerations required in specific operational environments.

Chapter 1: Introduction

Chapter 2: Seal Tactics

Section I. Basics

Section II. Tactics, Techniques, and Procedures

Section III. Support

Chapter 3: Jungle Operations

Chapter 4: Desert Operations

Chapter 5: Mountain and Arctic Operations

CHAPTER 2

SEAL Tactics

SECTION I BASICS

2.1 INTRODUCTION

This chapter is divided into three sections. Section 1 provides the basic platoon organization, duties and responsibilities of the individual members, basic combat patrol concepts, and techniques generally applicable to all SEAL patrol members and formations. Section 2 establishes tactical principles that can function as a foundation for further tactical development. Specific tactics are addressed as stepping-stones to more complex tactics and procedures that are presented later in the section. Various types of external combat support available to SEAL patrols are also discussed. Section 3 provides a framework for effective coordination with support elements.

2.2 PATROL ORGANIZATION

2.2.1 General. A routine patrol is organized for small unit SO. The basic operational unit is the SEAL platoon, consisting of 16 men with warfare skill specialties in combat diving, small unit tactics, air operations, diving, demolitions, communications, and small boat handling.

2.2.1.1 Tailoring the Patrol. The SEAL platoon usually consists of two officers and 14 enlisted men. The platoon is subdivided into two 8-man squads, but can be further subdivided into 4-man elements called fire teams (FT) or 2-man operation pairs. The composition of a SEAL element is tailored to fit mission requirements. Most land warfare missions, however, employ an 8-man squad size element. On occasions the platoon commander (PC) may not be the PL, so this manual is intended for use by anyone charged with the PL responsibility.

2.2.1.2 Larger Units. Although SEALs do not normally operate as more than a 16-man platoon, they should be prepared to conduct operations involving larger units. SEAL operations that require a larger force will normally be conducted by multiple SEAL platoons, although specific operations may involve other tactical units. Rules of thumb for larger unit operations are:

1. Conduct more detailed, mission specific planning, rehearsals, and realistic full mission profiles (FMPs).

2. Operate with the minimum number of personnel possible to enhance command, control, and communications, stealth, and the probability of success.

2.2.1.3 Patrol Augmentation Requirements. Various types of personnel can augment the SEAL platoon to meet unique mission requirements. Augmentees must be physically conditioned to the environment and capable of patrolling with SEALs or their presence may endanger the mission (see Section 2.2.5 for details regarding patrol augmentation).

2.2.2 Patrol Responsibilities. Assign every member on a patrol an area of responsibility (AOR). All personnel should be cognizant of aircraft, which can appear on the horizon quickly, boats drifting silently in waterways, people in the vicinity of the patrol, and wheeled vehicles approaching the patrol (particularly in open areas such as fields and meadows). If this happens, it will be necessary to seek concealment or cover as rapidly as possible to avoid detection and subsequent engagement. All personnel have the following common responsibilities:

1. Observing fields of fire
2. Perimeter security
3. Action upon enemy contact
4. Taking head counts
5. Passing hand and arm signals
6. Passing the word
7. Detecting danger.

2.2.2.1 Patrol Leader. The PL has the responsibility and authority for leading the patrol. This requires strong tactical knowledge and leadership skills.

1. Position. The PL places himself where he can best control the patrol during each phase of the operation. During patrol movement he normally positions himself near the front of the patrol formation, directly behind the point element.

2. Duties. The PL/APL plans the patrol and is responsible for its execution. His primary responsibility is to direct and control the patrol. During engagements, his first responsibility is to assess the situation and direct patrol actions. This takes precedence over all other actions, including returning fire. His duties are to:

   a. Plan the mission and direct all pre-mission preparation

   b. Direct all actions of the patrol

   c. Direct navigation of the patrol.

3. Comments. The PL:

   a. Makes decisions about the direction of travel, designates rally points (RP) and Lay-up Points (LUPs), conducts final planning after reconnaissance (recon) of the target, directs actions at the objectives, and directs actions upon enemy contact

   b. Constantly anticipates and considers actions if contact is made

   c. Is responsible to the NSWTG/TU

   d. Plans and initiates evasion and recovery (E&R).

2.2.2.2 Point Man. This is normally only one person; two men may be required, depending on the situation. When the PL determines the situation requires two men, the position is called the point element.

1. Position. The point man (PT) patrols ahead of the main body of the patrol. Although he may periodically be required to scout further ahead, he normally patrols within visual signal range of the rest of the patrol.

2. Duties. The PT is responsible for finding a safe route for the patrol by watching for danger areas, the enemy, and booby traps. He is not the primary navigator, but responds to directions from the PL. Proven navigation skills, however, are required assets for the PT; teamwork and mutual assistance between PT and PL while navigating are critical to the patrol maintaining its course. PT is part of a security team and his duties include the following:

   a. Assist the PL in route selection through careful map study

   b. Work with the PL to plan evasion and escape (E&E) routes

   c. Maintain the security of the patrol within his field of fire

   d. Alert the patrol to possible enemy movement if suspected

   e. Alert the patrol to obstacles and booby traps.

3. Comments. The PT or point element:

   a. Maintains communication with the PL

   b. Carries a compass to navigate, but does not act as the primary navigator

   c. Has good tracking and navigation skills

   d. Carries a lighter load than others for ease of movement and noise discipline

   e. May need frequent rest periods or replacement in order to reduce stress and fatigue

   f. Needs extra hydration since he will be more active than others in the patrol.

2.2.2.3 Radio Telephone Operator. Also known as the radioman, the radio telephone operator (RTO) is the patrol's main communicator.

1. Position. The RTO patrols closely behind the PL, except when the PL is away from the main element during a recon.

2. Duties. The RTO is responsible for ensuring that the patrol can send and receive all communications necessary for the mission. His duties include:

   a. Drafting communication plans based on mission requirements as identified by the PL and NSWTG/TU.

   b. Carrying primary radio and cryptographic equipment and acting as a rifleman/grenadier (R/GN or R/G).

   c. Conducting pre-coordination with the mobile communications team (MCT) and all supporting and supported elements included in the communications and electronics operating instructions (CEOI).

   d. Establishing and maintaining radio contact with the NSWTG/TU and supporting forces as required by the CEOI.

3. Comments. The RTO:

   a. May be read into some aspects of the mission not normally briefed to all patrol members.

   b. May require extra hydration because of his extra equipment weight.

   c. Should be trained in equipment repair and antenna theory.

   d. Should realistically test all equipment prior to going in the field.

2.2.2.4 Automatic Weapons Man

1. Position. Depending on the size and disposition of the patrol, the automatic weapons (AW) man is usually at or near the middle of the formation. The mission and situation will dictate the number of AW men required. A standard eight-man squad normally has two AW men.

2. Duties. The AW man is responsible for base or covering fire for the patrol. Patrol immediate action drills (IAD) and fire and maneuver depend heavily on his heavier sustained rate of fire. The AW man is expected to:

   a. Provide security cover for the patrol as it moves, or for elements moving into attack position

   b. Provide covering fire for maneuver elements

   c. Maintain a heavy sustained rate of fire during engagements

   d. Provide maximum suppressive firepower

   e. Designate point targets in stand-off operations

   f. Designate point targets for supporting elements (close air support (CAS), call for fire, etc.)

   g. Concentrate fire on weapons of greatest threat to patrol.

3. Comments. The AW men:

   a. Carries the heaviest platoon weapon, presently the M60 light machine gun and its ammunition load

   b. Requires extra hydration

   c. Requires exceptional physical strength, endurance, and tactical prowess that must be developed through realistic and intense training.

2.2.2.5 Rifleman/Grenadier

1. Position. Normally R/GNs are distributed evenly throughout the patrol to complement the placement of the AW man/men and the requisite security elements.

2. Duties. Routine patrol responsibilities. Will normally be assigned duties for special elements and teams (e.g., flank security, demo team, prisoner handler, etc.). The R/GN is expected to:

a. Carry M4 with the M203 grenade launcher and 40mm rounds.

b. Deliver accurate rifle fire against single targets when acting as riflemen. Normally he fires his weapon in semi-automatic mode.

c. Use special purpose munitions such as high explosive, illumination, smoke, signal, or CS rounds when acting as a GN.

3. Comments:

a. 40mm rounds require 14 to 28 meters from launch before arming themselves.

b. Depending on terrain and mission that can adversely affect movement or if weight is a concern, the PL may opt to have all personnel act as R/GNs, however, the platoon's ability to lay down an effective base of fire can be affected.

c. When using multiple R/GNs the PL should ensure they know who is primarily a GN and who is primarily a rifleman.

d. The PL should also ensure R/GNs know who is responsible for illuminating the target/enemy, covering/screening maneuver elements with smoke, covering withdrawal with smoke/CS, and providing signals.

e. The R/GN may also serve as demolitions carrier.

2.2.2.6 Rifleman

1. Position. Rifleman may be in any patrol position.

2. Duties. Routine patrol responsibilities. Will normally be assigned duties for special elements and teams (e.g., flank security, demo team, prisoner handler, etc.). A rifleman:

a. Will carry an M4, M14, or other selected assault rifle. He delivers accurate rifle fire against single targets in an assigned field of fire. Fires in semi-automatic mode.

b. May be demolitions, AT4, extra M18A1 Claymore carrier or prisoner handler, etc.

2.2.2.7 Hospital Corpsman

1. Position. The hospital corpsman (HM) patrols as directed by the PL.

2. Duties. The HM:

a. Provides first aid, preventative hygiene, and treatment of wounds and injuries.

b. Plans and coordinates medical evacuation (MEDEVAC) to the forward operating base (FOB) or safe area.

c. May act as rifleman, GN, prisoner handler, or demolitions carrier.

2.2.2.8 Assistant Patrol Leader. The APL assists the PL in all phases of coordinating and executing the patrol. He is usually positioned near the rear of the patrol.

1. Position. During movement, he patrols ahead of the rear security (RS) and in a different FT or element than the PL. He may be given a special job for each phase of the patrol. He helps the PL control the patrol by being where he can best take command, if required.
2. Duties. The APL is expected to:
   a. Take over if the PL is incapacitated; he is the second in command.
   b. Command the second FT (or second squad if in platoon formation) when the patrol is split.
   c. Assist with planning and preparation.
   d. Keep abreast of situation and status of mission.
   e. Know what the PL knows.
   f. Serve as a rifleman.
3. Comments. Commands the main body of the patrol during absence or incapacitating injury of the PL.

2.2.2.9 Rear Security

1. Position. The RS patrols as the last man in the patrol.
2. Duties. The RS is expected to:
   a. Be responsible for the security of the patrol from the rear.
   b. Pass all information to the PL regarding the situation behind the patrol.
3. Comments.
   a. The RS serves as a rifleman or GN.
   b. Observes overhead and to the rear.
   c. Does not walk backward, rather he stops to observe.
   d. Acts as the PT in situations when the patrol reverses direction.

2.2.2.10 Pacers. Pacers are men assigned the job of recording distance by pacing. There are usually two in each patrol, but there may be more. They use any of a number of techniques to keep track of the distance traveled. They should be separated in the patrol formation so that they will not influence each other's count. When the PL asks for the count, both pacers send up their count and an average of the counts is taken to get a closer approximation of the distance. The pacer count is always given in meters.

2.2.3 Patrol Command and Control. The essence of success in small unit tactics is the ability of the PL to command and control (C^2) elements of the patrol. Small unit tactics require a clear and defined chain of command. This enhances C^2 and the ability of the PL to make sound decisions quickly. Often patrol members mistake the chain of command as a one-way communication process when in fact the PL cannot make comprehensive decisions without accurate input from patrol members. This is especially true when an unexpected contact or ambush is made. Particular mention is required for direct action (DA) missions where follow-on actions are required, such as removal of captured enemy and material or destruction of the latter. This is accomplished through training and experience, using standard commands and procedures.

2.2.3.1 Patrol Integrity. If the patrol is ever reduced to individuals acting alone without coordination with the rest of the patrol, its tactical integrity has been lost and it is extremely vulnerable. In complex environments, this can be caused not only by the enemy, but also by the environment. The rule is **STAY TOGETHER**! Always be aware of those around you.

2.2.3.2 Patrol Chain of Command. Unless individual SEAL Team standard operating procedures (SOP) dictate differently, the PC will always be the PL when he is on the patrol.

1. Alternate PL. The PC will designate the PL for any patrols in which he does not participate.

2. APL. The APL is the SEAL next in seniority to the PL. He will become the acting PL if wounds or injuries incapacitate the PL.

2.2.4 Special Assignments. Patrol members normally have special assignments that they will be required to perform on command, as needed, or on a continuing basis throughout the patrol. These special assignments may be made to meet specific mission requirements, or they may be generally assigned during all patrols as routine or for contingency purposes. The specific assignments are normally given based on:

1. Mission requirements

2. Individual skills and/or abilities

3. Equal distribution of the workload throughout the patrol.

2.2.4.1 Security Elements. The security element is normally made up of one or more security teams that protect the patrol by blocking likely avenues of enemy approach, covering patrol movements, and giving early warning of an enemy approach.

1. Perimeter Security. Perimeter security is used throughout the patrol: during security halts, rest halts, at all RPs, and in patrol bases. Individuals or teams are assigned to provide visual and auditory coverage of their sector or AOR. The teams must be positioned so that complete 360-degree coverage is ensured.

2. Flank Security. Flank security can be one or more men positioned to the side of a formation. Their primary responsibility is to provide early warning against an enemy ambush or attack. Depending on the size of the flank security element/team and the situation, they may be able to deliver limited flanking fire on enemy positions. The use of the flankers requires solid C^2 from the PL to avoid engaging friendly forces. (Disadvantages to using flankers include limited fire and movement, and C^2 upon enemy contact)

3. Covering Force. A covering force is a security element positioned to cover an element's movement. It is positioned so that it can engage enemy forces to allow continued movement by another element that is attacking an objective (as in a raid) or retreating (e.g., breaking contact).

4. Blocking Force. A blocking force is a security element that is specifically positioned to defend against assault from a known or anticipated enemy threat. The blocking force is positioned so that in case the patrol is compromised, the blocking force can engage possible enemy reaction forces moving against the main body. If the blocking force is not utilized, it will usually rejoin the patrol at a predetermined time and location. The blocking force's employment is the same during an assault, i.e., to prevent the patrol from being engaged by enemy reaction forces.

2.2.4.2 Reconnaissance Elements. The recon element confirms, clarifies, and supplements information provided to the patrol about the routes and objectives from maps, aerial photos, and other sources. If a recon element is used to scout an objective, an objective rally point (ORP), or an ambush site, it will usually act as PT to lead the patrol to its new location.

1. Patrol Reconnaissance. During security halts the patrol recon element or team (normally the point element plus the PL) will reconnoiter ahead of the main patrol prior to moving into a patrol base, LUP, RP, ambush site, or assault position.

2. Target Reconnaissance. The point element and the PL usually conduct target recon. The patrol's tactical recon of the target is designed to confirm all the data that was previously provided in the target information package and on which they have developed their plan of attack and to identify any differences. The recon should be used to gather any of the specific information that the patrol needs, such as the location and movement of any guards, confirmation of the actual location and position of the target, and the layout of the facility. The recon team should pay particular attention to the points designated for attack and/or demolition charge placement, look for the best avenues of attack or the best way to attach a demolition charge to a target stress point.

2.2.4.3 Search Teams. After assaults or ambushes are conducted, the search team is responsible for seizing equipment or documents and for searching buildings, positions, dead, wounded, and prisoners. First, the team wants to ensure that there are no other enemy or potential threats to the patrol such as booby traps, unexploded ordnance, weapons, etc. Then their job is to collect information from the objective that can be added to the intelligence assessment of the area, the enemy, and his capabilities, strengths, and weaknesses. The types of searches are:

1. Ambush Search. The team will search for killed in action, wounded, prisoners, weapons, and documents.

2. Target Search. Depending upon the target, hard or soft, the team normally consists of one element. The PL must continuously monitor time on target (TOT) and what possible reactionary forces are doing. Teams must adhere to strict C^2 and maintain the ability to re-engage the enemy.

3. Prisoner/Personnel Search. Approach these searches with extreme caution. If a conflict occurred preceding the search, be aware of "sleepers". Severe trauma or shock may inhibit immediate responses from some personnel. A search is done on everyone, not just the ones that look important, using the "Five S's" standard:

   a. Search

   b. Segregate

   c. Silence

   d. Speed

   e. Safeguard.

2.2.4.4 Demolitions Teams. The demolition (demo) team is composed of men assigned to carry demolitions for specific target objectives or standard charges to be used for contingencies. The members will be assigned specific duties and responsibilities for team actions such as rigging explosives, running the detonation cord, tying into the trunk line, and pulling fuses. The demo team must be well briefed and its actions must be well rehearsed and coordinated. Demo team organization and duties are discussed in detail in the NSW Demolitions Handbook.

2.2.4.5 Prisoner Team/Handler. These are the men assigned to take prisoners or to take control of and be responsible for any prisoners that are captured during patrol operations. (Reference Section 2.3.12).

2.2.4.6 Field Interrogators. These are personnel specially trained and qualified in impact interrogation (i.e., interrogation immediately following combat action and capture). They quickly elicit tactical information (e.g., what reaction forces are in the area, their composition, direction of approach, time of response, etc.).

2.2.4.7 Swimmer Scouts. Swimmer scouts are assigned the task of conducting a recon of the selected beach-landing site (BLS) while the remainder of the patrol waits a safe distance offshore. After they scout the beach and determine it is clear of obstructions/obstacles and that there is no enemy in the immediate vicinity, they signal the swimmers or boats into the beach using pre-arranged signals. They also provide initial security for the swimmers or boats as they land. They may join the Security team guarding the boats or act as point to the objective on short raids if they scouted the objective earlier.

2.2.5 Patrol Augmentation Requirements. Patrol augmentees provide the patrol with skills not held by SEAL platoon members. These personnel may be a scout, interpreter, or other military personnel with specialized skills.

2.2.5.1 Scout/Guide. The scout or guide is a patrol augment who has an intimate knowledge of an area, or who can lead friendly forces to a known enemy location. Guides may be surrendered enemy personnel, personnel native to the area, or anyone else familiar with the area.

1. Position. The scout/guide generally patrols directly behind or with the PT/element, and in front of the PL.

2. Precautions. The scout/guide requires constant watch, even if there is little doubt of his loyalty. Verify his travel direction against the map, compass direction, and known positions so he doesn't lead the patrol into the wrong area. Carefully evaluate each guide to determine if he will carry a weapon and never allow him to walk point alone. Have him wear your type uniform if interaction with indigenous persons is not required.

2.2.5.2 Interpreter. The interpreter will provide language skills required in areas where SEALs do not have native language proficiency.

1. Position. Generally patrols directly behind the PL and in front of the RTO.

2. Precautions. Requires constant watch. More than likely, this individual will have little patrolling skill. It may be useful to provide patrol-skills training if this individual is to remain as platoon interpreter. Have him wear your type uniform if interaction with indigenous people is not required and only communicate with him when required.

2.2.5.3 U.S. Military Personnel. On some occasions explosive ordnance disposal (EOD), air/naval gunfire liaison company, U.S. Air Force Combat Control Team and other military personnel may accompany SEAL patrols to provide expertise not organic to the platoon.

1. Position. Other U.S. military personnel generally patrol directly behind the RTO.

2. Precautions. These individuals will require briefings on platoon SOPs and IADs prior to departure. Assign a coordinator before and during the mission and only provide the augmentee with mission information he must know for his assignment or patrol effectiveness.

2.2.5.4 Equipment. Patrol augmentees may be required to carry specialized equipment for their specific mission tasks.

1. Augmentees may require equipment to accompany SEALs; training may be necessary to assure proper use.

2. The question of whether or not to arm augmentees should be evaluated on a case-by-case basis.

3. Platoon personnel may be asked to carry specialized equipment for the augmentees. Calculate weights and how this would affect speed, mobility, and mission accomplishment.

2.3 COMBAT PATROL BASICS

2.3.1 General. Combat is an evolving art. SEALs must be well trained and capable of operating efficiently and effectively in all environments. Most of the patrol TTPs that are discussed in this NWP can be employed in all environments, varied terrain, and during daylight or at night. Although SEALs prefer to operate at night because of the additional concealment offered by the darkness, the tactics in this manual can be used in almost all lighting conditions. Those tactics and techniques that apply to finite circumstances are so indicated.

2.3.2 Guiding Principles. There are four basic areas of concern that SEALs must consider while preparing for and conducting combat operations. They are security, firepower, movement, and communications.

2.3.2.1 Security. Surprise and freedom of movement are essential to the success of SO. These vital factors are based on accurate and timely intelligence. Because of the nature of SEAL operations, all aspects of OPSEC should be rigorously observed throughout the planning cycle and during the conduct of operations. Information to friendly forces should be available only on a need-to-know basis. Negotiations with local political factions or military organizations necessary for the conduct of a SEAL operation should be carefully planned to preclude compromise. Security is accomplished by concentrating on the following areas of concern:

1. Maintaining OPSEC throughout all phases of planning and execution

2. Using cover, concealment, and fire discipline to avoid giving away the patrol's position

3. Establishing local security perimeters and zones of responsibility

4. Conducting reconnaissance

5. Protecting the patrol with preplanned fire support.

2.3.2.2 Firepower. Although SEALs employ only individual weapons, they are better armed than most elements of their size. Because of the number of AWs, grenade launchers, and other specialized weaponry normally carried by a SEAL patrol, they can deliver a disproportionate volume of fire for their relative size. This can be a short-lived advantage, however, since a SEAL patrol can expend all or most of its ammunition within the first few minutes during an intense firefight. In order to effectively utilize this firepower, a patrol must:

1. Establish immediate fire superiority through volume and accuracy of fire

2. Be able to accurately and quickly direct supporting fires

3. Use movement, position, and surprise to maximize the destructiveness of the fire.

2.3.2.3 Movement. Movement is necessary to reach mission objectives. It also is critical when avoiding unplanned enemy contact and when breaking contact. Movement of the patrol and its subordinate elements should be designed to put the patrol at an advantage and negate or minimize the enemy's capabilities. Effective movement is accomplished by:

1. Using cover and concealment to their fullest advantage

2. Conducting observation and recon prior to moving

3. Accurate land navigation

4. Using routes and movements that will not be expected by the enemy.

2.3.2.4 Communication. Communication is divided into two categories: communication within the patrol (see Appendix C) and external communications. In modern combat, verbal communications are required for coordinated mission support, to coordinate and direct patrol activities, and keep the chain of command informed. Verbal communications plans should:

1. Be easily understood by all members

2. Include a back-up communications net and lost communications plan in case of equipment failure

3. Minimize communication and consider enemy direction finding (DF) capabilities.

2.3.3 Basic Patrol Principles. The PL determines his patrol organization and patrol plan based on the following criteria:

1. Mission

2. Estimate of the situation based on criticality, accessibility, recuperability, vulnerability, affects on populace, and recognizability (CARVER)

3. Analysis of terrain by observation and fields of fire, cover and concealment, obstacles, key terrain, and avenues of approach (OCOKA)

4. Reconnaissance

5. Support available.

2.3.3.1 Mission. The mission is defined as those operations conducted by specially organized, trained, and equipped military and paramilitary forces to achieve military, political, economic or psychological objectives by non-conventional means in hostile, denied, or politically sensitive areas. SEAL missions primarily involve two types of combat patrols: Special Reconnaissance (SR) and DA.

1. SR Mission. SR complements national and theater intelligence collection assets and systems by obtaining specific, well defined, and time-sensitive information of strategic or operational significance. It may complement

other collection methods where there are constraints of weather, terrain-masking, hostile countermeasures and/or other systems availability. SR is a human intelligence (HUMINT) function that places U.S.-controlled "eyes on target" in hostile, denied, or politically sensitive territory. SEALs may conduct these missions unilaterally or in support of conventional operations. SEALs may use advanced recon and surveillance techniques, and/or sophisticated clandestine collection methods. Recons are usually referred to as either a point, area, or route recon. Complete details of SR are contained in NSW SR Operations and Reporting Tactical Memorandum (TACMEMO) XL-0080-7-89. Typical missions may include:

a. Contact with a resistance movement to assess resistance potential

b. Recon operations in advance of operations by conventional forces

c. Target acquisition of enemy C^4I systems, troop concentrations, lines of communications, special weapons, and other military targets of significance to the combatant commander and his operational force commanders

d. Collection and reporting of critical information about the movement of enemy forces in or adjacent to the main battle area

e. Location and surveillance of critical or sensitive facilities in hostile or denied territory

f. Geographic, demographic, and hydrographic recons to support specific land and maritime operations

g. Post-strike recon.

2. DA Mission. In the conduct of DA operations, SEALs may employ raid, ambush, or direct assault tactics. They may emplace munitions and other devices, or conduct standoff attacks by fire from air, ground, or maritime platforms. They may also provide terminal guidance for precision-guided munitions, and conduct independent sabotage. DA operations are normally limited in scope and duration and usually incorporate a planned withdrawal from the immediate objective area. SEALs may conduct these missions unilaterally or in support of conventional operations. They are designed to achieve specific, well-defined, and often time-sensitive results that have strategic, operational, or tactical significance. Operations typically involve:

a. Attack on critical targets (material or personnel)

b. Interdiction of critical lines of communication or other target systems

c. Location, capture, or recovery of designated personnel or material

d. Seizure, destruction, or neutralization of critical facilities in support of conventional forces or in advance of their arrival.

2.3.3.2 Estimate of the Situation. The SEAL combat patrol must have an accurate assessment of the operational situation in order to develop the best plan of action. The acronym METT-T (mission, enemy, terrain, troops, time available) identifies the primary areas of concern. Intelligence updates, maps, charts, photos, and sketches should be used during the planning and preparation phase. In addition, intelligence specialists should query SOCRATES (Special Operations Command Research, Analysis, and Threat Evaluation System) to ensure that the latest possible intelligence is obtained. Target selection is based on the following six CARVER factors:

1. Criticality

a. How valuable is the target?

b. What is the gain-to-loss ratio between the values of destroying the target and losing a SEAL platoon in the objective area?

2. Accessibility

a. Is the target a riverine or maritime SO target?

b. Is it accessible to the patrol considering:

(1) Insertion/extraction methods available

(2) Insertion/extraction areas available, given the nature of the environment

(3) The maximum range of the patrol, given its mobility in the environment

(4) The time constraints of the overall operation, given the limited mobility in the environment

3. Recuperability

a. Is the target repairable? How much force is required to do the following:

(1) Completely destroy the target and prevent any repair of its capabilities within a tactically significant period.

(2) Destroy sufficient parts of the target to prevent its use within a tactically significant period.

b. Can a SEAL platoon deliver sufficient force to prevent the target from using its capabilities within a tactically significant period?

4. Vulnerability

a. Is the target vulnerable to attack by SEALs?

b. Is the size of the installation and guard force such that the target is within a SEAL platoon's capabilities?

5. Effect on Populace

a. Given the rules of engagement (ROE) in effect, is the degree of force required consistent with respect to limiting damage to non-military targets?

b. Does target destruction have enough of a political-military impact on the populace to alter the ability of the enemy to wage war?

c. Are psychological warfare forces required in connection with the operation?

6. Recognizability

a. Is the target sufficiently recognizable that it can be rapidly classified and attacked without losing an unacceptable amount of time during identification?

b. Given the nature of the environment, is the target identifiable at sufficient range to allow a tactically advantageous attack?

7. Political Analysis

a. What effect will the operation have on the politics of the region?

b. Will it lean the government's politics one way or the other?

c. If a counter-guerrilla operation, will the population favor the actions taken or will it strengthen the resolve to support the guerrillas?

d. What are the present feelings within the indigenous government towards the U.S. involvement? Within the population?

8. Military Analysis.

a. How will the military be affected by the mission?

b. Will the military's stature within the country be altered?

c. What will be the resultant effect on the enemy's strategic or tactical capabilities?

2.3.3.3 Analyzing Terrain. The acronym OCOKA details the factors that should be used to analyze terrain. It is used as both a planning tool prior to the patrol as well as a working tool throughout the patrol.

1. Observation and Fields of Fire. The effects the terrain will have on observation and fields of fire influence selection of and movement over a route. The PL must constantly consider the observation and fields of fire available to his patrol and the enemy. This affects formations, rates and time of movement, and methods of control.

2. Cover and Concealment. For most SEAL patrols, stealth is extremely important to the success of the mission; therefore the patrol must always consider the proper use of cover and concealment in deciding the proper tactical approach or attack. Both cover and concealment are protection from enemy fire and may be either natural or artificial/manmade. The most important distinction between cover and concealment is that cover provides protection from some types of enemy fire while concealment merely hides the concealed element.

a. Cover. Cover can protect you from flat trajectory fire and partially protect you from high-angle, indirect fire and explosions. Even the smallest depression or fold in the ground may provide some cover when you need it most. Properly used, a 6-inch depression may provide enough cover to save your life under fire. Natural frontal cover is best. It is harder for the enemy to detect. The following principles apply:

(1) Natural cover includes logs, trees, rocks, stumps, depressions, ravines, hollows, reverse concave slopes, etc.

(2) Artificial/manmade cover includes fighting positions, trenches, walls, buildings, rubble, abandoned equipment, bomb craters, etc.

(3) When moving, use a route that puts cover between your patrol and the places where the enemy is known or thought to be.

(4) Use hills and wooded areas and other natural cover to hide or conceal movements. Ravines and gullies can also be used, but caution must be exercised to avoid channeling the patrol into possible ambush sites.

(5) Avoid open fields and "sky-lining" on hills and ridges.

(6) Make a habit of looking for and taking advantage of every bit of cover the terrain offers. This habit should be combined with proper movement techniques discussed in Section 2.5 and Appendix A.

b. Concealment. Concealment is anything that hides the patrol, its weapons, position, and equipment from enemy observation. Discipline in the control of camouflage, light, noise, and movement must be enforced. Camouflage, a very important aspect of concealment, is discussed in more detail in Appendix A. The following principles apply:

(1) Avoid unnecessary movement. Remain still; movement attracts attention. You may be concealed when still, yet easily detected if you move. Movement against a stationary background causes you to stand out clearly. When you must change positions, move carefully over a concealed route to the new position.

(2) Keep quiet. Noise caused by talking, changing a magazine, or cycling a bolt can be picked up by enemy patrols or listening posts (LPs). The noise created by loose equipment, particularly metallic sounds, carries over long distances and is easily discernable. Properly secure your equipment for noise discipline.

(3) Darkness alone does not hide you from an enemy with night vision, thermal, and/or motion sensors.

(4) Stay low when observing the area. Observe from a crouch, a squat, or the prone position. In this way, you present a low silhouette, making it difficult for the enemy to see you.

(5) Background is important, blend with it to prevent enemy detecting your position. Trees, bushes, grass, earth, and manmade structures forming your background vary in color and appearances, making it possible for you to blend with them. Select trees or bushes that blend with your uniform and absorb the outline of your figure.

(6) Natural concealment is provided by your surroundings and needs no change to be used; for example, bushes, grass, and shadows.

(7) The best way to use natural concealment is not to disturb it when moving into an area.

(8) Artificial concealment is made from materials such as burlap or nets, or from natural materials such as bushes, leaves, and grass that are moved from their original location.

(9) Be sure to consider the effects of changes of season or location on the concealment provided by both natural and artificial materials.

(10) Shadows help to hide you. Shadows are found under most conditions of day and night.

(11) Expose nothing that shines. Reflection of light on a shiny surface instantly attracts attention and can be seen for great distances.

(12) Keep off the skyline. Figures on the skyline can be seen from a great distance, even at night, because a dark outline stands out against the lighter sky. The silhouette formed by your body makes a good target.

(13) Alter familiar outlines. Military equipment and the human body are familiar outlines to all soldiers. Alter or disguise these revealing shapes.

3. Obstacles. Obstacles, both natural and manmade, must be considered in planning routes and throughout the patrol. They may stop, divert, or impede movement along a route or limit maneuver. Obstacles that may limit or restrict enemy action/reaction may be used to your advantage and should be exploited.

4. Key Terrain. Key terrain features are those that have a controlling affect on the surrounding terrain. Key terrain features must be identified and actions planned to take advantage of them. Similarly, the advantages offered to the enemy by key terrain features must be countered.

5. Avenues of Approach. The PL and team leaders must analyze the avenues of approach into patrol positions and when on the attack into the enemy's positions. Whenever the patrol moves into a patrol base, LUP, RP, or ambush site, proper defense and security of avenues of approach is a major consideration for establishing all around security.

2.3.3.4 Reconnaissance. The importance of recon cannot be overstated; it is essential in both the planning and conduct of patrols. It can be accomplished in several ways.

1. Map Reconnaissance. The PL, team leaders, PT/men, and navigator must make a thorough map recon of the terrain that the patrol must cover. The map study should include:

   a. A check of the map or chart's marginal data. Older maps or charts without recent updates should be closely evaluated for reliability. Insure correct data is entered to electronic navigation equipment.

   b. Consideration of the terrain in relation to all available information of known and/or suspected enemy positions, patrols, and previous ambush sites.

   c. Evaluation of the terrain from the enemy's perspective: where he could establish installations, patrols, ambushes, etc.

2. Aerial Reconnaissance/Over flights. Whenever possible, the PL should make an aerial recon. If an actual PL's recon is not feasible, the PL/Intelligence officer should consider overflights by local, theater, or national assets. The purpose of the recon is to obtain current and more complete information on the enemy situation, roads, trails, manmade objects, type and density of vegetation, and seasonal condition of rivers and streams than is available from a map recon.

   a. Indicators to look for:

      (1) Movement, or lack of movement in the area of operations (AO)

      (2) Smoke, indicating locations of campsites, patrols, or patrol bases

      (3) Freshly dug soil, indicating farmland, enemy positions or ambush sites

(4) Shadows, which may help identify objects

(5) Unusual shapes, sizes, shadows, shades, or colors which may indicate faulty camouflage

(6) Dead and/or unnatural vegetation.

b. Limitations. Some limitations of aerial reconnaissance include:

(1) It may warn the enemy.

(2) Structural strengths of roads, bridges, rail systems, and buildings or structures, cannot be determined.

(3) Terrain surfaces may be misinterpreted.

(4) Mines, booby traps, and well-prepared ambush sites may not be detected.

3. Sensor Reconnaissance. Take advantage of any sensor support that may be available, such as EA-6B aircraft or mobile inshore undersea warfare units, to gather additional intelligence about the enemy.

4. Tactical Reconnaissance. By their nature, SEAL operations generally require stealth to succeed. When approaching target areas or danger areas, a sound tactical recon is critical to avoiding premature compromise. Reconnaissance should be performed on all linkup sites, BLSs, helicopter landing sites, the patrol's rear, around patrol bases, and when the patrol is stopped for a rest break. Allotted time for a reconnaissance should be maximized to ensure thoroughness. Particular attention should be paid to the following:

a. Concealed routes for the patrol to and from the objective

b. Enemy position and routine

c. Avenues of approach for reaction forces

d. Danger areas and obstacles

e. Any additional OCOKA considerations applicable.

2.3.3.5 Patrol Routes. The patrol may be ordered to follow a specific route due to situation and mission requirements or it may be able to select its own route. OCOKA criteria, as discussed above, should be applied directly to assess patrol routes.

1. Selecting a Route. As far as orders and missions permit, routes should be selected in accordance with the following principles:

a. Avoid known enemy positions and obstacles.

b. Minimize chance of contact with local populace by avoiding manmade features and structures such as trails, buildings, wells, farms, campgrounds, etc.

c. In threat areas, take advantage of more difficult terrain such as swamps and dense woods.

d. Maximize use of identifiable re-set points to assist with navigation.

e. Avoid moving along exposed ridges. Move along the slope below the ridge to prevent silhouetting.

f. Avoid areas that may be mined, booby-trapped, or covered by fire.

g. For daylight movement, seek terrain that offers the most cover and concealment; avoid open areas.

h. Seek terrain that permits quiet movement at night.

2. Study the Route. Study maps, aerial photos, or sketches and memorize the route before starting. Note distinctive features (hills, streams, and swamps) and their locations in relation to the route. If possible, "box" the route in with terrain features to aid in navigation.

3. Alternate Routes. Plan an alternate route in case the primary cannot be used.

4. Route Offsets. In difficult terrain such as jungle or swamps, plan an offset to a known terrain feature.

5. Checkpoints. Establish checkpoints for the patrol routes. They are a means of control between the parent unit and patrol. These locations are decided upon and coordinated before the patrol leaves, so that both the patrol members and parent unit will know where the patrol is when the patrol reports its execution checklist items. The parent unit can follow the progress of the patrol without transmitting coordinates in the open. It is also an excellent means of coordinating the patrol route with a fire support plan.

6. Target Reference Point. A target reference point is an easily recognizable point on the ground, either natural or manmade, used for identifying enemy targets or controlling fires. These are used with the route plan and checkpoints in planning and controlling fire support.

2.3.4 Communications. The methods that the PL will use to communicate with elements of his unit, artillery, air or other supporting units, and higher headquarters need to be identified. During extended patrols it may be necessary to establish communication windows and relay sites (fixed/airborne). The paragraphs that follow provide guidance for communications in the field, both internal and external to the patrol. The NSW Communications Handbook should be consulted for more details on communications procedures.

2.3.4.1 Patrol External Communications. (Appendix B contains a matrix of available SEAL portable radio communications equipment.)

1. General Procedures

a. Think communications security (COMSEC) at all times. COMSEC is part of OPSEC.

b. Turn on communications equipment only when necessary.

c. Minimize radio traffic. Use brevity codes and burst transmissions whenever possible. Chances of interception by DF equipment increase with each transmission.

d. Don't transmit from the LUP site.

e. Use directional antennas whenever possible.

f. Be prepared to move after any transmission.

2. Satellite Communications (SATCOM)

   a. Uses ultra high frequencies (UHF) and is the best frequency band communications in most situations.

   b. Requires a clear "look" angle at the satellite, i.e., there should be no obstruction between the antenna and satellite, which is generally greater than 27 degrees above the horizon.

3. Very High Frequency (VHF) and UHF Communications. These frequencies are useable under most patrol conditions.

4. High Frequency (HF). HF communications are a useful long-range frequency band, but they are the most detectable and are therefore not the best frequency band for many missions.

5. HF Antennas. Overcoming adverse communications conditions demands the following:

   a. Use existing antennas correctly.

      (1) Keep the whip antenna vertical when transmitting.

      (2) Ensure a foreign object does not ground the antenna.

      (3) Locate the antenna as high as possible for best line of sight.

   b. Be prepared to utilize field expedient antennas. (Claymore firing wire can be useful in this regard.) The NSW Communications Handbook contains further details.

   c. Pour water or urinate on ground wires to improve grounding characteristics.

6. Considerations.

   a. When operating, don't leave plastic waterproofing on the handset:

      (1) Water condenses on the inside.

      (2) It makes noise.

      (3) It causes distorted voice transmission.

      (4) It shines at night when viewed through night vision electro optics (NVEO).

   b. Clean all contacts daily (use #2 pencil eraser).

   c. If the batteries for the AN/PRC-104 run low, they can still be used in the crypto unit, as this does not require as much power to operate.

   d. Warm climates reduce the communication range of man pack radios.

2.3.4.2 Patrol Internal Communications. Standard commands are the prerogative of the tactical commander, but use of the following standard procedures is encouraged to promote commonality.

1. Techniques. Use the following procedures:

   a. Pass all word without the aid of a radio, if possible (maximize use of platoon SOPs).

   b. Pass simple rather than complex instructions, giving time for the word to pass and then executing the instruction.

   c. Pay attention to the man in front, behind, and on either side. Ensure all signals are passed by all members of the patrol; therefore, each member must watch the signs of the others.

   d. Use silent hand signals to alert fellow patrol members. See Appendix C for standard hand and arm signals.

   e. Follow the lead of the other members because there may not be an opportunity for them to signal.

      (1) When the man in front gets down, all get down.

      (2) When the man in front freezes, all freeze.

2. Challenge and Reply. To ensure people are who they say they are, it is necessary to develop a series of challenge and reply codes.

   a. Use code names, first names or nicknames within the platoon.

   b. Use numbering system for others with a known total proving identity; e.g., the correct coded reply is "nine." If the challenge was "four," the other person should say "five."

      (1) Use odd numbers only for the prearranged total. This negates the likelihood that the challenged person can give back the challenger's number and get the correct answer.

      (2) The challenge is normally initiated by the returning element.

3. Develop signals using light, radio, or Morse code that cover the range of communication requirements. For example:

   a. All clear; come in

   b. Come in, but be extra careful

   c. Danger - enemy; don't come in.

2.3.5 Movement. The SEAL patrol spends more time moving than fighting. Moving carelessly may cause a patrol to make contact with the enemy unintentionally or when the patrol is unprepared. The goal of effective movement is to make contact with the enemy on your terms, rather than his. The fundamentals of movement include:

1. Using covered and concealed routes

2. Avoiding routes or actions expected by the enemy

3. Conducting observation and reconnaissance prior to movement

4. Accurately navigating

5. Maintaining all-around security (including air guard).

2.3.5.1 Patrol Movement. Consider all movements to be tactical movements.

1. Procedures.

   a. Do not move directly forward from covered positions.

   b. Avoid likely ambush sites and other danger areas (channeled terrain is especially dangerous).

   c. Enforce camouflage, noise, and light discipline.

   d. Take maximum advantage of weather conditions and ambient noises that can mask patrol noise and cover tracks.

2. Speed. Speed will always be limited by the necessity to avoid detection. In addition, the platoon's speed of movement is controlled by numerous factors including:

   a. The nature of the terrain

   b. The mission

   c. Time of day

   d. Environmental effects (storms, temperature, extremes, etc.)

   e. Personnel load

   f. Use of an experienced guide

   g. Booby trap or mine threat

   h. Requirements for security.

3. Load Bearing Equipment (LBE). NSW personnel are responsible for maintaining and carrying a great deal of equipment. Clothing, LBE, and rucksack designs should allow easy access during planned and emergency operations. Items frequently needed or changed should be located in the outer pockets or flap of the rucksack. Modifications may be required and the rucksack should have a separate component that can be removed and carried should the pack be ditched. One member of a unit may carry an item (e.g., tent) that is shared by two or more individuals. Individuals should stow equipment in a common manner, for easy location by other members of the unit. Equipment organization is described as follows.

   a. First Line. This equipment is used primarily for survival. It should include signaling devices, emergency rations, spare parts, etc. Survival items should be stored in the pockets of the outer garment.

   b. Second Line. These items should be carried by LBE, in a vest or on a belt. Second line equipment includes weapons, ammunition and enough supplies to survive should the third line (rucksack) be abandoned.

c. Third Line. This equipment is generally stored in the backpack. Items in the third line include the sleeping bag, ground pad, tent and other operational equipment.

2.3.5.2 Night Movement. Without a guide or unless familiar with the area, night movement may be difficult, but necessary to avoid detection.

1. Options. Moving on trails and up roads at night may provide better navigation; however, patrolling on trails and roads is dangerous. Off trail patrolling at night is preferred. Visibility at night in open or lightly forested areas is normally sufficient to see other patrol members because of starlight and moonlight. When moving at night in reduced visibility, the following procedures may be useful:

    a. Maintain visual contact between patrol members; hands-on contact may be required if visual contact is not possible.

    b. "Ranger tabs" or reflective glint tape attached to the inside fold of the collar or back of headgear can be effective for maintaining control.

2. Broken Contact. After a pre-planned time period, the patrol will conduct rendezvous procedures according to the plan promulgated at the PL order (PLO) briefing.

    a. The rear part of the patrol remains in place.

    b. The leading part retraces his steps to make up the separation, moving back on the route just patrolled.

    c. Patrol will rendezvous using a challenge and reply.

    d. If all else fails, proceed to the last enroute rally point. See paragraph 2.4.9.2 for procedures at enroute rally points.

2.3.5.3 Clandestine Individual Movement. Individual movement techniques are essential and require detailed thought. Movement occupies most of a patrol's time and can determine the success or failure of the mission. It can be a significant problem when patrol members become separated. The PL must always adapt speed of movement to that of the rear of the patrol. Responsibility for keeping together must be from front to rear. Rehearse and develop SOPs. Preparation for, and types of movement under varying conditions is presented in detail in Appendix A, Section A.3.

2.3.5.4 Noise Discipline. Noise discipline regarding voice and movement is essential at all times. Strict noise discipline is especially necessary when there is little vegetation present to absorb sound. Ensure all equipment is noise proofed by padding and taping. With practice, it is possible to move at a good pace in comparative silence.

2.3.5.5 Light Discipline. Exercise firm light discipline. The glow of a red lens flashlight may give away a position. Cigarettes are forbidden on patrol. A poncho should be used to shield flashlights when a map study is necessary on patrol. Light discipline includes shadow discipline. Patrol members should be aware of their shadows. Observation from overhead can easily detect shadows, which may not be visible from an observer on the ground.

2.3.5.6 Travel on Trails

1. Avoid movements on trails, except when speed is essential and the expected threat is very low.

2. Whenever possible, hide signs of movement to prevent leaving a trail. (This is especially important when crossing danger areas such as trails or beaches. Remember that standard issue boots leave a distinctive trail sign.)

3. Wear combat boots to prevent foot injuries.

4. Use operational deception (OPDEC) by never using the same trails or routes. This is especially true when entering or leaving the NSWTG/TU FOB or forward operating location.

2.3.6 Maneuver. An essential tenet of successful small unit tactics is the ability of the PL to organize the unit to maximize its maneuverability. Tactical maneuverability is the ability of the patrol to flexibly organize and reorganize to meet the current threat and conditions with minimum communication, while maintaining unit integrity and control. A competent patrol allows for innovation and ingenuity as well.

2.3.6.1 Use of Maneuver. Maneuver is used to describe the actions of a unit when one element is moving to a position to fight the enemy and another element is supporting that movement by fire. The objective of maneuver is to move an element into a position from which it is possible to maximize destructive firepower against the enemy. In order to assist movement to such a position, a separate supporting element provides covering fire. Maneuver can be used in an offensive manner (e.g., in an assault) or defensively (e.g., to break contact or to counter an ambush). Proper understanding and use of maneuver is essential to good patrol tactics. This definition of maneuver includes these key points:

1. Fire + Movement = Maneuver.

2. Maneuver is used ONLY to describe actions in contact with the enemy.

2.3.6.2 Maneuver Units. Patrol C^2 is exercised through the patrol maneuver units. The maneuver units for the 16-man SEAL platoon patrol are: platoon, squad, and FT.

2.3.7 Weapons Discipline. Weapons fire control is required by all when a patrol or element is fighting the enemy. The principles and uses of fire discipline discussed below must be second nature for all SEALs in order to be successful in combat. SEALs must know the characteristics of their weapons, how to fire them effectively and have an understanding of the different types of fire.

2.3.7.1 Fire Characteristics

1. Trajectory. Trajectory is the path of the projectile in its flight from the muzzle of the weapon to the point of impact. At greater ranges the shooter must elevate his weapon. This raises the height of the trajectory.

2. Danger Space. The danger space is the space between the weapon and the target where the trajectory does not rise above 1.8 meters (the height of an average man). At greater ranges, only a portion of the space between the weapon and the target is a danger space because the trajectory of the bullet must be elevated in order to cover the increased distance.

3. Cone of Fire. Each successive round fired from the same weapon at a target assumes a slightly different trajectory through the air. Taken together, these different trajectories form an elliptical impact pattern on the target that is called a cone of fire. The cone of fire is caused by variations in aiming and holding the weapon, vibrations, ammunition, wind and atmospheric conditions.

4. Beaten Zone. The beaten zone is the area where the cone of fire strikes the target or the ground. The beaten zone varies depending upon the type of terrain. For example:

   a. Over uniformly sloping or level ground the beaten zone will be elliptical.

b. If the ground slopes downward, the beaten zone will be longer.

c. If the ground slopes upward, the beaten zone will be shorter.

5. Casualty Radius. When high explosive projectiles are fired, they produce casualties by fragmentation and concussion. Casualty radius or fragmentation range is the area around the point of impact where personnel can be expected to be injured or incapacitated.

6. Kill Radius. The area around the point of impact where personnel will be killed. (Always smaller than the casualty radius.)

2.3.7.2 Types of Fire

1. Grazing Fire. In grazing fire the trajectory of the rounds is parallel to the ground and does not rise above one-meter from the ground (waist high).

   a. Because of the tendency to elevate fire after the first round, especially during an ambush, the firer must make a conscious effort to aim low; less than waist high (one meter).

   b. Effective grazing fire can be made from the prone position to a range of approximately 500 meters with rifles and AWs over uniformly sloping or level terrain.

   c. Low grazing fire, which is aimed at the foot of personnel targets, is especially effective. Not only does it wound directly and through ricochets, but also the dirt and debris thrown up effects the enemy's ability to see, shoot and exercise control.

2. Plunging Fire. Plunging fire is fire that has a trajectory path higher than a standing man. It strikes the ground at a high angle of impact so that the danger zone is practically confined to the beaten zone. The longer the ranges of fire, the more plunging it becomes because the angle of fall of the bullets becomes greater. Plunging fire can be considered "indirect" in that it does not have a flat trajectory.

3. Flanking Fire. Flanking fire is shot into the flank (side) of the target. It is normally considered to be delivered at a right angle to the target's line of march if moving, or orientation if fixed. When correctly employed, flanking fire can be especially devastating because it comes from an unexpected direction.

4. Frontal Fire. Frontal Fire is delivered directly into the front of the target, i.e., directly into its line of march if moving, or orientation if fixed.

5. Oblique Fire. Oblique fire occurs when the long axis of the beaten zone is oblique (approximately 45 degrees) to the long axis of the target.

6. Enfilade Fire. Enfilade fire occurs when the long axis of the beaten zone is the same as the long axis of the target. Enfilade fire can be flanking, frontal, or oblique. It is the best type of fire with respect to the target because it makes the best use of the beaten zone. An example would be firing at the front of a column of enemy soldiers or at the flank of soldiers on line. The M60 machine gun is an especially effective enfilade fire weapon against personnel targets because of its ability to cause damage through the length of a column.

7. Direct Fire. Fire directed at a target that is visible to the aimer. Sometimes referred to as Point fire.

8. Indirect Fire. Indirect fire is delivered at a target that cannot be seen by the aimer. The mortar is an example of a SEAL weapon capable of delivering indirect fire.

9. Area Fire. Area fire is distributed in width and depth so that overlapping fields of fire effectively covers a linear, column, or area target. This is the quickest and most effective method of ensuring that all parts of the target are brought under fire. It is especially effective in ambush, dense vegetation, or limited visibility.

10. Suppressive Fire. Suppressive fire is a type of fire that does not let the enemy see, shoot, or attack your position or one of your elements. It may be either direct or indirect fire that can screen squad or element movements.

11. Fire Superiority. A heavy, sustained, accurate volume of fire that renders enemy fire ineffective or causes it to cease completely. Fire superiority is usually the objective at the initiation of a contact with the enemy. Once fire superiority has been established, then the squads or elements will give suppressive fire to maneuver to kill the enemy or break contact.

2.3.7.3 Weapon Ready Position. Whether stopped or moving, carry weapons in accordance with platoon SOPs.

1. In a vehicle or a boat, the weapon should be pointed upward or outboard.

2. In a helicopter, the weapon should be pointed downward.

3. Carry the weapon with magazine in, round chambered, dust cover closed, and safety on. Ensure that spare magazines are readily accessible.

2.3.7.4 Commence Fire. A SEAL patrol should not commence fire (unplanned) without analyzing the situation. If it is receiving ineffective fire only, it is possible the enemy has not pinpointed the patrol's position. Opening fire prematurely will remove all doubt as to patrol location. Stealth is usually the best defense. Commence fire when:

1. The PL orders it.

2. The PL opens fire.

3. Per platoon SOPs (e.g., when ambushed or taking effective fire).

2.3.7.5 Fire Responsibilities

1. Riflemen will shoot point targets with effective fire.

2. The PL/APLs primary responsibility is to direct the actions of the elements, his secondary responsibilities are to engage the enemy.

3. AW men will shoot with controlled rates of fire.

4. GNs will:

   a. Place grenade rounds carefully to ensure grenades do not bounce back and injure the platoon

   b. Ensure illumination is kept constant, if used.

2.3.7.6 Verbal Communication. When firing, clear communications are essential. Yell:

1. "Change Magazine" when changing a magazine
2. "Out Of Ammo" if out of ammunition and requiring more
3. "Jam" if your weapon is jammed
4. "Cease Fire" called by the PL and passed by all personnel.
5. It is also possible to use a code system to increase security, such as:
   a. "Changing Magazine" might be "Green"
   b. "Out Of Ammo" might be "Black"
   c. "Jam" might be "Red".

2.3.8 Patrol Navigation. Constant training in map and compass skills in a variety of environments is required to maintain navigation proficiency. Some of the key principles to successful navigation are listed below:

1. Performing a thorough map study prior to going in the field
2. Using a navigation plan that includes enroute reset points, which can be used to verify the patrol's position
3. Maintaining an accurate pace count
4. Taking corrective action when unsure of position (e.g., stop and take resections, or patrol to an obvious reset point).

2.3.8.1 Navigator. Maintaining correct direction and knowing the distance traveled are critical to mission success. The PL is the primary navigator. Everyone in the patrol should keep track of their location in case they are separated during contact with the enemy.

2.3.8.2 Techniques. Unless a man pack global positioning system (GPS) is available, the patrol will navigate by "dead reckoning" (see paragraph 2.3.8.3 below.). If using GPS, dead reckoning should be used as a back-up. It is essential to verify GPS data with your map and compass because GPS can be inaccurate (e.g., undetected equipment malfunction, low batteries or inadequate number of satellites).

2.3.8.3 Direction. The compass is the only reliable and constant guide to direction unless in a region of magnetic anomaly. All members of a SEAL platoon must be able to use a compass with confidence and accuracy. They should also be proficient in land navigation. Maintaining direction during cross-country movement is extremely difficult, even when landmarks are present.

1. Selecting Approach Avenues and Infiltration Routes. Base this selection on a detailed survey of all possible sources:
   a. Recent photographs
   b. Updated maps and charts

c. Information from personnel familiar with the area.

2. Compass Techniques. When good landmarks are limited, the compass can be your best friend if used properly.

   a. Hold the compass at waist level.

   b. Walk the set azimuth direction.

   c. Determine a star along the azimuth and use it as a guide. This requires a high degree of concentration and constant verification with a compass. If using a star other than the North Star, verify the star's position approximately every fifteen minutes.

   d. "Dead Reckoning" can be used when a good landmark is available (a stationary object that is highly visible and readily identifiable from among other similar landmarks, and which is at or near the direction the patrol is moving). This involves sighting on it, noting the direction, and pacing toward it. Take a new compass reading on the landmark's direction at least every half-hour (preferably every fifteen minutes) and record the direction (avoid "wrapping around" a landmark using frequent compass sightings).

3. Drift and Direction. Most navigators have a natural tendency to drift to the left or right. Learn this natural drift and compensate accordingly.

2.3.8.4 Distance Traveled. Determine distance traveled by checking terrain features with an updated map, knowing the speed of movement, and pacing. Distances to key terrain features such as distinguishable mountains, manmade objects, and heavily forested areas should be committed to memory. Special attention should be focused on linear features such as roads, trails, and rivers as these are often unavoidable obstacles. (It is important to remember that very few maps show all roads, trails and streams - they can be confusing.)

1. Pacing. Pacing is the best way to keep track of distance traveled.

   a. Practice pacing in different environments.

   b. Determine how many paces it takes to patrol 100 meters in each type of terrain.

   c. Keep track of accumulated distance. Some methods that may be used to keep accurate count of the pace include: tying a knot in a line for every 100 meters, moving pebbles from one pocket to another, or using a mechanical/digital counter.

   d. Designate at least two men as pacers. Compare paced distances at least every half-hour. Determine the most reasonable distance covered given the terrain and by using available topographic landmarks.

   e. Pace count per 100 meters will increase with altitude and fatigue.

2. Offset. An offset is a planned deviation to the right or left of the straight-line azimuth to the objective. Each degree you offset will move you about 17 meters to the right or left for each 1,000 meters traveled. Planning an offset in your route helps to ensure that you know which direction the target is in when you arrive in its vicinity. The patrol can offset to bypass enemy positions, to go around obstacles and danger areas, or to ensure that the patrol does not unknowingly enter the immediate vicinity of the target.

2.3.8.5 Following Routes. Keep yourself oriented at all times. As you move along, observe the terrain carefully and mentally check off the distinctive features you noted in studying and planning the route. Many aids are available to help you check and double check where you are and where you are going.

1. The location and direction of flow of principal streams.

2. Hills, valleys, and peculiar terrain features such as swamps and barren areas.

3. Railroad tracks, power lines, roads, towns, and other man-made objects.

4. The fire of machine-guns, mortars, or artillery:

   a. If you know the general origin and direction of machine gun fire, you may be able to use these fires to orient yourself.

   b. Mortar and artillery rounds fired on known locations can guide you or help you orient yourself.

5. What to do if lost. This should never occur; especially with the use of handheld GPS. However, should you lose your way, do not continue until you have reoriented yourself with respect to the immediate terrain and position of the enemy and friendly forces. Keep calm and think the situation over.

   a. Use of a map and compass in correlation with the terrain is the best way to determine your location.

   b. If you do not have a map, establish the direction in which you were traveling (with a compass, or in relation to the sun, moon, or identifiable terrain features).

   c. Mentally retrace your route, recalling hills, trails, streams, and other terrain features you passed.

   d. If necessary, go back in the direction from which you came until you find features from which to orient yourself.

   e. If you do not have a compass, use the sun, the North Star, or the constellation Cassiopeia to find north. After finding north, you can move in any desired direction by remembering that, as you face north (0°/360°): (The constellation, Southern Cross, can be used to find south in the southern hemisphere.)

      (1) Northeast (45 degrees) is to your right front.

      (2) East (90 degrees) is to your right.

      (3) Southeast (135 degrees) is to your right rear.

      (4) South (180 degrees) is to your rear.

      (5) Southwest (225 degrees) is to your left rear.

      (6) West (270 degrees) is to your left.

      (7) Northwest (315 degrees) is to your left front.

2.3.9 Security Perimeters. Establishing effective security perimeters is key to patrol security. Defensive perimeter procedures are used:

1. Upon insertion/infiltration or exfiltration/extraction.
2. Prior to moving from the RP to the objective.
3. When stopping to rest or discuss plans.
4. When listening/watching for the enemy.
5. During any prolonged stop (for rest, eating, or sleeping).
6. At all designated RPs.
7. When stopping for communications.
8. After crossing obstacles and prior to proceeding. The perimeter should be incrementally set as each man/buddy pair crosses the obstacle.

2.3.9.1 Guiding Principles

1. Set perimeters with a minimum number of alterations in the patrol formation.
2. Use camouflage techniques and take maximum advantage of the existing cover and concealment. (Appendix A pertains.)
3. Position heaviest weapons (AWs and GNs) to cover most likely avenues of enemy approach. The PL may position each person as he moves into the perimeter.
4. Whenever possible, fishhook into perimeters in order to watch for enemy trackers.

2.3.9.2 Perimeter Configuration. Baseline perimeter radius should allow each man to see the men on either side of him. Size depends on illumination, cover, terrain, and situation. Perimeters can be set in various formations to take advantage of patrol composition and the terrain. Details of perimeters are contained in Section 2.4.8.2.

2.3.10 Breaks. Rest is a weapon; use it to your advantage.

2.3.10.1 Rest Management. Breaks must be called frequently to give opportunities for rest, observation, and listening. Terrain difficulties will usually dictate when to take breaks. "Halt" is synonymous with "rest" and "break."

1. Good physical conditioning and hydration are paramount to better cope with all the physical demands SEALs should anticipate.
2. The patrol can go only as fast as the slowest man. Remember the RTO and M60 men are carrying the greatest weight. It is better to take breaks early to maintain stamina rather than push men to fatigue.
3. Regardless of when you halt, get into a relaxed position, but remain alert. Halts are good times to compare notes, check your compass bearing, and get each other briefed on what you've seen and heard since your last stop.

2.3.10.2 Break Standard Operating Procedures. Breaks are vitally important to the accomplishment of the mission, but they cannot be haphazard. Solid working SOPs are a must for secure breaks. The following are some general recommendations for breaks:

1. All Stops

   a. Establish an "all around defense" posture, maintaining 50/50 security

   b. Maintain good concealment and noise/light discipline

   c. Each man face in his direction of responsibility.

2. On a Trail

   a. Move away, well clear of the trail (fishhook, if possible)

   b. Adopt an ambush position.

3. Eating/Sleeping

   a. Allow only a limited number to take the time to eat or sleep while the others maintain security. Use the buddy system. Always maintain 50/50 security whether eating, drinking, or sleeping.

   b. Use of a "tug" line is advisable to signal the patrol members without noise. Use 550 cord wrapped around a finger.

   c. If men cannot stay awake while on watch and at breaks, have them pair up and kneel rather than lie down. Do not allow personnel to rest their heads on anything, as they will definitely fall asleep.

2.3.11 Handling of Wounded and Dead. The nature of the operations being conducted may dictate the appropriate method for handling wounded personnel. Whatever the method used for handling wounded, dead and prisoners, mission accomplishment should be the foremost consideration. The PL should analyze all options before deciding to abort an operation due to casualties. Planning should include contingencies for casualties and define the minimum number of healthy personnel required to complete the mission. General principles for handling wounded and dead follow.

2.3.11.1 Wounded. The first response must be to suppress enemy fire and then remove the wounded from the immediate area before applying first aid. Administering first aid during contact causes more casualties. When tactically feasible conduct triage, determine a casualty's status, and determine follow-on action based on mission and situation.

1. Walking Wounded. The following options are generally available:

   a. MEDEVAC.

   b. Accompany patrol.

   c. Conceal them for later pick-up. An unwounded man should be left with the wounded man or men.

   d. Return on their own to friendly areas. An unwounded man should accompany him or them.

2. Seriously Wounded. The following options are generally available:

   a. MEDEVAC.

   b. Be carried by the patrol. This is generally practical only when the patrol is returning to a friendly area.

   c. Be concealed for later pickup. An unwounded man should be left with the wounded man.

2.3.11.2 Dead. The dead may be handled the same as the seriously wounded except that, when concealed for later pick-up, no one is left with the body. While it is true that SEALs have never left their dead in the field, it is imperative to suppress enemy fire prior to taking action to recover a body.

2.3.11.3 Prisoners. Prisoners are bound and gagged and may be blindfold as appropriate. They may then:

1. Be returned, under guard, to friendly areas

2. Be evacuated by air, taken with the patrol, or concealed for later pick-up, in the same manner as seriously wounded.

2.3.12. Enemy Prisoners of War, Documents, and Material

2.3.12.1 Prisoners. The handling of prisoners of war is governed by international agreement (Geneva Convention Relative to the Treatment of Prisoners of War, 12 August 1949). Taking prisoners when that is not the specific mission of the patrol demands at least one member's full attention that effectively eliminates him from other patrol functions as well as slowing down and jeopardizing the patrol.

1. Preparation. The mission could be to abduct key personnel or a situation could come up where you may have to take a prisoner such as a target of opportunity. The responsibility for abduction and detailed handling instructions of the prisoner should be delineated in the PLO. Planning and practice with restraint equipment are a must to ensure the proper gear is brought on the operation.

2. Types of Prisoners.

   a. A compliant prisoner could be a downed pilot or a hostage.

   b. Compliant to non-compliant prisoners are possibly friendly, but their attitude may change.

   c. Non-compliant prisoners are combative and openly hostile.

   d. Non-combative and non-compliant prisoners could be stupid, scared, or deaf.

3. Sequence of Events for Prisoner Handling. If the PL decides to take prisoners, the directed handlers will comply with the 5 "Ss."

   a. Seize the prisoner using sentry stalking or hand-to-hand methods. During abduction take care not to injure the prisoner to the point where he can't walk or else you'll be giving up two or more patrol members to carry him. Maintain control. Indicate by appropriate force, speech or sign language that you want him quiet. Don't let him talk, move around, look back, or distract you.

b. Secure the prisoner's sight, sound, and freedom of movement as required to ensure they won't escape or compromise security. Safeguard the prisoners as you take them to the rear. Make sure they arrive safely. Do not allow anyone to abuse them, but also do not allow anyone to give them cigarettes, food, or water (unless required for mission completion).

c. Search prisoners for weapons and documents as soon as you capture them. Take weapons to prevent resistance and documents to prevent the prisoners from destroying them. Prisoners, from whom personal property is taken, including personal documents, should be given a written receipt for the property as soon as it is practical to do so. Tag documents and other personal property taken so you know which prisoner had them.

d. Silence is essential. Do not allow prisoners to talk to each other. This prevents them from planning escape and from cautioning each other on security. Report anything a prisoner says to you or attempts to say to another prisoner.

e. Segregate them into groups: officers, noncommissioned officers, privates, deserters, civilians, females, and political indoctrination personnel. This prevents the leaders from organizing for mass escape and from making the rest of prisoners security-minded. Keep the prisoners segregated as you evacuate them to rear. The information they have does no good until it is obtained by an interrogator and processed.

2.3.12.2 Documents. The importance of captured documents cannot be overstated. Timely and accurate information is vitally important to successful combat operations.

1. Prisoner Documents. Documents taken from prisoners are tagged with the name of the prisoner, date, time, place of capture, and the unit making the capture. All documents are sent or carried to the rear with the prisoner escort. In the rear area the prisoner is questioned about the information in the document.

2. Loose Documents. Documents found on the ground, in enemy command posts, or similar places are identified to show where and when they were found and the unit that found them. Tag these documents and give them to your squad or PL. He ensures that the documents are given or sent to the TG/TU intelligence officer. The intelligence officer gets whatever information he can use and forwards the documents to the next higher headquarters.

2.3.12.3 Material. Captured or observed materials help to build an accurate intelligence assessment of the enemy.

1. New Materials. Report any new type of weapon or equipment you find to your squad or PL. If it is light enough to be carried and you are certain it is safe and not booby trapped, note everything you can observe about its condition, recover it, and report it when you have the opportunity. Make notes or sketches to help describe what you saw. Photos of the objects are excellent when practical. If you cannot carry the item, report it to your squad or PL on scene so that it can be addressed at the debrief.

2. War Trophies. Captured documents and material are the property of the United States. Do not attempt to keep unauthorized items. They are needed to help your unit accomplish its mission.

SECTION II TACTICS, TECHNIQUES, AND PROCEDURES

2.4 PATROL FORMATIONS AND TACTICS

2.4.1 General. SEAL patrol tactics are of vital importance to mission accomplishment and patrol survival. Specific patrol tactics vary throughout the NSW community, however, and can be a source of controversy. What worked for one platoon in a Vietnam firefight may be different from what worked for another platoon. Often, the only obvious commonality is the fact that all participants feel strongly about the tactics that worked for them.

2.4.2 Understanding Tactics. Every platoon member must have a solid understanding of SEAL tactics, platoon SOPs, and platoon IADs. PCs must ensure that each member understands platoon SOPs and IADs and is able to act instinctively and in concert with other patrol members.

2.4.2.1 Team Concept. Building team concept in tactics is essential. Nobody wins alone. It is imperative that each patrol member's actions be coordinated with the rest of the patrol.

- **TEAM:** **T**ogether **E**veryone **A**chieves **M**ore

2.4.2.2 Standard Operating Procedures. SOPs are procedures practiced by a SEAL platoon over and over, until they become automatic responses to various situations encountered when on patrol. SOPs and IADs must be designed to get platoon members into some pre-planned actions with minimal or no orders.

2.4.2.3 The Basics of Tactics. The fundamentals of SEAL tactics include the following basic concepts:

1. Maneuver to attack when the enemy is off guard (surprise).
2. Detect the enemy before he detects you (reconnaissance and surveillance).
3. Use stealth and surprise to their maximum advantage. Doing the unexpected is a small unit's most effective asset.
4. Apply disciplined, accurate, and concentrated firepower.
5. Take and maintain the initiative; take decisive violent action when attacking.
6. Use the terrain and environment with greater flexibility, more stealth and cunning, and superior aggression than the enemy.

2.4.3 Formations and Movement Techniques. Formations are ways to place and organize patrol members in relation to one another. Movement techniques differ in that they encompass various methods that may be used from a variety of formations. Both formations and movement techniques provide the PL options for the different tactical situations he may encounter. The following paragraphs provide basic information about formations and movement techniques.

2.4.3.1 General Considerations. Because of their small unit size (i.e., platoon or smaller element), SEALs normally move as a single unit in a file formation (see Figure 2-1). Depending on a variety of factors such as the enemy situation, likelihood of contact, terrain, size of the patrol, and time available, there may be times when it is more appropriate to use one of the infantry movement techniques or formations discussed in the following sections. In any case, a full understanding of these techniques will increase a SEAL's patrolling expertise.

The most important factor in selecting a movement technique or formation is training the patrol members. Patrol SOPs and IADs must be rehearsed extensively for each anticipated technique/formation.

2.4.3.2 Interval. Once a formation is selected, the distance between each person within the formation will depend on such factors as overall visibility (day or night, thick brush, fog, etc.), the amount of control required based on enemy intelligence, and the terrain. In a file, members should be able to see hand signals passed by the men directly to the front and rear. Avoid bunching the patrol together; when this occurs the patrol becomes a much easier ambush target.

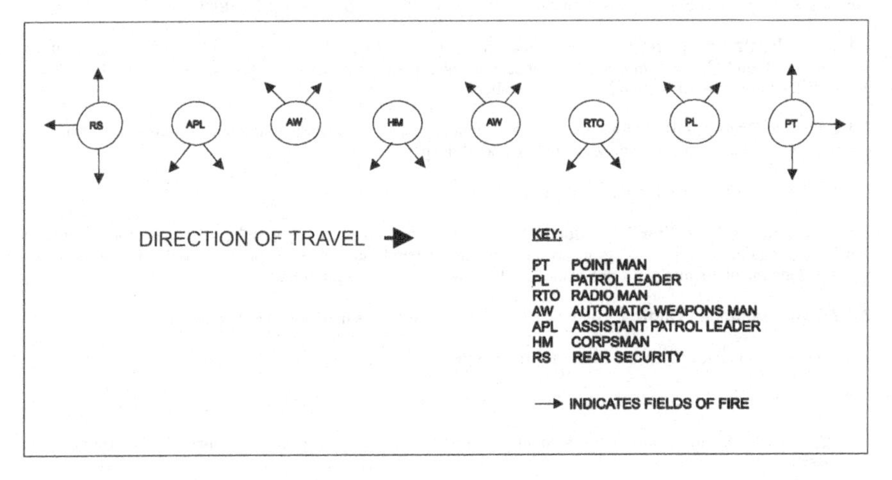

Figure 2-1. File Formation (Fields of Fire Depicted by Arrows)

2.4.3.3 Fields of Fire. The term "Field of Fire," as currently used in NSW, defines a specific area of coverage assigned to an individual or element within a formation. Both fire and visual surveillance are included. Fields of fire are designed so that the responsible individual or element can effectively fire at the enemy without endangering other patrol members or friendly personnel in the vicinity.

Fields of fire are normally unrestricted except by patrol fire coverage assignments and the presence of other friendly forces. At times, some personnel will have a restricted field of fire. Figure 2-2 shows the restricted sectors of the RS and AW man because of the location of friendly boats.

Fields of fire should overlap to provide maximum coverage of a given area and efficient use of individuals and elements.

Although heavy forestation or hilly terrain may sometimes interfere with visual surveillance, all personnel must remain vigilant in their assigned sectors and to ensure a patrol maintains 360-degree security.

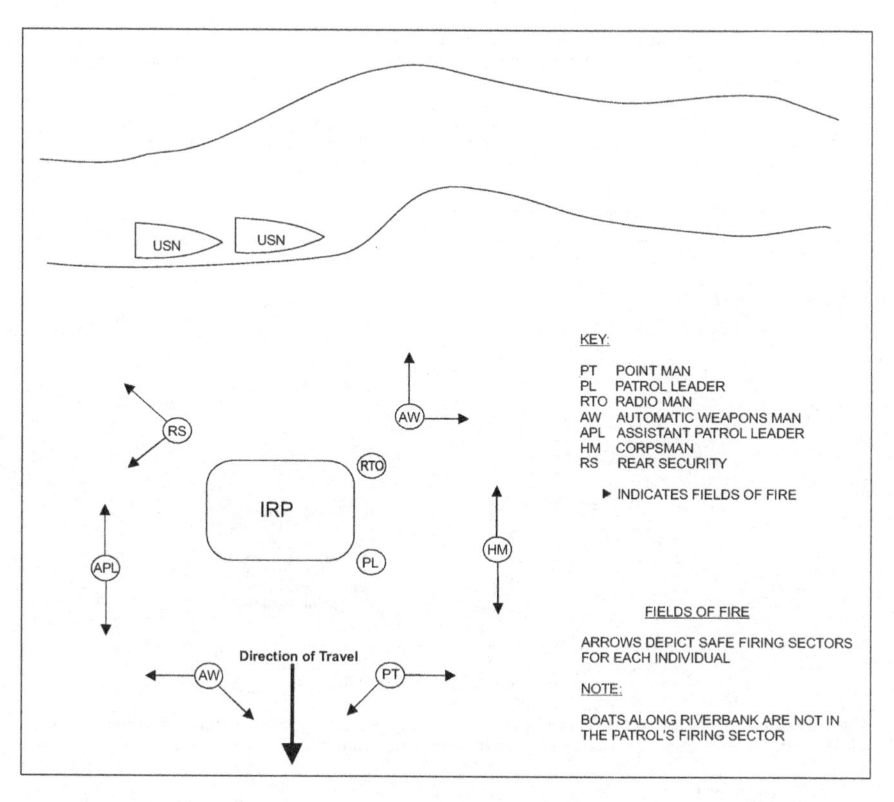

Figure 2-2. Fields of Fire

2.4.3.4 Movement Techniques. The following techniques can be used in various formations. The descriptions below refer to distances between individuals, FTs, and squads. All distances are approximate and will vary with the terrain and situation. The techniques described involve separating the patrol into sub-units. This greatly increases the chance of inadvertent contact between friendly units unless there is extensive training and clear SOPs are developed.

1. Traveling. Traveling is a patrolling technique used when contact with the enemy is not expected or probable, and when speed of advance is important. Traveling requires the patrol to be split into two units. When traveling as a platoon, the two squads will be separated by approximately 20 meters, depending on the terrain. See Figure 2-3.

2. Traveling Overwatch. Traveling overwatch is the technique used when contact is possible, but speed is still necessary. The platoon is divided into two squads (or the squad into two FTs). The trailing team drops about 50

meters behind the lead team and is prepared to support the lead squad. If the lead squad receives fire, the trailing squad is far enough away so that it shouldn't be hit by the same enemy fire, yet close enough to maneuver in support of the lead squad.

3. Bounding Overwatch. Bounding overwatch is used when contact is probable. This technique reduces the speed of advance. One squad moves while the other remains in a good fire position, covering the moving squad. The key to this movement technique is using terrain features for cover and concealment. A bound is normally about 100-150 meters forward of the overwatching squad (depending on terrain and visibility). This technique is also referred to as the leapfrog method or "leapfrogging".

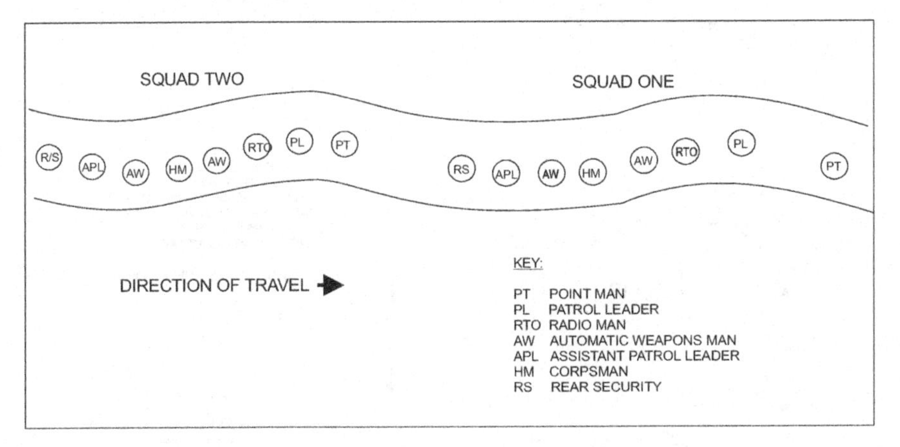

Figure 2-3. Traveling (In File)

2.4.3.5 Selection of Movement Techniques. The primary consideration in selecting a movement technique should be the threat of enemy contact. In addition, the following factors should be considered:

1. Speed requirements
2. Dispersion inherent to each technique
3. Control requirements
4. Security requirements.

ENEMY CONTACT	MOVEMENT TECHNIQUE
Not Probable	SEAL FILE
Not Probable	Traveling
Possible	Traveling Overwatch
Probable	Bounding Overwatch

2.4.4 Types of Formations. Formations are common to all military units. They provide frameworks for tactics for fighting or withdrawing and optimize the security and firepower of the patrol. Formations may be offensive or defensive. Details of the basic formations are discussed below:

2.4.4.1 File Formation. The file formation is the most common formation used by SEAL patrols. It may also be referred to as a column. In infantry terms, it offers little fighting advantage and is generally reserved for low threat areas. Although a SEAL patrol may decide to use it due to its compatibility with small units, the file's strengths and limitations need to be understood completely.

1. Formation Basics. The best time to use the file is during hours of limited visibility and in thick vegetation. It is also appropriate for narrow trails. (SEAL patrols should avoid trails, however, unless absolutely necessary for mission accomplishment.) This is the most basic of all combat formations and the one from which all other formations originate.

   a. Normally the PL is the number two man in the file. In situations presenting a higher probability of enemy contact, an AW man can move into the number two position ahead of the PL. This places the AW man in a better position to immediately engage the enemy with concentrated firepower while the rest of the formation comes on line.

   b. Once in contact and the AW man has expended his first rounds, the patrol should, if feasible, engage the enemy with an M203 grenade launcher. A grenade can have a significant and immediate impact on the enemy, often providing a better maneuver opportunity than engaging the enemy with only rifle fire.

2. Advantages. The file formation has the following advantages:

   a. Good C^2 (unit integrity is maintained)

   b. Stealth

   c. Ease of movement at night and in thick vegetation

   d. Good for quick movement.

3. Disadvantages. Unless moving through urban or well-traveled areas, the greatest threat from inadvertent contact with the enemy generally comes from the front of the patrol where the file has the least amount of fire coverage (the PT only). For this reason, more conventional units usually discourage use of the file in medium to high threat areas. The file has the following additional disadvantages:

   a. It leaves a trail if the patrol is not careful. (Following the previous man's footsteps will help to minimize footprint detection and cause confusion as to how many personnel are in the patrol.)

   b. The file can become too dispersed in open areas, especially when operating as a full platoon. This can make C^2 difficult.

4. Effects of Fire

   a. Frontal, enfilade fire is most effective against a file formation.

   b. Grazing fire at the PT requires windage correction only, since elevation rises naturally, increasing the depth of the beaten zone.

c. The file formation represents an area target when exposed to frontal fire.

d. Flanking fire requires point target corrections in windage and elevation.

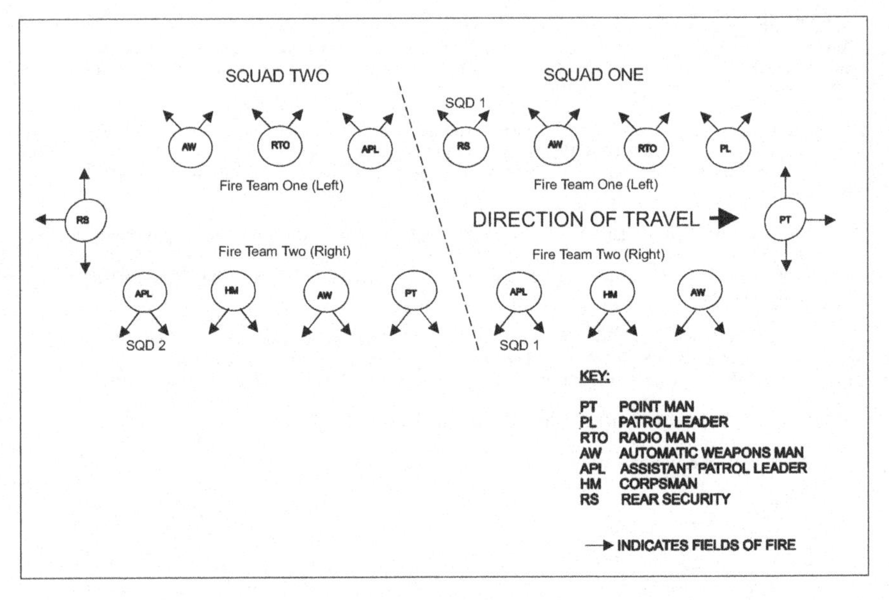

Figure 2-4. Platoon Staggered File (Fields of Fire Depicted by Arrows)

2.4.4.2 Staggered File Formation. The staggered file (sometimes referred to as a staggered column) uses all the strong points of a file formation but adds depth. It is best used on trails, paths, and cuts where a file formation would be less appropriate. It can be used in vegetation of medium density.

1. Formation Basics. The standard file is usually preferred for squad or smaller formations; however, a staggered file shortens the distance between the PT and RS, thus increasing C^2. The simplest way to form this formation is from a standard file: each man steps toward his field of fire and then resumes patrolling. (The number three and four men, the RTO and AW men, can step opposite from their fields of fire in order to keep the RTO next to the PL). See Figures 2-4 and 2-5.

2. Advantages. The staggered file formation has the following advantages:

   a. Gives the front of the formation more firepower than a standard file.

   b. Good when the platoon is operating together as two squads.

c. Good C^2.

d. Ease of movement on wider trails/areas where file would be too dispersed.

e. Accentuates individual fields of fire.

f. In certain urban environments it can be used to cover both sides of a street at the same time.

g. Can quickly move into the wedge or "V" formation.

h. Aids in movement of prisoners or rescued personnel.

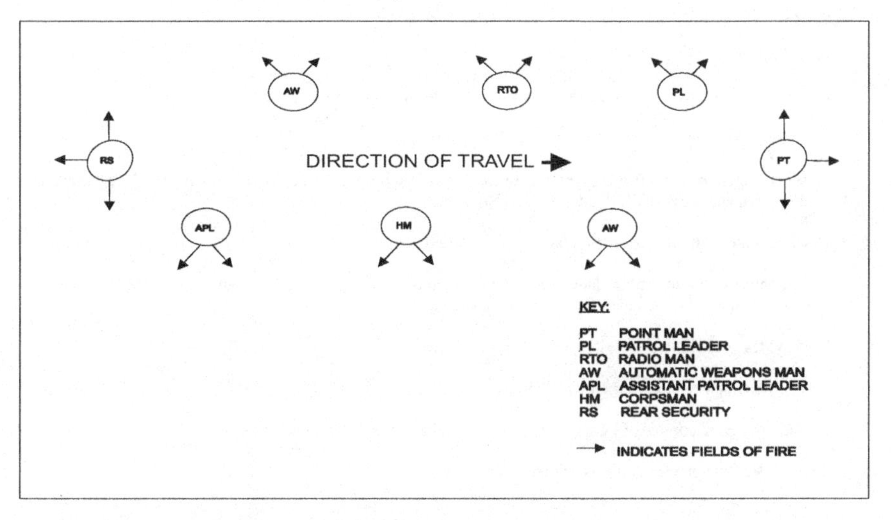

Figure 2-5. Squad Staggered File (Fields of Fire Depicted by Arrows)

3. Disadvantages. The staggered file formation has the following disadvantages:

a. Makes more noise than a file.

b. Requires clearing of two adjacent trails, increasing the booby trap threat and the possibility of being tracked.

c. Additional SOPs and IADs are necessary to ensure instinctive reaction if contacted.

4. Effects of Fires.

a. Increases the number of people in the kill zone from flanking or oblique fire.

b. Same effects as file formation.

2.4.4.3 Line Formation. The line formation is normally used to make the final rush on an objective. The line must move steadily and guide on a base man or team.

1. Formation Basics. In this formation, patrol members are positioned in a line perpendicular to the direction of movement. See Figure 2-6. This formation is sometimes used when crossing linear danger areas and in some IADs. It is also known as a skirmish line, line abreast, or being on-line.

2. Advantages. The line formation has the following advantages:

   a. Provides for maximum firepower in the direction of the threat or contact.

   b. Fastest for crossing a linear danger area.

   c. Facilitates good control and allows for immediate movement or maneuver.

   d. Allows for an immediate leapfrog forward or rearward.

   e. When contacted from front or rear, this formation allows immediate lateral movement out of the danger area through a side peel left or right.

3. Disadvantages. The line formation has the following disadvantages:

   a. The entire formation enters danger areas simultaneously, thus exposing the whole patrol to unknown threats (instead of a point element only) and decreasing stealth.

   b. Provides very limited security to the flanks and rear.

   c. More difficult to control than a file formation.

   d. Requires good control of fields of fire and good awareness of men on either side.

   e. Can't be used on narrow trails or in heavy vegetation.

4. Effects of Fire.

   a. Exposes the entire formation to enfilade fire from the flanks. Same effects of fire as the file formation given flanking fire.

   b. Movement left or right in the danger area reduces each man's ability to fire at the enemy since the fields of fire are reduced.

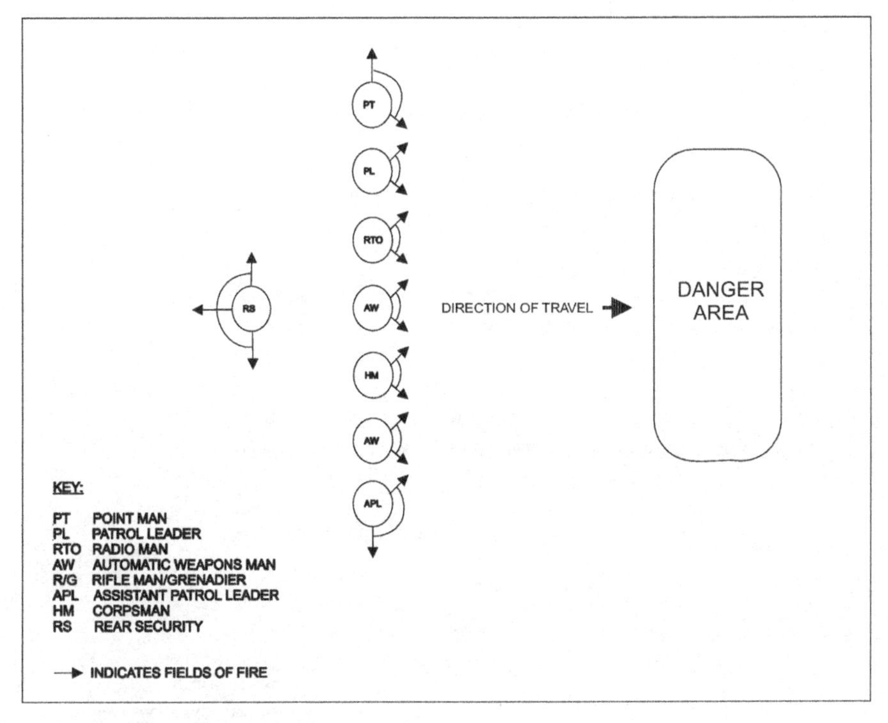

Figure 2-6. Line Formation

2.4.4.4 Wedge Formation. The wedge formation is best used in relatively open terrain such as farm country, or sand dunes near a beach or desert.

1. Formation Basics. The wedge is an excellent formation to break out of an enemy encirclement or when approaching a probable danger situation. The wedge can also be employed when in open terrain. See Figures 2-7A and 2-7B. The double wedge, also known as a flying wedge, is two wedge formations in a column; it is a good option for a full platoon patrol. See Figure 2-8. These require extensive practice; they can become confusing in the field.

2. Advantages. The wedge formation has the following advantages:

   a. Concentrates firepower in primary movement direction with good fields of fire.

   b. Easy transitions to overwatch and bounding overwatch movement techniques.

c. Provides good C^2.

d. Good dispersal of personnel.

e. Allows for ease in coming on line or returning to file.

f. Distances between FTs can be easily increased or decreased.

3. Disadvantages. The wedge formation has the following disadvantages:

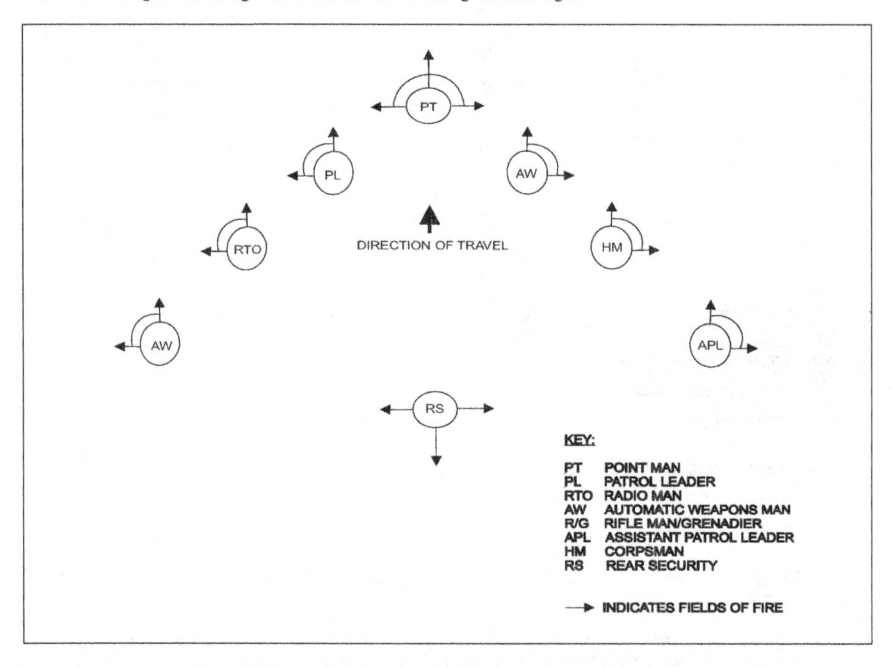

Figure 2-7A. Wedge Formation (Fields of Fire Depicted by Arrows)

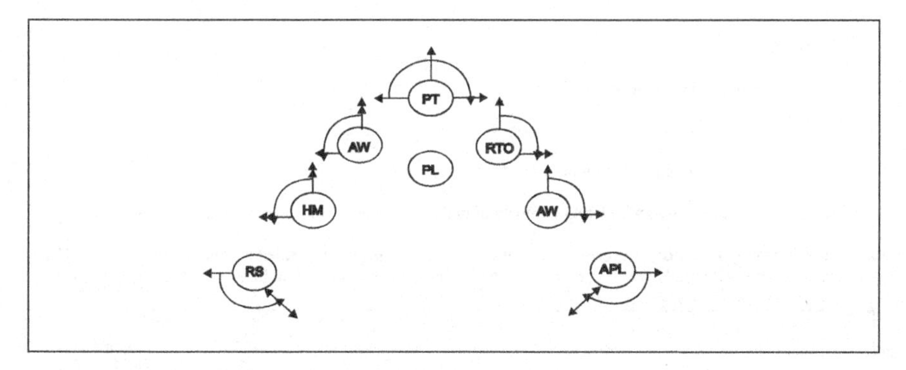

Figure 2-7B. Alternate Wedge Formation (Field of Fire Depicted by Arrows)

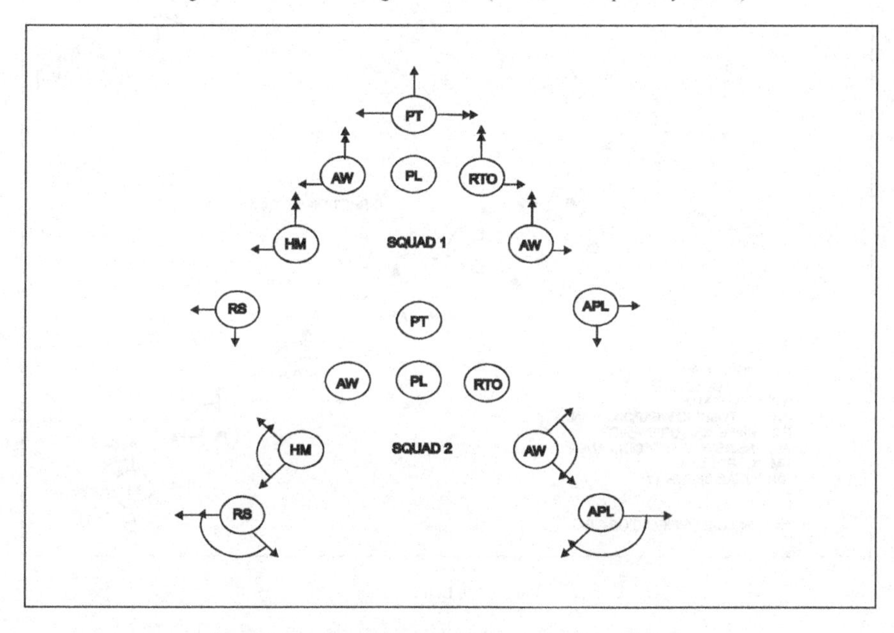

Figure 2-8. Double Wedge (Also Known as the Flying Wedge)

a. Requires revised SOPs/IADs and extensive training

b. Noisier than a file

c. Requires good control of fields of fire

d. Can't be used on narrow trails or in heavy vegetation

e. Leaves the patrol vulnerable in the rear

f. Hard to control if used with larger numbers of personnel (more than 5 or 6 persons).

2.4.4.5 Echelon Formation. The echelon formation has many of the advantages of the line formation, yet still has a point element. It can be used over short distances where contact is anticipated but the PL wants to move as one unit.

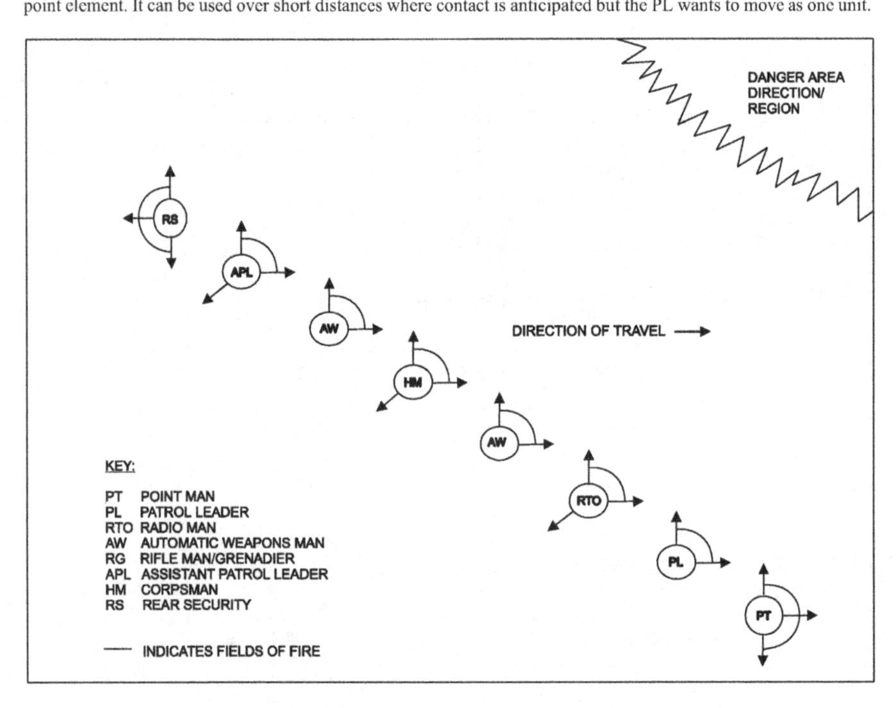

Figure 2-9. Left Echelon

1. Formation Basics. This formation can be configured either echelon right or echelon left. The decision to employ this formation is dictated by terrain and enemy situation. It can be used to cross large, open danger areas or on final approach to the target. See Figure 2-9.

2. Advantages. The echelon formation has the following advantages:

   a. Is good over short distances where control is important and the requirement is to keep the entire patrol together

   b. Allows the firepower of the patrol to be instantly employed in all directions

   c. Provides good C^2

   d. Can easily come on-line to conduct IAD upon contact with the enemy in any direction.

3. Disadvantages. The echelon formation has the following disadvantages:

   a. Not good for thick vegetation

   b. Not good for long durations, as it produces individual strain and increased fatigue

   c. Not good in mined or booby-trapped areas

   d. Requires good control on each individual's field of fire.

4. Effects of Fire.

   a. Frontal fire is less effective since fewer personnel are in the danger area.

   b. Matching oblique enfilade fire will be highly effective against this formation since the beaten zone matches the formation.

2.4.4.6 Circle and Diamond Formations. The circle formation (Figure 2-10) has limited useful applications during patrol movements, but it can be used in specific circumstances. It is useful only for short distances and in relatively open terrain. See Section 2.4.8.2 for procedures for getting into the circle formation. The diamond formation (see Figure 2-11) has similar characteristics and uses.

1. Formation Basics. The circle formation is useful for the security of hostages, friendly POWs, VIPs, or another unit that you are trying to protect in a "hot" area. It is possible to ring them in a defensive circle or diamond and move them a short distance for extraction. The formation can be expanded large enough to hold a landing zone (LZ) at the center.

2. Advantages. The circle and diamond formations have the following advantages:

   a. There is continual 360-degree security both while patrolling and while stopped.

   b. Firepower is immediately available in any direction.

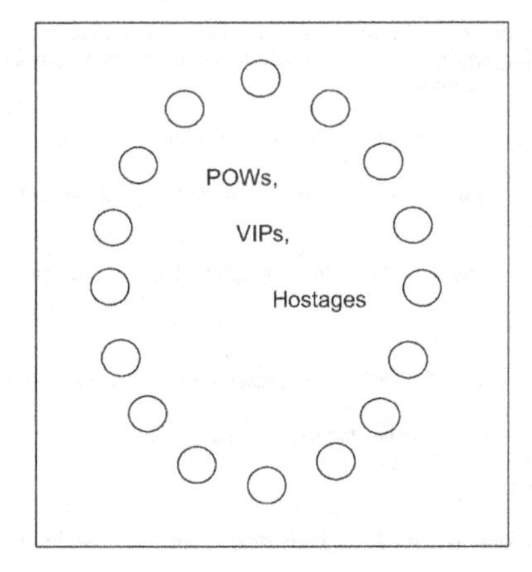

Figure 2-10. Circle Formation

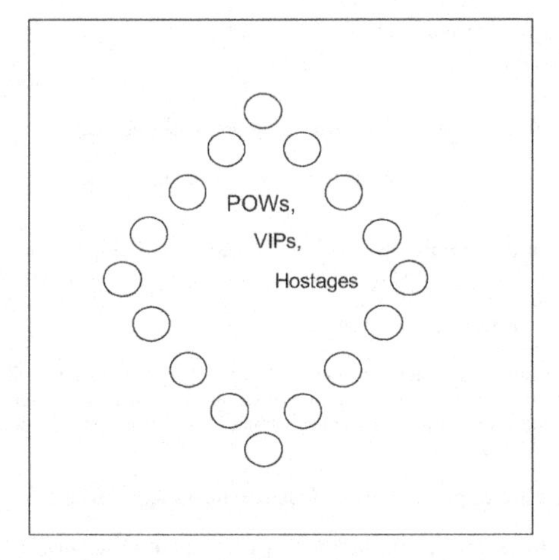

Figure 2-11. Diamond Formation

3. Disadvantages. The circle and diamond formations have the following disadvantages:

   a. They are difficult to maintain for large groups.

b. The formations are not good for long durations, as they are slow and produce individual strain and increased fatigue.

c. They are not good in thick vegetation.

d. Additional SOPs and IADs are necessary to ensure effective instinctive reaction if contacted.

4. Effects of Fire. The formations represent an area target.

2.4.4.7 "V" Formation. The "V" formation is best used in open terrain such as sand dunes or beach.

1. Formation Basics. The "V" formation is excellent for approaching a probable target as it concentrates fire on a point target.

2. Advantages. The "V" formation has the following advantages:

a. It provides good C^2.

b. There is good dispersal of personnel.

c. It concentrates firepower in direction of movement.

d. There are good fields of fire with established skirmish lines if contacted from the rear or flanks.

e. It allows for ease in coming on line or returning to a file formation.

f. Distances between FTs can easily be increased or decreased.

g. There is an easy transition to over-watch and bounding over-watch movement techniques.

3. Disadvantages. The "V" formation has the following disadvantages:

a. It requires extremely good control of fields of fire.

b. It cannot be used on narrow trails or in heavy vegetation.

c. The formation leaves you vulnerable in rear.

d. It is hard to control large numbers or teams larger than 5 or 6 persons.

4. Effects of Fire.

a. The formation minimizes effects of frontal or flanking enfilade fire.

b. Oblique fire can represent a more significant threat to this formation.

c. Indirect fire, grenade fire, or 40mm fire will increase the number of casualties since this formation presents an area target.

2.4.5 Basics of Immediate Action Drills. IADs are actions taken by patrol members upon inadvertent contact with the enemy. They should be practiced with regularity and applied with consistency so that each patrol member instinctively knows what his counterparts will do when contacted. The objective for each SEAL platoon should be the ability to instinctively react to contact, even if the PL is unable to direct the patrol's actions. In order to achieve this goal, consistent training in all possible scenarios is necessary. This section will summarize the fundamentals of successful IAD performance.

2.4.5.1 Immediate Action Drill Fundamentals. Tactically sound IADs should include the following:

1. Maximum utilization of available cover and concealment. Personnel conducting IADs should minimize exposure to enemy fire. It is not realistic to run, stand, or kneel when taking effective fire. IAD training should reflect this fact.

2. A means to suppress enemy fire. This is done by concentrated application of accurate firepower. While all patrol members play a part in this, it is often quite difficult for members caught in a fire zone to provide their own suppressive fire.

3. A means to quickly move patrol members out of the enemy's fire zone. This is done by personnel movement, usually in element or squad units, covered by suppressive fire and making use of available cover/concealment and patrol equipment (smokes, white phosphorous (WP) grenades, etc.).

4. An effective way to consolidate and account for personnel and equipment after contact is broken.

2.4.5.2 Immediate Action Drill Movement. There are a variety of ways to move patrol personnel during contact, most of which fall into one of the following four categories: on-line, leapfrog, peel, or flanking movement. Specific IADs may use any or all of these methods. Each is described below.

1. On-line. When using this method, the patrol operates as one unit, maximizing weapons brought to bear against the enemy. Since it maintains unit integrity, this method offers the most control of the patrol. By itself, however, it has very limited use since it offers no ability to provide covering fire for moving personnel (all patrol members are either shooting or moving simultaneously). Figure 2-12 depicts a technique that may be used against small, lightly defended objectives. The main reason for bringing the patrol on line is to transition into another phase of an IAD (e.g., to consolidate forces at the end of a contact or to shift from leapfrogging into a peel).

2. Leapfrog. In order to perform leapfrog, a patrol divides into two units, usually squads or FTs (see Figure 2-13). Much like the bounding over-watch movement technique described in Section 2.4.3.4, one FT moves while the other provides cover fire. In contrast to bounding over-watch, movement distance for each bound is severely restricted. Each movement must use the often limited cover fire offered by the other FT. In addition, each FT must take advantage of all available cover and concealment both while moving and while stopped. A key safety factor is the dead space between the FTs, shown in Figure 2-13, which helps to minimize the chance of casualties due to friendly fire.

3. Peel. In a peel, rather than moving the patrol as squads or FTs, individual patrol members move in sequential order. There are four methods of performing a peel: the side peel, left or right (Figure 2-14), the standard peel back (Figure 2-15), and the center peel-back (Figure 2-16). The side peel provides significant firepower against the enemy. The standard and center peels, in contrast, are very limited in this regard - only one or two weapons can fire simultaneously due to field of fire restrictions.

4. Roll. In a roll, patrol members move as FTs, squads, or larger elements rather than moving as individuals or pairs. When the command is given to roll right or left, the base element will maintain fire superiority while the moving element moves to the appropriate side and comes back on line with the base element. See Figure 2-17.

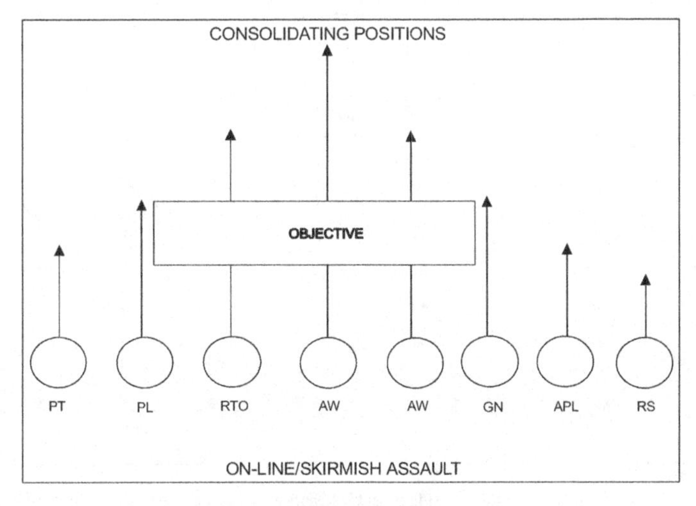

Figure 2-12. Online/Skirmish Line Assault

5. Flanking Movement. A flanking movement can be part of a leapfrog or can lead into a leapfrog or peel. It is a movement performed by the members of the patrol not receiving effective fire to position themselves to provide covering fire for those caught in the fire zone (see Figure 2-18). The formation optimizes the patrol's firepower.

6. Fire and Maneuver. This is a common infantry tactic that uses a supporting FT positioned in front of the target. The maneuver element sets up on the flank of the target for an assault through the target. This formation provides maximum firepower. It also provides for ease of movement. This formation is difficult to control and terrain often limits its employment.

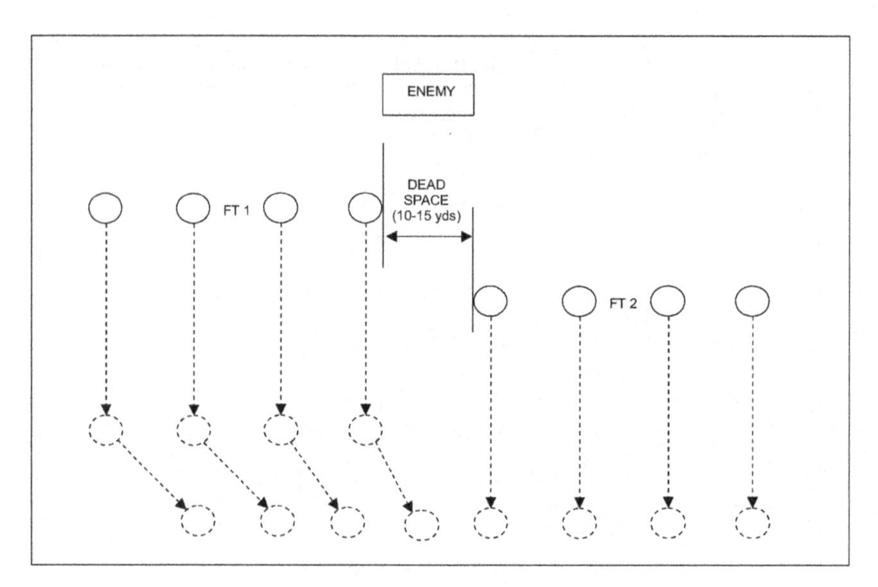

Figure 2-13. Leapfrog

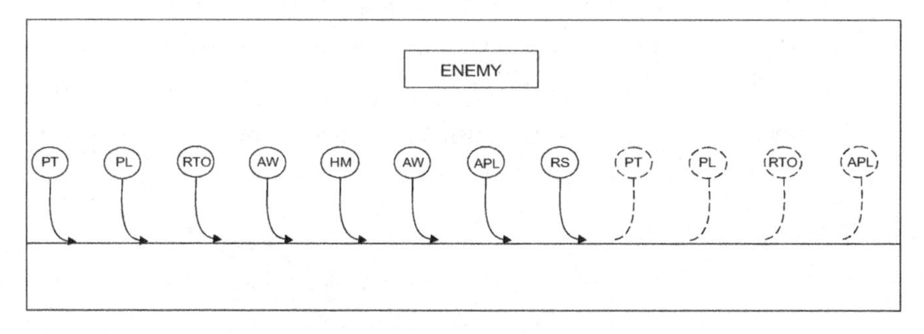

Figure 2-14. Side Peel

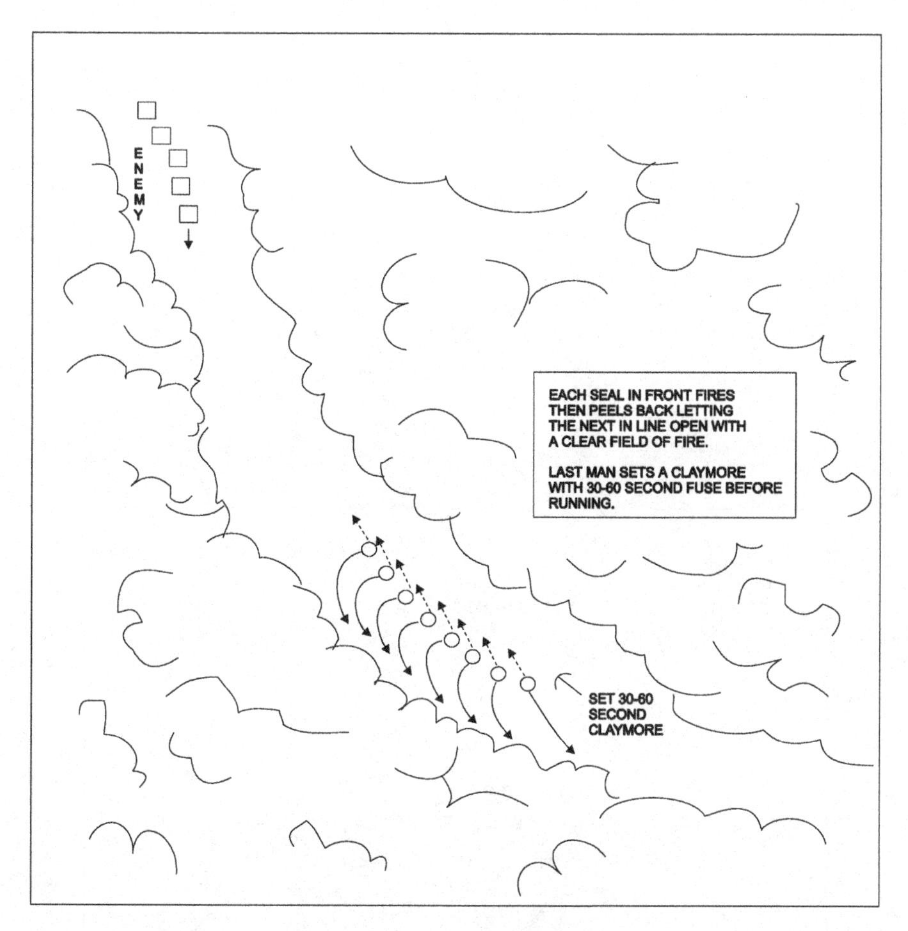

Figure 2-15. Standard Side Peel-back

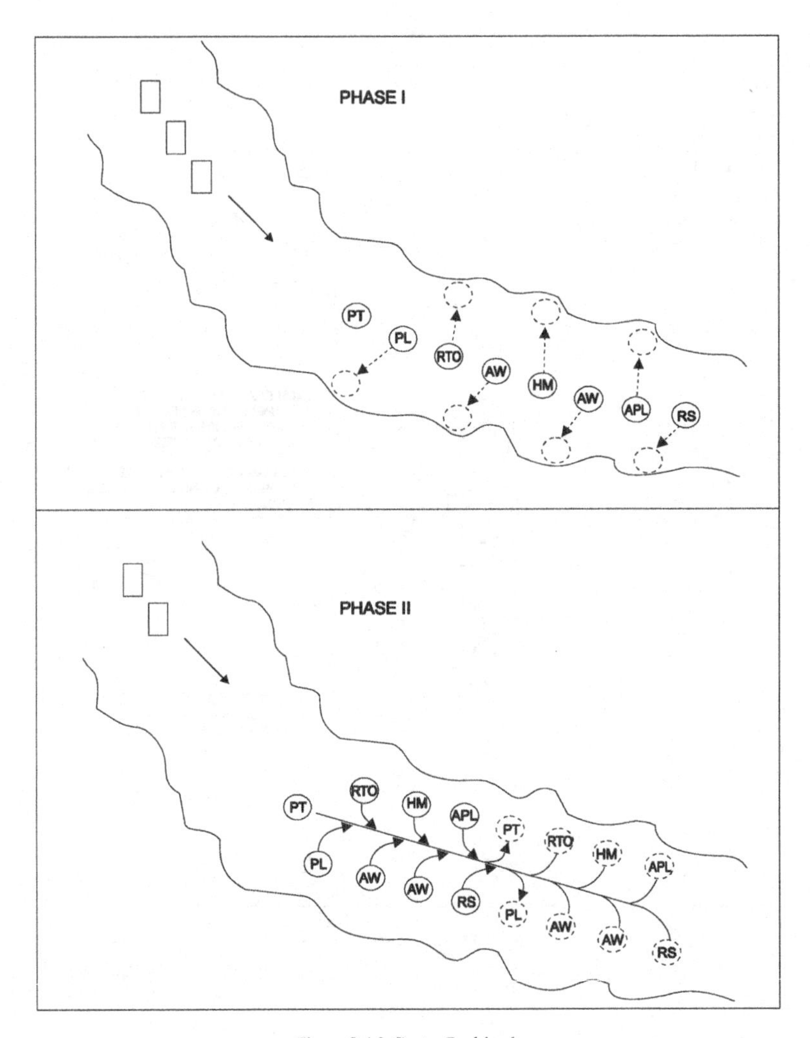

Figure 2-16. Center Peel-back

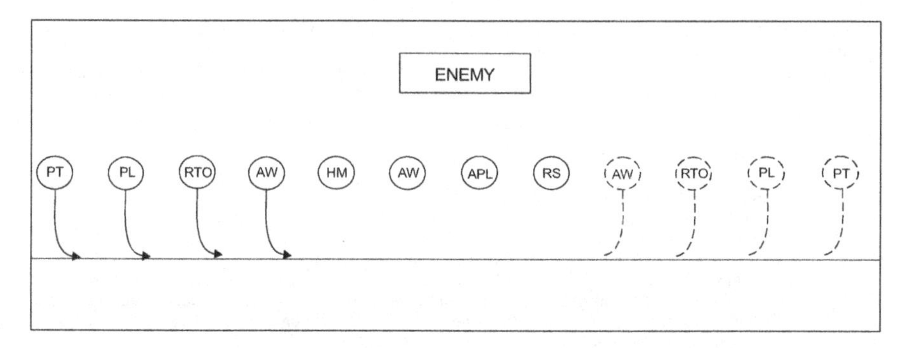

Figure 2-17. Roll

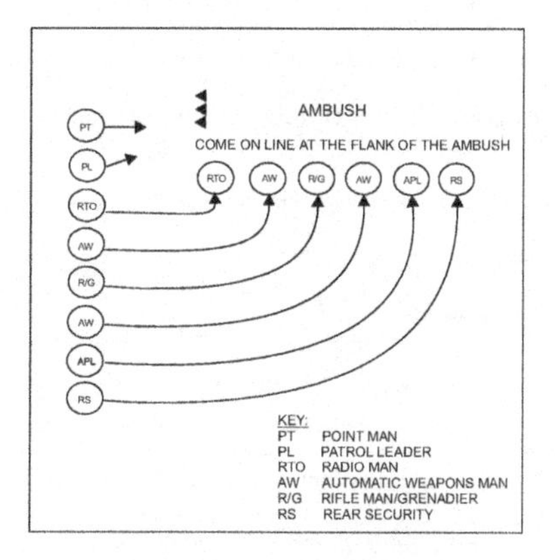

Figure 2-18. Flanking Movement

2.4.6 Immediate Action Drill Procedures. The variety of tactical opinions within the NSW community can be overwhelming. This is especially true regarding IAD procedures. This section will summarize the various drills currently used.

2.4.6.1 Counter Ambush Procedures. SEAL patrols should be particularly concerned about ambushes when patrolling on low ground in the vicinity of high ground or in any channeled areas that limit maneuverability. Where there is an increased threat of ambush, patrols should continuously monitor available cover in their immediate vicinity and be ready to call whatever fire support is available. The immediate tactical goal when ambushed is to survive the initial engagement through effective use of cover and concealment. Counter maneuvers can then be performed as feasible.

Any of the IADs discussed can be attempted to counter an ambush. An ambush has the potential to quickly devastate a small unit like a SEAL patrol; maneuvers by individuals or pairs may at times be the only defense possible. As a last resort, if all other maneuvering is impossible, the PL may resort to assaulting the ambush position from the kill zone. See Figure 2-19. A successful assault from this situation is highly unlikely and should be considered only as a last means of survival.

1. Individual actions when ambushed.

   a. Seek cover and return fire as accurately as possible.

   b. If possible, ambushed patrol members should attempt to fall back and get on line with the rest of the patrol.

   c. In a near ambush (jungle trail) within 10 meters and in which the whole patrol is involved, a rush to break into a skirmish line may be the only individual action each patrol member can make to counter the ambush. A rush into the ambush will remove the patrol from the danger area (based on fires) and remove the ambushers' ability to support each other's field of fire.

2. PL's action when ambushed. When ambushed, and the patrol is returning fire, the PL should:

   a. Immediately assess the enemy's effectiveness of fire and strength.

   b. If practical, determine the condition of the element ambushed, including the number and seriousness of any wounded members.

   c. Make the decision to assault or break contact. This must come as soon as possible.

   d. Use fire support if available.

3. Action when entire patrol is ambushed.

   a. Seek cover.

   b. Patrol members commence intense fire and violence of action into assigned areas of responsibility.

   c. Commence fire against selected ambush positions (look for muzzle flash and smoke as an indicator).

   d. PL should assess the enemy strength and decide to assault or break contact.

   e. A successful assault against an ambush is unlikely and should only be considered a last means of survival.

   f. Fire and maneuver to assault or break contact, as required.

   g. Regroup.

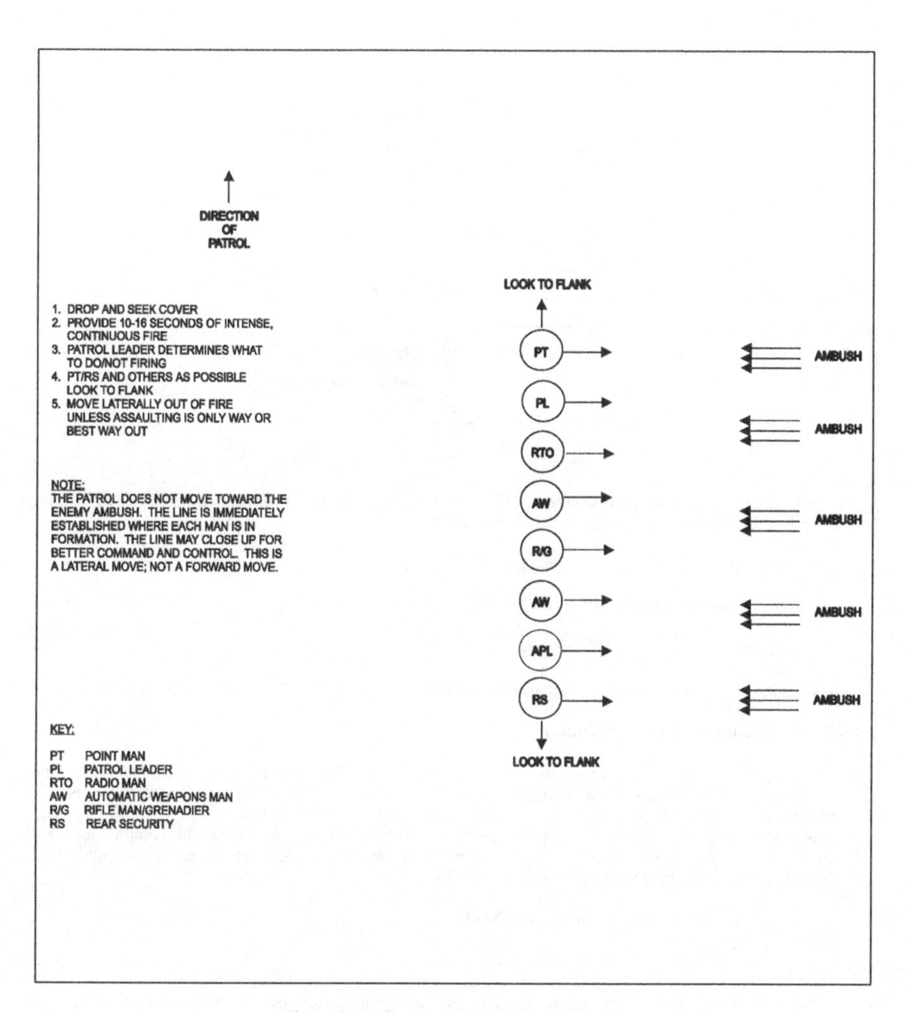

Figure 2-19. Counter Ambush (Entire Patrol)

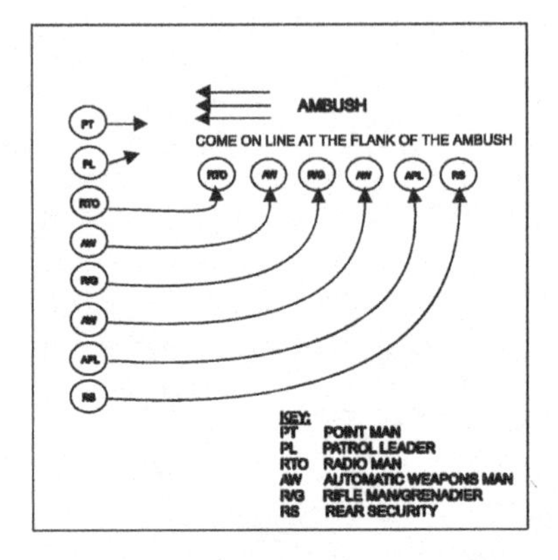

Figure 2-20. Counter Ambush (Point/Lead)

4. Action when point is ambushed. See Figure 2-20. When only the lead man in the patrol is ambushed by flanking or oblique fire, the sequence of events are as follows:

   a. Call the command "On Line" to define the skirmish line.

   b. Call the command "Flank Left" or "Right".

   c. Immediately flank and attack the enemy.

   d. Gain immediate fire superiority; suppress enemy fire.

2.4.6.2 Procedures on Enemy Sighting

1. Unintentional Meeting. For a well-trained patrol that maintains strict field discipline and sound tactics, this situation should be the most common type of unplanned contact with the enemy. Unless there are unusual circumstances warranting a different approach, the patrol should endeavor to break contact without giving away its position through fire or noise. It is important, however, to maintain readiness for possible compromise at any time. Performing a *silent* leapfrog or peel will allow cover fire to be immediately available if compromised. Procedures for responding to this situation are as follows:

   a. Point freezes in position; signals "enemy ahead."

   b. Patrol freezes in position with weapons at the ready.

   c. When the patrol and enemy are on the same trail, the point should take cover on the right or left, the remainder of the patrol should move together to the same side.

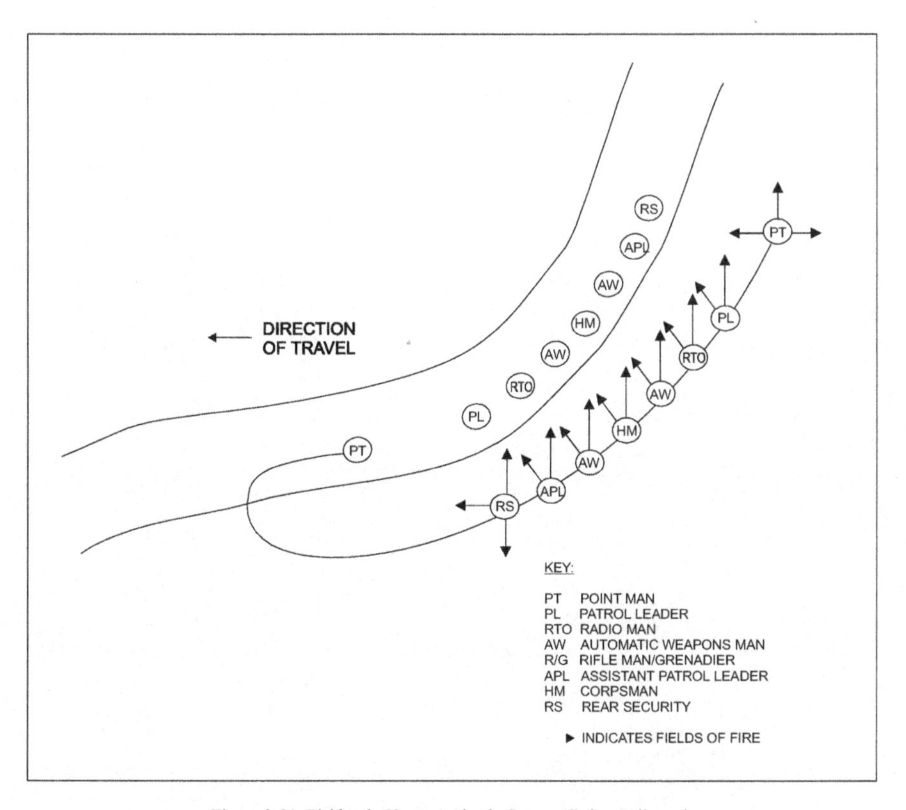

Figure 2-21. Fishhook, Hasty Ambush, Suspect Being Followed

d. PL assesses enemy strength, patrol mission and decides on a course of action: avoid enemy, proceed with mission, or maneuver to hasty ambush positions.

e. If avoiding contact use silent IADs and take advantage of terrain features.

f. The decision for a hasty ambush needs to be based on enemy strength and dispersal, ROE, and guidance for targets of opportunity, as well as the expected impact the ambush will have on the accomplishment of the mission. Figure 2-21 shows a hasty ambush when the patrol suspects it is being followed. Figure 2-22 shows a hasty ambush for meeting an oncoming enemy.

g. Back into ambush position when meeting an oncoming enemy.

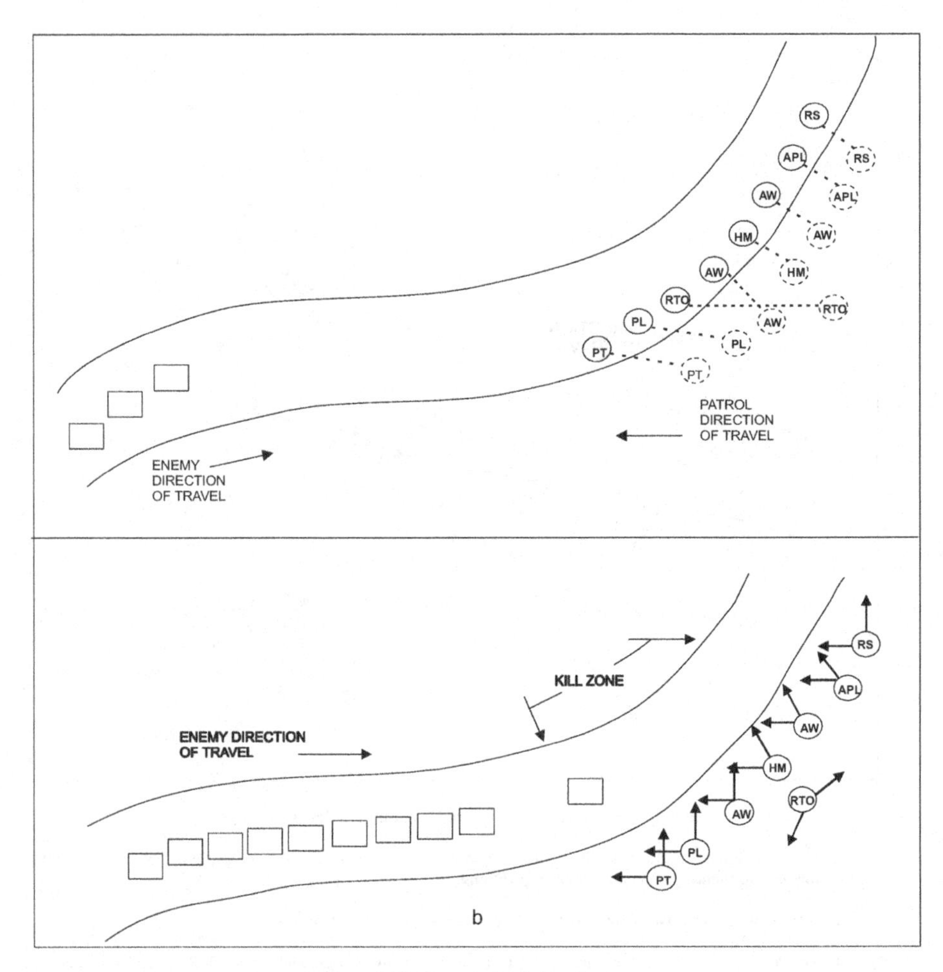

Figure 2-22. Hasty Ambush Ahead

h. Allow as many enemy personnel as possible to enter the kill zone before initiating fire.

i. The PL opens fire when the enemy is in the kill zone; however, if the enemy spots this hasty ambush, any patrol member can initiate fire.

j. Allow enemy patrols that are too large to pass by, if possible.

k. If time permits, use Claymores.

2. Simultaneous Meeting.

a. The first man to see the enemy will immediately engage the target and indicate direction "enemy front, rear, left, or right" (relative to the patrol's direction of travel).

b. PL orders "Come On Line" in the direction of the contact.

c. Commence immediate high volume fire to gain fire superiority.

d. PL assesses enemy strength.

e. PL must control the distance of the assault (if an assault is conducted).

3. Simultaneous Meeting When Out-manned. In a field or open area, it may be best to come on line and use a leapfrog to break contact. This method is shown in Figures 2-23A and 2-23B. All personnel execute it immediately coming on-line. Usually the personnel not in direct contact maneuver to those engaged. Once on-line, the patrol begins a leapfrog maneuver to break contact. The advantages of this method are that the patrol remains together; making C^2 easier, and all weapons are on-line for maximum firepower. The disadvantages to this method, however, are significant. As can be seen from Figure 2-23A, in order to get on-line, the maneuvering unit must move forward, into enemy fire, to the area where the point and PL have established the line. In an intense firefight, this move will often be impossible. The other significant potential pitfall of this method can also be seen in Figure 2-23A. Notice that the leapfrog goes directly away from the enemy, the patrol remains in the fire zone. With the range of some small arms in open terrain being 1,000 yards and greater, the leapfrog will have to continue for a great distance or until suitable cover is found. The procedure for this method follows. There are three phases to this maneuver:

a. Phase One (Initial contact procedures)

(1) Point makes contact front and shouts "Contact Front".

(2) Point fires into the enemy position on full automatic.

(3) Squad steps into fields of fire. PL and RTO fire 3 to 5 round bursts.

(4) PL gives command "Get On Line". FT or squad one moves to on-line position and picks up fire.

(5) Second FT moves to an on-line position on the opposite side, leaves a dead space if possible, and commences heavy sustained fire into the enemy position.

b. Phase Two (Break contact, get out of the field of fire)

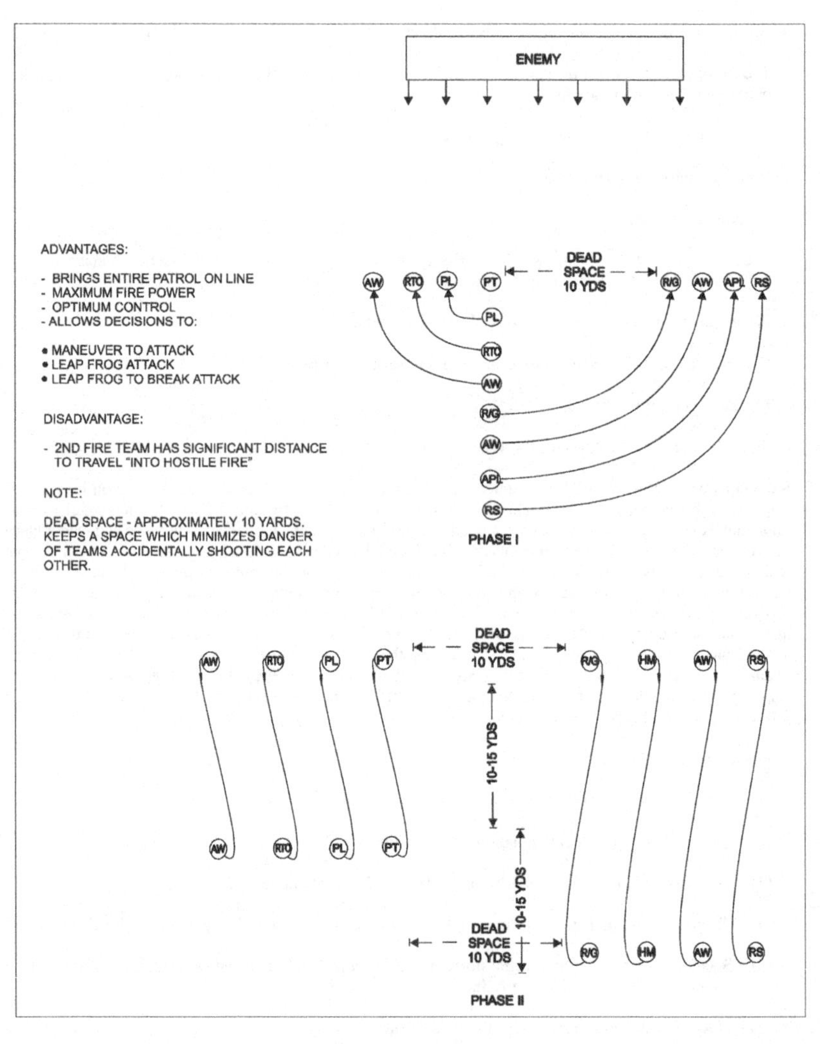

Figure 2-23A. Open Terrain Online Method

(1) The platoon is on line, laying down a sustained base of fire.

(2) PL gives command "Leap Frog Rear!" "Ones Back!". All patrol members pass commands.

(3) Base element (FT 2) picks up rate of fire. Maneuvering element (FT 1) moves on the PL's command "Ready Up."

(4) Observe muzzle discipline.

(5) Maneuvering element (FT 1) moves back to cover, establishing dead space between FTs, gets down, and picks up fire immediately.

(6) If you were unable to change magazines on the line, do so on the move, placing expended magazines inside shirt.

(7) Movement must coincide with sustained rate of fire from base element to provide "cover fire" when up.

(8) In some cases crawling may be required. Make sure members are still with you before you fire.

(9) Base element (FT 2) hears or sees (FT 1) sustained rate of fire, determines effectiveness of enemy fire, and prepares for movement back.

(10) Like the previous FT, FT 2 uses arm and verbal commands for movement back.

(11) This fire and movement process continues until patrol is in suitable cover.

c. Phase Three

(1) The FT that reaches suitable cover or is no longer taking effective fire will give the command "Consolidate!". This FT provides cover for the team moving to consolidate.

(2) The moving FT follows all the previous rules for moving, but instead of moving past the base FT, this time takes up position on line with the base FT.

(3) After assessing the best method of withdrawal, the PL or APL gives the command to "Peel Left/Right! PT (or RS), Take Us Out!".

(4) Move 90 degrees to contact and get a head count.

(5) Move 100-200 meters or as required, form a hasty perimeter, conduct a head count, treat the wounded, redistribute ammunition, and reorganize for follow-on contact.

(6) Patrol moves out quickly, but tactically.

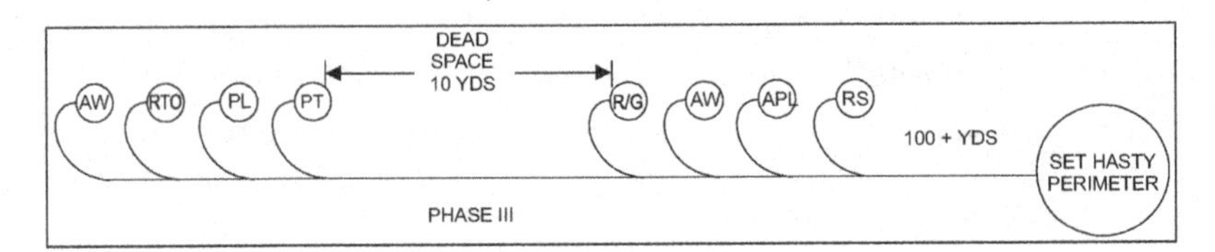

Figure 2-23B. Open Terrain Online Method

d. Effects of Fire.

(1) Frontal Fire. Initially the file or staggered file formation that the patrol is in will match the beaten zone. It is important that the patrol move as quickly as possible to a line formation so the effects of fire are minimized as much as possible. Each element should look for cover when moving on line in order to shield from the effects of grazing fire.

4. Simultaneous Meeting When Out-manned in a Channelized Area. Channeled areas represent some of the most hazardous situations in which a patrol must fight. The inability to move left or right 90 degrees from the contact creates a funnel effect on the platoon. Effective reaction in close and channeled terrain requires extensive training. In this environment, squad and FT movement can be severely limited by a commanding enemy position. In these cases smaller groups or even individuals may be required to maneuver with little or no direction from the PL. IADs must be rehearsed well enough for all individuals to know what is required of them in all situations. The three methods of peeling from this type of contact (see Figure 2-14, 15, and 16). The peels are very similar except that the center peel (also known as the Australian peel) provides more firepower towards the enemy (two weapons employed simultaneously, vice one). The standard peel is performed as follows:

a. Phase One

(1) PT yells “Contact Front!” and fires full automatic into the enemy’s position.

(2) “Peel Back!” is called by the PL.

(3) Patrol members step away from berm to create a clear path to peel back in, weapons pointed away from the berm.

(4) PT turns away from patrol into the berm, taps out, withdraws to the rear, and looks for an out. If there isn’t an out, he turns on line.

(5) The next man will not fire until each peeling man has cleared his field of fire.

(6) This continues for each patrol member until out of the immediate threat or an out is found.

b. Phase Two

(1) Any member can determine the out. For best control the PL should give the command, “Out Here.”

(2) As members withdraw, the PL or APL will pull the AW man into position to cover the out and direct the rearward movement to a lateral movement 90 degrees to the contact, reestablishing the firing line as a line.

(3) Each member moves to the out and starts a side peel left or right.

(4) The object of the out is to find cover and move away from the contact. It can also be used to maneuver and reengage the enemy by flanking.

(5) As the patrol peels into the out, they should not fire back towards the contact until all members are on line.

(6) As the last man peels out of the channel, a head count is immediately conducted to ensure all patrol members are out of the channel.

(7) PL at this time determines whether or not to move out or continue the fight.

c. Phase Three

(1) If everyone is accounted for, the PL tells the PT/RS to take the patrol out.

(2) Patrol moves out as quickly as possible.

(3) Move 100-200 meters, or as required, form a hasty perimeter, conduct another head count, treat the wounded, redistribute ammunition, and reorganize for a follow-on attack.

(4) Patrol moves out of the area quickly, but tactically.

Although the peel method is the most commonly taught IAD for the scenario, it has some significant limitations. First, it provides very limited firepower employed against the enemy; a maximum of three weapons simultaneously, and for most of the drill only two. Second, especially in a channeled area, the patrol does not move laterally with respect to the enemy, but stays in the enemy's fire zone throughout the drill.

With very limited firepower for cover, each patrol member must move down a path directly in the enemy's cone of fire. All fighting is done from within the enemy's fire zone; there is no attempt to break out and take back some initiative by offering fire from a different sector.

With all IADs the objective is to suppress fire and neutralize the threat by either overwhelming the enemy or moving to cover. If practical, a patrol may elect to move from a channeled area by filing out instead of moving 90 degrees or laterally from the contact. In areas that provide cover from direct fire or grazing fire this method may be effective. If indirect types of fire are encountered, movement laterally may be the only tactical means of removing the platoon from the danger area.

5. Contact from the Rear in a Channeled Area

a. The same responses and IADs are used to withdraw. Only individual responsibilities change.

b. Effects of Fire

(1) Frontal fire. Initially the file or staggered file formation that the patrol is in will match the beaten zone. It is important that the patrol move as quickly as possible to break the contact so the effects of fire are minimized as much as possible. Each man should look for cover when peeling back, as cover minimizes the effects of grazing fire by creating dead space.

(2) The use of this tactic in open terrain will increase the patrols' exposure to the effects of enfilade fire.

6. Simultaneous Meeting when Out-manned on a Jungle Trail. Trails are a common means of movement and navigation in densely vegetated environments. Withdrawal from enemy contact is difficult at best. The added difficulties associated with the dense vegetation makes normal withdrawal by peel or leapfrog almost impossible, and provides only minimal fire support. The following modification of the center peel can provide beneficial characteristics of both on-line and leapfrog movements. It is especially important that all platoon members are thoroughly indoctrinated and trained prior to executing this IAD or method of withdrawal.

a. Phase One

(1) PT initiates "Contact Front." Steps to right/left side of trail and simultaneously fires on full automatic into the enemy position. The PL moves immediately on line with the PT, establishing a two-man element, and commences firing. The RTO moves to the opposite side of the PL/PT on the trail. The AW moves up and on line with the RTO to form a second two-man staggered element and commences firing. The second FT also forms staggered two-man elements with the AW man, HM/GN, and APL/RS. "Peel Back By Twos" is called by the PL/APL.

(2) Patrol members move to staggered two-man positions and prepare to peel to the rear, creating a corridor or clear path down the center of the trail to peel back in. Weapons are pointed in established fields of fire away from the opposite two-man elements.

(3) PL/PT turn into the trail and away from the men behind them, they tap the AW/RTO and withdraw to the rear, looking for cover or an out. If no out is located, both the PL/PT turn on line in a staggered position behind the RS/APL.

(4) The opposite two-man element waits until they receive the tap before they move. This is mandatory in dense vegetation since visual contact is not always possible.

(5) Each element will not fire until the peeling element has cleared their field of fire.

(6) This continues for each element until out of the immediate threat or an out is found.

b. Phase Two

(1) Any member can determine the out and give the command "Out Here."

(2) As elements withdraw, the PL/APL will place a pair (AW optimal) into position to cover the out and direct the rearward movement 90 degrees to the contact, reestablishing the firing line to an on-line.

(3) Each element moves to the out and starts a side peel left or right as individuals/pairs.

(4) The object of the out is to find cover for moving away from the contact. It can also be used to maneuver and reengage by flanking.

(5) As the patrol peels into the out, the members should cease fire unless otherwise directed by the PL or receiving effective fire.

(6) As the last element peels off the trail, a head count is taken to ensure all patrol members are off the trail.

(7) The PL at this time determines whether or not to move out or continue the fight.

c. Phase Three

(1) If the head count is good, the PL tells the PT/RS to "Take Us Out."

(2) The patrol moves out as quickly as possible.

(3) The patrol moves 100-200 meters or as required, forms a hasty perimeter, conducts a head count, tends to the wounded, redistributes ammunition, and reorganizes for patrol movement out of the area quickly, but tactically.

(4) If the PL finds that the immediate threat is no longer a factor and withdrawal can be expedited by filing out from the two-man peel, he need only tell the PT to "Take Us Out."

d. Advantages

(1) Provides maximum use of firepower on contact and while withdrawing

(2) Expedites the withdrawal

(3) Minimizes movements of patrol

(4) Improves control of patrol

(5) Improves accountability of patrol members

(6) Provides clear fields of fire.

e. Disadvantages

(1) Requires substantial platoon training.

(2) Requires the PL to make the call immediately.

(3) Requires a high level of fire control to ensure safe fields of fire.

f. Effects of Fire.

(1) When receiving frontal fire, the file or staggered file formation of the patrol will initially match the beaten zone. It is important that the patrol move as quickly as possible to break the contact so the effects

of fire are minimized as much as possible. Each man should look for some cover when peeling backing in order to shield from the effects of grazing fire.

(2) The use of this tactic is not recommended in open terrain because it will increase the patrol's exposure to the effects of enfilade fire.

7. Open Terrain/Flanking Method. As with the on-line method above, this method also uses a form of the leapfrog movement. Initial actions on contact, however, are significantly different in that the FT or squad not taking effective fire attempts a flanking maneuver. Once the flanking FT begins firing, the FT in contact can begin movement similar to leapfrog, except that its orientation to the enemy will be canted to one side. The advantages of this method are that the flanking movement utilizes the strengths of a small unit (surprise and quick movement) and the flanking maneuver reduces time spent in the enemy's effective fire zone.

a. Phase One (see Figure 2-24A)

(1) PT makes enemy contact, drops to the ground shouting: "Contact Front! Contact Front!" and fires into enemy (full auto).

(2) PL drops into his field of fire, assesses strength and direction of fire, and then gives command, "Flank Left" (or right as appropriate, depending on enemy and terrain).

(3) PL then brings FT 1 on line as best he can with available cover and the team returns fire. (Although PL can fire at point targets and to pick up lulls, his main concern is C^2.) Cover is not sacrificed in order to come on line; the main objective is to provide covering fire for the maneuvering element.

(4) FT 2 breaks to the flank side called by the PL and quickly moves to the best fire position available. This must be done swiftly in order to provide immediate suppressive fire to protect FT 1. Often only a partial flanking maneuver can be performed, but this will still help the personnel in the fire zone.

(5) The flanking FT should initiate fire when directed or as per platoon SOP.

b. Phase Two (Figures 2-24B-24D)

(1) The patrol is now ready to begin a leapfrog maneuver with FT 2 providing a base of suppressive fire.

(2) FT 1 members move on the PL's command, "ONE Up!" FT 1 members guide on FT leader as he leads them back past FT 2, keeping dead space. FT 1 turns, gets down, and picks up the fire immediately. Magazine changes should be performed while moving. As in the on-line method described previously, it may be necessary for the FT to crawl to its next position, depending on how effectively fire is suppressed.

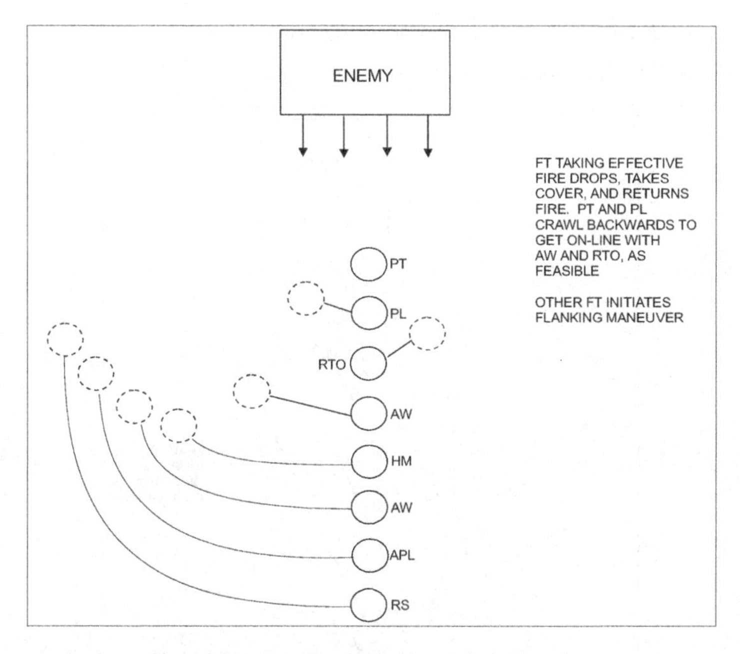

Figure 2-24A. Open Terrain Flanking Method - Phase One

c. Phase Three (Figure 2-24E)

(1) The leader of the FT that reaches suitable cover will give the command "Consolidate! Consolidate!" This FT provides effective fire at the enemy as necessary to allow the other FT to move.

(2) The moving FT follows all the previous rules for moving, but instead of falling back past the base FT, this time takes up position on the base FT.

(3) The PL now has his patrol consolidated. He assesses the situation and decides on the direction for subsequent movement. If cover is adequate, he may order, "PT, Take Us Out!" or "RS Take Us Out!" as appropriate. If enemy fire is still a threat, he can order, "Peel Left! (Or Right)." Another good option if still in an open area but not receiving enemy fire is to continue the leapfrog, maintaining maximum readiness to return fire as necessary.

(4) A head count is taken while on the move. Check for wounded.

(5) The patrol moves 100-200 yards and quickly forms a perimeter.

(6) The head count is verified, an ammunition count is performed, and the wounded are accounted for.

(7) The patrol moves out of area quickly, but tactically.

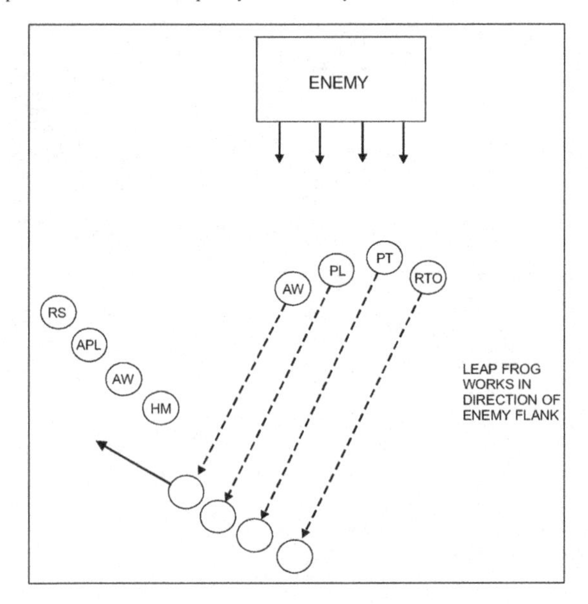

Figure 24B. Open Terrain Flanking Method - Phase Two

The above procedures are described for a "contact front" situation; they are easily modified for side and rear contacts.

In a rear contact, the exact same procedures are used with FT 1 and FT 2's roles being reversed. In a side contact near the front (or rear) of the patrol (Figure 2-25), the situation is ideal because members in the fire zone are already on-line relative to the enemy. The other FT can maneuver quickly to flank the enemy.

The other situation is a side contact in the middle of the patrol. The counter ambush technique for that is discussed in Section 2.4.6.1.

8. Close Terrain/Flank Method. In close terrain, especially canalized areas, the flanking method described above is more difficult to employ but is still usable. In thickly vegetated areas, the flanking FT may be required to bulldoze through considerable brush or jungle to reach a suitable fire position. While this is not ideal, as long as it is done quickly the drill remains the same.

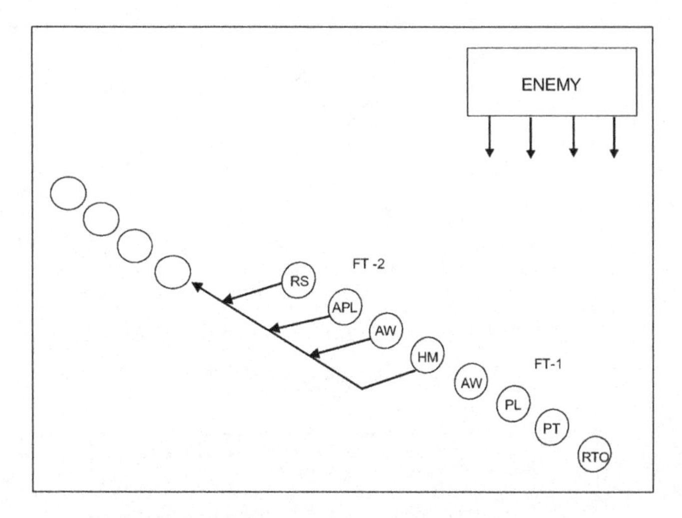

Figure 24C. Open Terrain Flanking Method - Phase Two Continued

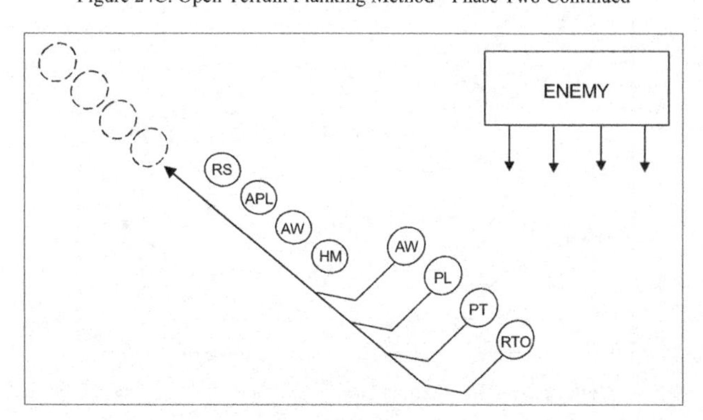

Figure 2-24D. Open Terrain Flanking (FT roll) Method - Phase Two Option

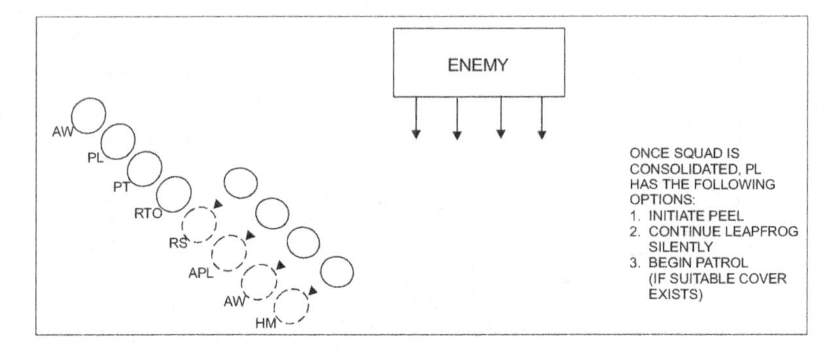

Figure 24E. Open Terrain Flanking Method - Phase Three

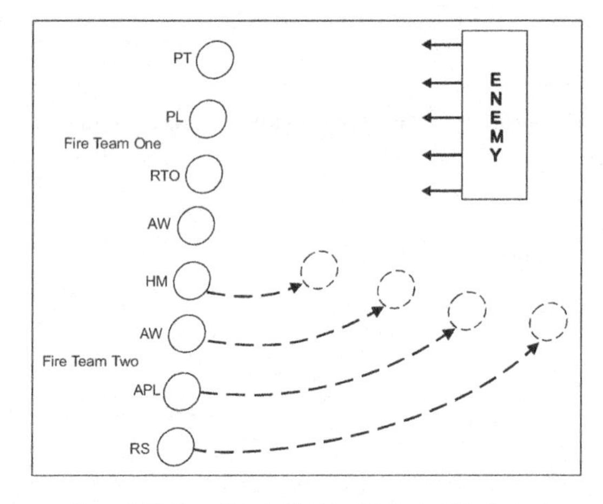

Figure 2-25. Open Terrain Flanking Method - Side Contact

In a canalized area, the patrol is very vulnerable to enemy fire since maneuverability is restricted. However, even if only one or two men are able to flank (by climbing out of the channel or utilizing its farthest side), this still offers a distinct advantage over returning fire only from within the enemy's fire zone. Figures 2-26A-C show one example of this method. Each engagement will look different depending on the situation, but the flanking principle remains the same. Although the method described here offers less control and is not choreographed like most IADs, it employs the strengths of a SEAL patrol; the element of surprise and individual initiative.

Successful tactics in close and channelized terrain require extensive training. In this environment, squad, and FT movement can be severely limited by a commanding enemy position. Instead, smaller groups or even individuals may be required to maneuver with little or no direction from the PL. IADs must be rehearsed well enough that all individuals know what is required of them in all situations.

9. Actions in Response to Indirect Fire

a. Exposure to indirect fire such as artillery or mortar fire may come at any time during the mission. Specific tactics for minimizing exposure to this type of fire vary based on availability of cover and extent of fire. All tactics stress the same thing; remove the patrol from the impact area as quickly as possible unless adequate cover is available.

b. Reaction to indirect fire while moving is as follows:

(1) Any member of the patrol detecting incoming indirect fire gives the alert: "Incoming".

(2) Patrol members assume prone positions immediately.

(3) Once the indirect fire stops, the PL gives the patrol the direction and distance to move.

(4) The patrol moves quickly out of the impact area as directed.

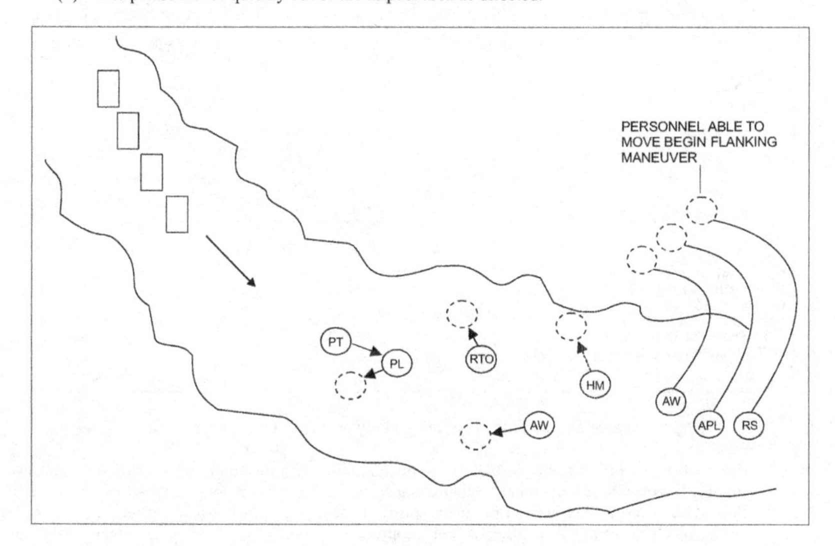

Figure 2-26A. Close Terrain Flanking Method - Phase One

10. IADs Supplemental Information

a. PLs must immediately determine the type of fire being received. A key consideration in this situation is the availability of fire support. Accurate fire support from gun ships, artillery, or naval gunfire can quickly and significantly change the tactical situation. Since almost all fire support involves some time delay before it can be directed accurately against an enemy, the patrol's immediate actions on contact are of great importance.

(1) Effective fire will cause the patrol to seek cover for protection. This situation calls for immediate effective action. Effective fire directed against a small unit can have devastating effects. Whereas a large unit can withstand a number of casualties and retain much of its fighting capability, one or two casualties to a SEAL patrol will immediately affect firepower and maneuverability. More numerous casualties can degrade the patrol's ability to fight or survive.

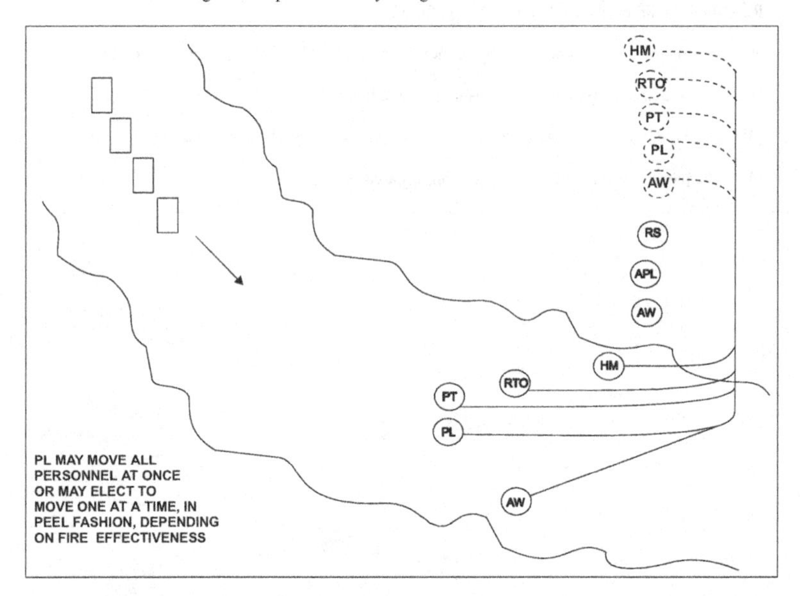

Figure 2-26B. Close Terrain Flanking Method - Phase Two

(2) Ineffective fire doesn't impede movement of the patrol. Fire discipline is essential for small units. Enemy fire may be indiscriminate reconnaissance by fire, a sentry response to noise, or target practice. Responding to such ineffective fire can compromise the patrol's position and possibly cause a firefight with a much larger unit. A leapfrog or peel is appropriate to break contact while maximizing the patrol's ability to respond if pinpointed. Danger areas are undefined, but fire may become effective at any time.

b. Control of the patrol is the first priority, followed by fire discipline.

c. Recognize, analyze, react, and then make a determination as to whether to assault or break contact.

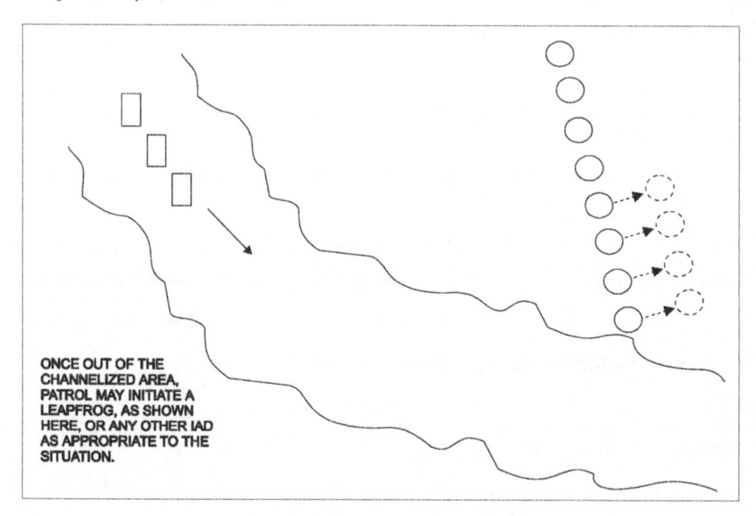

Figure 2-26C. Close Terrain Flanking Method - Phase Three

d. Initiate an accurate and heavy base of fire, keeping in mind at all times the positions of all patrol members.

e. Listen for commands from the PL/APL.

f. Rehearse inserts/extracts IADs from boats, helicopter, and vehicles.

g. FT integrity is most important when a contact is made. All IADs discussed represent the most effective application of this concept. Once the original contact is made, it is a constant struggle to maintain FT integrity. It is most affectionately referred to as "organized chaos." This is why standardized IADs should be used throughout the NSW community, and any variation must always start and end with this objective. Proper patrolling techniques will assist in maintaining FT integrity.

2.4.7 Danger Areas. Danger areas can range from low to high risk. When dealing with danger areas, it is important that the patrol tempo remain as constant as possible, otherwise the mission's time window may close before it can be completed.

2.4.7.1 Types of Danger Areas. A danger area is any area that exposes the patrol or forces it to channel itself into an area where its security is threatened (e.g., roads, open areas, fences, river crossings, extraction points (EP), etc.). All obstacles should be treated with caution and crossed tactically. These are the places a patrol could easily be ambushed. Three types of danger areas are:

1. Linear Danger Areas: roads, rivers, streams, beaches, large power lines, pipelines, fences, etc.

2. Small Open Areas: a field or clearing

3. Large Open Areas: a large field or clearing.

2.4.7.2 Guiding Principles

1. Avoid danger areas whenever possible. The number and types of known danger areas should be key considerations during route selection.

2. Known danger areas should be planned similar to actions on the objective. Training and rehearsals for navigating through, near, or over them are critical.

3. Perform thorough reconnaissance prior to entering a danger area.

4. All members must maintain security; IADs for compromise during a crossing should be rehearsed.

5. Minimize time spent in danger areas. Training and rehearsals enhance smooth and quick movement.

2.4.7.3 Techniques for Crossing Danger Areas. Figure 2-27 is a deliberate danger area crossing.

1. Hasty.

   a. Pass danger area sign to patrol.

   b. PT moves up to edge of danger area and stops; conducts recon of the other side by sight and sound.

   c. Second man covers left and right respectively.

   d. When ready, PT crosses danger area, stops, and conducts reconnaissance of far side by sight and sound. He waits for the second man to come across and bump him off of his position so he can travel on. This process continues until all men are across. A head count is then conducted.

   e. PT signals "All Clear" to patrol.

   f. Unit crosses one man at a time: a man should not cross until the man ahead of him has crossed and is ready to accept him. The next man behind him must quickly move up into position to be the next to cross. This is a very quick, low threat crossing. The platoon should never lose its overall momentum.

2. Doubles; used in medium threat areas.

   a. Pass danger area sign to patrol.

   b. As a pair crosses, one man will look left and the other right.

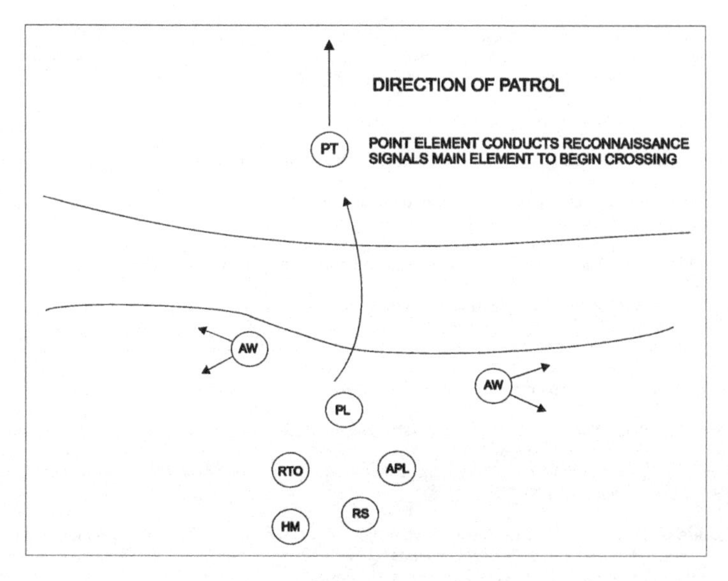

Figure 2-27. Deliberate Danger Area Crossing

3. Deliberate danger crossing; used in high threat, highly traveled areas.

   a. Pass danger area sign to the patrol.

   b. PT and second man move up to the edge of the danger area and recon the far side by sight and sound.

   c. AW and/or GN next in line set up security positions left and right of point element (the crossing point), within hand grenade range. These can be one- or two-man positions. RS is set behind the element to cross.

   d. Point pair crosses to far side and recons using a diamond pattern. When cleared, a signal is passed to the unit. It is wise not to let the PT go by himself on the recon. A three-man team can cross if it is desired to set a security man on the other side, while the far side recon takes place.

   e. Upon return from the recon, the recon personnel may be repositioned on the flanks. Otherwise the first two men across take the far side flanks.

   f. The point recon is not conducted for a large area. The purpose is to find adequate perimeter area for patrol personnel.

   g. This danger area crossing is lengthier because it is used in heavily traveled, high threat areas. Personnel will have to be staged in a concealed perimeter area while awaiting the completion of the crossing.

   h. On-line, if used, will be initiated by verbal whisper or "On line" hand signal. All personnel get on line and move across together. This technique is used only where it is obvious that the threat level is very low. This will maximize patrol exposure, so it must cross quickly but cautiously. The patrol then moves on.

4. Actions at Small Open Areas.

   a. Point element halts the patrol and the command element moves to its location.

   b. Patrol sets up in a security halt posture. Command element determines if it is necessary to negotiate the danger area.

   c. Patrol is informed and moves to adjust their line of march, such as spacing intervals.

   d. The easiest method to clear a small area is to skirt it.

5. Actions at Large Areas.

   a. Point element halts the patrol and the command element moves to its location.

   b. Command element determines if it is necessary to negotiate the danger area. If so, crossing points are located.

   c. Overwatch (security) position is set up; the rest of the patrol crosses the danger area using terrain and patrol formations to their best advantage.

2.4.7.4 Roads or Trails. There are various acceptable methods of crossing a trail or road depending on the situation. Whatever the method, basic principles of reconnaissance and security presented in the previous paragraph apply. Additional guidelines for road or trail crossings are as follows:

1. Avoid crossing at any road/trail junction.

2. Form up and cross one or two at a time until all are across. Move quickly and quietly in each other's steps. Be aware that passing enemy may easily detect tracks. The last man should brush the trail.

3. At the first sign of an approaching vehicle or patrol (a cloud of dust, headlights, or the noise of engines or personnel are good indicators), take the best available cover or concealment.

4. The main point of consideration in any road crossing is control of the patrol.

2.4.7.5 River and Stream Crossing. Site selection is based on security and safety. Method selection is based on width, depth, and current of the river or stream.

1. Guiding Principles. In addition to the guidelines set forth in Section 2.4.7.2, the following principles apply:

   a. Time is critical when crossing rivers and streams. Without proper planning, training, and rehearsal, crossings can become long evolutions, exposing the patrol for much longer than needed.

   b. Use the simplest method possible.

   c. SEALs should always have the ability to conduct waterborne operations.

2. Site Selection. In crossing a river or stream, the objective is to get across quickly, quietly, and as safely as possible. During initial planning, do a detailed map study of the river or stream to be crossed to select a potential crossing point.

3. Methods.

   a. Fording or Wading. This is the most typical method for shallow rivers or streams. Wear a deflated flotation device in case you fall. Use the same procedures as for crossing a road or trail.

   b. Swimming. This is the simplest method in deep rivers or streams with very slow current. Orally inflate flotation devices and wear fins if conditions dictate (i.e., in a medium to fast current).

   c. Use of a Line. Deep rivers and streams with swift current require the use of a line.

      (1) Prior to patrolling to the riverbank, set perimeter security for a danger area and conduct a PL/PT recon.

      (2) Select a narrow site with cover. Current, debris, and bends should be taken into consideration.

      (3) Return to the patrol, where the PL briefs the squad/platoon. Patrol members prepare river crossing equipment and patrol to the crossing point.

      (4) Security elements should position themselves to observe the river crossing as well as boat traffic or obstacles coming up or down the river.

      (5) PT and PL or APL will swim the line across. Fins are recommended if the current is strong or the river is wide. The element enters the water upstream from the intended landing point, or it enters the water across from the desired landing point and then patrols back upstream once it has landed on the opposite side.

      (6) One man does the recon and provides security while the other secures and tends the line.

      (7) The running end of the line should be secured a few inches below the water. It must be easily untied (or cut) in case of approaching craft or emergency. The rope is a guide only, not a replacement for flotation.

      (8) Personnel begin moving across once the point element signals "All Clear."

      (9) Inflate lifejacket and unbuckle LBE.

      (10) Snaplink waterproof buoyant rucksacks and bulky gear to the line. Position weapons across rucksacks.

      (11) Tell the man following that you are entering the water prior to crossing the river.

      (12) Enter water and cross on the downstream side of the line using the hand over hand method.

      (13) When one man is half way across, the next man enters the water.

      (14) Take security position after exiting water as soon as possible.

      (15) Downstream security must be aware of last man and line crossing.

(16) PL insures all personnel are accounted for and gear is stowed and ready before moving out.

4. Equipment.

a. Lines for river crossings should be easy to untangle, lightweight, strong, and easy to stow and carry. Different lines are used as follows:

(1) Shroud line from a cargo chute

(2) Tubular nylon (20 foot 1/2" preferred)

(3) Braided Sampson lines (used for lanyards)

(4) Rappelling line

(5) Polypropylene line.

(6) Static climbing rope (used for over water crossing because when using green or gold rappel line, it is difficult for the carabineer to slide across the lay of the rope when it is weighted).

b. Snaplinks used for securing the river crossing line and for securing gear to the line can be locking or non-locking.

c. The bag or pouch used for carrying the river-crossing line can be a Claymore bag or a demo haversack.

d. Improvised Flotation Devices. Used for floating heavy and bulky gear across water.

(1) The preferred method is haversack flotation devices and closed cell foam

(2) Waterproof plastic bags

(3) Wrap in poncho with foliage to create air gaps

(4) Clothing, when used to trap air in the manner taught in lifesaving and water survival classes

(5) Air mattress.

5. Considerations.

a. A buoyancy check must be conducted with a full load of ammunition and equipment prior to any operation that could require a river or stream crossing. The number of UDT life jackets and floatation devices (either ether-foam filled or bladders) needed to float the man and his load out must be predetermined.

b. Waterproofing. Ensure equipment is adequately waterproofed both inside the rucksack and on the body.

c. Carry an extra life jacket for an intended prisoner, down pilot, or wounded person.

d. AW and RTO will need extra flotation.

e. Get out of water slowly, to minimize noise.

f. Clothing and equipment should have drainage grommets. Trousers should be unbloused or loosely bloused to allow quick drainage.

g. Weapons. Tape or use covers over the end of the barrel. Remember to drain water and ensure weapons are free from mud. Ensure weapons are ready to fire once across.

h. Hook line to PT/swimmer where he has easy access to undo it.

i. Angle line downstream for fast flowing rivers.

j. Keep the line from flapping on water as much as possible.

k. Over the beach (OTB) SOPs.

2.4.7.6 Booby Traps, Mines, and Improvised Demolitions. There are many devices and methods of employment, including non-explosive options.

1. Primary Uses. These include:

   a. Protecting an area from intrusion by enemy forces

   b. Slowing the progress of enemy forces moving through an area

   c. Inflicting damage on personnel and equipment

   d. Demoralizing enemy forces through fear

   e. Engaging the enemy with minimum risk to employing personnel

   f. Instilling fear in the local populace and government

   g. Sabotaging installations.

2. Scope. A booby trap is a device that will injure or kill a person who disturbs an apparently harmless object or performs a presumably safe act. A mine is an explosive device designed to destroy or damage vehicles, ships, boats, or aircraft, or to wound or kill personnel. It may be detonated by the action of its victim, by the passage of time, or by a controlled means. These devices can vary from the simplest to fairly complex items. Deception is often used to lure personnel into the areas where they are employed. They are usually disguised carefully, requiring personnel to conduct a thorough and patient check of the route to preclude triggering them.

   Quite often it is possible to obtain information from the local populace who generally know where the traps are located. All intelligence sources should be exploited to ascertain types of traps in use and where they are located. Ensure the patrol knows the positions of booby traps laid by friendly personnel.

3. Indicators

   a. Shiny objects

   b. Flat areas

c. Differences in color of vegetation

d. Trip wires

e. Anything that appears out of place.

4. General Precautions. When booby traps and mines are a possibility, do the following:

   a. Proceed slowly and carefully. As the name implies, booby traps are used to catch careless personnel.

   b. When sweeping an area, remember both explosive and non-explosive booby traps may be employed with mines.

   c. Be suspicious of all objects that appear to be loose. When checking captured factories, supply dumps, or material, watch for booby traps, as there may be more than one.

   d. Never cut a trip wire. Before disturbing any object, check it for wires and pressure release devices.

   e. If you find explosive booby traps, mark them and leave them alone unless it is absolutely necessary for you to move them. Let EOD personnel take care of removal and disposal.

   f. Do not take anything for granted. An object may appear to be innocent, but if it is found in suspicious circumstances, suspect a trap.

   g. Common sense factors should not be overlooked. The enemy marks most mined areas in some way to alert their own troops. Learn to read the signs indicating mines or booby traps.

2.4.8 Breaks and Perimeters

2.4.8.1 Breaks. A break is usually a brief halt of only a few minutes duration. Breaks must be used to keep the patrol fresh so its overall effectiveness won't deteriorate. The patrol can only go as fast as the slowest man. A break is an enroute control measure that can be used for many purposes; actions are as follows:

1. Set Security. The PL should pre-brief the perimeter formation that the patrol will use during the halt. The emphasis will be on maintaining 360-degree coverage of the patrol's position. Various types of perimeters will be discussed in the following sections. Whenever possible, the patrol should fishhook into its perimeter in order to observe its trail for enemy trackers.

2. Pass the Word. This is an opportune time to disseminate new or updated information and instructions to all the patrol members. It is also important to ensure that the patrol members have an opportunity to give any information that they have on what they may have seen or heard during the move from the last halt.

3. Map Check. Every time that the patrol stops for a brief period, the PL should conduct a map check to pinpoint the patrol's exact location. The position should be confirmed by using the portable GPS, the pace count, resection of prominent terrain features, and time of travel. All of the above data should correlate to provide a very accurate position plot. The PT, RS, APL, navigator, and assistant navigator should all be a part of the process.

4. Reconnaissance. If necessary, a quick area reconnaissance can be conducted by selected patrol members to ensure that the area is clear. At a minimum, a RS element can be sent back along the patrol's trail to determine

whether or not the patrol is being tracked. If an individual or element leaves the patrol's perimeter, follow the "GOTWA" formula (explained below).

a. Ensure that all patrol members know what your plans and intentions are (GOTWA).

(1) Going - where you are going and what you are going to do.

(2) Others - who you are taking with you.

(3) Time - maximum amount of time you will be gone.

(4) What if - what the person in charge will do if you do not return on time.

(5) Actions - actions to be taken by each element upon enemy contact.

5. Communications. Radio communications can be performed more easily when halted than when moving, especially if satellite and burst communications are being used. Make communications the final event before the patrol moves out to reduce the enemy's ability to DF your location. Never make your last communication of the day from your night patrol base or LUP.

2.4.8.2 Perimeters

1. Establishing Perimeters. This is a key ingredient of patrol security measures. Defensive perimeters are used when:

    a. Inserting or extracting

    b. Moving from objective RPs

    c. Stopped or resting

    d. Listening for or watching for enemy

    e. Prolonged stops to eat, rest, or sleep

    f. At all RPs

    g. Stopped to communicate

    h. Negotiating danger crossings.

2. Setting Perimeters. This entails dispersion of the patrol to provide observation in every direction. Usually the PL and RTO are together in the center. Figure 2-28 represents typical perimeter positions. SOPs for setting perimeters are as follows:

    a. Accomplish with minimum changes to the patrol formation.

    b. Use cover and concealment, stay low or prone.

    c. Position AW and GN to cover most likely positions of enemy encounter.

3. Perimeter Size/Radius. The perimeter should allow each man to see the man on either side. Factors to consider include:

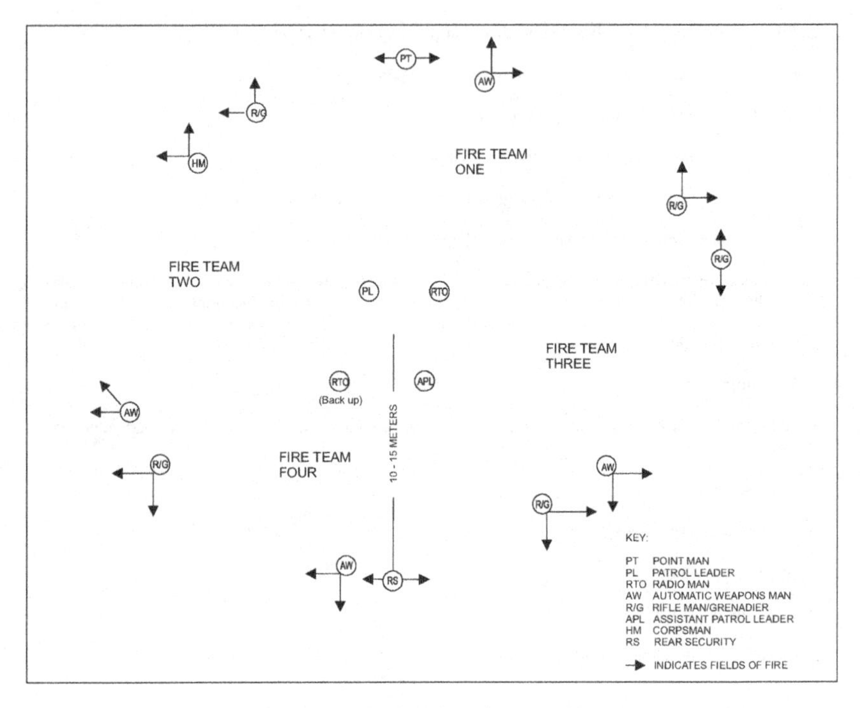

Figure 2-28. Security Perimeter Positions

a. Illumination

b. Cover

c. Terrain

d. Situation.

4. Types of Perimeters.

a. Cigar perimeter. The cigar perimeter is a simple and very effective security perimeter. It is easy to establish from any movement formation and easy to move out from. Each man has a clear field of fire in formation.

Movement within the formation perimeter is relatively easy using the center aisle. The perimeter conforms well to terrain such as ridgelines and fingers. The perimeter is good for virtually any size patrol. See Figure 2-29. Procedures for formation are:

(1) Be in or shift to a file or staggered file formation.

(2) On signal, each man steps a few paces into his field of fire, gets down, and uses available cover and concealment.

(3) To resume the patrol, reverse the procedure.

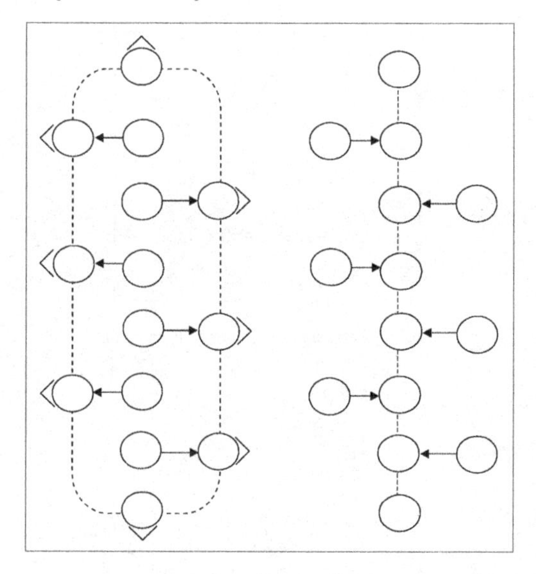

Figure 2-29. Cigar Perimeter

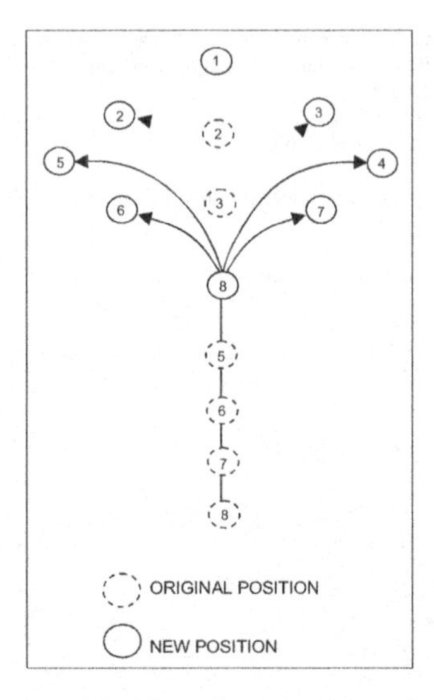

Figure 2-30. Diamond Perimeter from a File

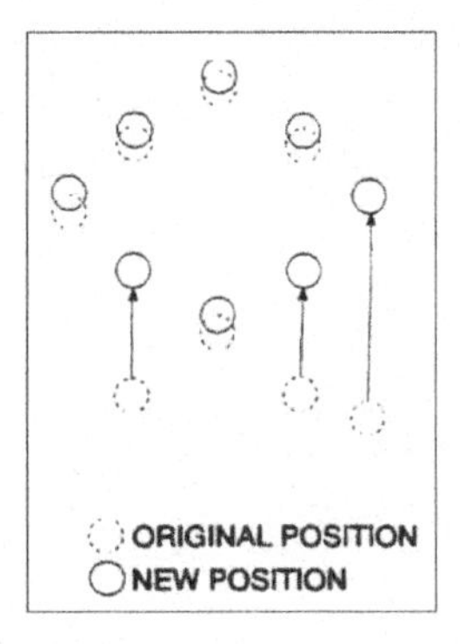

Figure 2-31. Diamond Formation from a Wedge Formation

b. Circle and Diamond Perimeters. The circle and diamond perimeters are other common perimeters. They give good, all-around defense with 360-degree coverage. They are relatively easy to establish and move out from if rehearsed. They can also be used for limited movement as discussed in Section 2.4.4.6. Two basic methods for putting the patrol into the circle perimeter are discussed below. The diamond perimeter is very similar and is shown in Figures 2-30 and 2-31.

Procedure One. See Figure 2-32. This procedure is best for a squad-size patrol. Procedures follow:

(1) Halt the patrol. If not in a file, get into file formation.

(2) Establish one man as 6 o'clock for the circle. Placing the AW man in that position in a squad-size patrol is a good SOP because it gives more firepower to the rear.

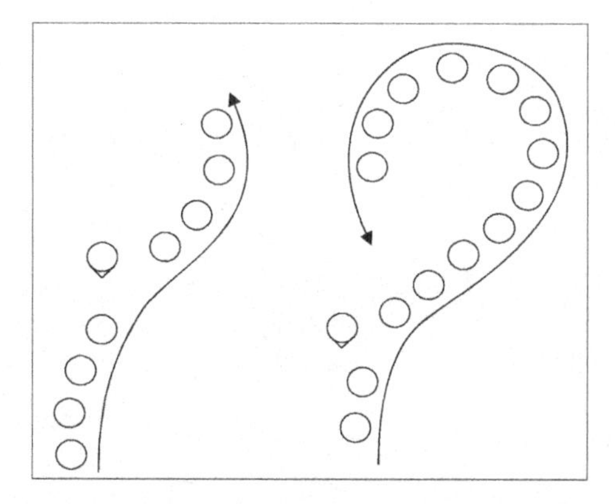

Figure 2-32. Circle Perimeter Procedure One

(3) Everyone else files around in a circular pattern, clockwise or counterclockwise, until the lead man reaches the 6 o'clock again.

(4) Personnel halt, face outboard, establish a tactical spacing, get down, and use available cover and concealment.

(5) To resume the march, collapse the circle into the desired patrol formation, ensuring rear or direction of most likely threat is covered.

(6) Assume proper line of march and move out.

Procedure Two. See Figure 2-33. The second method for establishing a circle perimeter can be used by a squad-size patrol and is easier for controlling a platoon-size patrol. The method is:

(1) Halt the patrol. If not in a file, get into file formation.

(2) Assign FTs or pairs to clock positions 2, 4, 8, and 10 o'clock, or as terrain dictates.

(3) Personnel face outboard, establish a tactical spacing, get down, and use available cover and concealment.

(4) To resume the march, collapse the circle into the desired patrol formation, ensuring rear or direction of most likely threat is covered.

(5) Assume proper line of march, and move out.

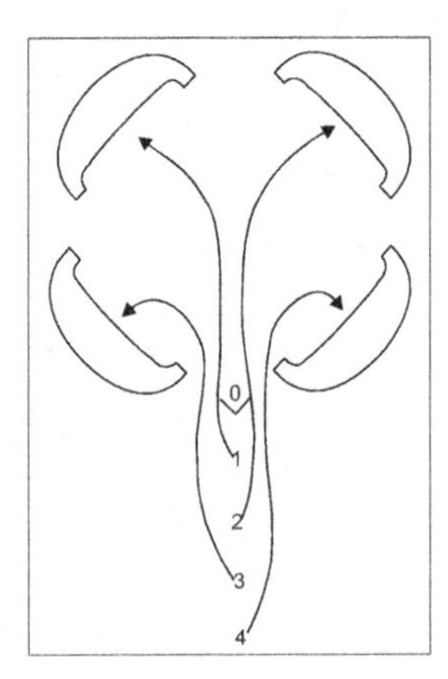

Figure 2-33. Circle Perimeter Procedure Two

2.4.9 Rally Points. RPs are a control and security measure. A RP is a place where a patrol can reassemble, reorganize, and prepare prior to actions at the ORP. The PL should select RPs in planning and during the patrol. They consist of initial, enroute, objective, and extraction RPs. RPs should:

1. Be large enough for the patrol to assemble in
2. Be easily recognized
3. Have cover and concealment
4. Be defensible for a short time
5. Be located well away from normal routes of movement.

2.4.9.1 Initial Rally Point. The Initial Rally Point (IRP) is used immediately after insertion. It should be an easily recognized point or terrain feature within about 100 meters of the insertion point. The patrol's actions immediately upon stepping off the insertion platform are:

1. Set initial security.

2. Quickly get into patrol formation and move out in the preplanned direction and distance.

3. If there are no prominent features, move for no more than five minutes in a preplanned direction.

4. Set perimeter security. STOP, LOOK, LISTEN, and SMELL (SLLS) to attune yourself to the sounds, sights, and smells of your immediate surroundings.

5. Establish communications with the insertion platform in case immediate extraction is needed. This on-call extraction should be in effect for only a short time or a set distance into the patrol in order to reduce the potential threat of compromise to the insertion platform and the patrol.

6. Make sure you are ready to move out: conduct final equipment preparations.

7. Once the patrol moves out and feels secure, detach the insertion platform.

2.4.9.2 Enroute Rally Points. An enroute RP is where a patrol rallies if dispersed or separated enroute to or from its objective. If separated or dispersed by a chance contact, the patrol members go back to the last designated enroute RP, regroup, and reorganize. Prominent, but not obvious, terrain features should be used.

1. Actions at enroute RPs after a contact are:

   a. Approach and observe the RP.

   b. Cautiously enter.

   c. Use challenge and reply procedures.

   d. Set security.

   e. Help guide other patrol members into the RP.

   f. Consolidate ammunition and ordnance.

   g. Treat wounded/injured.

   h. Stay only for the preplanned and briefed time.

   i. The senior man present is the acting PL.

   j. PL (or acting PL) determines whether to:

      (1) Continue patrol (depending on patrol status and mission requirements).

      (2) Follow or modify briefed rally duration if patrol members are missing. It is usually better to not exceed the briefed duration unless good information about the missing personnel is available.

      (3) Send out search team for missing personnel.

      (4) Proceed to next enroute RP.

(5) Patrol to planned XRP.

(6) Call for hot or emergency extraction.

k. Patrol out as an organized unit, maintaining strict patrol discipline.

l. Sterilize RP on leaving.

2. Considerations. In planning operations, the PL should always plan to use a series of enroute RPs. These are especially useful for helping to guide the patrol to the XRP. If the patrol is separated for tactical reasons or as a result of a chance contact with the enemy, the series of RPs gives the patrol multiple opportunities to link up again with the main patrol element.

   Pre-plan enroute RPs by making a thorough map and imagery study. Whenever possible, confirm the information by an aerial reconnaissance. Ensure that all patrol members study the RPs on the maps and photos carefully and understand the rally plan prior to beginning the mission.

2.4.9.3 Objective Rally Point. An ORP is where the patrol halts to prepare for actions at its objective. It is also where the patrol will return after its action at the objective (as required). In general it:

1. Must be located near the objective.

2. Should be far enough away from the objective so that the enemy will not detect the patrol at the OPR.

3. Must be far enough from the objective so that it will not be overrun if the patrol is forced off its objective. The patrol's actions on nearing and establishing an ORP are:

   a. Approach cautiously because the target may have moved, enemy security may have changed, or you may not be exactly where you think you are.

   b. Make a security halt and reconnoiter the ORP and vicinity before bringing the patrol in.

   c. Move in by fish hooking into the ORP to cover your trail.

   d. Set security and maintain a high alert status. Ensure any obvious approach routes are well covered.

   e. Leave subordinate team leaders to make preparations at the ORP while the recon team reconnoiters the target.

   f. Follow the GOTWA formula.

   g. Return to the ORP and conduct a final brief to ensure that any changes or modifications to the original plan are thoroughly briefed to all, especially those affected, and that everyone acknowledges and understands the changes. Security should be maintained during the brief, which may require two or more briefs.

   h. Prepare equipment being left at the ORP for quick pick-up upon return.

   i. Organize into the proper elements to conduct the actions at the objective.

   j. Sterilize the ORP.

k. Conduct actions at the objective.

2.4.9.4 Extraction Rally Point. The XRP is used as the final staging point for regrouping and reorganizing the patrol and preparing a reception for the extraction platform. However, the mission is not over. If the enemy has been trailing or tracking you, this is the most opportune time for him to attack. Not only does he expect that you are tired, hungry, and somewhat relaxed with your extraction imminent, but it is an excellent opportunity to take out the extraction platform as well. Low hovering or landed helicopters and beached boats are easy targets. Actions on nearing an XRP are:

1. Make a security halt and reconnoiter the XRP and vicinity before bringing the patrol in.
2. Move in by hooking into the XRP to cover your trail.
3. The actual extraction site should be easily accessible from the XRP.
4. Lay up and wait, maintaining 360-degree security.
5. Initiate communications or signals for the extraction platform.
6. Reorganize into proper extraction loads for helicopters, aircraft, boats, or other craft.
7. Prepare equipment as required.
8. Sterilize the RP.
9. Move to extraction site and prepare it if necessary.
10. Maintain a security perimeter. Be extremely alert.
11. Board extraction platform quickly and efficiently.

2.4.10 Lay-up Points and Posts. LUPs and the posts needed for the safety of the platoon and observing designated areas or targets should be planned, rehearsed, and used in all combat patrols.

2.4.10.1 Lay-up Points. A LUP is a position established when a patrol halts for an extended period outside a friendly controlled area. The patrol must take active and passive measures for security while in such a vulnerable position. It is most effective to plan, select, occupy, and organize an area, which, by its location and nature, provides passive security from enemy detection.

1. Purpose of a LUP. The typical purposes of a LUP include:

   a. A place for concealment when disadvantageous visibility conditions require a patrol to stop moving.

   b. Hiding the patrol during a lengthy, detailed reconnaissance of the objective.

   c. Preparing food, maintaining weapons and equipment, and resting after extended movement.

   d. Reorganizing after a patrol has infiltrated the enemy area in small groups.

e. Providing a base from which to conduct several consecutive or concurrent operations, such as ambush, raid, reconnaissance, or surveillance patrols.

2. Selecting a Site. An initial LUP is one that is usually selected by map or aerial reconnaissance, or prior knowledge of the area during the patrol planning. It remains tentative until its suitability is confirmed and it has been secured for occupation. The PL must also select an alternate site, a rendezvous point, and a RP.

3. Passive Security Measures. The PL should select terrain that is considered of little tactical value to the enemy. When feasible, the area should be remote from human habitation, but near a water source. Difficult terrain impedes foot movement. Dense vegetation (e.g., bushes and trees spread out close to the ground, etc.) impedes foot movement. The PL should avoid known or suspected enemy positions, built-up areas, and ridgelines and topographic crests. He should also avoid roads, trails, and natural avenues of travel and wet areas.

4. Active Security Measures. Plan for:

   a. Observation post (OP) and LP systems that cover avenues of approach into the area.

   b. Establish a radio communications net to provide early warning of an enemy approach.

   c. Select an alternate area for occupation if the initial area is found unsuitable or compromised.

   d. Possible withdrawal, to include multiple withdrawal routes.

   e. Enforcement of camouflage, noise, and light discipline.

   f. Conducting necessary activities with minimum movement and noise.

5. Establishing an LUP. The establishment of an LUP is usually part of an overall plan for a patrol's operation. In some circumstances, however, establishing an LUP may be an on-the-spot decision.

   a. The maximum time an LUP may be occupied depends on the need for secrecy. In most situations, it should not exceed 24 hours. In all situations, an LUP is occupied for the minimum time necessary to accomplish its purpose. The same LUP is not usually used again.

   b. In guerrilla operations, secrecy of the LUP is mandatory. An LUP is evacuated if discovery is even suspected.

6. Occupying an LUP. An LUP may be occupied in either of two ways:

   a. By moving to the selected site and expanding into and organizing it in the same manner as an on-the spot establishment.

   b. By halting near the selected site and sending forward reconnaissance forces. The method used must be thoroughly planned and rehearsed. Use of the LUP drills in these methods will assist in swift and efficient establishment. Occupation methods are shown in Figures 2-34 and 2-35.

      (1) The PL halts the patrol at a suitable position within 200 meters of the desired LUP. He establishes close-in security. Previously designated individuals join the PL.

(2) The PL designates the point of entry into the LUP location as 6 o'clock, assigns areas by the clock system, and designates the center of the base as patrol headquarters. The reconnaissance team reconnoiters the area for suitability and returns to the patrol. See Figure 2-34.

(3) After the tentative LUP has been reconnoitered, the patrol moves forward to the center of the base as shown in Figure 2-35.

(4) Once the LUP is established, each FT reconnoiters its sector and reports indications of enemy or civilians, suitable OP and LP positions, RPs, and withdrawal routes (see Figure 2-35).

(5) The PL designates RP, positions for OPs and LPs, and withdrawal communications.

7. Security Considerations. Security is of prime importance during operations. The following principles are vital to the security of every man moving in and around the LUP.

a. Use only the point of entry and exit. Camouflage and guard it at all times.

b. Stoves and heat tabs are used only when necessary and, as a general rule, only in daylight.

c. Accomplish noisy tasks only at designated times and as early as possible after occupation, but never at night or during the quiet periods of early morning and late evening. When possible, perform noisy tasks when other sounds will cover them, such as the sounds of aircraft, artillery, or distant battle noises.

d. Movement, both inside and outside the LUP, is restricted to the necessary minimum.

e. When enough personnel are available, man LPs with at least two, preferably three, men so they can help each other stay alert.

f. One hundred percent alert is observed 1/2 hour before and 1/2 hour after light in the morning, and 1/2 hour before and 1/2 hour after dark in the evening. This ensures that every man is acclimated to changing light conditions, and is dressed, equipped, and ready for action. The PL will designate a security watch as the situation dictates for all other times.

g. Each man makes certain that he knows the location of men and positions to his flanks, front, and rear, and that he knows the times and routes of any expected movement within, into, and out of the LUP.

h. Permit rest and sleep only after all work is done. Stagger rest periods to maintain proper security.

i. Send two people for water, but only when needed. Make sure everyone knows when a man leaves, his expected time of return, and his directions for leaving and returning.

j. Take out only necessary gear and put it away after its use. Keep web gear, pack, and weapon ready to go immediately. A man's gear, especially his weapon, should never be more than an arm's reach away.

k. If the patrol is to be resupplied by air, locate the flight path and drop zone so that neither the base nor possible objectives are compromised.

8. Defense. Defensive measures must be planned, but a LUP is usually defended only when evacuation is not possible.

a. Do not construct elaborate firing positions.

b. Stress camouflage and concealment.

c. Plan for artillery and mortar fires if available. Place early warning devices on avenues of approach. If the base definitely will be defended, place mines, trip flares, and booby traps on avenues of approach and in areas that cannot be covered by fire. The value of these devices must be weighed against the fact that their discovery automatically compromises the LUP.

d. Establish an alert plan. This includes plans for evacuation and defense. All members must know the plans, signals or orders for implementation. Base defense plans should include pursuit and destruction of the attacking force.

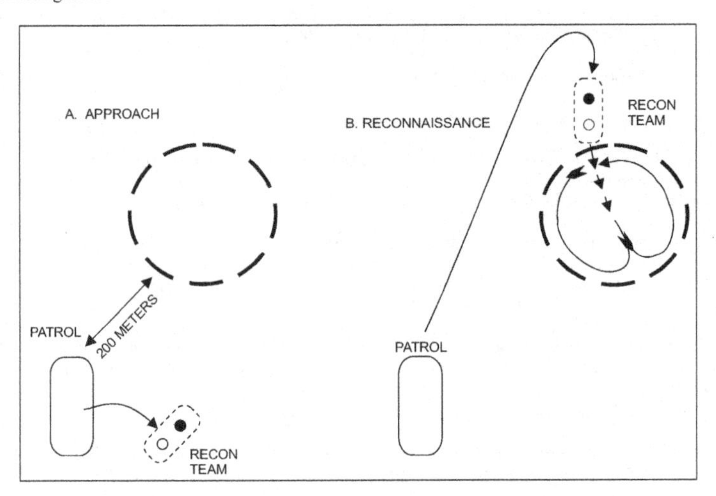

Figure 2-34. Lay-up Point Establishment - Reconnaissance

9. Communication. Communications are established with higher headquarters, and with OPs or LPs as required. A system must provide for every man to be alerted quickly and quietly.

   a. Hand and arm/voice communications are an excellent and preferred means of alerting everyone.

   b. Tug or pull wires may be used for signaling and waking people who are asleep.

10. Departing an LUP. Remove or conceal all signs of the patrol's presence. This may prevent the enemy from learning of the patrol's presence, may prevent pursuit, and may prevent the enemy from learning the patrol's methods for operating LUPs. The patrol evacuates as a unit when possible.

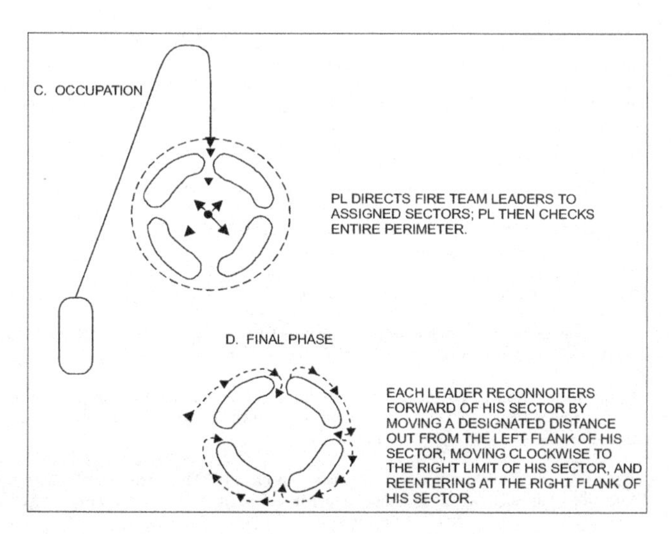

Figure 2-35. Lay-up Point Establishment - Occupation

2.4.10.2 Observation Posts. OPs are positions from which you watch and listen to the enemy activity in a particular sector. OPs can be used offensively and defensively. OPs are used primarily during the hours of daylight, while LPs are used during the hours of limited visibility caused by darkness or weather conditions. LPs are positioned forward of and on exposed flanks of the unit position and generally along probable avenues of enemy approach to provide early warning, as well as to listen and observe activity. Their mission will often require location in an exposed position. They are operated in reliefs except when movement to and from positions would reveal their locations or endanger the personnel. Otherwise, LPs are established and operated in the same general manner as OPs.

1. Offensive OPs. The offensive OP is used to provide eyes on the target. This OP can be conducted as a mission itself, such as a SR mission. It can also be used for tactical observation of the target for the patrol to conduct a follow-on raid of the target or to ambush forces moving or leaving the target area being observed.

2. Defensive OPs. The defensive OP is used to observe the activities that occur within its sector and provide early warning of enemy approach. The number and location of OPs established depend on the patrol's mission, AOR, and the degree of observation permitted by the terrain and weather.

3. Selecting an OP. The site should provide:

   a. Maximum observation of the desired area

   b. Cover and concealment for the occupant

   c. Concealed routes to and from the OP.

4. General Considerations. Observation is the primary consideration whether or not the above conditions exist at a site. Normally, the best location for an OP is on or near the military crest of a hill. The topographical crest should be avoided because you may be skylined when occupying the OP. You must remember that selecting a position for an OP is very similar to selecting firing positions for direct fire weapons, such as a rocket launcher or recoilless rifle. It may be appropriate to establish the OP well down the forward slope when the terrain restricts observation.

5. Establishing and Operating an OP. The procedures for manning and operating an OP are as follows:

   a. The PL should take the men who are to man the OP to it, especially for the initial establishment of the OP. He will show them where to set up and give rules and instructions.

   b. OPs should have communications. Radio antennas should be well camouflaged to avoid detection.

   c. OPs should consider using any portable sensor devices that are available, such as NVEOs, infrared (IR), and thermal and seismic (motion) sensors, to enhance their observation capabilities.

   d. Personnel going to and from the OP must move carefully so that movement does not reveal the location to the enemy. Separate routes to and from the OP are established. When natural concealment is not adequate, the OP is camouflaged.

   e. OPs are operated in reliefs of a minimum of two men for each relief. The frequency of reliefs depends on the physical condition of the men, weather, number of personnel, the patrol's next operation, and enemy situation or threat.

   f. OP observers should rotate jobs while on duty. One observes while the other records and reports information observed. The observer and the recorder must alternate about every 30 minutes because an observer's efficiency decreases rapidly after that period.

   g. OPs and LPs should be within effective small arms range of friendly personnel and supported by other means when possible.

   h. Those manning OPs should shoot at the enemy only in self-defense or to cover their withdrawal. Their mission is to observe and report.

2.4.11 Tactical Innovation. No amount of planning, drilling, or use of standard techniques and procedures will prepare a patrol for all eventualities. A skilled and competent SEAL platoon knows when its IADs are honed correctly and when innovation and ingenuity must be employed. It is not possible to predict when this will occur. If the platoon is a competent team and knows and understands the basics of small unit tactics, however, it will adjust its actions quickly.

2.5 AMBUSH TACTICS, TECHNIQUES, AND PROCEDURES

2.5.1 General. The ambush is one of the most lethal combat tactics available to small units, especially in NSW. A thorough understanding of ambush principles is essential for field survival. This section will discuss the tactical principles, techniques and procedures associated with both conducting and surviving ambushes.

2.5.1.1 Definition. An ambush is a surprise attack characterized by firepower and violence of action from a concealed position upon a moving or a temporarily halted target. Ambush operations may involve actual closure with the enemy, or they may involve fire only. Ambush operations are highly effective when used in conjunction with DA missions to destroy hard targets.

2.5.1.2 Ambush Objectives. Ambush operations seek to reduce the enemy's overall combat effectiveness by destroying or capturing troops and material, and by damaging the enemy's morale through harassment. A senior headquarters may adopt an ambush strategy for all SOF units to blanket an entire conflict zone with ambushes. Such a strategy can have a significant effect on the enemy.

1. Destruction of Enemy Forces. Ambush operations, if properly conducted, can produce a significantly disproportionate impact on the enemy relative to the number of SOF involved and ammunition/munitions expended. In ambushes, firepower can be carefully concentrated for maximum damage to the enemy.

2. Capture of Enemy Personnel and Material. Ambush operations may be conducted for the express purpose of capturing enemy personnel and equipment for intelligence purposes. In such cases, the ambush force must plan carefully for protecting and moving all captives to the rear immediately. Also, as much intelligence as possible must be obtained from the ambush.

3. Harassment. Although harassment is a secondary purpose of the ambush, SEALs can cause the enemy to divert critically needed personnel from other support or combat operations. Ambushes create a defensive mind-set within enemy ranks. Since most ambushes are conducted at night, the enemy may avoid night operations, thus decreasing his overall combat effectiveness, and his war-fighting psychology may change.

2.5.1.3 Advantages of Ambush Operations. Ambush operations are a favorite tactic of guerrilla forces because they do not require that ground be seized or held and they enable small forces with limited weapons and equipment to harass or destroy larger, better armed forces. Ambush operations are also effective counter guerrilla tactics because they force an enemy guerrilla force to engage in decisive combat in disadvantageous circumstances.

2.5.2 Ambush Fundamentals

2.5.2.1 Surprise. There must be surprise. This allows the force to seize and retain control of the situation. If an ambush does not achieve surprise, it is not an ambush, just another form of attack. Careful planning, preparation, and execution are the keys to the surprise, as is rehearsal of squad member's points of performance, responsibilities and actions prior to the ambush.

2.5.2.2 Coordinated Fire. The ambush force leader should position all weapons, including mines and demolitions, and coordinate all fire and fire support such as artillery, CAS, and naval gunfire support (NGFS) to achieve the following:

1. Surprise delivery of a large, accurate, volume of highly concentrated fire into the kill zone to cause maximum damage in the least amount of time.

2. Isolation of killing zone to prevent escape or reinforcement.

2.5.2.3 Control. The ambush force leader must control the entire SEAL element, especially when approaching, occupying, and withdrawing from the ambush site. Control measures must provide for:

1. Early warning system in place to alert the ambush force to an approaching enemy.

2. Fire discipline. This must be maintained to ensure that fire begins only when the proper signal is given with the enemy in the kill zone. Appendix H may be used to inform all personnel of weapon responsibilities during the evolution.

3. Initiation of appropriate IADs if the ambush is prematurely detected.

4. Proper coverage of after-ambush search teams while searching.

5. Quick withdrawal to the planned RP.

2.5.2.4 Reconnaissance. Proper reconnaissance of an ambush site is critical. The ambush force must have a thorough knowledge not only of the terrain at the ambush site, but also of the surrounding area. Potential approach and escape routes for both the enemy and the ambush force must be carefully surveyed.

2.5.2.5 Security. Proper security is essential to a successful ambush operation. Security must cover both the ambush site and expected routes into and out of the area. Poor security will often lead to a premature initiation of contact with the enemy, which can put the ambush force into an extremely vulnerable position.

2.5.3 Types of Ambushes. There are two types of ambushes; the hasty ambush and the deliberate/point ambush. The hasty ambush is discussed as an IAD in Section 2.4.6.2. There is little or no planning or rehearsal, and members must react according to platoon SOPs. The deliberate ambush is planned and rehearsed with respect to terrain and actions at the objective.

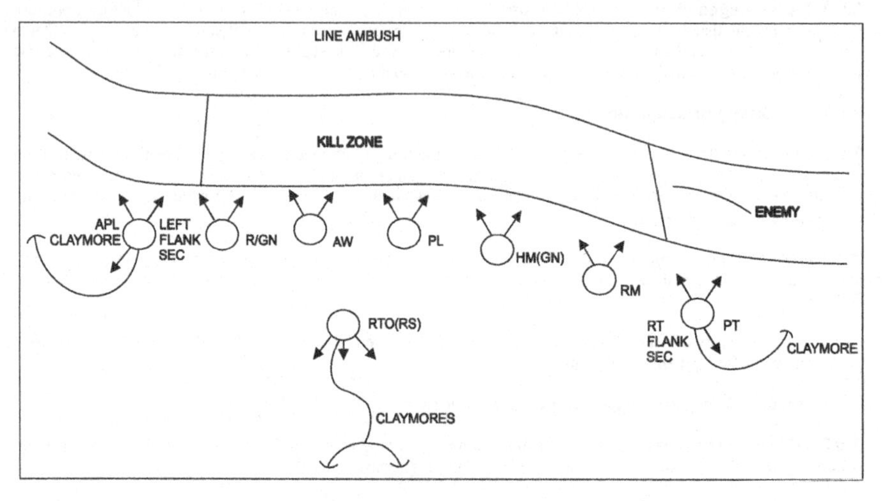

Figure 2-36. Line Formation

2.5.3.1 Deliberate Ambush. The deliberate/point ambush involves a single kill zone. This is the type of ambush primarily used by SEALs. The formation of a deliberate/point ambush is a carefully considered function of the terrain, visibility conditions, forces involved, weapons employed, and the overall enemy situation in the AO. The most common formations for hasty and deliberate/point ambushes are the line and L-shape.

1. The Line Formation. In this formation, the ambush force deploys parallel to the enemy's route of advance through the ambush zone. (see Figure 36). This is the ideal ambush formation. With the ambush force parallel to the enemy's route of movement, maximum flanking fire is possible. This formation is the easiest to control, but there is always the chance that lateral dispersion of the target may be too great for effective coverage. The line formation works best in close terrain, which restricts target maneuver, and in open terrain where one flank is restricted by natural or other obstacles. Selection of the ambush site should be based on the following criteria:

   a. Type of ambush desired/available. The line formation is best suited to sites where the enemy direction of approach can be well predicted. They work best along trails or roads or in areas where the direction of movement is restricted by terrain features such as rivers, cliffs, and gorges.

   b. Ability to exercise C^2. The line formation is the easiest to control since all patrol members respond to one ambush force leader.

   c. Sectors and ranges available for clear fields of fire. The line formation provides for the simplest and most direct organization of fields of fire. All fields of fire are directed ahead of the formation.

   d. Likely avenues of approach of the targeted group. The line formation is the least adaptable to an enemy approach from a direction other than that for which the ambush was set. It is difficult to rapidly coordinate an undetected swing of an entire line along a new direction of approach.

   e. Likely defenses of the targeted group. If the target is predicted to have a high rate of firepower, particularly firepower capable of rapid area saturation, the line formation is vulnerable. Once the ambush is executed, the ambush positions are compromised and the patrol must rely on its own firepower to overwhelm the unprepared enemy. If the enemy is felt to have the capability for concentrated area saturation firepower that can react quickly, another type of ambush should be considered.

   f. Number and type of targets within the targeted group. The line formation is designed to attack an enemy formation of relatively soft targets that can be mostly contained in the kill zone. If there are a number of hard targets (such as tanks or armored personnel carriers) another type of ambush should be considered (unless the patrol is equipped with anti-tank standoff weapons and/or anti-tank mines/demolitions).

2. The L-Shaped Formation. The L-shaped formation is a variation of the linear formation. The long side of the ambush force is parallel to the killing zone and delivers flanking fire. The short side of the ambush force is at the end of, and at right angles to, the killing zone. The job of those on the short side is to deliver enfilade fire, which interlocks with fire from the killing side. (see Figure 37). This formation is very effective at the bend of a straight road, stream or trail; however, C^2 is critical in order to ensure that the patrol does not fire on itself. Selection criteria for the "L" shaped formation is as follows:

   a. Type of ambush site desired/available. The "L" formation is effective for an ambush site in which enemy movement is funneled by natural bends or curves such as in a road, trail, river, or stream. In these cases the "L" should bend to conform to the geographic feature. Use the regular "L" with one avenue of approach and the inverted "L" when there are two possible avenues of approach.

   b. Ability to exercise C^2. The "L" formation is more difficult to control than the line formation because each leg of the "L" has a different direction for fields of fire. Set up the C^2 so the PL exercises command through two squad or FT leaders.

c. Sectors and ranges available for clear fields of fire. The "L" formation provides for the most complete coverage of the kill zone through overlapping fields of fire. Fields of fire from each leg cover the same area, but from different directions. This requires a high degree of fire discipline to avoid firing into the other leg.

d. Likely avenues of approach of the targeted group. The "L" formation provides for more flexibility with regard to knowledge of the enemy approach direction. It provides a formation that can conduct an ambush against an enemy approaching from within a ninety-degree sector.

e. Number and type of targets within the targeted group. Like the line formation, the "L" formation is designed to attack an enemy formation of relatively soft targets that can be mostly contained in the kill zone. If there are a number of hard targets (such as tanks or armored personnel carriers) another type of ambush should be considered (unless the patrol is equipped with a number of anti-tank standoff weapons). If the target is of such a size that a majority of it is not in the kill zone at the time of ambush execution, another type of tactic should be considered.

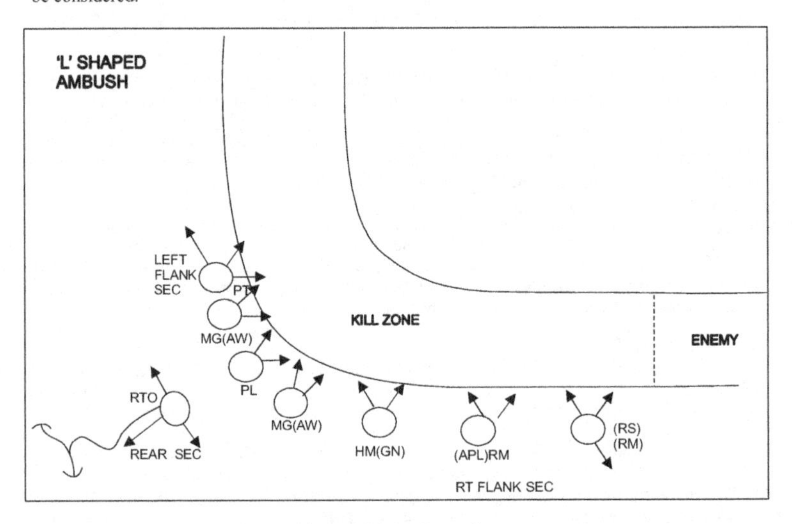

Figure 2-37. L - Formation

2.5.3.2 Demolition Ambush. Demolition ambush tactics are discussed in the following paragraphs. Figures 2-38 and 2-39 are two examples of a demolition ambush. See Appendix A for more detailed discussion regarding demolition procedures. When engaging numerically superior forces, a demolition ambush can be initiated remotely from a safe OP out of effective fire range.

1. Characteristics. A demolition ambush can use mines, Claymores, fragmentation grenades, concussion grenades, and detonation cord. Improvised demolitions can also provide excellent results. This type of ambush can deceive the enemy, who might not know whether they tripped a booby trap or if troops are present.

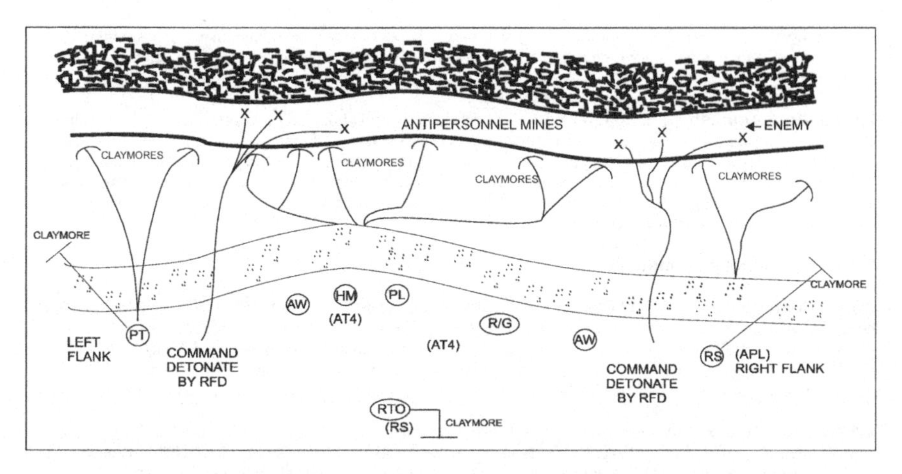

Figure 2-38. Demolition Ambush, Example One

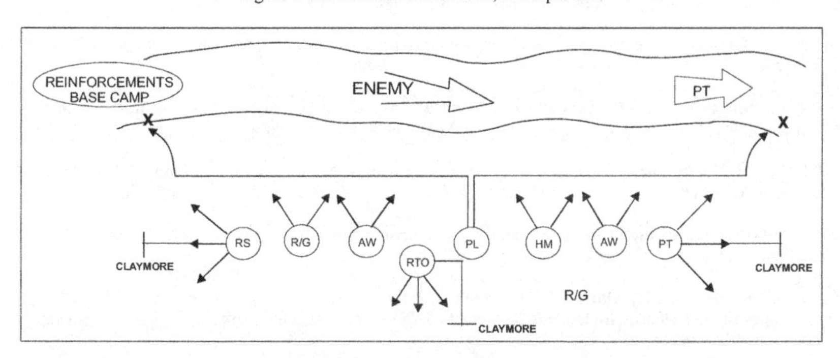

Figure 2-39. Demolition Ambush, Example Two

2. Selection Criteria. The demolition ambush is effective in an ambush site in which enemy movement is funneled by natural bends or curves in a road or trail, river or stream. It can be very effective against targets in narrow streams or canals where the enemy is forced by the terrain to pass through a potential kill zone. The demolition ambush is not effective unless the enemy approach direction and intended path is known with a high degree of certainty. Careful site selection is necessary to avoid friendly casualties from fragmentation devices such as hand grenades.

3. Ability to Exercise C^2. The demolition ambush is easy to control as there are minimum patrol positions involved. It provides for detonations within a defined area and the only enemy defense against it is to detect the concealed charges.

2.5.3.3 Night Ambush. Most SEAL ambushes are conducted during hours of darkness. The general principles of day ambushes also apply to night ambushes. Concealment is easier at night, but target detection is more difficult and weapons fire is less accurate. All AWs, particularly the M60 light machine gun, should have their left and right arcs of fire fixed to prevent shooting friendly patrol members. Patrol members must never change their positions once set up, as movement could be regarded as hostile. In night ambushes, clear orders, precise fire control instructions, clear RPs, and easily identifiable signals are critical.

2.5.4 Personnel Assignments

2.5.4.1 Command and Control Element. This element will usually consist of the PL, RTO, and HM. The PL will be in overall control and must have a clear vantage point of the kill zone and entire area. The RTO will provide illumination to the kill zone when the ambush is initiated and ensure illumination is maintained throughout the ambush (if illumination is used). Most often, he is located behind the PL and may serve as RS for the ambush. The HM will participate in the ambush as directed by the PL. He will join the search/prisoner handling team, if directed.

2.5.4.2 Security Element. The PL designates the RS and flank security. Flank security is used to signal if someone or something is approaching, and to cover the flanks of the ambush site. Flank security should be positioned several yards out from the main assault element ensuring visual contact is maintained. The duty of flank security is to watch for any possible enemy flanking maneuver and to keep a lookout for any reinforcements.

2.5.4.3 Rear Security. The RS force is designated by the ambush force leader and protects the rear of the ambush site from enemy approach. This is usually the RTO in a small ambush.

2.5.4.4 Assault Element. Members of this element will assault and control the kill zone at the ambush force leader's direction. Members will shoot point targets. Weapons should be fired on semi-automatic.

2.5.4.5 Search Team. The search team, comprised of riflemen, is pre-designated by the ambush force leader from the assault element. The search team will disarm and search the downed enemy after the ambush.

2.5.4.6 Prisoner Handling Team. If the ambush yields prisoners, pre-designated search team personnel should be prepared with handcuffs, blindfolds, gags, and line.

2.5.4.7 Perimeter Security Element. This element will fire during the ambush until "Cease Fire" is called. It will then provide perimeter security for the search element while its teams search the kill zone. AWs are assigned to the perimeter security element.

2.5.5 Procedures Prior to Initiation. Pre-initiation preparation is of critical importance to a successful ambush. A well set-up ambush will not only surprise the enemy, but also minimize his ability to react.

2.5.5.1 Selecting Ambush Method. The method of ambush used depends on the following:

1. Ambush site selected.

2. Ability to exercise C^2.

3. Sector and ranges available for clear fields of fire.

4. Likely avenues of approach of the targeted group.

5. Likely defenses of the targeted group. Does he have fire support and reactionary forces?

6. What are the SOPs of the targeted group? What is his training and does he flank?

7. Number and type of targets within the targeted group.

2.5.5.2 Ambush Objective Rally Point. The ambush ORP should be established close enough to the ambush site to be useable, but not too close, i.e., usually at least 500 meters away from the ambush site. The PL and the PT should leave the patrol at the ORP to conduct a reconnaissance of the ambush site. For long duration ambushes or during hours when the ambush may be sprung unintentionally, the SEALs will use the ORP as a LUP to rest or as a fall back position. The LUP is also used to hold any person who accidentally discovers the ambush.

2.5.5.3 Patrol Leader's Reconnaissance. The PL's reconnaissance of the ambush site is conducted before the SEALs occupy the ambush positions. The reconnaissance should identify concealment opportunities offered by the terrain. Try to maintain high ground advantage and view the ambush site from the enemy's perspective. If the site is to be occupied at night, care must be taken to ensure that positions will remain concealed during daylight. The reconnaissance should also identify where to place demolitions and mines and to establish a good centralized location for the PL to provide C^2.

2.5.5.4 Ambush Site Selection. The site for the ambush is selected based on the following:

1. Routes. The estimated enemy approach routes to the ambush site must be determined. All reasonably possible approaches to the ambush must be covered. The route into the site should be from the rear, avoiding the kill zone if possible.

2. Kill zone. The intended kill zone must be carefully selected. The kill zone is the area in which the target is located when the primary ambush is initiated. The kill zone should be divided into fire sectors based on ambush positions. Overlapping fire sectors should cover the entire kill zone. The depth of the ambush must also be considered. When the firing begins, the enemy may immediately scatter; exit points should be covered with fire (and demolitions if possible). The ambush site should be laid out so the enemy will probably have to go through the kill zone even if the situation changes. The PL may determine not to execute the ambush, especially if a larger than anticipated enemy force arrives.

2.5.5.5 Positioning Patrol Personnel for Ambush. Early warning of the enemy's approach is vital. Security should be positioned first to prevent surprise and protect the flanks of the ambush site. The PL should position personnel with the following criteria in mind:

1. Preparing the Individual Ambush Position. The PL will set a time deadline to prepare and inspect all positions. Noise and light discipline must be maintained. If possible, items that could cause noise later (such as dried branches and leaves) should be removed. Movement within the ambush area should be kept to an absolute minimum. Each position should be tied in carefully to adjoining positions to provide supporting fields of fire and to facilitate emergency communications. Escape routes should also be carefully identified and coordinated within the ambush force.

2. Security Elements. Positions should be established on the flanks and to the rear of the ambush force. Quiet means of communications must be established with these elements. It is critical for flank security to see who is approaching and in what numbers in order to defend against enemy flanking movements.

3. Weapons. AW men should be placed as soon as possible. AW men are often placed in the center of the ambush in order to sweep the entire kill zone. The PL should consider, however, positioning at least one AW to provide enfilade fire, i.e., fire from one end of the kill zone through the length of the target. The effectiveness of a machine gun in this position can be devastating. AW men must be ready immediately while the remainder of the platoon sets the mines and demolitions and prepares for the ambush.

2.5.5.6 Booby Traps. Booby traps should be used as a "trip wire" to signal an approach of an enemy from an unexpected sector. Booby traps should also be set along estimated enemy lines of retreat, in which case they should be designed to prevent the enemy from regrouping and mounting an effective counterattack. One or two command-initiated Claymores monitored by RS should be set up to the rear to allow the patrol an exit path clear of booby traps.

2.5.5.7 Ambush Discipline. All SEALs must exercise maximum discipline when in the ambush site. Personnel should train themselves to maintain a high state of visual and mental alertness and a minimum of physical movement. The patrol should be prepared to remain in the ambush position as long as the situation dictates. Some discipline considerations are listed below:

1. Perform head calls prior to taking up an ambush position, if possible.
2. Ensure cologne, chewing tobacco, cigarettes, gum, and candy are not used prior to and during the operation.
3. Equipment must be secured to prevent noise.
4. Turn off alarms on wristwatches and cover watch faces.
5. Personnel should hydrate before going into the ambush position.
6. All movement should be severely limited. Slowly move different body parts every 15-30 minutes to prevent numbness and cramping.
7. No lights should be used once in position.
8. To avoid bug and insect bites or annoyance, use a well-camouflaged head net.

2.5.6 Ambush Execution

2.5.6.1 Signaling Sighting of the Target. The first person to notice the approaching enemy, usually the flank security, should alert the other personnel. In accordance with platoon SOPs, this may be done by hand and arm signals, tug line, or radio. Upon receiving the alert, the signal is passed up and down the group. Unless the enemy makes eye-to-eye contact with a member of the ambush patrol, everyone remains quiet and waits for the PL to initiate.

2.5.6.2 Initiating the Ambush. The PL should wait and see if more enemy forces are approaching. If not, he should wait until the target is well within the kill zone, taking care to distinguish between a PT or an advance element and the main body, and then initiate the attack by a pre-planned method. The initiation method should always produce casualties.

The optimum time to initiate the ambush is when the main body of the target fills the kill zone. This may necessitate allowing a PT or an advance force to pass through the kill zone. If this occurs, the PL should assign an element to track the PT or element to ensure it is not able to flank the patrol position or cut off patrol retreat from the ambush site. Premature initiation of the ambush can result in a failure of the ambush, as well as placing the patrol in significant danger.

2.5.6.3 Actions During the Ambush. Once the ambush is initiated, come off safe, and deliver a heavy and accurate volume of fire. Individual actions are:

1. Riflemen should hit point targets ensuring "one round, one hit". The flanks must continue to watch for reinforcements or flanking maneuvers.

2. AWs should concentrate fire against massed targets and provide suppressive fire across the entire kill zone.

3. GNs should have their weapon trajectories figured in advance to place rounds on the trail/route, along enemy routes of retreat, and against any enemy flanking attempts.

4. The radio operator provides RS and should provide illumination or calls for extraction, or fire support, as directed by the PL.

2.5.6.4 Actions after the Ambush. The ambush should be terminated when the enemy is destroyed, routed, or in retreat. The PL terminates the ambush with a pre-arranged signal. Individual actions are as follows:

1. Command is given "Cease Fire, Search Teams Out".

2. The search team moves through the kill zone, penetrating a short distance beyond in order to check for enemy who may have been wounded and attempted to crawl off.

3. Flank men must be aware of their 180-degree responsibility during movement.

4. Security members on the flanks deploy far enough to give advance warning, but not so far away that the PL loses control of the situation.

5. Illumination may be used during ambush and kept up during the search phase.

6. The RTO should announce elapsed time in 30-second intervals.

7. The search team keeps the PL updated as to what it is finding and, when finished, informs the PL by saying "Search Team Coming In".

8. As the platoon consolidates, a head count is gathered and the platoon moves out as per SOP.

2.5.6.5 Departing the Ambush Site. The patrol should depart from the ambush site expeditiously in order to avoid enemy reaction forces. Fuzed Claymores used as per SOP can be used to cover the withdrawal and discourage pursuit. Other time delay munitions and booby traps can also be set at the ambush site or along the exfiltration route to disrupt enemy reaction forces. The SEALs should initially withdraw to the ORP to regroup and redistribute ammunition. Departure from the ORP should not be by a previously used route. If available, supporting close air, naval, and other fires should be called in to hit potential enemy quick reaction force routes.

2.6 RAIDS

2.6.1 General. SEALs may conduct raids on selected targets, which may vary from lightly defended mobile soft targets to permanent or semi-permanent hard targets with established defensive fighting positions. Raids are a subset of DA missions. As ambushes are used primarily against moving targets, raids are directed primarily against fixed targets. A raid is usually a small-scale operation, involving a swift penetration of hostile territory to secure information or property, confuse the enemy, or destroy his installations. It ends with a planned withdrawal upon completion of the mission.

2.6.1.1 Raid Principles. The principles of surprise, concentrated and effective firepower, rapid attack, and planned withdrawal are all applicable in a raid. Whenever possible, the fire and maneuver assault technique should be employed. As discussed subsequently in Section 2.6.4, the concept of a covering force providing suppressive and diversionary fire while an assault force approaches the target is a proven one.

2.6.1.2 Raid Objectives. The objectives of a raid may be to accomplish one or more of the following:

1. Destroy or damage enemy installations.

2. Secure information.

3. Confuse and harass the enemy.

4. Destroy or capture weapons, ammunition, equipment, and supplies.

5. Eliminate or capture enemy personnel. Such missions require detailed intelligence, planning and preparation, and firm control during execution. Section 2.6.6 discusses this operation in more detail.

6. Liberate friendly personnel, i.e. hostages and POWs. Due to its small size, a SEAL platoon will normally conduct hostage/prisoner of war (POW) rescue only in *in extremis* circumstances. Most hostage/POW liberations are best suited to units specifically designated and trained for that mission.

7. Divert attention from other operations.

2.6.1.3 Methods for Conducting Raids. There are basically two operational methods for conducting raids: clandestine and overt. The method selected determines the tactics the patrol will use in conducting the mission.

1. Clandestine Raid. The clandestine raid is characterized by stealth throughout the patrol. Successful execution of a clandestine raid allows the patrol to insert, infiltrate, approach the target, conduct its actions at the objective, and exfiltrate (move to a tactically secure location or extract/return to friendly lines) before the enemy knows the patrol was there.

2. Overt Raid. The overt raid is simple and direct. It is a DA mission in which the patrol uses stealth during its insertion, infiltration, and approach to the target. Once readied for the attack phase, immediate violent action is used to secure and conduct its actions at the objective.

    a. The overt raid is an extremely violent tactic designed to kill or rout a defending enemy and destroy his equipment by overwhelming him with surprise, aggression, accurately concentrated firepower, and demolitions.

    b. It might also be used if the primary mission objective is to secure information held at the target. In this case, the enemy is destroyed or routed and the target is searched for intelligence.

    c. Without modification, it is not a suitable tactic for personnel abduction, hostage, or POW liberation missions. These missions often can be better accomplished using a variety of other mission-specified tactics discussed in this section.

2.6.2 Planning Considerations. In planning any combat operation, an essential element of mission analysis is to determine an effective means for mission accomplishment that offers the least amount of risk to friendly personnel. PLs should recommend appropriate options to higher authority despite pre-conceived notions for execution that may already exist. Fixed targets can often be effectively attacked with standoff weapons such as 60-mm mortar, .50 caliber sniper weapons, and shoulder launched munitions. Standoff weapon employment is discussed in the NSW Standoff Weapon Tactical Guide. The PL should also consider using fire support, (e.g., naval gunfire, artillery, missiles, or attack aircraft) against hard targets, with SEALs in supporting roles as spotters, laser designators, etc.

When planning raids, consider the following:

1. Mission objective.

2. Target location and immediate environment.

3. Ability to exercise command, control, and communication.

4. Fields of fire.

5. Avenues of approach to the objective (with specific attention to cover and concealment).

6. Likely defenses at the objective, including booby traps, early warning devices, and command-detonated explosives.

7. Number and type of specific objectives within the target objective.

8. Avenues of enemy approach, reaction, time, and size of reaction force.

9. Effects the mission will have on the overall campaign against the enemy. Is the risk worth the gain?

10. Target critical nodes.

11. Weapons to achieve minimum required success with acceptable degree of reliability.

12. Ability to extract given the actions at the objective.

2.6.2.1 Phases. Raids are normally planned in phases. These phases are addressed in more detail in Section 2.6.4. Insertion and extraction are discussed in Section 2.7.

1. Infiltrate to ORP

2. Establish ORP

3. Conduct reconnaissance and surveillance of objective

4. Move to assault positions

5. Execute the raid

6. Secure the objective and, if possible, collect intelligence

7. Exfiltrate.

2.6.3 Raid Force Organization. The size of the SEAL raid force depends on the mission, the nature of the target, its location, and the enemy situation. A platoon would be the normal size for most SEAL raid missions, but fewer personnel should be used if the target permits, since stealth and C^2 will be enhanced. Raids requiring a larger force are not usually considered SEAL missions and should only be undertaken with extensive training and rehearsals. Due to the limited size of a SEAL patrol, personnel are usually assigned multiple tasks. A SEAL platoon conducting DA missions normally consists of C^2, base/fire support, assault, and security elements.

2.6.3.1 Command and Control Element. The C^2 element is usually a part of one of the base elements, but it may be a separate element. It may or may not be assigned specific duties in other elements. The C^2 element typically consists of the PL and primary RTO.

2.6.3.2 Base or Fire Support Element. The base element consists of one or more teams that provide supporting fire for the maneuvering and/or assaulting elements.

1. It delivers neutralizing fire on the objective. The element will lift or shift fire when the assault is initiated.

2. The base element may need to provide covering fire for the assault element as it withdraws from the objective.

3. Accurate small arms fire is extremely important. It is imperative that the raiding force plans its fire initiation sequence carefully to immediately eliminate key threats such as guard stations and AW posts.

2.6.3.3 Assault Element. The assault element will close with and assault through the objective, moving far enough through it to place fire on withdrawing enemy elements in order to protect against counterattack. The assault may employ standoff weapons. Once the objective has been achieved, the assault element establishes security and protects the special purpose teams while they conduct their actions at the objective.

Most assaults are conducted with an assault element and a base element, as this will maximize surprise. However, assaults may be conducted on-line. (see Section 2.4.5.2.) Small units are not well suited to conduct on-line assaults against any target because they lack supporting fire, and the entire unit is vulnerable to exposure during the assault. If used, the on-line assault should be limited to very small, poorly defended targets. The assault element usually includes a number of special purpose teams charged with various responsibilities such as:

1. Eliminating Sentries. Lone sentries should be dealt with preferably by suppressed weapons, and at the last possible moment. See additional information on sentry stalking in Section 2.6.7.

2. Breaching Obstacles. As much preparation of breaching materials as possible should be done at the forward operating base. Breaching materials need to be configured quickly at the ORP. Refer to the NSW Demolitions Handbook (NWP 3-05.23) for more details.

3. Initiating Diversions. Conventional artillery or air attacks on adjacent targets can be excellent diversions. The PL should also coordinate with the NSWTU commander to consider the feasibility of coordinating diversionary attacks. Coordination with the conventional force commander to conduct conventional force patrols is another option. The purpose of diversions is to make the enemy believe that reconnaissance is being conducted at another target for possible strike. The more forces diverted from the actual target the better.

4. Setting Security. Security is the responsibility of the security element and is addressed in Section 2.6.3.4. However, due to the small size of SEAL patrols, assault element personnel are usually required for security as

well. This is accomplished by designating assault team members to take up perimeter security positions on target, immediately after the assault.

5. Searching the Target. Once the target has been secured, the search team takes over. Care should be taken to identify and avoid booby traps. Target searches should take no more than 10 minutes. In addition to searching for documents, one man should be designated to photograph all enemy dead. Searchers must pay attention to name tags and rank devices. Equipment should be photographed if left behind. If possible, a post-destruction photo of the target should be taken for battle damage assessment purposes.

6. Setting Demolitions and Booby Traps. Dual prime each Claymore for both electric and non-electric firing. The time fuse should be pre-cut for 30, 60, 90, or 120-second delay, for pursuit or break contact situations. WP grenades have a great psychological effect against enemy troops. Using CS grenades in conjunction with WP grenades doubles this effect, if CS has been approved by the ROE.

7. Prisoner Seizure, Search and Handling. If a prisoner is ambulatory he should be handcuffed with plastic or metal handcuffs. Prisoners will move faster if handcuffed with their hands in front. If the prisoner is wearing a belt, rotate the belt so that the buckle is to the center of his back. The cuffs can then be looped around the belt, which now serves as a restraining "belly band". If several prisoners are involved, a length of shroud line can be put through the same arm (right or left) on each individual prisoner. For added security, one loop around each prisoner's arm above the elbow will ensure that the enemy prisoners maintain pace or the circulation in their arms will be affected. Two SEALs at each end of the line can control several prisoners in this manner.

2.6.3.4 Security Element. The security element protects the patrol. To best accomplish this task it may be divided into teams. Duties and requirements for the security element will vary according to the situation and may include:

1. Secure RPs
2. Watch for an enemy's approach
3. Block enemy avenues of approach
4. Prevent enemy from reaching the target
5. Provide cover during withdrawal.

2.6.4 Basic Raid Tactics. Basic raid tactics are outlined and discussed by phase in the following paragraphs.

2.6.4.1 Phase I, Patrol to the Objective Rally Point. The procedures for reconnoitering, moving into, and occupying an ORP are discussed in Section 2.4.9.3.

2.6.4.2 Phase II, Establish Objective Rally Point. The ORP is the final RP prior to the assault and is used as a secure position from which to conduct target reconnaissance and issue final orders. The ORP is where the final attack preparations are completed before conducting the actions at the objective.

1. Set up the ORP as close to the target as possible, but with sufficient offset from the target to avoid compromise.
2. Ensure security and OPs are adequate.
3. Establish a recall system between the reconnaissance team and the squad or platoon.

4. Establish a challenge and reply system (such as a number system or the use of first names).

2.6.4.3 Phase III, Conduct Reconnaissance and Surveillance of Objective. The patrol's tactical reconnaissance of the target is designed to confirm all the data that was previously provided in the target intelligence package and which was the basis for the plan of attack. The reconnaissance should be used to gather any of the specific information that the patrol needs, such as the location and movements of any guards, confirmation of the actual location and position of the target, and the layout of the facility (paying particular attention to the points designated for the attack). Also look for potential dangers such as obstacles that must be breached, the most likely avenues of approach or attack, and reaction force location and capabilities. The information that the reconnaissance team returns with will be the basis for any modifications or adjustments to the plan prior to the attack.

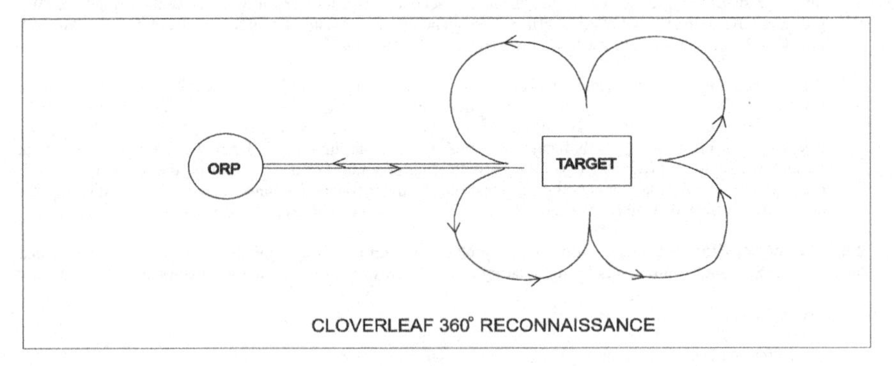

Figure 2-40. Cloverleaf Reconnaissance

1. Procedures. The following are general procedures for the reconnaissance or surveillance of the objective:

   a. The PT from each element (maneuver and support) and the PL conduct the reconnaissance. Surveillance should continue up to the time of attack.

   b. Brief GOTWA and use radio if possible to notify elements of delays.

   c. Do a 360-degree reconnaissance of the objective, if possible. A "cloverleaf" pattern, as depicted in Figure 2-40, is useful since it minimizes lateral movement near the target. Close in lateral movement is easier to see than a direct approach.

   d. Locate all sentry positions.

   e. Locate positions and routes for maneuvering and fire support elements.

   f. Take a close look at the maneuvering element's avenue of approach. Look for pits, barbed wire, fences, barricades, and other obstacles.

   g. Locate blind spots, dead space, and escape routes.

h. Locate gun and other weapon positions.

i. Determine the number of personnel on target, if possible.

j. Observe target routines, especially those of sentries.

k. Confirm or discount intelligence reports; update target intelligence.

l. Conduct target surveillance up to the time of assault, if possible.

m. Be alert for animals, particularly dogs and domesticated geese that may give warning to the enemy. If these or other animals are present, plan the reconnaissance and assault to avoid alerting the enemy prematurely.

n. Personnel not involved in the reconnaissance will make final preparations to required weapons and equipment.

o. When the recon is complete, the PL will issue final orders for the assault.

2 Final Assault Brief/Checks. This is the PL's responsibility and his last opportunity to ensure that the latest information and plan of attack have been disseminated and are understood by all the patrol members. He should check all equipment and ensure it is ready.

2.6.4.4 Phase IV, Move to Assault Positions. The PL should designate the route and formation to move the patrol from the ORP and into attack positions with the optimum patrol security during this critical phase. Once the patrol is in place, all elements wait for the pre-designated signal to initiate the attack.

1. General. Whenever possible, all elements should approach the target as a single patrol. This allows the PL to maintain control and ensure all elements are properly positioned. The PL can then "drop off" each element into position.

2. Base or Fire Support Element. The base element is placed first.

   a. The PL positions the unit where the best field of fire is available, shows it the target location, and points out the maneuvering element's avenue of approach.

   b. Each member of the base element should be able to see the men immediately to his right and left.

   c. The base element should include heavier weapons, such as AWs and grenade launchers, and standoff weapons, such as .50 caliber sniper rifles and AT-4s, to suppress the enemy and prevent his use of NVEO as required.

   d. Provides cover for the maneuvering element in case it is spotted before the assault is initiated.

   e. Provides illumination, if used, during the assault.

3. Maneuver Element.

   a. The maneuver element moves from the base element location to its location on the left or right flank of the objective, perpendicular to the base element.

b. The PL usually leads the maneuver element.

c. If the maneuvering element is detected enroute, the roles of "base" and "maneuver" elements are then automatically switched since the original base is now the only undetected element which can pose a surprise for the cross-fire and assault.

d. The maneuver element identifies the position from which the final assault will be initiated and provides the base-of-fire element with the shift fire signal to be used, such as, voice, whistle, smoke, or flare.

2.6.4.5 Phase V, Execute the Raid. The tactics for the assault will depend on whether the patrol conducts a clandestine or overt attack. See Figures 2-41 and 2-42.

1. Clandestine Assault. The clandestine assault is characterized by stealth throughout; i.e., there are no discernable violent actions heard or seen. The following considerations apply:

   a. Base element. The base element is positioned in the best vantage point to give covering fire to the attacking element throughout the entire attack phase. The base element may use a sniper pair as the primary covering force. Depending on the size of the patrol, this pair may be the entire base element.

   b. Movement security. The second or maneuver element provides its own moving security. This element moves in with the assault team and covers their actions throughout the attack. Their actions must be coordinated with the base element coverage to ensure on-site security throughout the actions at the objective.

   c. Withdrawal. The assault element's withdrawal from the objective is covered throughout by the base element until it reaches the predetermined point or gives the signal for the base element to also withdraw.

2. Overt Assault. The overt assault is the same as the clandestine assault up to the point of initiation (i.e., stealthy). Upon initiation, however, overt assaults are characterized by violence of action. The following sequence applies:

   a. The base element commences fire at a predetermined time or on a pre-arranged signal from the PL. The members attempt to kill all visible enemy personnel and provide a heavy volume of concentrated fire to suppress, kill, or disperse any remaining enemy.

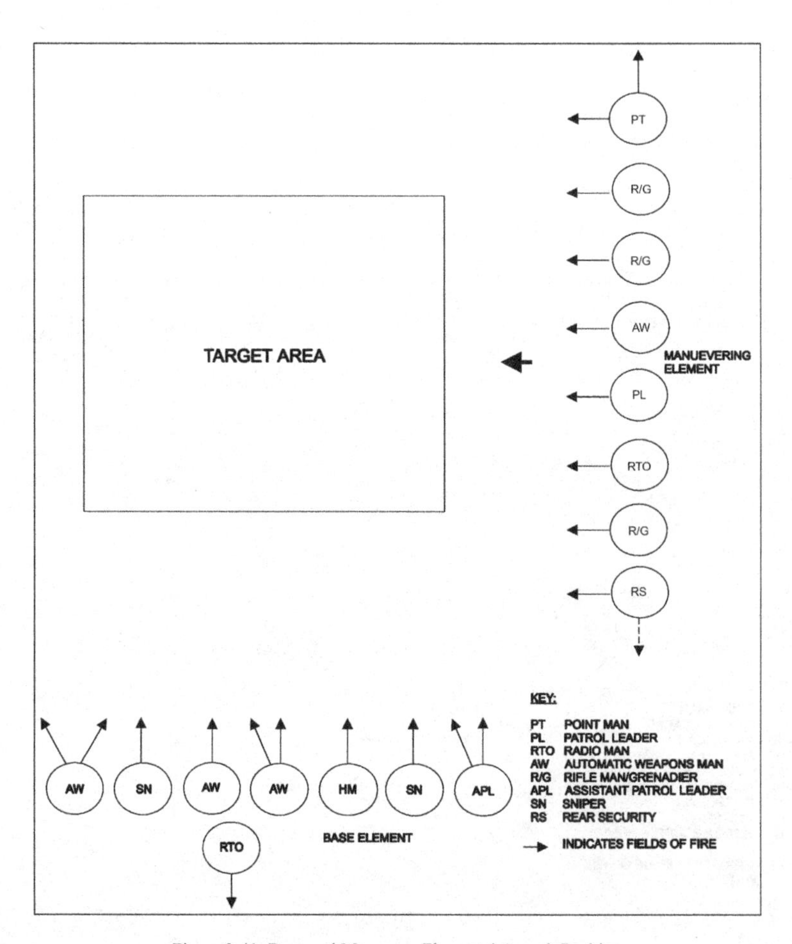

Figure 2-41. Base and Maneuver Elements' Assault Position

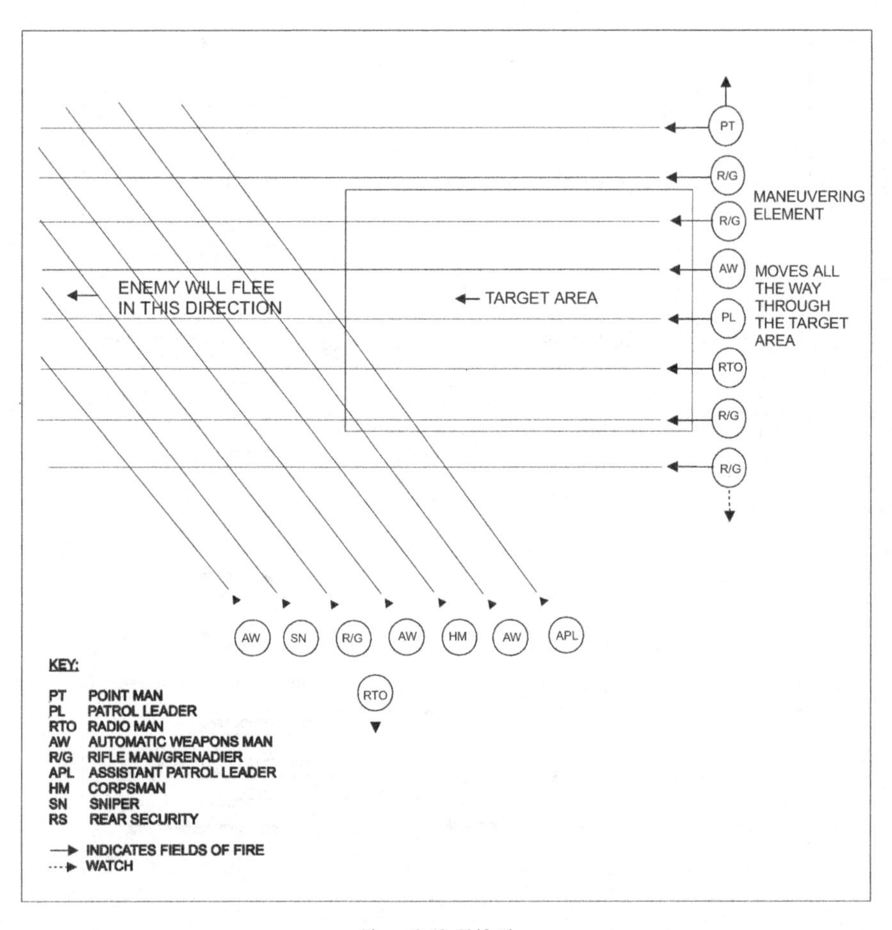

Figure 2-42. Shift Fire

b. Using available cover and concealment as much as possible, the maneuver element clandestinely moves as close as possible to the target under fire support of the base element.

c. The maneuver element then opens fire, catching the target in a lethal crossfire. Once the target has been successfully neutralized, the maneuver element begins assaulting through the target. They use any cover afforded them while bounding through the target and engaging point targets.

d. Once the maneuver element has approached a pre-determined flanking point, a clear signal to shift base element fire will be initiated by the PL. This is essential to avoid casualties from friendly fire. The most effective signal is the parachute flare/star cluster because audible signals may not be heard. All signals during action must still be backed up by voice, however, and repeated loudly by all personnel.

e. Once the shift fire has occured, the maneuver element advances through the target to the designated limit of the assault and the PL calls a cease-fire. Both base and maneuvering elements must strictly observe their fields of fire during the assault.

2.6.4.6 Phase VI, Secure Objective. See Figures 2-43A and 2-43B.

1. Set Security

a. The maneuvering element, base element, or members of each may set security. If the base element will be brought onto the objective, the members move only when verbally told to do so. The maneuvering element should maintain security and minimize movement until the base element arrives. All patrol members should be made aware of any friendly movement into and out of the objective.

b. The base element either comes into target site the same way the maneuvering element entered (since any perimeter booby traps would have been found or set off) or they come into the target single file guiding on voice communications.

c. The team reports status of ammunition, casualties, and equipment (ACE report) to the PL.

2. Search Objective, Set Demolitions and Prepare Prisoners for Exfiltration.

a. Special elements, such as search and demolition teams move out to individual target areas.

b. The predetermined timekeeper continually updates all with TOT. Search/demolition teams inform the PL when they have completed their tasks.

c. The RTO contacts supporting forces and provides information as required.

d. If pre-planned, place booby traps or set up a stay-behind operation to catch the enemy reaction force as it moves in to inspect or protect the target.

e. PT verifies his compass bearing for patrol exfiltration.

f. The PL then directs his people to pull off their security positions and vacate the area. The PL directs the PT to head out.

g. PT then physically motions to let everyone know what the direction of travel is.

h. A head count is conducted as personnel are pulling out of the target area and designated persons are pulling fuses.

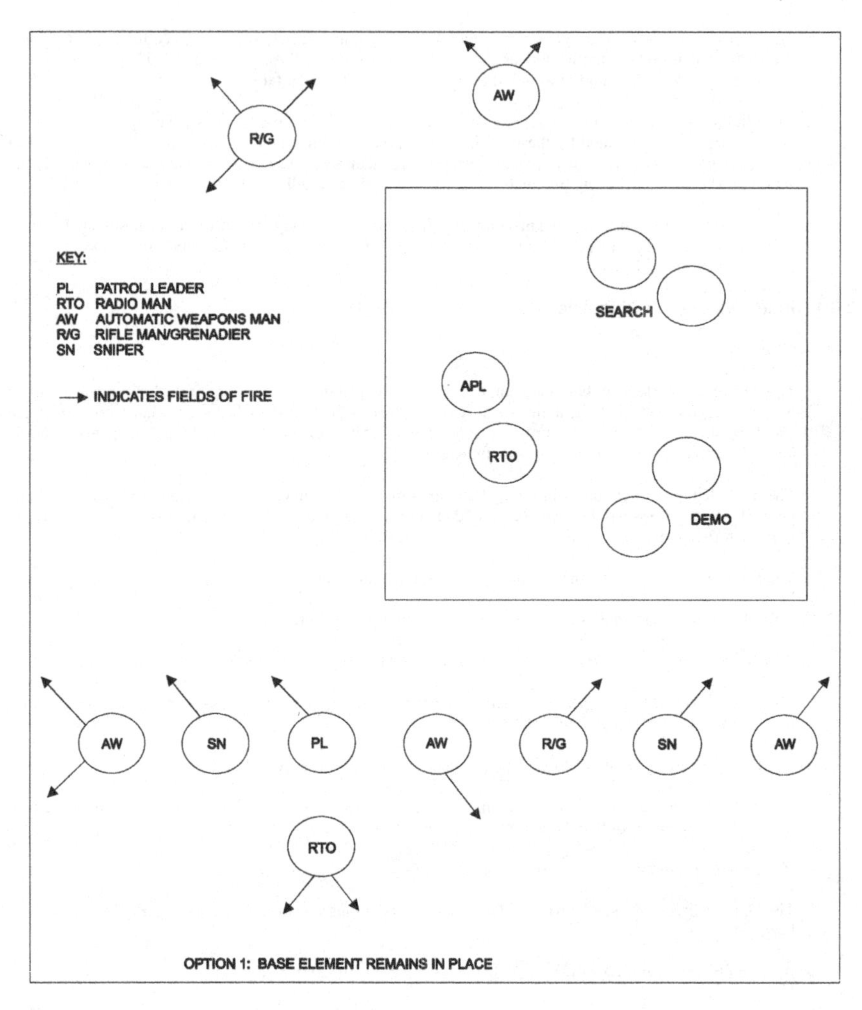

Figure 2-43A. Securing the Objective - Option One

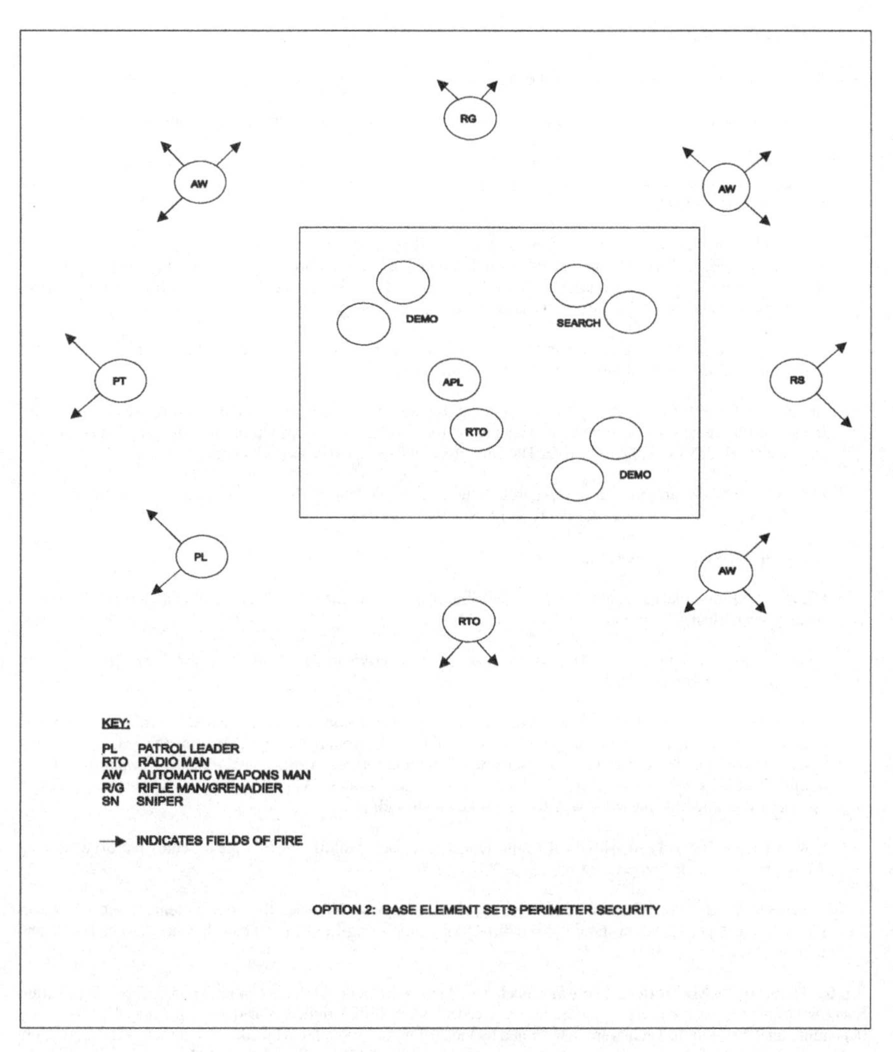

Figure 2-43B. Securing the Objective - Option Two

2.6.4.7 Phase VII, Exfiltrate

1. Set a quick perimeter out of range of the demolitions that were set on target.

2. Get another head count, determine personnel status, conduct ammunition checks, and redistribute ammunition.

2.6.5 Commands. Verbal commands will be difficult to distinguish and must be as clear and concise as possible. All verbal commands should be visually reinforced. Besides hand and arm signals, visual aids and pyrotechnics should be used as much as possible.

1. Commence Fire. If possible this signal should be initiated with a weapon that will produce accurate and concentrated fire that will produce enemy casualties and take him by surprise. The idea is to initiate together. If the tactical situation permits, a set time period should be allowed to concentrate heavy fire into the objective before the assault team begins its final assault.

2. Assault. The assault team picks up and begins its final attack to sweep through the objective. This maneuver doesn't stop until the attacking force has gone to the far side of the target.

3. Lift and Shift Fire. If a supporting FT is part of the mission, this command tells them to shift their fire ahead of the assault force and off the objective. This allows the assault force to sweep through the target and continue suppressive fire towards the objective. The supporting FT can then engage fleeing enemy.

4. Hold. Once the assault force has swept through the objective it halts its forward movement. It continues to fire from hasty positions, driving the enemy fully from the objective.

5. Cease-fire. All elements cease-fire.

6. Clear. Each individual assault element member should check the immediate area around him and yell "Clear" if he is, in fact, clear.

7. Search Teams, Demo Teams, Medical Teams In. Respective teams break off from hasty positions and accomplish their assigned tasks.

8. Consolidate. Pre-planned actions are taken to develop security on the objective and prevent an immediate counterattack. If the plans call for the supporting FT to assist on target, it quickly joins the assault team and assists in securing the objective. All assault team members must be fully aware that the supporting FT is approaching the objective and from which direction they are coming. When possible, the supporting FT should approach the objective the same way the assault element did.

9. Search Teams, Demo Teams, Medical Teams Out. Team leaders inform the PL (and the entire patrol) when they are returning from the target area.

10. Point Take Us Out. When all actions on target are complete, PT leads the assault element out. Whenever possible, the assault element should go out through the supporting element, which will conduct a head count and fall in behind.

2.6.6 Personnel Abduction. There are specialized tactics that will aid in the capture of a live, healthy prisoner. Successful abductions require extensive planning, rehearsal, and experience. Various tactics may be employed, depending on the situation. Tactics are determined by the designated abductee's location, whether in a fixed position, such as sleeping quarters or place of work, or moving along a road or trail. Preparation and methods follow.

2.6.6.1 Preparation

1. Intelligence. Intelligence is vital during preparation. The platoon preparing to capture an enemy must have as much information as possible on the designated prisoner (s) including appearance, defenses, bodyguards, and habits. Positive identification is of utmost importance. Other required essential elements of information (EEI) are listed in the NSWMPG.

2. Rehearsals. Knowing exactly what to do and when is paramount. The sequence of events for prisoner handling is discussed in Section 2.3.12. Rehearse everything, including:

   a. Binding, gagging, and searching

   b. Carrying the prisoner (in case he/she is wounded)

   c. Control throughout all phases

   d. Contingencies.

3. Designated Duties. Each patrol member must have designated duties, including:

   a. Who will bind the prisoner?

   b. Who will search the prisoner?

   c. Who are the prisoner handlers?

   d. Who will field interrogate the prisoners?

2.6.6.2 Methods

1. Silent or No Fire Capture. This method is the ideal method of capturing an enemy, but it requires easily penetrable defenses. The abducted is taken alive with no disturbance so the enemy will not immediately realize he is gone.

2. Silent Assault and Capture. This method requires the use of suppressed weapons. It is useful for capturing an enemy who is poorly protected and may be used against an individual in a fixed position or moving along a road or trail. It prevents enemy reaction forces from reacting immediately, although the fact that the patrol has been at the site cannot be hidden.

3. CS Powder and Gas. This method requires the SEAL patrol to use gas masks. The CS powder will cause noise when dispersed and CS powder clinging to the patrol's clothing and equipment may affect extraction forces, such as helicopter or boat crews. SEALs require realistic training with CS in order to use this method. On the positive side, it quickly incapacitates the abducted and requires a reaction force to use gas masks, which may reduce their effectiveness.

2.6.6.3 Small Building Search Procedures. These procedures could be used for a boat or room.

1. Reconnaissance. Conduct a 360-degree reconnaissance of the building and surrounding area.

2. Set security.

3. Building Entrance. A single-room building may require only two personnel. Use the procedures described in the NSW Close Quarters Combat Manual (NWP 3-05.24) for searching and moving through the room.

2.6.7 Sentry Stalking. The following procedures are recommended for silently killing a sentry prior to a raid.

2.6.7.1 Situation

1. Study the situation and identify sentry positions.

2. Avoid sentries whenever possible.

3. Identify the uniform. Is he wearing protective clothing such as body armor, flak jacket, or helmet?

4. Identify weapon type. Is he manning a machine gun in a protected emplacement or carrying an assault rifle slung over his shoulder?

2.6.7.2 Sentry Movements. Study the sentry's movements and routes of patrol.

1. Does he move into the dark?

2. Does he stay in well-lighted areas always?

3. Does he look at the lights?

4. Does he rove a perimeter?

5. Does he go out of sight of the guard shack?

6. Is his platform off the ground?

7. Does his vision extend beyond the lighted area?

8. Is he outside or inside?

9. Does he appear to be well trained and alert as a sentry?

2.6.7.3 Approach Route

1. Identify the best approach route and beware of mine fields, booby traps, and other obstacles. Approach routes are usually also covered by machine guns.

2. Identify cover and concealment, including noise cover and dark areas.

3. Identify positions for security/cover man or sniper team.

2.6.7.4 Techniques. The recommended actions are:

1. Use a sniper team to shoot the sentry with a suppressed weapon.

2. Use a two-man team, with one man killing the sentry with a suppressed pistol (subsonic round) at close range and the other covering the first with a suppressed weapon.

3. Knives, garrotes, and hand-to-hand techniques should only be used *in extremis*, as they have great potential for compromise and/or failure.

2.6.7.5 Pointers. The following is a general list of pointers for stalking:

1. Supersonic rounds, even from a suppressed weapon, will make a loud crack.

2. When feasible, use a suppressed pistol firing a subsonic round.

3. Untrained sentries may "spook" easily and may fire randomly into the dark.

4. Just because a sentry fires his weapon does not mean the patrol has been compromised. Study the situation carefully - do not react rashly.

5. The primary shooter must avoid obscuring his cover man's line of fire.

6. The cover man must always have his weapon trained on the sentry.

2.6.7.6 Sentry Reporting Routine. Be aware that sentries may report to a Sergeant or Corporal of the Guard. If time permits, use the leader's reconnaissance to determine the frequency with which the sentry reports.

2.6.7.7 Domestic Animals. Be aware of domesticated animals. Dogs and geese can compromise a patrol/individual. A dog can smell you long before he sees you. Geese are very alert and territorial. Once alerted the noise factor of domesticated geese cannot be ignored. Be aware of this when establishing your ORP. A possible mission compromise might occur form something as simple as being upwind of an alert dog.

2.6.7.8 Hand-to-Hand Combat. One-on-one hand-to-hand combat to eliminate a sentry is not recommended.

2.7 INSERTION AND EXTRACTION TACTICS AND PROCEDURES

2.7.1 General. Insertions and extractions are essential parts of every SEAL mission. Consequently, SEALs and the personnel who operate the various insertion/extraction platforms must have a full understanding of insertion/extraction tactics and procedures. While there are several options available to the SEAL platoon for the type of insertion/extraction platform, certain methods have proven more effective than others, depending on the specific operational situation. This section provides the TTP for insertion and extraction operations under various conditions and with various platforms.

2.7.2 Insertion Point Selection. The insertion point marks the beginning of the infiltration phase. Criteria for selecting an insertion point relative to the target area are:

1. Primary Insertion point

   a. The need for the unit to have time to acclimate once inserted

   b. Mobility restrictions caused by terrain

   c. Water sources enroute to the target and on the subsequent route to the EP

d. The need for a clandestine insertion. Considerations:

(1) Points where livestock are watered can be good because tracks are masked naturally.

(2) Inserting at high or slack tide is better than at low tide.

(3) Inserting near bluffs/higher elevations should be avoided when possible.

2. Alternative Insertion Points. If advance reconnaissance is conducted, identify all possible insertion points, LZs, and water sources within and adjacent to the AO. LZs may be for boat or helicopter.

a. In addition to the primary LZ, identify two or more alternative LZs; these alternative LZs should be in the direction of any E&E route.

b. The final primary and alternate extraction LZs should be easily identifiable rendezvous points capable of supporting selected extraction craft.

2.7.3 Tactics and Procedures to the Insertion Point. While the platforms may vary as to type and method for insertion into an AO, the tactics to the insertion point are relatively standard despite the variety of environments that can be encountered.

2.7.3.1 Routing

1. Route Planning. Routes are developed and based upon tactical and technical factors. The following characteristics will not always be factors, but should be considered during planning:

a. Routes should be as short and as easy to navigate as possible, consistent with other considerations.

b. Terrain should be considered in terms of its potential for masking enemy observation, direct fire weapons, radar acquisition, etc., while not masking friendly fire support.

c. Routes should avoid known enemy units and air defense positions as well as built-up areas.

d. Use a different route leaving the insertion point than when entering.

2. Dissemination of Route Information. Maps or overlays containing route information should be disseminated to the NSWTU commander in the event there is a problem along the infiltration path and assistance is required. This will also help prevent friendly-on-friendly actions.

3. Supporting Fire along the Route. Fire missions along the route are planned to suppress known or suspected enemy positions, if necessary. Support may consist of smoke, chaff, or countermeasures to prevent the enemy from acquiring the friendly insertion platform, or fire support to suppress and/or destroy enemy weapons.

During night operations, the use of illumination or tracer munitions requires detailed planning since these can interfere with night vision goggles/devices being used on the insertion platform and cause unsafe conditions.

4. Security Concerns when Clearing Insertion/Extraction Routes and Operational Areas. When submitting these clearances, there is a possibility that information about where the SEALs will be operating could fall into the wrong hands and be compromised.

a. Clear the areas for a longer period of time than the actual mission will take, with the mission time frame not necessarily in the middle of the clearance window. This prevents the enemy from knowing exactly when the SEAL mission will take place.

b. Clear more than one area and multiple routes to those areas. This means that if the enemy is planning to ambush the patrol either enroute to or in the operational area, he will have to divide his forces between several places at the same time to ensure finding the SEAL patrol. Some area commanders do not like to approve requests for multiple operating areas and insertion/extraction routes as it restricts other operations during the period, but safety of the SEALs and SEAL support personnel is the most important consideration.

5. OPDEC. Every effort should be made to deceive the enemy as to the actual location of the insertion point and the timing of the operation. Ways to accomplish this may include:

a. Multiple insertion points

b. False insertions

c. Diversionary actions such as fire missions, air attack, and psychological operations

d. False intelligence leaks

e. Deceptive routes to make the enemy believe that the operation will take place in an area other than where it actually will.

2.7.3.2 Securing the Insertion Point Area. Immediately upon insertion, the patrol must set 360-degree perimeter security. Usually the PL and RTO are in the center. Initial perimeter security should be set as close as possible to the actual patrol formation. It also should be done taking maximum advantage of patrol weapon types/numbers, the existing cover, and concealment and camouflage techniques. How tightly the perimeter is set depends on the same factors as setting interval:

1. Illumination

2. Cover

3. Terrain

4. Situation.

2.7.3.3 Actions at the Insertion Point. Communications should be established by radio check if called for by the communications plan. If the radio check does not establish the required communications, the insertion vehicle, depending on the plan, may be required to extract the patrol immediately.

After the insertion platform has departed the patrol should conduct a head count, move out to the IRP, reset security, and conduct SLLS. Ensure all personnel are ready to begin the patrol and all equipment adjustments have been made.

2.7.3.4 Departing the Insertion Point. Once the patrol has adjusted to the environment, it should assume the patrolling formation (pre-planned) best suited to the terrain and threat, and proceed. The APL should automatically send a head count forward. Any noise from loose gear or equipment should be identified. It will take a few moments for the patrol to get into the routine of movement.

2.7.4 Tactics and Procedures to the Extraction Point. Extreme caution should be exercised when approaching the EP due to the possibility of compromise. To avoid risk to both the SEAL patrol and the extraction platform, the PL should send a recon element ahead to recon the EP. Upon determination that the EP is safe, the PT and the patrol should move into the area.

Any time the patrol is split, even for EP reconnaissance, the risk of mistaking the friendly element for the enemy is very real. Splitting a SEAL unit should only be done if the situation is well understood, has been rehearsed, and most importantly, if the communications between the two units is such that there can be no problem identifying friendly forces from enemy forces.

2.7.4.1 Securing the Extraction Point. Immediately upon entering the EP, security should be established so that the approaches to the position are covered and there is covering fire for the extraction platform. The positions of the personnel around the EP should be such that the personnel can be quickly "folded in" for embarkation on the extraction platform, yet still maintain the security of the position. Section 2.4.8 discusses security perimeters.

2.7.4.2 Extraction Point Departure Tactics. There are a variety of tactics that may be used by the patrol and the extraction platform as it departs the EP.

1. Extract in a direction not expected.

2. Lay a smoke cover to mask the extraction.

2.7.5 Insertion and Extraction Platforms. While there are several options available to a SEAL platoon for insertion/extraction platforms, certain methods have proven more effective than others, depending on specific situations.

2.7.5.1 Helicopter. The helicopter is an excellent combat insertion platform and is frequently used for SEAL operations, but bad weather, aircraft breakage and availability, flight ranges, and enemy air defenses may preclude its use. The helicopter is extremely vulnerable to small-arms fire and heat-seeking missiles. Under normal conditions helicopters can climb and descend at steep angles, allowing them to fly from and into confined and unimproved areas. In the absence of suitable LZ in an intended AO, the helicopter's hover capability affords the SEALs various methods of insertion and extraction. Helicopters can provide significantly enhanced C^2 capabilities in some circumstances. Helicopters configured for SO missions (i.e., those with night vision, special navigation packages, and in-flight refueling capability) are best suited for insertion or extraction of SEAL patrols.

1. Helicopter Limitations. Helicopters are limited in operational range and cargo load. They are further affected by various environmental factors such as precipitation, visibility, temperature, altitude, and humidity. Helicopters may lose over 40 percent of their lift capacity in high temperature and humidity situations such as those in the Middle East maritime regions.

   LZ requirements vary based on the helicopter type and situation at the site. The surface condition of the LZ should be firm enough to support the weight of the landing helicopter. It must be such that excess dust, snow, or debris will not be blown when a helicopter lands. Approach and departure directions should be into the wind if possible. The slope of the ground should not exceed eight percent. If it does, the helicopter should hover while disembarking/embarking SEALs. LZ requirements are addressed in NSW Air Operations Instruction 3000.3A.

2. Types of Helicopter. There are four types of helicopters that may be employed. All can provide lift for SEAL personnel, although only the utility version is usually used, as the others are not configured for insertion and extraction missions. They should only be employed during *in extremis* situations.

a. Observation helicopters. These are normally used to provide C^2, aerial observation, reconnaissance, and target acquisition.

b. Utility helicopters. SEALs use these most often. They are used to transport SEALs, provide C^2, resupply, act as a MEDEVAC for dead or wounded, and various support roles. SO helicopters are the most versatile of this type as they are configured for long-range, low-level flight and their crews are trained to perform insertion/extraction missions.

c. Cargo helicopters. These provide transport, resupply, and recovery of equipment and personnel from the field.

d. Attack helicopters. They provide overwatch, destroy point targets, provide security, and suppress air defense weapons.

3. Helicopter Loading. Personnel should load helicopters in reverse of the order they will exit the aircraft. SEALs should carry their weapons with the muzzles pointed down in order to prevent an accidental discharge from hitting the rotors or engines.

4. Helicopter Insertion/Extraction Methods. The most common and preferred means of using helicopters for insertion/extraction is to land for boarding and debarking. There are other methods of inserting and extracting SEALs from a helicopter without actually bringing the helicopter close enough to the ground to jump from the aircraft. The Maguire rig and special procedures insertion/extraction rig methods allow several men at a time to be raised or inserted in a short period of time. Fast roping, rappelling, and parachuting are also possible from helicopters. Each of these methods is described in more detail in the NSW Air Operation Manual.

2.7.5.2 Fixed Wing Aircraft. There are several types of fixed wing aircraft that can be used to insert or extract SEALs, but there are only two ways of actually conducting the extraction or insertion: parachuting into the intended LZ or landing and disembarking/embarking.

1. Parachuting. Because of the uncertainties associated with this means (e.g., personnel injuries, malfunctions, wind effects), this should be used for SEALS only after other possible means of insertion are determined to be unfeasible.

2. Landing. Fixed wing aircraft can only extract by landing. Short takeoff and landing aircraft can be used in unprepared fields.

2.7.5.3 Water Craft. Because most SEAL operations are conducted in a maritime region, maritime platforms are primary SEAL insertion/extraction vehicles. Several factors can limit boat support for insertion/extraction operations: coastal radar/patrol craft, craft range, and weather/sea state. In addition, riverine environments offer additional limiting factors:

1. Heavy vegetation along the banks of inland waterways can offer excellent concealment for enemy ambush actions.

2. Steep, slippery riverbanks, shallow mud flats, and heavy vegetation can also complicate insertions or extractions.

3. Navigation on riverine waterways can be difficult since much of the terrain often looks alike and because tides and currents can be erratic. In such circumstances, consideration should be given to using aerial photos to augment available maps and charts.

2.7.5.3.1 Planning. Plans for waterborne operations must be detailed enough to give all participants complete mission information, yet flexible enough to be modified as the tactical situation changes. The warning order and PLO should provide all information required. The planning process should ensure that all SEAL, Special Boat Unit (SBU), and mission attachment personnel review the appropriate charts, tides, sun/moon time periods, weather, and weapon requirements. The "scheme of maneuver" portion of the overall operational plan should include:

1. A transportation and support plan for the embarked SEALS and others involved in the operation
2. A plan to conduct or support blocking actions as necessary
3. A direct and indirect fire support plan, if such support is available
4. A MEDEVAC plan.

The plan should designate the sequence, time, and place of the landing. The size and composition of the waterborne force is based upon:

1. Mission.
2. Enemy situation and capabilities (enemy order of battle).
3. The influence of predicted weather, tides, and visibility.
4. The availability of fire support assets.

The plan must take into account SOPs and contingencies. The following considerations must be addressed:

1. Actions to be taken if ambushed before or after the insertion.
2. Selection of alternate insertion points if the mission permits.
3. Determination of actions while onboard the insertion/support craft.
4. Development of an emergency extraction plan.
5. Determination of where each SEAL and personal/unit equipment are to be placed on board the craft while enroute to the insertion point, or where equipment, wounded, and prisoners are to be located following an extraction.
6. Identification signals, challenge and reply signals, and designated fields of fire during extraction and insertion.
7. SEAL body and equipment weight is crucial during extraction; the added weight can suddenly ground a craft, preventing quick departure from a possibly hostile EP. Further, if the same personnel with the same equipment come back onboard, they can still weigh significantly more just from being soaking wet with water.

2.7.5.3.2 Insertion/Extraction Craft. There are numerous types of waterborne platforms employed by SEALs for insertion/extraction operations. These include:

1. Combat rubber-raiding craft (CRRC)

2. Rigid hull inflatable boat (RIB)

3. Patrol craft such as Mk V Special Operations Craft (Mk V SOC), Patrol Boat Light (PBL), etc.

4. Ships such as the CYCLONE Class Patrol Coastal Ship, major combatants, Military Sealift Command vessels, etc.

5. SEAL Delivery Vehicles (SDV) and Advanced SEAL Delivery Systems

6. Submarines

7. Indigenous craft such as dhows, sampans, barcas, etc.

8. High-speed boats.

The value of using indigenous craft for clandestine operations, transits, and other missions in host nation (HN) waterways cannot be overemphasized. Indigenous craft can be tailored to support missions and may move with less risk in the areas where the enemy is known to operate.

2.7.5.3.3 Small Craft in SEAL Operations. Perhaps the most common SEAL craft is the CRRC. See Figure 2-44. This small, inflatable craft can be propelled by various sizes of outboard motors. Generally it has a rigid floorboard placed over an inflatable keel and is surrounded by the main inflatable tubes. Whether using this craft or one of the other small craft such as the RIB or PBL, certain techniques remain constant. Factors regarding loading, operation, defense, and embarking/disembarking should be kept in mind. Also see the NSW TACMEMO XL-2080-2-89, Combatant Craft TACMEMO (U) for general guidance.

1. Loading. Personnel and equipment should be positioned to evenly distribute the weight on both sides of the centerline, but also so the fore/aft center of balance is approximately one-third up from the stern. This will maximize stability in all seas and create optimum performance with most horsepower engines and average hull characteristics. See Figure 2-44. In a heavily loaded CRRC, personnel may need to shift forward to allow the boat to get "on step." Once it is up and running at speed, weight can then be shifted back to the optimum point. All equipment should be secured by lanyards, snaplinks, or lashed and netted to the craft in such a manner that if the craft is knocked about badly by the seas or capsized, the equipment will remain secure.

2. Procedures while Embarked. While embarked in the insertion platforms, certain precautions must be taken:

   a. Weapons should be carried in the proper attitude and on safe.

   b. The first and second lines of equipment should be worn. If the route is long, the second line may be secured until entering the threat area. All equipment must be easily accessible.

   c. Night vision should be protected at all times.

   d. Complete gear checks and rattle checks should be conducted prior to boarding the insertion platform.

   e. Adequate personnel flotation devices (PFD) should be worn.

   f. Appropriate body armor should be worn if required for the operation.

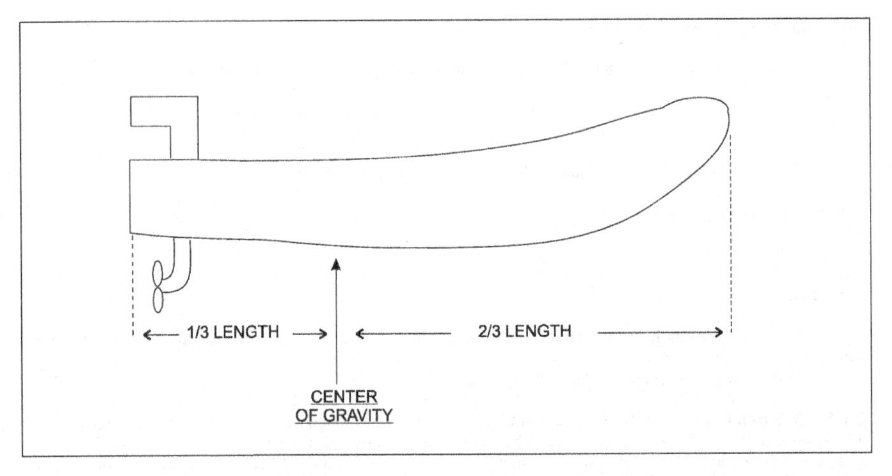

Figure 2-44. Combat Rubber Raiding Craft with Proper Weight Distribution

3. Craft Selection Factors. The number of personnel involved, the amount of equipment required, the distance to be traveled, and the enemy situation all impact on the type and number of craft needed for a particular operation. Generally, these craft should be operated in groups of two or more to provide maximum safety and security. This will:

   a. Distribute the weight, personnel, and equipment over the two hulls

   b. Provide increased craft performance under lighter loads

   c. Create a back-up in case of engine or hull problems/failures

   d. Enhance security during insertion/extraction and transit.

4. Craft Operation. The following factors are critical to successfully employing small craft in SEAL operations:

   a. Engines. Engines should be equipped with propellers of proper diameter and pitch ("propped") to maximize the craft's performance, given the projected speed and load carrying requirements. When using boats with more than one engine in a riverine environment, varying engine use and engine speed can provide deception regarding the number of boats.

   b. Fuel. Contingency movement plans must also be considered in fuel planning. Each engine/boat combination must be run under operational loads and expected sea states to determine its fuel efficiency. Using different size engines on the same craft, "propping" variations, and varying personnel or equipment loads will all affect fuel consumption.

   c. Tools and spare parts. A tool kit, repair or patch kit, and adequate spare parts should be secured on each craft..

d. Radar Cross-section. The radar signature of the loaded craft is of primary importance and must be considered carefully in load plans. In some cases, personnel and equipment may produce more of a radar signature than the craft itself. The NSW Vulnerability and Signature Database should be consulted to estimate detection ranges versus specific threats.

5. Craft Handling. During transits, craft operators should maintain throttles at 80 percent or less to promote fuel efficiency (if this can be done while staying on step). Engine cooling water should be monitored carefully and engines shut down and repaired immediately if the flow is interrupted. When nearing danger areas, shallow water, reef areas, and the objective area, speed should be reduced to a slow idle to maintain the best possible noise discipline. In transit, craft operators should quarter into the seas instead of going directly into them to improve the ride and safety of the transit. This may require adjustments to the navigation plan.

6. Small Craft Navigation. Craft operators should use a compass, timer, LORAN, GPS, and any other aid available for accurate navigation. The GPS is currently the best system available. LORAN is a secondary choice, but not available in all areas of the world, and its coverage is likely to be reduced in the future. All electronic means of navigation should be backed up by dead reckoning using a compass, chart, timer, and knowledge of currents and speed. If the boat does not have a speedometer (pit log), the engine rpm (or engine setting) should be calibrated to an estimated speed. Marking approximate speeds on the engine throttle beforehand can be helpful beforehand to ensure accuracy.

7. Defense. If possible, craft should be positioned with one craft ahead and to the side of the other. See Figure 2-45. This allows the lead craft to act as point element with the trail craft acting as the RS and "shotgun" for the lead. During transit in a craft with little or no organic firepower (CRRC, RIB), patrol members must maintain security for assigned fields of fire. The PL should assign these responsibilities. While one craft is inserting, the weapons of the other will cover it and then they will reverse. There may be times when only one craft can be used. In such cases, the personnel in the craft will provide defense/security for those inserting.

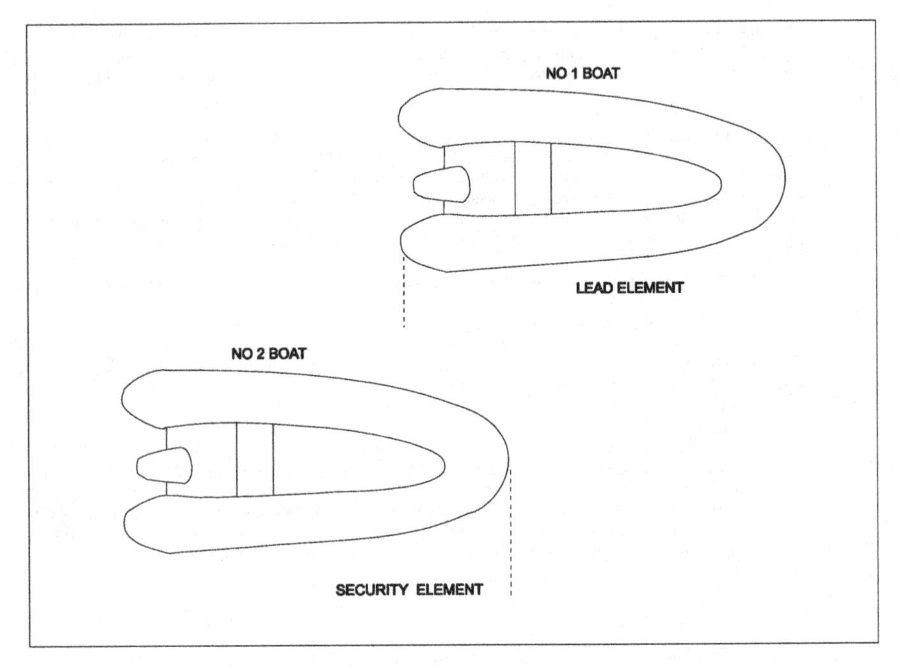

Figure 2-45. Boats in Tandem

Once on shore, the personnel will take up their defensive positions. See Figure 2-46. In this manner, there will always be individuals providing cover and security for those unable to do so at the moment.

If using a RIB, there will be at least a two-man crew attached to the craft that will remain onboard during and after insertion. One individual is assigned to coxswain or pilot the craft, while the other serves as the engineer and weapons man. They may loiter in the vicinity to await the return of the patrol, move to a new site to set security or conduct interdiction efforts, or return to a safe zone away from the AO.

8. Small Craft Formations. Craft movement to the BLS can be accomplished in a variety of formations depending on environmental conditions and requirements for C^2.

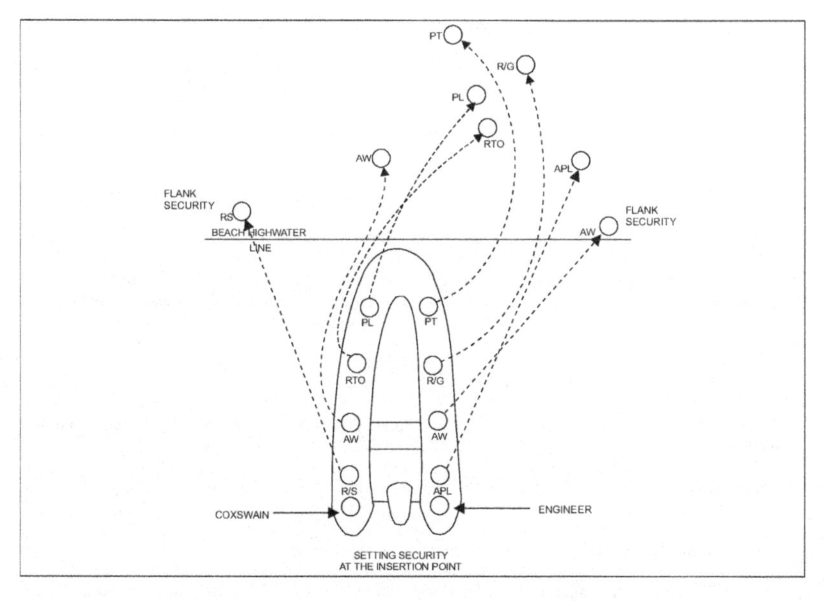

Figure 2-46. Setting Security at the Insertion Point

a. File. This formation is used in tight areas and when expecting contacts from flanks. The file presents a small radar signature to the BLS.

b. Wedge/diamond. This formation provides good all around defense and can be moved into a diamond formation with four boats.

c. Echelon left and right. This formation provides for all around flexible defense, but it is highly visible and needs a very open area.

d. Boat pool. This is a planned or unplanned stop to gather all boats into a group. The officer in charge (OIC) can give last minute orders, confirm location, launch or recover swimmers, or perform repairs. Boat pool planning and considerations as follows:

(1) Location must be briefed and all personnel must know when, where, and how the boats are to pool up.

(2) Noise discipline must be a major consideration at all times; stay tactical. If tactically practical, keep the motors running as this makes less noise than restarting them.

(3) Visual signature of the pool to enemy tracking devices.

(4) Weather conditions can make signaling between pool and support or scouts difficult.

(5) Signals need to be established for pooling up and the means of passing signal (radio, hand signal, or preset time) need to be considered.

(6) Personnel in boats should have second line equipment on but unbuckled, and primary weapons clipped to the boat with a quick release.

(7) Each boat should have a set location in the pool and know whom to tie up to, who will anchor, etc.

9. Movement of Equipment and Personnel to the Beach. Some equipment can be permitted to get wet, but other equipment that must remain dry must be waterproofed prior to departing on the mission. Equipment may be swum ashore by personnel or kept in the boat, which can be brought ashore to off-load. Bringing a boat up on the beach with water and/or equipment onboard is not easily done by hand. Besides concern over security, the craft is often too heavy to lift. The following factors are important to transporting equipment in small boats:

a. The equipment must be properly secured by lanyards and/or netting to prevent its loss in the event of capsizing.

b. As the boat approaches the beach, heavy surf or currents may make the equipment hard to control unless it is well secured.

c. The most vulnerable point of the insertion is when the boats or personnel are nearing the beach. If the patrol is attacked at this time, the following courses of action are possible and should be pre-planned.

(1) Exfiltrate to sea. Seek another insertion point, or abort the mission.

(2) Infiltrate by swimming, retreat the boat, and attempt to reach the first LUP while avoiding the enemy.

10. Tactics while Disembarking. Personnel should depart the insertion platform in the order of patrol with the PT and PL (with his RTO) moving out of the insertion platform in the planned direction of the patrol. The operational plan must cover the contingency of being attacked at this point and everyone must be prepared to execute IADs as necessary.

11. Boat Disposition. Decide what to do with the boats during the land phase of the mission. The three primary options are:

a. If the boats have an assigned crew (such as the RIB), send them away from the insertion point to stand off and await recall to pick up the patrol, or return the boats to base and use another method for extraction.

b. Scuttle the boats and use another means of exfiltration.

c. Anchor or cache the boats near the BLS and plan on retrieving and using them for exfiltration. Consider leaving a small security force with the boats. This may not be feasible if it will be difficult to link up with them upon return and it also reduces the number of personnel available for the mission. Upon return to the cache site, extra caution will be needed to determine if the boats have been compromised and if they are safe to embark for exfiltration.

12. Tactics while Embarking on the Extraction Platform. Combat loading (i.e., embarking in the reverse order of the insertion sequence) is utilized whenever a SEAL element loads craft for a mission. When embarking the craft as

part of an extraction, personnel should use the reverse order of the original offload sequence. The PL should maintain the head count and board last, along with the security element. Security and fields of fire should be maintained while embarking. Some of the security should be shifted to the personnel operating the platform. Some platforms have gunners assigned to assist with mounted weapons. The personnel still on the ground will cover the embarking personnel. The personnel already on the platform will cover for the personnel still on the ground and those embarking. Equipment should be rapidly re-secured. All personnel should wear PFDs. If there are any prisoners or agents picked up, there should be additional PFDs for them onboard or with the patrol. Also, any agents or prisoners should be given a specific position to sit in the craft (where they can be closely observed and kept secured). All weapons should be pointed upward to prevent an accidental discharge from penetrating the hull or electronic systems of the craft.

Depending upon the type of craft being used in a waterborne extraction, it is helpful to have a cargo net draped over the forward part of the craft to assist the SEALs getting back aboard, especially when exfiltrating from a riverbank or muddy delta area where footing is difficult. This same netting can provide handholds for swimmers to be pulled away from the danger area and out to safety before boarding, if required.

2.7.6 Over the Beach Operations with Boats. OTB operations in NSW require exceptionally detailed and careful planning. This section will present the fundamentals of OTB operations to include planning considerations and tactical methodologies. OTB operations using small boats offer many advantages and are planned when:

1. The BLS is a long distance from the mother craft
2. The SEALs may be accompanied by attachments that are weak swimmers
3. Heavy equipment or large quantities of equipment must be transported.

2.7.6.1 Purpose of Over the Beach Operations. OTB operations are conducted for a variety of operational reasons. Among these are:

1. Beach reconnissance
2. SR inland
3. Inland raids
4. Maritime exfiltration of downed pilot (combat search and rescue (CSAR))
5. Exfiltration of political prisoner/agent
6. Coastal target designation/verification.

2.7.6.2 Personnel and Equipment Preparedness. Personnel should be appropriately dressed for the environment. Wetsuits, dry suits, Mustang suits, raingear, and any other systems available should be used to keep the personnel as comfortable as possible. Temperatures easily withstood on dry land may quickly reach dangerous levels on the water due to humidity, sea spray, and wind evaporation on the skin and clothing. Even when conducting landings from small craft (e.g., CRRC), the possibility of the boat capsizing must be planned for and appropriate uniform and equipment preparations must be made. All personnel must wear PFDs capable of supporting them and their personal equipment. Non-inflatable types (e.g., sterns vest) are recommended, since they will still float even if hit by rounds or shrapnel. This section addresses appropriate uniform and equipment for OTB operations.

Embarked personnel should prepare equipment against the environment by using waterproof containers, and plastic covers and sealants in key areas to prevent electrical problems. Fuel bladders should be used to allow the fuel to be consumed without creating an air space in the cell, thereby increasing risk of gas explosion if a round or shrapnel penetrates the cell. There are self-sealing fuel cells being produced commercially that can be extremely valuable.

1. Clothing. The basic military battle dress uniform (BDU) is not a good choice for OTB operations. When wet, this uniform becomes heavy and cumbersome, and it is slow to dry. The "lightweight" BDU is better. The rip-stop material of these uniforms dries more quickly and is more comfortable. In terms of color, however, the woodland and olive drab (OD) fatigue stand out against the beach. In this regard, uniforms with the desert camouflage pattern are better.

   The optimum OTB uniform solution has generally been the older OD green utility uniform. This uniform is dark enough for the water and on patrol, but is generally faded enough to facilitate beach crossings. Turning BDUs inside out for the beach-crossing phase is also a good option.

2. Thermal Protection. Thermal protection is required even for operations in warm water. 80 degrees Fahrenheit water is still almost 20 degrees Fahrenheit lower than core body temperature and prolonged exposure to such a temperature differential can produce hypothermia. Consider wearing a "cheater top" or "half farmer john" wetsuits at minimum. A detachable hood is recommended. Up to 70 percent of body heat is lost through the head and neck region. In extremely cold conditions, dry suits should be worn.

3. Footwear. Most OTB operations will require significant movement over land. Boots should be worn for this. In colder conditions, a pair of thin wool or silk sock liners under heavy wool dive socks work extremely well to keep feet warmer and to prevent blisters.

4. Flotation. Personnel and equipment should be neutrally or slightly positively buoyant. Flotation can be provided by inflatable means or by non-inflatable means. Non-inflatable materials such as ethafoam provide reliable flotation. Lifejacket vests are also helpful but they are bulky and difficult to manage once ashore. On the other hand, inflatable equipment can be deflated and stored once ashore. Inflatable equipment, such as a UDT lifejacket, can be punctured, however, and rendered inoperable at sea with serious consequences. Generally, some combination of the two flotation methods should be employed.

5. Fins. Each man must be able to secure his fins on his gear and take them off within seconds without assistance. With weapons, ammunition, and other operational gear, the swim to the beach is often a slow and cumbersome process. Fins with large pockets to accommodate footgear are recommended. Some SEAL operators turn the fin upside down and wear boots. Adjustable heel straps are required. Spare heel straps should always be packed.

   Fins should be carried on second line gear whenever possible. One method is to place a small plastic tie-tie loop on the back of web gear between the shoulder blades. The loop should then be secured with rigger tape. The fins can be secured to the tie-tie loop with a snaplink on the fin heel strap. A loop of 550 cord through the heel strap can facilitate donning or doffing the fins in seconds.

2.7.6.3 Weapons and Ordnance. As with clothing and equipment, the surf and sand associated with OTB operations require special consideration in the choice of SEAL weaponry.

1. M203 40 mm Grenade Launcher. The M203 can be fired right from the surface of the water with little weapon preparation. Also, M203 reports do not give away your exact position, especially at night. The M203 can also have a significant impact on enemy morale and can impede enemy tactical maneuvering.

2. Hand Grenades. Explosive grenades are heavy, bulky, difficult to waterproof, and lack the range of an M203. Smoke or CS grenades, however, can cover a withdrawal when conditions are appropriate. See Section 2.8 for details on use of smoke and grenades.

3. Assault Rifles/submachine guns. Extra effort must be taken to waterproof and sand proof assault rifles and submachine guns to ensure they will be ready to fire right out of the water. In OTB operations, the majority of sand will be encountered in the final few feet of surf. Various types of waterproof bags may be used.

2.7.6.4 Selecting a Beach Landing Site. As with any clandestine activity, security is of utmost importance. Points to consider when selecting a BLS:

1. Know the major obstacles on the boat's general route that could cause an accident.

2. Know the enemy's sonar, surface radar, air radar, night vision, noise enhancement, and thermal capability.

3. Know the tides and currents.

4. Know the navigational aids.

5. The site should be as far as operationally possible from civilian or military installations, activities, and villages.

6. The BLS should not be easy to access and secure or be easily seen from a high point.

7. Make a horizon sketch of the site with bearings to confirm the location from the boat pool.

8. Divide the site into sections for planning and contingencies: red, yellow, and green sections.

2.7.6.5 Approach to the Beach Landing Site. Caution must be exercised during the final approach to the BLS. This is an extremely vulnerable time. Many SEAL patrols have encountered enemy personnel and/or ambush at the insertion point due to security leaks, poor intelligence, or compromise during the approach. Remember:

1. Slow movement is best.

2. Noise must be kept to an absolute minimum. The bow of the boat must be kept pointed towards shore to minimize engine noise from the beach. Use insulated cowlings on engines. Paddle when necessary.

3. Silhouettes should be kept as low as possible.

4. Light discipline is essential.

5. The cover of darkness and inclement weather conditions should be used to the best possible advantage. Be aware of ambient light conditions that silhouette forces. A moon or light source behind the approaching craft can create an excellent opportunity for anyone on the beach to observe and target the approaching craft.

2.7.6.6 Swimmer Scout Reconnaissance. The BLS must be scouted prior to inserting the main body. The swimmer scouts (generally two, but may be more) will approach the beach by swimming. Purpose, equipment, planning considerations, and tactics are as follows.

1. Swimmer Scout Purpose.

a. Confirms that the enemy doesn't occupy the site, that it is suitable, and the surf is passable.

b. Signals boats to the beach and provides security while coming ashore.

c. Directs landing party to cache site, transition site, and RP.

2. Swimmer Scout Equipment. The swimmer scouts should have their weapons, signal devices, communications equipment (if required), first and second line equipment, and fins over their patrolling uniform. The rest of their equipment should remain in the insertion craft. The following considerations apply:

a. The swimmer scouts should be neutrally buoyant and armed with assault rifles and side arms. The use of suppressed weapons should be considered.

b. Thermal protection such as a wet suit or dry suit appropriate for the type of climate should be worn. The scout needs to consider that he may get inserted off the BLS and spend more time than anticipated in the water.

c. Fins are secured after use.

d. Lifejacket, knife, and flare are needed. Scout web gear should allow quick movement, stealth, and enough firepower to break contact and swim to the pickup point.

e. A directional IR light or strobe with helmet is optional.

f. E&R kit with rations for 24 hours.

g. Map and compass.

h. Waterproofed radio and NVEO should be secured to the body.

3. Swimmer Scout Planning Considerations.

a. Scouts are designated during the planning phase so plans can be formulated and drills rehearsed. Unforeseen problems may require the scouts to be ashore longer than anticipated.

b. Alternates should be assigned in case there are unforeseen personal problems.

c. The situation will dictate the composition of the team.

d. Contact drills, casualty recovery, and recovery must be rehearsed in detail.

e. Before deploying, a detailed briefing must be conducted. All "what ifs" and "actions on" must be covered and briefed back so there are no misunderstandings.

f. Scouts must have a finite time to achieve their goal and a contingency plan if that time is exceeded. If scouts take longer than the designated time, they should make radio communications back to the boat element and apprise them of the situation. (A secure visual signal can be used.)

g. If scouts don't return, wait for briefed drop-dead time and move to the alternate rally point for rendezvous at contact times.

4. Employing Swimmer Scouts and Beach Reconnaissance.

   a. After forming the boat pool, testing the current and drift, and confirming the location, one boat may break away from the pool and either move closer to the beach to launch the swimmer scouts or take them all the way into the beach.

   b. The PL releases the scouts. They enter the water on the seaward side of the craft, and give the "GOTWA" to the PL.

      (1) Going - where you are going and what you are going to do.

      (2) Others - who you are taking with you.

      (3) Time - maximum amount of time you will be gone.

      (4) What if - what the person in charge will do, if you do not return on time.

      (5) Actions - actions to be taken by each element upon enemy contact.

   c. The scouts will swim a compass bearing or to a fixed point on land. They will log time or kicks to the beach and maintain all around security.

   d. Move through the water with an underwater recovery stroke and facing each other. Don't break the surface of the water with your fins. Swim perpendicular to the reconnaissance area. The human eye is less equipped to detect motion in the near/far plane than in the side view.

   e. Observe the surrounding water for obstacles that may pose a threat to the boat element. Before committing to the surf zone take a short stop to conduct SLLS.

   f. At waist deep water remove the wet suit hood and conduct SLLS again. Look for concealment, beach composition, potential ambush sites, obstacles, and signs of human inhabitants.

   g. If all clear, make weapons ready and remove fins. One man at a time will remove fins with the support and cover of the other.

   h. When ready to move inland, the senior member of the element considers the shortest distance to cover and where the fewest tracks will be left.

   i. Once the beach has been surveyed and appears safe, secure fins to facilitate quick movement.

   j. After the beach looks, sounds, and feels safe, start the reconnaissance using the leapfrog method of movement. Use a NVEO to assist viewing the area surrounding the beach site for dangers and enemy presence. Look for signs of activity such as footprints, booby traps, etc. While moving try to avoid walking all over the beach and move from cover to cover with weapon at the ready. Go all the way to the hinterland to make sure there are no surprises. If the CRRC is to be cached, look for a possible site. Select a good landing site that is close to concealment.

   k. At completion of the reconnaissance, three things can happen:

      (1) Have the rest of the patrol swim ashore

(2) Signal the boats to the beach.

(3) The swimmers can swim back to the boat pool.

5. Tactical considerations of swimmer scouts. Remember, you are alone. You have limited firepower and you make all the decisions. This is the most vulnerable point of the mission. Try to avoid enemy contact or observation and think of contingency plans.

6. Scouts IADs.

a. Swimmer scouts get split up.

(1) Designate right and left swimmer.

(2) Move to waist deep water.

(3) Designated left swimmer turns to the right and starts looking for buddy up to 50 meters. Designated right swimmer turns left and does the same. If linked up, continue with mission; if not, swim out to the CRRC for pick-up.

b. Scouts contacted in the water. If the enemy compromises the swimmer scouts, they should return to the insertion craft under the covering fire of the personnel in the craft. Fire at the enemy only if necessary to suppress their fire while retrieving the swimmer scouts.

c. Scouts contacted by enemy on approaching land or during reconnaissance. Contingency planning requires:

(1) Break contact, find cover, move to the secondary BLS, and call for pick-up.

(2) Break contact, scramble into the water, swim out, and signal the CRRC for pick-up.

d. Scouts contacted while signaling CRRC. SOPs are the same as the previous action.

e. Scouts find the beach occupied or it is a bad area. The scouts have the option of radioing the PL and moving the site up or down the beach or swimming out to the CRRC and moving the site.

f. E&R or link-up plan.

(1) If the scouts take longer than the designated time, they should make radio communications back to the boat element and apprise them of the situation.

(2) If scouts don't return, wait for the briefed drop-dead time and move to the alternate RP for rendezvous and contact times.

(3) Follow arranged E&R plan. Abort mission.

2.7.6.7 Calling in the Patrol. Calling in the patrol can be done by various means. The most common way is to use a unidirectional light-signaling device such as a directional beam red flashlight or strobe light. An alternative clandestine method is to use an IR signaling light such as a flashlight with an IR lens to notify the personnel in the craft it is safe to come in and to designate the exact insertion point. The signal plan must be an integral part of the PLO.

There is a possibility the boats can drift out of position from where they dropped the swimmer scouts. Further, each party can become disoriented as to where the other is on a dark night. Either anchor the boat or establish some other means of knowing that they are holding in position. If possible, identify two fixed objects ashore to use as a "range." If they remain in line or in the same relative position, this indicates lack of lateral movement by the boat.

1. Swimmer Scout and Boat Pool Signaling. Boats should only land after receiving a pre-arranged call or recognition signal from the swimmer scouts. Signaling methods between scouts and boats are:

   a. Primary means is light signaling with a directional red lens flashlight or IR device.

      (1) The signaler should face the waterline on the back bearing to the boat pool. Signal once to the back bearing and then slightly to the right and left of the bearing. Signaler must be aware of his vertical position relative to the pool.

      (2) The signaling should be repeated until the scouts see the lead boat. Scouts must ensure the light does not reflect off the water.

      (3) Boats coming in to the beach should never signal the scouts.

   b. Secondary means are radios. The scouts should initiate communications whenever possible, and have some type of headset so no one else will hear transmissions. A back-up radio should always be taken.

2.7.6.8 Combat Rubber Raiding Craft and Rigid Hull Inflatable Boat Insertion and Extraction Methods. Planning for insertions and extractions must include the Special Boat Detachment. Signals, radio frequencies, grid coordinates, and support information must be briefed in detail to minimize confusion. Boat loading needs to be preplanned. Seat assignments, weight distribution, and weapons positioning affect the transit. This section presents procedures for different craft.

1. OTB Insert by RIB.

   a. During transit at five nautical miles off shore the coxswain will slow the RIBs to reduce the signature. At 2,000 yards out from shore the boats will stop.

   b. The lead RIB will continue toward the beach and drop swimmer scouts as tactically close as possible. The OIC will note insertion time and grid coordinates and then return to the boat pool to wait for the scouts' signal.

   c. If the beach is good, and upon receiving the scouts' predetermined signal, the craft will proceed to shore, following range markers to the insertion point.

   d. If the beach is bad, a pro-word will be passed. One pro-word will be used if the scouts will return to the craft and another will be used if they will search for another site. If there is no word from the scouts in one hour the force will follow prearranged contingency plans.

   e. For a dry insertion the RIBs will be in an on-line formation, with all platoon members manning the sponson to provide security. The lead boat will advance as far as surf conditions allow. (RIBs cannot beach in surf greater than two feet.) Platoon members should have positive control of their equipment when leaving the boat.

f. For a wet insertion, the RIBs will assume an abreast formation and approach the beach as close as tactically possible. The platoon will have control of their rucksacks and lie on the sponsons facing out, with weapons at the ready. On command, the platoon goes over the side tactically and forms a swimmer pool. The RIBs will then back away a set distance and remain on station until the platoon leader determines a hot extraction is unnecessary.

2. OTB Extraction by RIB.

   a. When the platoon is moving towards extraction it will pass approximate time periods to the task force commander via pro-words. A prearranged pro-word will be passed to the task force commander to let him know when the platoon is three hours and one hour out from extraction time. The extraction window begins one hour after the one-hour pro-word is received and lasts for 30 minutes.

   b. To insure the RIBs make it to the beach the platoon must do a reconnaissance with two men to chest deep water.

   c. The RIBs will approach the BLS using the pro-word indicating they are ready for range markers. The platoon must be prepared for radio and light signals between them and the RIBs.

   d. The lead craft will approach the beach while the trail boat stands off and provides cover. One SEAL will meet the boat and hold the bow and keep it aligned perpendicular to the incoming surf.

   e. Loading will be done from both sides. The first man to board will hand his pack to the crewmember in the bow and be assisted aboard. He will move to the opposite side of the bow and start assisting other platoon members. Each man boards and moves to amidships on the boat. This will keep the weight centered.

   f. The last man will send up the head count and shoulder the boat off the beach before being pulled into the boat by the two men in the bow.

   g. Once the lead boat is away the trail boat will repeat the process.

3. OTB Insertion by CRRC. The boats and coxswains may not need to come ashore or go on the operation. The boat can drop off the platoon and return to sea.

   a. The CRRCs will approach at least 1,000 yards off BLS, form a boat pool, and prepare swimmer scouts.

   b. The CRRCs will then move as close as tactically possible to launch the scouts. Prior to launch, they conduct a drift test.

   c. Scouts will move to the beach and conduct a recon per platoon SOPs.

   d. Upon receiving a good signal from the beach, the boats will proceed to the beach in a file formation using the scout signal as a guide.

   e. Everyone gets into a tactical position.

   f. The lead boat will beach with the others landing to the right and left, or use the same SOPs as with a RIB.

4. OTB Extraction by CRRCs. This will depend on whether the boats were cached or the platoon was dropped off.

a. If dropped off, the patrol forms a perimeter at the extraction RP and sends out a reconnaissance team to reconnoiter the beach.

b. The radioman communicates with the CRRC and the patrol moves to the beach and forms a perimeter.

c. While the patrol is moving to the beach, the CRRCs are moving as close as tactically possible to the beach.

d. The scouts initiate the signal to the boats and the platoon members don their packs.

e. As soon as the boats are almost to the beach the patrol will move in file formation across the beach.

f. The first person to ankle deep water will count the patrol. As they enter ankle deep water they will break off into FTs and move to their respective boats. The last two men to each boat ensure it can move off the beach.

2.7.7 Swimmer Across the Beach Operations. There are two primary methods of conducting across the beach operations by swimmers: the parallel and the perpendicular methods. The parallel method can be used for units of all sizes. It is recommended for situations where compromise by the enemy may be possible and it is prudent to be able to focus the maximum firepower from the water onto the beach. The perpendicular method is used for small units and offers the best chance for concealing tracks across the beach. The differences between these two methods are in the insertion; the extraction methods are the same.

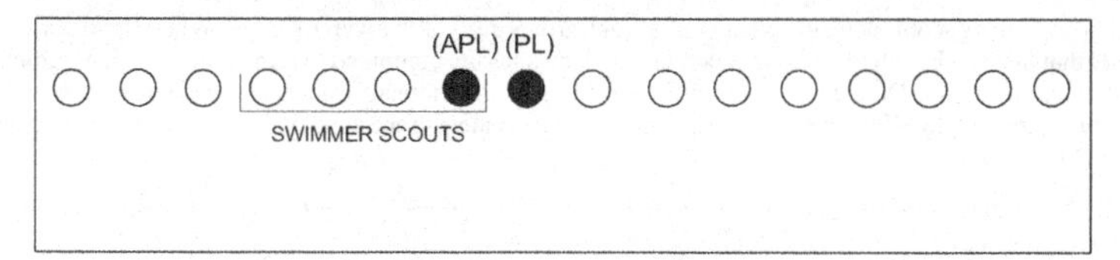

Figure 2-47. Parallel Method; Movement to the Beach

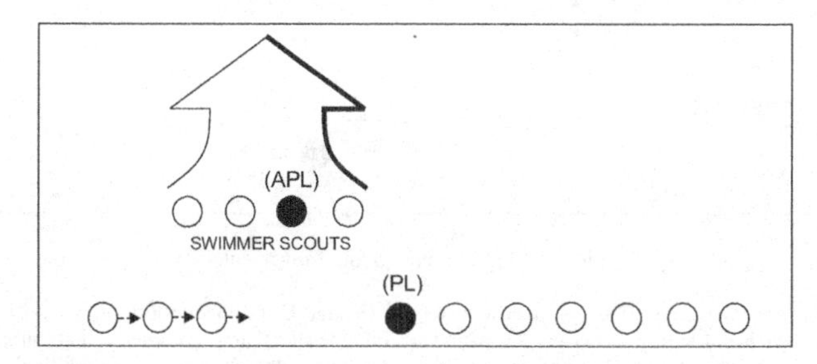

Figure 2-48. Swimmer Scout Deployment

2.7.7.1 Parallel Method. Swimmers are generally launched approximately 600-1,000 meters offshore. Actual launch distance will depend on the threat, visibility, sea state, and currents. A range finder should be used while still aboard the mother craft when possible. Operational planning should address the effect of tide and currents on the OTB force.

1. Movement to the Beach. Swimmers should surface swim in a wedge or file formation or on-line and parallel to the beach. See Figure 2-47. The wedge or on-line methods are easiest to control but the file formation is the least visible from the beach. The swimmers should be approximately one meter apart. The unit leader should occupy a position in the center of the line/wedge for C^2. The maximum number of swimmers manageable in a single line is about 16. If the unit experiences difficulty swimming on-line, the PL should swim out in front. The rest of the element should then form a slight wedge. This will make it easier for the PL to maintain control of the swimmer line.

2. Hand and Arm Signals While in the Water. Appendix C contains the standard hand and arm signals. Additional signals or variations to those in Appendix C will be developed in accordance with platoon SOPs.

3. Swimmer Scout Deployment. Approximately 100-300 meters from the beach the main swimmer line halts. Up to this point the swimmer scout element has been positioned toward the center of the main swimmer line. The swimmer scout element consists of the second in command and the center, left, and right flank security. The swimmer scouts are located next to the unit leader in the main body for ease of control. As the swimmer scouts proceed toward the beach, all other swimmers close down the formation gap by swimming toward the unit leader (Figure 2-48).

   The swimmer scout element remains in a tight group until at the water's edge. When the determination is made that the beach is clear, the scout element leader issues the command to remove fins, which is done in pairs to maintain security. On command from the swimmer scout element leader, the element then moves inland, separates, and deploys left and right. SEALs must remain alert to the scout element leader at the center of the BLS (Figure 2-49).

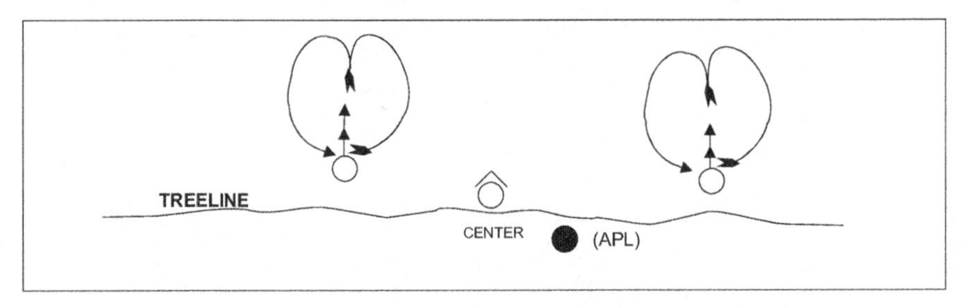

Figure 2-49. Swimmer Scout Movement

4. Beach Zone Reconnaissance. The swimmer scout element leader should conduct a close-in reconnaissance of the immediate beach area to inspect berms, dunes, and any other terrain features of interest in the area immediately overlooking the actual beach. A useful reconnaissance pattern is to have the men on the flanks move inland towards beach center as a pair, then reconnoiter and return to beach center. This is called a heart-shaped beach reconnaissance (Figure 2-49). If discovered at this point, the element leader and those with him must be prepared to support the reconnaissance element, link up, and return to sea.

5. Swimmer Scout Weapons. Careful consideration should be given to the type of we the swimmer scout element. Swimmer scouts must be lightly armed to be as mobile be prepared so as to be immediately ready for firing upon leaving the water. Conside using indirect fire weapons such as the M79 grenade launcher and pre-planned air and/or grenades can also be effective in covering a hasty withdrawal in some circumstances. Swin engage lone targets, such as sentries, with suppressed weapons. Use of the M4 with supp considered.

6. Signals Between the Swimmer Scout Element and the Main Body. Radio communications and red lens (directed out to sea) are the primary methods used by the swimmer scout leader to signal the main b swimmers. IR signaling devices can also be used as long as the lead swimmer in the main body has a n vision scope capable of acquiring the signal. Generally, a red light is the most effective signal for guidan purposes. Another feature of red light is that it is at the edge of the light spectrum visible to humans. It is difficult to see from any great distance and if one is not looking for it, chances are one won't see it.

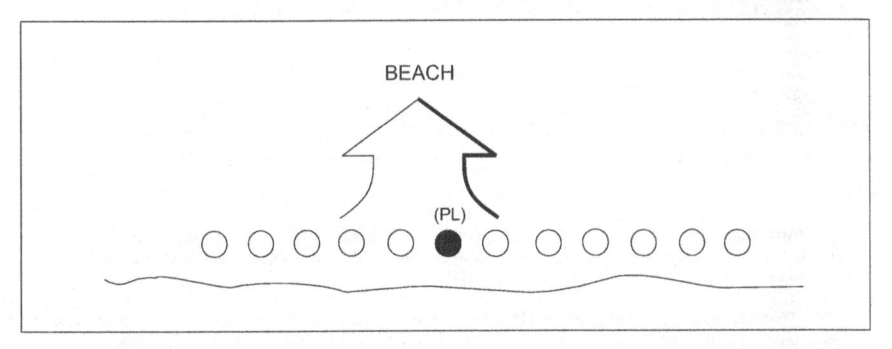

Figure 2-50A. Main Body Movement

7. Main Body Movement to the Beach. Once signals are passed and acknowledged, the main body swims cautiously ashore. The swimmer line stops in chest-to-waist-deep water. On command from the PL, and while the swimmer scouts provide coverage, all hands remove their swim fins and put them on a wrist. The unit leader gives this signal by removing a single fin and holding it out of the water (still maintaining a low profile). Each man indicates that his fins are off by signaling a "thumbs up." The signal begins outboard and is passed inboard. This signal is never to be sent by any individual until he is actually ready to move. On command, all hands cautiously advance on-line across the beach toward cover with weapons at the ready. The entire line guides on the PL located in the center of the line. Depending upon how wide the BLS is, it is best for the group to quickly cross to the first available cover. (see Figure 2-50A)

As the main body passes the locations of the individual swimmer scouts, the scouts then move with the main body. If swimmer scouts are located on a dune or berm line further inland, they wait to be signaled before joining the main body.

A second method, which provides better track discipline, is for the line to collapse onto the PL (or onto the flank) and then cross the beach in a file formation. (see Figure 2-50B)

ach RP. At the initial beach RP the entire group conducts SLLS for signs of enemy activity for a brief Once the "all clear" is passed, the unit leader signals to secure fins. The entire group secures fins in pairs. one man covers, the other secures his fins. At this time, any other gear preparation is accomplished (e.g., king out radios, donning boots, etc.).

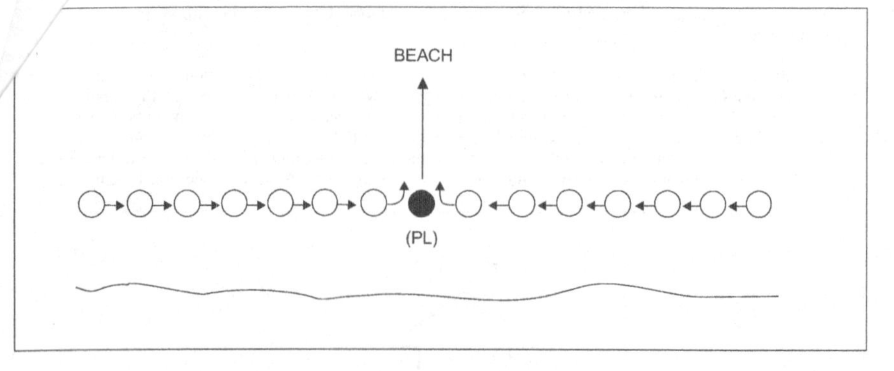

Figure 2-50B. Main Body Movement (Options)

2.7.7.2 Perpendicular Method. As in the parallel method, swimmers should be launched approximately 600-1,000 meters offshore. A range finder should be used while still on the mother craft when possible. Operational planning should address the effect of tide and currents on the OTB force.

1. Movement to the Beach. Swimmers should surface swim in a wedge or file formation towards the beach. The swimmers should be approximately one meter apart. The unit leader should occupy a position in the center of the wedge for C^2.

2. Hand and Arm Signals While in the Water. Signals for water operations will be developed in accordance with individual platoon SOPs.

3. Swimmer Scout Deployment. Approximately 100-300 meters from the beach, the main swimmer line halts. The swimmer scout element is sent forward to recon and secure the beach. The swimmer scout element consists of two or more swimmers. The swimmer scouts are located next to the unit leader in the main body for ease of control. As the swimmer scouts proceed toward the beach, all other swimmers close down the formation gap by swimming toward the unit leader.

   The swimmer scout element remains together in a tight file until at the water's edge. When the determination is made that the beach is clear, the scout element leader issues the command to remove fins. On command from the swimmer scout element leader, the element then conducts its reconnaissance as previously planned. The weapons and signals used by the swimmer scouts are the same as in the parallel approach.

4. Main Body Movement to the Beach. Once signals are passed and acknowledged, the main body closes its wedge into a single line perpendicular to the beach and swims cautiously towards shore. The beach is treated the same as a danger area on a land patrol. Each swimmer will, when signaled by the man ahead, cross the beach

individually in the same location. Once ashore, a security perimeter is established in the nearest available cover. The PL, who is the lead swimmer, is the first ashore after the swimmer scouts. After all swimmers are ashore and within the security perimeter, the remainder of the operation proceeds as described in the parallel method. Because there has been only one path made across the beach into the initial beach RP, the footprints can be easily obscured before proceeding on the mission.

2.7.7.3 Establishing Security Ashore. As the personnel exit the water or the craft at the beach, they should immediately form a defensive perimeter that sets security while the equipment and boats are secured. A fire support plan should be considered in advance in case the SEALs need supporting fire. Direct fire can be provided by fixed-wing aircraft, helicopter gunships, or by surface close-in support such as small combatant craft with Mk 19 40 mm grenade launchers and machine guns. Indirect fire support can be provided by major surface combatants or land-based artillery.

2.7.7.4 Patrol Preparation. While security is set, a designated group can prepare their personal equipment and also do whatever is planned with the unit equipment. This means:

1. Cache any equipment that is to be left at the BLS. The craft may have to be camouflaged for retrieval upon return or scuttled if alternate extraction platforms are planned.

2. Derig waterproofed gear to be used during the rest of the mission.

3. Leave any equipment in its waterproof packing if there is a chance it will be submerged or subjected to heavy rainfall.

4. Change out of wetsuits/wet gear into dry patrolling equipment and operational clothing.

5. Ensure that the three lines of personal patrolling equipment have been planned into the equipment requirements and that each patrol member is familiar with the status of the other patrol members' equipment.

2.7.7.5 Extraction Phase. For extraction the patrol moves to a beach RP to prepare for the swim to sea. The patrol should form a perimeter within available cover. Fins are then readied and put on wrists. Hand and arm signals are the same as when coming ashore. Any other gear preparation is also accomplished at this time. Regardless of the type of withdrawal, everyone will have a swim buddy and all personnel will be accounted for. Each man in the group should have the coordinates of the beach EP memorized. All hands should have an escape azimuth to "kick, stroke, and glide." If separated during extraction or in heavy surf, the azimuth should be followed for the prescribed time. When one group is picked up, the others should be in the same general area regardless of currents.

1. Withdrawal - No Duress. This is a clandestine extraction to sea. The presence of the team is not compromised and there is no known enemy in the vicinity. The group faces inland. The last cover positions are occupied by FTs numbered from left to right. (see Figure 2-51.)

   The PL is located in the center of the formation. He sends one FT to sea while the rest provide cover. Prior to leaving the beach, the PL ensures that all men in each FT are accounted for. The FT then moves to waist-deep water, dons fins, and swims to sea. See Figures 2-51 and 2-52.

   As the first FT leaves the beach for sea, the PL signals the second FT by either radio or red lens light. The second FT moves to the PL's location and passes the head count. The entire second FT crosses the beach, dons swim fins, and enters the water. This procedure continues until the last FT begins swimming to sea. The PL follows with his swim buddy at the rear of the last FT as the last pair to leave the beach.

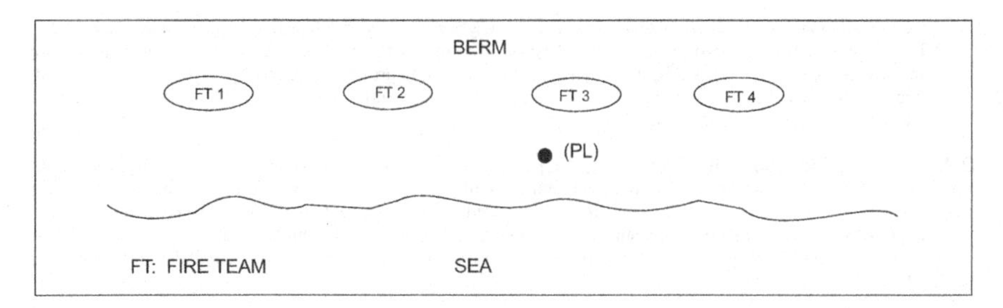

Figure 2-51. Withdrawal - No Duress: Cover Positions

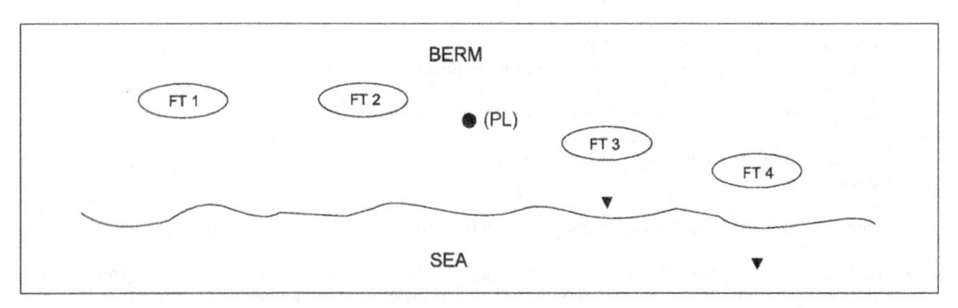

Figure 2-52. Withdrawal - No Duress: Movement to Sea

2. Withdrawal - By Pairs. This technique is used when the group's posture remains clandestine but the presence of enemy in the immediate vicinity is probable. The group is positioned in the last available cover facing inland. From left to right the FTs are numbered 1, 2, 3, and 4. The PL begins tapping pairs out of the center. The FT leaders leave with the last pair in their FT. Doing that, each FT leader informs the PL that their respective FT is all accounted for. While one FT is tapped out to sea all others remain alert. See Figure 2-53.

   Maintain unit integrity at all times. Each FT links up as they enter the water. FTs attempt to link up with other FTs as they swim to sea. Once seaward of the outermost line of surf and beyond immediate danger, FTs link up and swim to sea.

3. FT Leapfrog. This method is used when the unit is under fire but their position is not in danger of being overrun. The patrol withdraws to the final position of cover and either attempts to conceal its exact location from the enemy or engages the enemy. The PL issues loud commands for a specific FT to depart, e.g., "Three Out!" The command is repeated several times and passed by all hands. See Figure 2-54.

   At this point, smoke and/or CS may be employed. It can be dispensed to obscure the enemy's field of vision and assist in withdrawal. The designated FT sprints to the water and swims to sea together. While moving toward the water they must be aware of other members who may trip or be wounded. The FT does not leave the beach without a full head count.

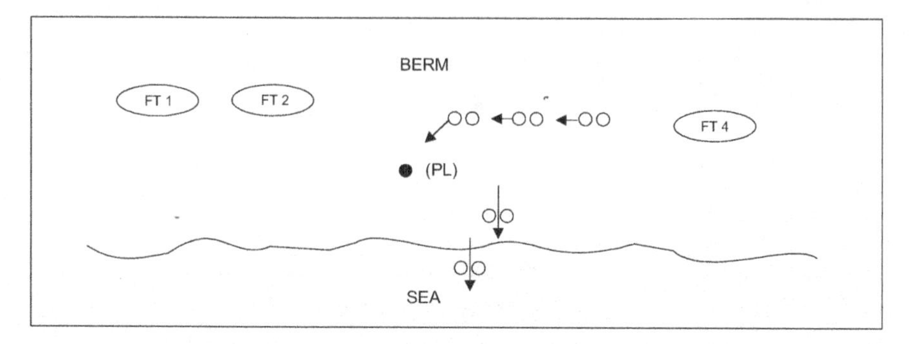

Figure 2-53. Withdrawal - By Pairs

Each FT will be sent to sea in a similar manner. If engaging the enemy, the covering FT must increase its volume of fire as more personnel leave the beach for the water (40mm is especially effective for this). The PL and his swim buddy depart with the last FT. They ensure that no one is left on the beach. The withdrawal of the last FT can be covered with fire from the shallow surf zone (from previous FTs); this requires training and rehearsals. Fire should not be attempted once through the surf zone and swimming. As they swim to sea, FTs link up and head counts are immediately passed to the senior man. Stay on the evasion/pick-up heading and "time to kick" amount of time (determined by PL in the planning phase) required to reach a predetermined point.

4. Everybody Out. This technique is a last resort. If the unit is under intense fire and in danger of being overrun, this method may be used. See Figure 2-55.

The entire unit fights in a retrograde maneuver (usually leapfrog) to the last cover available on the beach. While continuing to engage the enemy, listen for the PL's commands. When he issues the command, "EVERYBODY OUT!" all hands unleash a final brief volume of fire by emptying their magazines, firing 40mm grenades, and throwing screening agents. If Claymore mines are available and time permits, use them. After the ordnance has been expended, all hands turn and sprint for the water. See Figure 2-56. Even a determined enemy will hesitate rushing a position from which heavy fire was just received. During this hesitation, you can enter the water and be lost in the darkness. Each man stays with his swim buddy. Each FT attempts to stay together.

Swim to sea by swim pair. Link up by FT once beyond the surf zone and clear of enemy small arms fire. Swim to sea on your escape/pick-up azimuth and adhere to your "time to kick". Remember that it is difficult for someone on the beach to identify a swimmer in the water at night. Do not reveal your position by firing at the enemy from the water. Your probability of hitting a target while bobbing in the water is minimal and you risk firing a weapon with water in its barrel, which will result in self-injury.

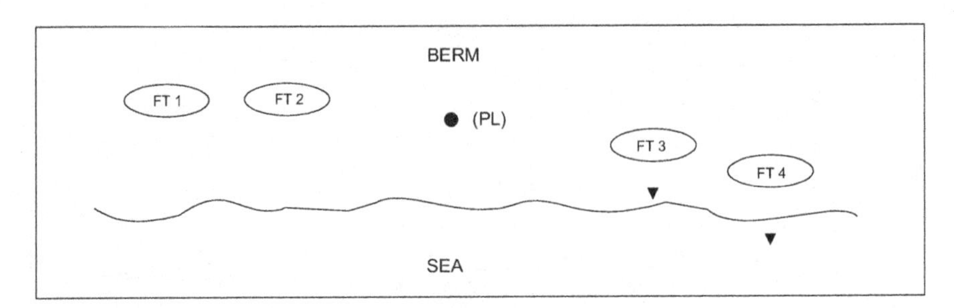

Figure 2-54. Fire Team Peel-off

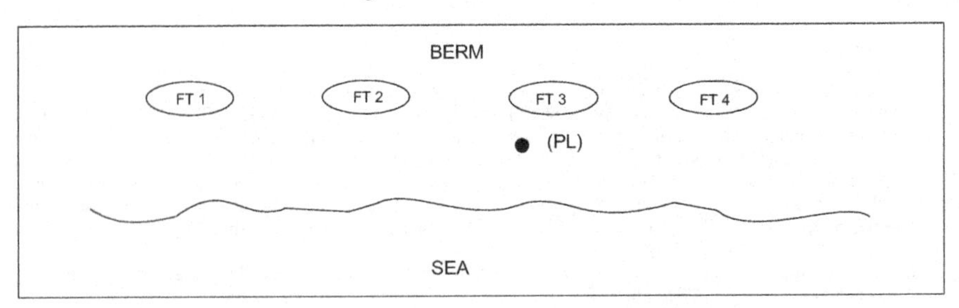

Figure 2-55. Everybody Out; Retrograde Maneuver

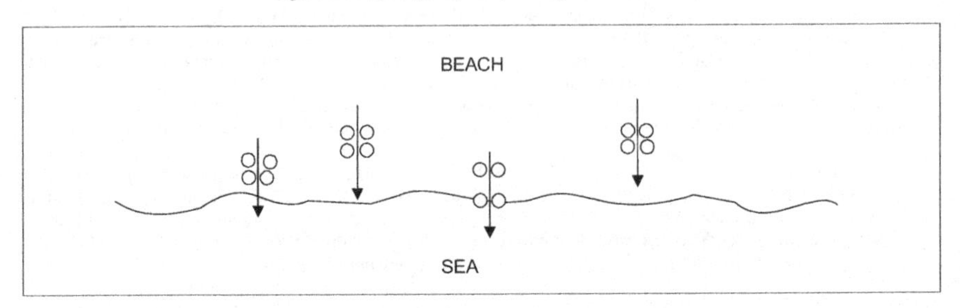

Figure 2-56. Everybody Out; Movement to Sea

2.8 HAND GRENADES, BOOBY TRAPS, and EXPLOSIVES

2.8.1 General. The basic SEAL patrol, operating as a small unit well forward and/or independent of major friendly forces, needs to use grenades and booby traps effectively. With little or no supporting arms at its immediate disposal, the patrol must be able to prosecute any enemy contact to its best advantage. The many types of hand grenades and booby traps available give the patrol useful and versatile weapons that can help the patrol succeed in its situation. The purpose of this section is to address considerations and preparations for the use of hand grenades, booby traps, and explosives. For further clarification refer to NWP 3-05.23, SEAL Demolition Manual.

2.8.2 Grenade Uses. The basic uses for the hand grenade are offensive, defensive, signaling, and harassment.

2.8.2.1 Offensive Uses. Depending on the type of grenade, its basic application is to produce enemy casualties. It can also be used as a diversion, to initiate an assault, or destroy enemy equipment.

1. Casualties. The fragmentation and concussion (overpressure) produced by a hand grenade detonation can be effective against personnel, vehicles, and light armor. Using grenades against such targets would normally be done in an assault or firefight.

2. Diversion. Smoke, WP, and CS grenades are good for creating a diversion, attracting the enemy's attention to a point or location to your advantage, or covering your moves or intentions.

3. Initiating an Assault. The blast and fragmentation produced by grenades can help to give a decided edge to a patrol that initiates action on an enemy contact. If the assault is in the open, the attackers will not achieve complete surprise unless the grenades are pre-rigged on trip or pull wires with instantaneous fuses. Otherwise the enemy will hear the "pop" of the grenade spoons releasing.

4. Destroying Equipment. The thermite grenade is ideal for destroying all types of equipment. The intense heat that it generates can fuse, melt, and warp metal as well as start fires in flammable materials. Fragmentation, offensive, and WP grenades can also destroy equipment. The dense smoke from smoke grenades may have some effect against sensitive electronic equipment such as computers and communications transceivers in enclosed spaces.

2.8.2.2 Defensive. The hand grenade can be used defensively for all the same reasons and in the same manners as discussed above for offensive uses. Hand grenades can help defenders beat back an assault by massed troops, delay pursuit, defend perimeters, cover movements, and destroy friendly equipment to deny its use by the enemy.

2.8.2.3 Signaling. Smoke grenades are primarily used for signaling and marking positions. Such signals can include marking LZs and drop zones (DZ), cease or shift fire, and forward air control (FAC) spotting for close air or gunfire support.

2.8.2.4 Harassment. Harassment will almost always be in the form of uncontrolled grenade booby traps; that is, they are not command- or time-detonated.

2.8.3 Methods of Control. For all practical purposes the grenade is an uncontrolled weapon. The only exception to this is a grenade with an instantaneous fuse on a pull wire that a patrol member will initiate as a designated time. Otherwise they are all uncontrolled detonations. Think the action through, because once the grenade has been thrown, control is gone.

2.8.4 Patrol Load-out and Handling of Grenades. A SEAL patrol should carry an assortment of hand grenades because of their usefulness and versatility. How many, what kinds, where in the patrol, and where on the members will be governed by the mission, the enemy threat, and SOPs. How and when to best use grenades is very important to every SEAL. The following are some guidelines.

2.8.4.1 Load-out. The patrol should carry a mixture of offensive, fragmentation, CS, and WP grenades for the following reasons:

1. Offensive grenades are good for work around water and in close proximity with the enemy.
2. Fragmentation grenades are good for inflicting casualties.
3. CS grenades are ideal for stopping or slowing down enemy troops or dogs pursuing your team. In wet weather, CS grenades work better than CS powder.
4. WP grenades have a great psychological effect against enemy troops and can be used for the same purpose as CS grenades. The use of CS and WP at the same time will more than double their overall effectiveness.
5. Smoke grenades should be carried in an easily accessible location on the web gear or harness and used to provide masking when breaking contact with the enemy. (HC white smoke works best.)
6. Each patrol should carry at least one thermite grenade for destruction of either friendly or enemy equipment.

2.8.4.2 Grenade Tips. The following tips are provided for consideration in the use of hand grenades:

1. When on patrol make continuous daily checks on all grenades to ensure that the primers are not coming unscrewed.
2. Do not bend the grenade pins flat, otherwise the rings are too hard to pull when needed.
3. Leave cotter pins in place and do not carry the grenade with straightened safety pins.
4. Fold riggers tape through the rings of grenades and tape the ring to the body of the grenade. This also keeps the ring open for your finger, stops noise, and prevents snagging.
5. Use black or OD spray paint to camouflage smoke, CS, and WP grenades.
6. Don't carry rubber baseball-style CS grenades; they are designed for riot control on city streets and are inadequate in the field.
7. Don't carry grenades on the upper portion of your harness. Always try to keep your shoulders clear for shooting and ease of movement.
8. Tie extra grenades on the rucksack between pockets.
9. Rubber band or paper-tape grenade spoons as added insurance against friendly casualties. (The last one-half inch of tape should be folded back so it can be easily identified and ripped off).
10. Look for corrosion and inspect the fuse to ensure it has not been tampered with or substituted for an incompatible type, e.g., the previous holder of the device may have inserted and instantaneous fuse to use the grenade for a booby trap. Refer to NSW Demolitions Manual for fuse nomenclature.

2.8.5 Grenade Tactics. Every time that a grenade is used, the use should be based on sound tactical thought, thorough knowledge, intense training, and a conscious decision. Never, never throw a grenade simply as an SOP! The following paragraphs discuss how to use grenades tactically. Keep in mind these two points.

1. The average man will normally throw a grenade about 35yards or 100 feet under training range conditions; that is, a clear overhead, no ground obstacles, standing up, with no time constraints, and no one shooting at him. This is not the same distance he can throw a grenade under combat conditions.

2. The casualty, fragmentation, and burst radii of most hand grenades are approximately 15 meters casualty, and up to 35 meters burst and fragmentation.

2.8.5.1 Offensive Tactics. Offensive tactics for using hand grenades are discussed below:

1. Fixed Site in Open Terrain

    a. The advantages in this situation are:

        (1) The enemy is fixed in his position and will tend to stay put and defend his territory.

        (2) The attacker can localize the enemy.

        (3) The attacking force can maintain a safe range from the intended point of detonation. (The overhead should be clear of obstructions that might interfere with the flight path of the grenade.)

        (4) A well-thrown grenade with good hang/air time limits the enemy's ability to get away from it or return it once it lands.

    b. The disadvantages of this terrain are:

        (1) The relatively short range of a thrown grenade

        (2) The defender's advantage of firing from fixed positions when he has:

            (a) Cover and concealment

            (b) Clear line of sight

            (c) Predetermined fields of fire

            (d) Range staked/marked perimeter.

2. Open Terrain Engagement. The rationale for using a grenade is basically the same as for assaults against fixed positions. The key point to remember is to always be aware of force dispositions in order to avoid the possibility of causing friendly casualties.

3. Fixed Site in Close Terrain

    a. Grenades can be used effectively in close terrain for the following reasons:

        (1) The enemy is fixed in his position; he will have a tendency to stay put and defend his territory.

(2) The attacker can localize the enemy.

(3) The terrain may allow the attacking force to move in close enough to deliver hand grenades accurately and effectively.

(4) The trees, bushes and scrubs will reduce much of the fragmentation hazard to friendly troops.

b. The disadvantages of using hand grenades in close terrain are:

(1) The trees, bushes, vines, shrubs, and other terrain features are dangerous obstructions to a clean throw.

(2) Ricochets, deflections, and bounce-backs hinder the ability to reach the target and increase the likelihood of friendly casualties.

(3) The dense vegetation reduces the casualty radius of the grenade.

4. Close Terrain Engagement. The rationale for using a grenade is basically the same as for assaults against fixed positions. The key point to remember during engagements involving moving forces in close terrain is that one must always be aware of force dispositions and obstructions to a throw path in order to minimize the possibility of friendly casualties.

5. Urban/Village Engagements.

a. Grenades can be very useful in the close-in environment of urban warfare. This includes:

(1) Building and room clearance. Offensive concussion grenades are preferred. Fragmentation will do, but they are more dangerous to friendly personnel. CS is appropriate if trying to take prisoners.

(2) Although there are generally a number of obstructions such as aerial wires (power and telephone), light and power poles, and awnings, the throwing space is usually relatively clear because such obstructions are easily seen.

b. The disadvantages/precautions are:

(1) One must always be aware of the type of structure that the grenade is being thrown into. Village hootches of straw or clapboard siding (as well as drywall structures) do not provide adequate protection to the people on the outside.

(2) One should not take cover behind doors or under glass windows.

(3) Grenades used in roads and alleys are easily seen and allow enemy action/reaction time.

6. Ambushes

a. Grenades can be used very effectively during an ambush for the following reasons:

(1) They can be used to initiate or augment the violence of an attack on an unsuspecting enemy force.

(2) If the ambush has been sufficiently pre-planned, grenades can be used to command initiate using a pull trigger wire and instantaneous fuse or can be used as the demolitions for pull-type booby traps when the enemy is coerced into that kill zone.

b. Throwing hand grenades during an ambush should be done with caution. Grenades should only be thrown if:

(1) It is a pre-planned ambush and there is enough time and visibility for the patrol to find good positions from which to throw grenades (adequate protection).

(2) No overhead obstructions, branches, trees, or bushes are in the intended path of the throw.

c. The disadvantages of using grenades in an ambush are:

(1) If the grenades are thrown to initiate fire, the sound of the spoons flying will alert the enemy and may give away the ambush position.

(2) Throwing grenades in close terrain reduces air/hang time and increases the enemy's chances of taking cover or returning the grenades.

2.8.5.2 Defensive Tactics. Grenades are used defensively for perimeter security and breaking contact.

1. Perimeter Security While on Ambush, at OPs, or at LUPs (in order to engage the enemy without disclosing exact position by firing). One of the more common uses for hand grenades in perimeter security is as a simple booby trap device. It is usually called a "grenade in a can" or "grenade trip wire". Whichever way it is used, the basic principal is the same.

   a. Place grenade in a can with an open top (the original grenade shipping can is excellent) or some simple means of controlling the spoon.

   b. With enough of the grenade exposed, yet allowing the can to restrain the safety lever, pull the pin on the hand grenade.

   c. Anchor or wedge the can in place.

   d. Attach trip wire to the neck of the grenade.

   e. Anchor the bitter end of the trip wire to a fixed position on the opposite side of the trail.

   f. Use the slack wire or tangle foot wire across the trail.

   g. Ensure that the wire is not obstructed and the grenade will pull completely out.

2. Another method:

   a. Fix the grenade in place by taping it low on a sapling or a similar fixed object.

   b. Leave the spoon free.

   c. Straighten the pin and ensure that it will pull out by the trip wire easily.

d. Fix the opposite end of the trip wire across the target path or trail.

e. Fix the running end to the grenade pin.

f. Use a slack wire or tangle foot technique so that you do not have tension on the wire while trying to fix it to the grenade pin.

3. Breaking Contact. Some general guidelines to break enemy contact and delay his pursuit without disclosing the exact position of the patrol are:

a. When the enemy discovers your ambush position, OP, or LUP at night:

(1) Detonate any Claymores set out (don't wait too long to do this or the enemy may pass through the kill zone).

(2) Use fragmentation grenades (explosions are disorienting and don't always give away your immediate position).

(3) Use M4/M203 and machine gun fire (the muzzle flashes will pinpoint your location; AWs usually draw maximum return fire).

b. If the enemy is pursuing you:

(1) Deploy grenades and/or Claymores with fuse delays of 60-120 seconds.

(2) Throw CS and/or WP grenades to your rear and flanks. (These actions will help to give the enemy every reason and/or excuse to quit.)

c. If the enemy is searching for you at night:

(1) Consider using CS grenades prior to using Claymores or grenades. This may cause the enemy to panic and will not give away your position.

(2) If you must, use time delayed Claymores or WP; firing weapons should be a last resort only after the patrol's position is compromised.

(3) If time permits, leave trip wire grenades along an exfiltration route to delay or deter pursuing enemy forces.

2.8.5.3 Signaling. This is a vital capability to all small unit operations. All signals must be clear and easily seen and understood by supporting units.

1. Visual Signals. The following are some recommendations for visual signals using grenades:

a. Violet and red are the smoke colors most visible from the air, but they are not good in dense jungle or wet weather.

b. Use HC smoke to signal aircraft in dense jungles and wet weather.

c. Notify aircraft before signaling with WP; gun ships or fighter-bombers may mistake it for a marking rocket from an enemy position and attack.

d. Use smoke, flares, pencil flares, and tracers as a last resort for marking own position.

e. Pilots should identify the color of the smoke used by teams on the ground after it has been thrown. The team does not identify the color to the pilots. Violet and red smokes are the best colors for use.

f. If contact is made with enemy while in dense jungle, use white HC smoke grenades to mark own location for the FAC. The red and violet smoke grenades are normally not sufficient.

2. Marking Location. There are several methods for marking own location at night; flashlight, strobe light, flares, or illumination grenades.

a. The flashlight should be placed inside the M203 barrel and aimed directly at the aircraft. This shields the light from enemy observation.

b. Use IR strobe. If there isn't any IR capability, paint or riggers tape the sides of the strobe to make it more directional and to reduce the illumination from the side.

c. Notify the aircraft before firing a pencil flare since it resembles a tracer. Never fire it directly at the aircraft.

d. WP grenades make effective night markers.

2.8.5.4 Throwing Techniques. For a hand grenade to be useful, the SEAL operator must be able to put it on target. The following are basic methods for throwing hand grenades.

1. Standing. This method is usually used for initial training.

a. Stand with feet approximately shoulder-width apart.

b. Half-face the target.

c. Hold the grenade in your throwing hand chest high with its lever against the palm.

d. Pull the pin with the index finger of your opposite hand using a twisting motion.

e. Throw the grenade with a free and natural motion, (similar to throwing a football or baseball) with proper follow through to aid in accuracy and distance and to reduce strain on the arm.

f. A good throw is 35 yards with a release angle of 45 degrees.

2. Kneeling

a. Kneel with the knee opposite the throwing hand being closest to the target.

b. Use the same techniques of grip, pull, and throw as in the standing method.

2.8.6 Booby Trap Basics. A booby trap is a concealed or camouflaged device designed to be triggered by an unsuspecting action of the intended victim. Booby traps are uncontrolled and indiscriminate. They will detonate on any unsuspecting act or person that satisfies the minimal criteria to trigger the device. This could be counterproductive should civilian or noncombatant personnel be killed or property destroyed.

2.8.6.1 Booby Trap Types. For the purpose of this manual, a booby trap is an explosive device contrived to be fired by an unsuspecting person who disturbs an apparently harmless object or performs a presumably safe act. There are two types of booby traps in use.

1. Manufactured. Factory-made from materials that are relatively easy and safe to set and may be disarmed.
2. Improvised. Homemade from readily available materials by the user. They lack the standard safety features of manufactured devices which may make them more dangerous to the user.

2.8.6.2 Initiation Methods. There are basically two methods for initiating booby trap firing devices: mechanical and electrical.

1. Mechanical Initiation. A mechanical firing device is one in which mechanical motion produces percussion or friction that initiates the explosive train. There are four modes of operation for initiating a booby trap mechanically.
   a. Pressure. The pressure that a man's foot or vehicle puts on the device causes a mechanical switch to trigger a firing pin that will initiate the firing train, nonelectric cap, or time fuse.
   b. Pressure release. This type of device actuates the firing train when a release of pressure triggers the firing device. It can be set to trigger when opening a door or picking up an item.
   c. Tension pull. A simple pull of a wire or object connected to a switch triggers the release of the loaded firing pin, which initiates the firing train of an explosive device. A trip wire or drag wire is set up so that an unsuspecting enemy pulls on the wire and initiates the device.
   d. Tension release. When the armed switch is under tension, the firing train is triggered by a release of the tension. An unsuspecting enemy breaks or otherwise cuts a taut trip wire, releasing the striker and initiating the device.
2. Electrical Initiation. A booby trap designed with an electric power source that, when armed, has an open circuit. When the circuit is completed, electric contacts are closed, and the firing system initiates the firing train.

2.8.6.3 SEAL Patrol Uses. Booby traps, when used appropriately, can help the patrol succeed in its mission. The basic uses for booby traps are:

1. Offensive Uses. The very nature of the booby trap does not lend itself to normal offensive actions such as direct assaults and raids because the weapon is neither controlled or time detonated. However, it can be useful as a trip wire to initiate an ambush.
2. Defensive Uses. Booby trap devices are very useful for perimeter defense, early warning, and enemy pursuit delay or deterrent.
3. Harassment. This is the most common use for booby traps because they are designed to be triggered or initiated by unsuspecting actions of personnel at an unspecified time. Booby traps are most effective when placed behind enemy lines and set up in locations that will hinder or disrupt the enemy's normal activities or operations.

2.8.7 Booby Trap Tactics. The use of booby traps should be based on sound tactical thought, thorough knowledge of the device and enemy SOPs, and intense training.

2.8.7.1 Offensive Tactics. The best offensive use of the booby trap is for signaling or augmenting the firepower of a patrol set up in an ambush.

1. Trip Wire. The booby trap can be used in the trip wire (pull or tension release) mode to:

   a. Initiate a signal device that alerts the ambush element that the enemy is in the kill zone.

   b. Initiate Claymore mines, hand grenades, or explosives that are positioned to cover the kill zone with deadly force.

   c. Initiate the "trap" mines, grenades, or explosives when the enemy is maneuvered into a predetermined position by the "bait" action.

   d. Laying the trip wire for such tactics requires sound principles of concealment and camouflage in order not to alert the enemy to the presence of the booby trap or the ambush site. The trip wire can be used on roads, paths, trails, etc. Some general guidelines are as follows:

      (1) Roads. This is the most difficult location to conceal the trip wire itself. It is recommenced that the trip wire be used on roads at night and against vehicles because:

         (a) Their rate of movement will reduce the probability of detecting the trip wire.

         (b) Their rate of movement may be too fast for the ambush to be initiated by sight or signal. That is, it may pass through the kill zone before small arms weapons fire can inflict adequate damage.

      (2) Paths and trails. This is a preferred location to place booby traps. It is more difficult to accomplish this satisfactorily in open terrain than close terrain, but with imagination and ingenuity it can be very effective. Some general thoughts on trip wire placement are:

         (a) Use the natural surroundings to help conceal or camouflage the wire, firing device, and demolition.

         (b) In close terrain consider stringing the wire vertically. Most point men are trained and accustomed to looking for something running horizontally across the trail low or close to the ground.

         (c) Attach the trip wire to a small branch or other obstruction in the line of movement that would trigger the device when pulled, pushed, kicked, or otherwise moved.

2. Pressure Modes. The booby trap can be used in the pressure or pressure release mode to:

   a. Initiate Claymore mines or explosives that are positioned to cover the kill zone with deadly force.

   b. Prevent or disrupt the enemy from regrouping and mounting an effective counterattack.

   c. Initiate the "trap" mines or explosives when the enemy is maneuvered into the predetermined position by the "bait" action.

d. Setting the pressure/pressure release devices for such tactics requires sound principles of concealment and camouflage in order not to alert the enemy to the booby trap or ambush site. The booby trap device can be used on roads, paths, trails, etc. Some general guidelines are as follows:

(1) Roads. Unimproved roads are the best location for placing and concealing pressure actuated devices. The booby trap can be placed in a hole with a light board covered with sufficient dirt or brush to disguise. It can be used on roads during the day or night and against personnel and vehicles. The rationale for using such devices on ambush is the same as discussed for the trip wire mode.

(2) Paths and trails. This is a preferred location to place booby traps. It is more difficult to accomplish this satisfactorily in open terrain than close terrain, but with imagination it can be very effective. Some general thoughts on booby trap placement are:

(a) Use the natural surroundings to help conceal or camouflage the firing and demolition device.

(b) The pressure-activated mechanism can be placed immediately on either side of a log, downed tree, ditch, footbridge, or other obstruction to the line of movement. Look for channelized terrain or places where movement is naturally funneled when placing booby trap devices.

(c) For pressure release devices use obstructions or material that the enemy must or will be readily inclined to move, kick, or pick up.

2.8.7.2 Defensive Tactics. Booby trap devices can be used quite effectively to help protect or defend a SEAL patrol's LUP, OP, or ambush site. However, they may be counterproductive if activated by someone or something other than the enemy or found by the enemy, thereby compromising the patrol's position or intentions. All the tactics for hiding and concealing booby traps for defensive use are the same as for offensive use.

1. In perimeter security situations where patrol personnel cannot see defensive booby traps on the perimeter, it is recommended that signaling devices such as trip flares or grenade and artillery simulators be used instead.

2. Booby traps should be used as a "trip wire" to signal the approach of an enemy from an unexpected sector.

2.8.7.3 Harassment Tactics. Booby traps are ideal for harassing the enemy because they can be placed for unexpected activation. The main tactical purpose for such harassment activity is to keep the enemy off balance and degrade his will and desire to fight.

SECTION III SUPPORT

2.9 COMBAT SUPPORT

2.9.1 General. Combat support enhances the combat power and effectiveness of the SEAL patrol. There is normally only limited support that is organic to the NSWTG/TU organizations and, for the most part, combat support elements are not organic to the SEAL patrol. The NSW task organizations that support operations or contingencies through theater Special Operations Commands will often have access to air, ground, and naval supporting assets. Operational planners must understand the employment of the various types of units assigned to support the joint or combined task force to make the best use of their capabilities.

This section will discuss the various types of external combat support that may be available to SEAL patrols and will provide a framework for effective coordination with those elements. This support may involve the establishment of blocking forces, provision of fire support, and/or air support.

2.9.2 Coordination. Coordination is continuous throughout planning and preparation. The NSWTG/TU may coordinate some things and leave others for the PL. For example, the NSWTG/TU may arrange for the patrol to be guided to the point of departure. Likewise, the NSWTG/TU may coordinate the patrol's reentry of friendly areas through another unit. If this help is not given, the PL will have to do it. Even though some coordination is done for him, the PL must ensure that nothing is overlooked.

2.9.2.1 Operations and Intelligence. The operations and intelligence requirements for conducting combat patrols are vital to mission success. The coordination for them is the responsibility of the PL as well as the NSWTG/TU operations and intelligence sections. Examples of things a PL must coordinate are listed below. A PL should prepare a checklist and carry it during his coordination to keep him from overlooking anything vital. Some items may need to be coordinated with more than one staff section.

1. Operations. The operations section should keep the PL updated and informed of any significant operations or activities that may affect his mission. The following items are among those that require coordination or discussion:

   a. Changes in the friendly situation that may affect route selection, LZ selection, etc.

   b. Link-up procedures

   c. Transportation

   d. Resupply operations (in conjunction with supply)

   e. Signal plan: call signs, frequencies, code words, pyrotechnics, and challenges and passwords

   f. Departure and reentry of friendly lines (see below)

   g. Other friendly forces in the area, and attachment of specialized personnel (demolition team, scout dog team, forward observers, interpreter, etc.)

   h. Rehearsal areas

      (1) Terrain similar to the objective

(2) Security of the area

i. Use of blanks, pyrotechnics, or live ammunition

j. Fortifications available

k. Time the area is available

l. E&R nets available.

2. Intelligence. The intelligence section should keep the PL informed. It is essential that any last minute updates be given to the patrol. All available information must be used in planning and conducting operations. This includes the information listed below. See the NSWMPG for a more complete EEI list.

a. Known and/or suspected ambush sites

b. Weather and light data

c. Reports from units or patrols that have recently operated in the area

d. Size, location, activity, and capabilities of the enemy forces

e. The attitude of the civilian population and the extent to which they can be expected to cooperate or interfere

f. Changes in the enemy situation

g. Special equipment requirements.

2.9.2.2 Fire Support. All combat patrols should develop a fire support plan to be put into effect from the time the patrol departs friendly lines until it returns. See Section 2.9.4 for a more detailed discussion of fire support for SEAL patrols. Minimum coordination should include:

1. Mission and objective
2. Routes to and from the objective (include alternate routes)
3. Time of departure and expected time of return
4. Pre-planned targets enroute to and from the objective, and fire on and near the objective
5. Communications (primary and alternate means, emergency signals, and code words).

2.9.2.3 Friendly Forward Unit. The PL must coordinate closely with the friendly forward units the patrol will pass through. If no time and place have been set for this coordination, the PL should establish them when he coordinates with operations. He must talk to someone at the forward unit who has the authority to commit that unit to help the patrol pass. This is normally the unit commander. Coordination entails a two-way exchange of information.

1. PL gives:

a. Identification (himself and his unit)

b. Size of patrol

c. Times(s) of departure and return

d. Area of the patrol's operation (if it is within the forward unit's area of operation).

2. Forward unit gives:

a. Information on terrain

b. Known or suspected enemy positions

c. Likely enemy ambush sites

d. Latest enemy activity

e. Detailed information on friendly positions

f. Obstacle locations

g. Fire plan

h. Support the unit can furnish (e.g., fire support, litter team guides, communications, reaction units, etc.)

i. Signal plan including the signal to be used upon reentry, and the procedure to be used by the patrol and guide during departure and reentry

j. Location(s) of insert point, IRP, departure point, and reentry point.

2.9.2.4 Adjacent Patrol. The PL should check with other PLs who will be patrolling in the same or adjacent areas and exchange the following information:

1. Identification of the patrol
2. Mission
3. Route
4. Fire plan
5. Signal plan (challenge and reply are critical)
6. Planned times and points for departure and reentry
7. Any information that other patrols may have about the enemy.

2.9.3 Employment of Blocking Forces. In any AO, blocking forces are often established to prevent the escape of enemy forces from a designated target or objective area or to prevent enemy reaction forces from entering the area to support other enemy forces.

2.9.3.1 Blocking Forces. When a SEAL patrol conducts assault operations, additional ground forces may be necessary to seal off potential enemy ingress routes for reaction forces and/or egress routes for enemy escaping the objective. As in any joint or combined force operations, extensive coordination is necessary between the supporting and supported units.

1. General Operational Coordination. The SEAL PL and the ground blocking force commander must have a thorough understanding of the overall operational plan and its support requirements. The PL must ensure that the blocking force commander understands the patrol's operational sequence of events and the planned positions of all elements of the patrol at each and every stage of the operation. He must also coordinate the actual blocking force positions to ensure (1) they are in the most effective locations to support the operation, and (2) friendly fire casualties are avoided.

2. Communications. The PL must ensure that the communications plan includes the blocking force and that the blocking force receives copies of the plan in a timely manner. Call signs, emergency signals, brevity codes, passwords, and scheduled event reporting for the supporting units should be included.

3. Fire Support. The coordination of fire support (air, ground, and naval) between the PL and the blocking force commander is critical. (Refer to Section 2.9.2.2).

2.9.4 Employment of Fire Support. In almost any AO, fire support may be available from a variety of sources. The PL must understand the unique capabilities of each fire support source and be able to effectively employ them. To be effective, fire support must be thoroughly planned and coordinated well in advance of an operation. Pre-planned calls for fire should be planned not only at the objective, but also along the infiltration and exfiltration routes, on potential enemy escape routes, and at designated points on friendly withdrawal (or E&E) routes as necessary. The PL is responsible for establishing fire support requirements and for developing the fire support plan using the most effective mix of fire support available.

2.9.4.1 Types of Fire Support. Fire support in the form of naval gunfire, field artillery, and aerial gunships may have other applications in the support of operations. Illumination of the enemy at night should not be overlooked as a pre-planned fire support mission. Each type of fire support listed has a unique application and capabilities regarding illumination that must be considered in the planning phase.

1. Naval Gunfire. Conventional naval gunfire may have limited range in coastal, littoral, and river delta areas. The ability of the NGFS ships to take up adequate fire support positions may be restricted by extensive mud flats and sandbars that extend offshore for several miles.

2. Field Artillery and Mortars. Depending on the vicinity of friendly units, artillery and mortars may provide considerable fire support to a SEAL patrol. Artillery planners should consider the following:

   a. Lack of adequate firing positions may make it impossible to use the quantity and caliber of artillery justified by enemy strength and area characteristics.

   b. Absence of positions in defilade, lack of cover and concealment, and/or positioning in an insecure area require artillery units to use direct fire techniques and antipersonnel ammunition.

   c. Artillery may not be able to deliver accurate fires without adjustment (registration) because it lacks survey and meteorological data.

3. Aerial Gunship Support/CAS. These will be addressed in the next section dealing with air support.

2.9.5 Air Support. Air support can be very important for SEALs for reconnaissance, mobility, resupply, and fire support. This can involve both fixed- and rotary-wing aircraft.

2.9.5.1 Air Interdiction. If enemy aircraft attack the patrol, friendly air interdiction support should be requested. This will normally involve fighter aircraft or attack helicopters capable of air-to-air combat (i.e., missile systems, cannon, and gunfire). When requesting air interdiction support, the call for assistance must include your position, the type of attacking aircraft, and the urgency of the support. At the same time, evasive maneuvers and return fire must be executed. It may be prudent not to return fire in some situations, however, as the attacking enemy aircraft may not have an accurate "fix" of your position. Return fire will reveal the patrol's position. See Joint Publication 3-01.2 Joint Doctrine for Theater Counterair Operations for further details.

2.9.5.2 Close Air Support. CAS is air action against hostile targets in close proximity to friendly forces. Enemy antiaircraft capabilities directly affect the type of air support used.

1. Mission types

   a. Pre-planned

      (1) Executed to arrive at a pre-designated TOT

      (2) Rare, because it requires that a specific target and time be identified well in advance.

   b. On-call

      (1) Aircraft is pre-loaded for a target type and placed on alert status.

      (2) Mission is launched at supported unit's request.

      (3) Detailed mission planning and pilot brief are essential.

      (4) Minimum communications are required to effect final coordination.

   c. Immediate. Air strike is directed at an emergent target. Mission coordination is often accomplished while CAS aircraft are enroute. The FAC will brief the attack aircraft, so successful execution of immediate missions depends on reliable radio communications. In a high intensity, jamming environment, immediate missions may be impossible.

2. Fixed-wing CAS. In a high threat anti-air environment, high performance fixed-wing aircraft will be most effective. Their weapons can be employed during rapid passes over the target. In lower threat environments, propeller-driven fixed-wing aircraft may offer the best CAS. Specialized aircraft such as the AC-130 that provide a variety of CAS weapons options and sensors for pinpoint accuracy are a good option in low threat environments. The use of FACs, if available, is strongly encouraged. RTOs should be trained in CAS radio procedures in case a FAC is not available. Effective, safe CAS requires prominent identification marking of friendly elements. Marking options include glint tape, panels, pyrotechnics, flares, and lights.

3. Rotary-wing CAS. Helicopters are useful in lower anti-air threat environments due to their flexible maneuverability and ability to loiter on target. Rotary-wing aviation can furnish suppressive fire prior to landings or debarkations, and while the patrol is moving to or leaving an area of operation. This type of fire discourages small arms and AW fire against the aircraft by forcing the enemy to take cover. Helicopters do not replace other types of supporting fire, but are coordinated with them. They are also coordinated with the operations of friendly

units in the area. Marking friendly elements prominently should be SOP before any armed helicopter fire support is delivered.

2.9.5.3 Medical Evacuation. An air MEDEVAC capability is an essential element in the planning and conduct of any combat operation. Helicopters provide the fastest method for transporting wounded to treatment facilities. MEDEVAC communications procedures, LZ selection (with alternates), and LZ security requirements are critical planning elements for MEDEVACs.

2.9.5.4 Command, Control, and Communications. Airborne Command, Control, and Communication (ABCCC) may be employed during SEAL operations. Use of ABCCC, particularly during an operation involving a large number of assets and/or a dispersed AO, offers the tactical commander definite advantages. There are numerous aircraft (AWACs, E-2C, P-3, and command configured helicopters) that can provide the mission commander with rapid mobility, a full communications suite, and the ability to achieve "eyes on" understanding of fluid and evolving tactical situations.

2.9.5.5 Aerial Reconnaissance. Aerial reconnaissance is the use of aircraft to provide observation, directions, and warning for operational forces. In addition to recon conducted in advance of an operation, aerial reconnaissance may be conducted during its execution. An aircraft ideally suited to this task, in a low anti-air threat environment, is the AC-130 with its suite of sensors and weapons. Air reconnaissance:

1. Collects multisensor imagery of areas of interest.

2. Provides and maintains surveillance of enemy activities or areas of interest.

3. Conducts airborne electronic reconnaissance.

4. Supports the direction and adjustment of artillery and naval gunfire. In some cases (e.g., AC-130) this may also include direct fire support with onboard weapon systems.

5. Provides intelligence collection capabilities.

6. Affords the SEAL PL the ability to sustain coverage in areas of operation and provides rapid and current information on enemy composition, disposition, activity, installations, and terrain.

2.9.6 Resupply

2.9.6.1 Resupply Responsibilities. The primary responsibility for resupply planning and coordination lies with the PL and the NSWTG/TU.

1. PL Responsibilities. The PL is responsible for selecting the primary and alternate DZ and the recommended track and azimuth of approach to the DZ. Map reconnaissance and imagery will accomplish this while in isolation. If OPSEC allows, DZs for resupply should be confirmed as soon as possible after a unit infiltrates into an area. The DZ selected should offer the least possibility of compromise. DZs can be changed during the course of operations should the tactical situation dictate. The PL should ensure that:

   a. The resupply schedule and procedures are thoroughly briefed and coordinated within the patrol.

   b. All bundle load contents meet the mission requirements and the equipment is properly prepared for aerial drop.

c. The bundles are the correct size, weight, composition, and marked appropriately.

2. NSWTG/TU Responsibilities. The NSWTG/TU will make the necessary coordination with logistics/supply support for procurement and packaging of required supplies. The actual packing and rigging of the bundles will be done by the patrol while in isolation. The NSWTG/TU will activate an emergency resupply mission if a predetermined number of scheduled communications contacts are missed.

2.9.6.2 Aerial Resupply. Unless otherwise specified, airdrop equipment requirements and computations for SEAL operations will be based on the following:

1. Airdrop of equipment and supplies will be by parachute, with a maximum weight of 500 pounds and a minimum of 250 pounds. (If extra weight is needed, sandbags can be used).

2. There will be no salvage of airdrop support equipment.

3. The PL will be responsible for packing pre-planned resupply bundles with the assistance of the mission coordination support team.

4. The PL will coordinate with the logistics/transportation officer for the correct size and weight of the bundles.

5. To minimize transmission time, supplies will be requested using prearranged bundle codes. The patrol will receive codes while in isolation.

6. Cache locations will be briefed in isolation. See Appendix A for a more detailed discussion of caching.

2.9.6.3 Pre-planned Resupply Procedures. Equipment losses that may occur during insertion, infiltration, and operations can greatly reduce the combat effectiveness of the platoon. During the planning phase of the operation the PL will make arrangements for resupply of the platoon should the requirement arise. Resupply contingencies used can be on-call, automatic, or emergency. The NSWTG/TU schedules and coordinates resupply procedures.

2.9.6.4 On-call Resupply. Once communications have been established with the NSWTG/TU or headquarters that is exercising mission control, the PL can request on-call resupply based on operational needs. These supplies are usually items that are not consumed at a predictable rate. The PL determines the quantities and types of supplies to be included in the on-call request and requests them using a pre-planned code. Automatic and on-call DZs will be marked as specified by the NSWTG/TU during the isolation phase.

2.9.6.5 Automatic Resupply. The automatic resupply is used to resupply the patrol until routine supply or resupply can be instituted. The time, location, and contents of the automatic resupply are pre-planned by the patrol while in isolation. The PL will determine the type and quantity of supplies to be included in the automatic resupply. It may be modified by the PL depending on the initial area assessment and status of the patrol after insertion into the AO and/or infiltration to the target area.

1. Automatic resupply is scheduled for delivery shortly after the patrol has been committed. It is prearranged as to time, delivery site, and composition of load.

2. This resupply is delivered automatically unless the PL cancels or modifies the original plan.

3. Automatic resupply is planned to replace lost or damaged items of equipment and to augment the patrol with equipment that could not be carried during the insertion/infiltration.

2.9.6.6 Emergency Resupply. Emergency resupply procedures are for restoring the operational capability of the unit. Prior to infiltration an emergency resupply procedure and DZ is determined and coordinated through the NSWTG/TU with the supporting air arm, based on all available intelligence on the operational area.

1. Purpose of Bundles. The contents of the emergency resupply bundle(s) normally will be communication, survival, ordnance, and medical equipment designed to:

   a. Restore the patrol's operational capability.

   b. Provide the patrol with sustenance and medical supplies in the event of injuries and loss of equipment.

2. Emergency Resupply Upon Initial Entry. Emergency resupply bundles should be planned if the insertion method may result in loss of essential operational equipment.

   a. The emergency resupply drop is flown when the operation has not established contact with the NSWTG/TU within a designated period of time after insertion.

   b. Infiltration will be considered completed when the initial entry report is made by the insertion platform or SEAL patrol, as appropriate.

   c. If infiltration has been completed successfully, a request to cancel the emergency resupply drop may be submitted with the initial entry report. The DZ data will be maintained on file at NSWTG/TU, however, until the unit confirms or modifies the data, or cancels the DZ.

3. Emergency Resupply During Patrol. The requirements and procedures after infiltration include:

   a. Emergency resupply DZ(s) are selected and reported to the NSWTG/TU as soon as practical after the initial entry report is submitted and the patrol is established in the operational area.

   b. The emergency resupply is prearranged as to time and composition of load.

   c. Only members of the SEAL patrol should know the location of the emergency resupply DZ(s) and contents of the load.

4. Bundle Contents. The contents of the load will consist of those items necessary to maintain and/or restore the operational capability of the patrol and may include:

   a. Patrol operational radio, communication, and signaling equipment

   b. Survival and medical equipment as required (climatic)

   c. Weapons and ammunition sufficient for the patrol to defend itself and/or continue the mission.

5. Loss of Communications. The emergency resupply procedure is initiated and based on the loss of communications between the operational unit and the NSWTG/TU for a pre-designated number of contacts. There are several contingencies in which the loss of communications may occur, such as:

   a. Communications equipment becomes inoperative.

b. Enemy pressure has negated the patrol's capability to continue operations, forcing them to become highly mobile or to remain in hiding and maintain radio silence.

c. If the SEAL patrol is in a situation that requires constant movement, emergency DZs should be selected and reported at the first opportunity. If a predetermined number of radio contacts are missed, then the resupply is initiated and the equipment is dropped on the last reported DZ.

6. LZs/DZs. The PL will select a primary and an alternate LZ/DZ for emergency resupply and submit the appropriate LZ/DZ reports while in isolation. This information will be passed to the NSWTG/TU for further processing.

a. The patrol in isolation selects its desired LZs/DZs.

b. After infiltration, the patrol is responsible for selecting, reporting, and marking additional LZs/DZs.

c. A LZ/DZ that does not meet all criteria for the safety of infiltrating personnel, but does meet all air safety criteria may be used for material.

d. Situation permitting, the emergency resupply DZ will be marked. A single light (preferably flashing) in the center of the DZ will usually suffice, but it should be coordinated in advance. If the emergency resupply TOT is unknown, or if the situation will not allow a light to be used, the emergency resupply will be a blind drop.

(1) In the event the mission is flown, the primary LZ/DZ will be used unless the pilot and/or crew of the aircraft observe signs of enemy activity on the LZ/DZ.

(2) If there is enemy activity, the aircraft will immediately fly to the alternate LZ/DZ.

7. Executing the Resupply. The emergency resupply plan will be activated as pre-planned.

a. The emergency resupply mission will usually be flown between the end of evening nautical twilight and beginning of morning nautical twilight (BMNT). Drop time must be scheduled to allow the mission aircraft time to enter and depart the AO during darkness. The emergency resupply mission may have a 24-hour weather/maintenance delay.

b. It is activated when the operational unit has missed a predetermined number of radio contacts consecutively. The dates and number of contacts missed will vary based upon many factors (i.e., contact schedule, type mission, etc.) and must be established by the unit commander while in isolation and coordinated with the NSWTG/TU. Two consecutive missed contacts are usually programmed to allow for radio operator error. If three contacts are missed consecutively, the emergency resupply is then dropped at the next pre-planned window.

c. A SEAL patrol should always pre-plan its emergency resupply after the planned target is hit. Even if it cannot communicate, this will allow it to be resupplied without alerting the enemy to its presence prior to actions at the objective. With the radios in the emergency resupply bundle, the patrol can confirm their EP if they have accomplished their mission. If they have not accomplished the mission, the NSWTG/TU can give them additional instruction based on their reported situation.

d. Aerial resupply operations within denied areas of the world may not be feasible during the early stages of hostilities or SO. If this situation exists, the PL should consider rigging and transporting the necessary emergency equipment into the AO on the insertion aircraft. This equipment can be cached for later use as required and a cache report submitted to the NSWTG/TU. See Appendix A for more details on caching procedures.

e. If the patrol has requested both an emergency resupply and an automatic resupply, the resupply plan should address which bundle, or both, should be dropped in the event contact is missed.

f. The PL should tailor his emergency resupply bundle load to fit the specific needs of the mission, terrain, and climate that the patrol is operating in. If additional items of supply and equipment are required, he requests these during scheduled communication windows, to be forwarded to the appropriate NSWTG/TU for coordination and action.

2.9.7 Rendezvous Procedures. A rendezvous is a predetermined meeting at a specific time and place. The procedures that govern the actions of both parties and how to authenticate each other during rendezvous should be prearranged. Rendezvouses are sometimes called link-ups when referring to friendly operational forces meeting in the field. The proper use of rendezvouses is fundamental to SEAL operations. Rendezvous procedures give the control mechanisms that will be used for pre-planned and emergency rendezvous and linkups. But rendezvouses and link-ups can be dangerous to carry out. They must be planned and executed with care. There are two main categories:

1. General rendezvous procedures will always be arranged for any mission. They include the general patrol SOPs: pick-up, alternate pick-up, emergency shore rendezvous, and evasion rendezvous.

2. Special rendezvous procedures are mission-specific and include the DZ, LZ, and rendezvous link-up procedures with another patrol.

2.9.7.1 Patrol Rendezvous Procedures. These are the SOPs that will be used at the RPs that have been pre-planned and those selected as the patrol moves along its route. As discussed in Section 2.4.9, pre-planned and enroute RPs are used should the patrol become separated for any reason. In order to prevent lengthy backtracking, sufficient RPs should be identified and they should be easily recognizable. General patrol rendezvous procedures are:

1. RPs should be open for a specified amount of time only.

2. Make use of RPs, both pre-selected and those chosen enroute. Make sure all patrol members are aware of enroute RPs and time windows.

3. When required, the XRP should be open for a specific period depending on the mission. It is used when the patrol has not been able to fully reorganize at the enroute RPs.

4. Positive recognition signals must be prearranged for all rendezvouses.

2.9.7.2 Link-up Rendezvous Procedures. If two patrols plan to link up, the following procedure must be prearranged:

1. One patrol is designated as "host" and the other as "guest."

2. The "host" patrol arrives at the RP well before the rendezvous time, establishes security, sets up observation of the RP, and conducts an area reconnaissance.

3. Once the RP is determined safe, the entire "host" patrol may or may not move into the RP. Depending on the threat and situation, they may elect to send a reception team into the RP and hold the remainder of the patrol in a security position. The security team maintains eyes on the reception team during the entire evolution.

4. The "guest" patrol ensures it does not arrive at the RP before the rendezvous time. If necessary, it lays up nearby. The patrol should find a position from which to observe the RP for a short period and provide security for the team that will conduct the link-up.

5. A team or one man from the "guest" patrol moves into the RP from a pre-planned direction and is challenged by the "host" patrol.

6. They exchange near and far recognition signals and challenge and reply.

7. The reception team and guest rendezvous and return to the host teams security position, consolidate, and move out on a predetermined azimuth.

2.9.7.3 Alternate Extraction Points . These are necessary when the enemy's presence and/or time prevent the use of the primary EP. Alternate EPs should:

1. Have a set rendezvous time and a clear time limit

2. Have recognition signal authentication.

2.9.7.4 Emergency Shore Rally Point. This should be used when the extraction craft does not arrive or cannot be reached. The RP serves as an emergency contact point and as an evacuation point. Emergency shore RP considerations include:

1. It should be easy to identify, especially if the extraction craft was not included in the pre-operational planning and preparations.

2. It is possible that the extraction craft may have to be directed by radio until it acknowledges visual signals.

3. The emergency shore RP may be open for an extended time (days or weeks). Exact times will vary according to operational needs.

4. Factors affecting the choice of the emergency shore RP are:

   a. The need to extract the patrol as soon as possible to decrease the risk of capture and prevent physical and mental deterioration.

   b. The inability of the patrol and the extraction platform to guarantee arriving at a specific time on a specific night. This leads to the need for both sides to keep trying as long as possible, until either contact is made or the period of the emergency shore RP runs out.

   c. If the type of vessel making the extraction is a submarine or indigenous craft, it may stay in the area for several days and attempt the emergency shore RP each night.

   d. The possibility of unforeseen delays holding up the arrival of the extraction craft.

5. Authentication and Recognition Procedures. The authentication procedures at the emergency shore RP should be preplanned. Lights, radio, or verbal signals may be used. A sample authentication procedure follows:

   a. Extraction craft approaches the emergency shore RP close enough to be clearly seen from the shore.

b. One man from the shore party comes forward and identifies himself by challenge and password.

2.9.7.5 Evasion Rally Point. This is the final fallback. Its use becomes necessary after capture and escape or the failure of the emergency shore RP. It should be a definite location, but one that has ample cover nearby for prolonged laying up and one that can be safely and fully observed before moving into it. It should be in a location that precludes accidental use by an innocent third party. This RP could take the form of a safehouse. It is open for weeks or months.

2.9.7.6 Loads and Signs. These provide means of communicating with other members of the patrol or in the evasion network. Loads and signs are visual mechanisms that can be observed from a relatively safe distance that tell the observer the RP is, or previously has been, occupied or the status of a dead drop. The loads/signs must:

1. Be simple
2. Blend in with the surroundings
3. Be at the specified times and locations
4. Be known by both parties.

2.10 LOGISTICS SUPPORT

SEAL platoons depend on the theater Navy component or the Joint Special Operations Task Force commander for logistic support. Being small and highly mobile, SEAL units carry tents, communications, weapons, diving, parachuting, and combat equipment, but they require support for land transportation, messing, fuel (e.g., high-octane gasoline for NSW support craft), water, sanitation facilities, spare parts and maintenance support for non-SOF peculiar items, and administrative consumables. Units operating from Navy platforms receive support from the host via logistics requests for medical, messing, communications, supply, and personnel support.

Operations ashore involve larger and more complex requirements. During the initial stages of an operation it may be necessary to request logistic support from the U.S. Navy. Ships, with the exception of a few special types, are neither designed nor equipped to give logistic support to ground forces. Limited support may be available if it is adequately coordinated in advance.

2.10.1 Supporting Force Requirements. Consider the following when planning support:

1. Location of Assets. Are the assets available in the AO or will they have to be imported?
2. Sufficiency of Assets. Are there enough assets and back-ups to handle the mission requirements?
3. Support Facilities. Are there adequate facilities available to the supporting forces during the operation? These include:

    a. Messing

    b. Berthing

    c. Equipment storage, hangar space, pier space, hazardous material storage (including demolitions and ammunitions), and vehicle lifting equipment (cranes, etc.).

d. Refueling and watering facilities

e. Communications

f. Electronics.

4. Base Posturing. Good base posturing is mandatory. Effective base perimeters must be established to protect against sappers and terrorist raids. See the NSW Ashore and Afloat Basing Manual for specifics.

2.10.2 Combat Service Support Team. One Combat Service Support Team (CSST) is assigned to each NSW Group and may be employed to coordinate and conduct a full array of forward logistics support for SEAL platoons, SPECBOATDETs, and Task Units/Groups. The CSSTs are outfitted with state-of-the-art expeditionary support equipment, Combat Engineer Support Equipment, advanced field utility systems, NBC decontamination gear, and a full compliment of defensive weaponry. The CSSTs supervise embarkation operations and load planning, coordinate movement of forces and cargo per the Time Phased Force Deployment Data, organize contract and funding requirements, secure approvals for land use, contract for vendor services, liaison with host nation support agencies, coordinate defensive planning and preparations, build expeditionary camps, manage supplies, fuel and ammunition, and ensure all other logistics requirements are satisfied.

After launch from a FOB or ship, SEAL operations are typically of short duration. Accordingly, tactical logistical support requirements are usually minimal, as load out requirements generally do not exceed those that can be carried by the individual operator. In most situations, sustainability factors such as food, water, and ammunition do not become an issue.

If the SEALs are to remain in the field longer than 24 to 48 hours, resupply may be required either by pre-staging a cache of required supplies, or by delivering resupply items by air or boat. For mobile task unit operations, intermediate FOBs may be required to support advancing operational forces. In either of these cases, CSSTs may be employed to coordinate necessary arrangements and to execute critical field support. Care must be taken with either method to ensure sufficient advance planning and security measures.

2.11 MEDICAL INTELLIGENCE

Medical intelligence is that intelligence produced from the collection, evaluation, and analysis of information concerning the medical aspects of foreign areas that have immediate or potential impact on policies, plans, and operations. Medical intelligence also includes the observation of the fighting strength of enemy forces and the formation of assessments of foreign medical capabilities in both military and civilian sectors.

All medical intelligence is provided to the medical planner by intelligence organizations. The medical planner must identify the intelligence requirements and provide them to the supporting intelligence organizations. In an emergency, up-to-date medical intelligence assessments can be obtained by contracting the Armed Forces Medical Intelligence Center (AFMIC), Fort Detrick, Maryland 21702-5004. The message address is DIRAFMICFTDETRICKMD. Medical intelligence elements and AFMIC produce Medical Capabilities Studies, Disease Occurrence, Worldwide Reports, Foreign Medical Materiel Studies, the Disease and Environmental Alert Report, the Foreign Medical Facilities Handbook, Scientific and Threat Intelligence Studies, Foreign Medical Materiel Exploitation Reports, Quick Reaction Responses, and the AFMIC Wire. The medical planner must use all available intelligence elements to obtain needed intelligence to support the military operation. Sources of medical intelligence are listed in Appendix N.

There should be a supporting intelligence element at some point in the medical unit's chain of command. This element, whether military or civilian, will be the primary source for the medical planner to access necessary intelligence. The medical personnel must develop a feedback system with the supporting intelligence element to provide and receive intelligence updates. To develop medical intelligence, information is gathered, evaluated, and analyzed on the following subjects:

1. Endemic and epidemic diseases, public health standards and capabilities, and the quality and availability of health services.

2. Medical supplies and blood products, health service facilities, and the number of trained medical personnel.

3. The locations of specific diseases, strains of bacteria, lice, mushrooms, snakes, fungi, spores, and other harmful organisms.

4. Foreign animal and plant diseases, especially those diseases transmissible to humans.

5. Health problems relating to the use of local food and water supplies.

6. Medical effects of radiation and pro-phylaxis for chemical and biological agents.

7. The possible casualties that can be produced by newly developed foreign weapon systems such as directed energy weapons.

8. The health and fitness of the enemy's force and his special use of antidotes.

9. Areas of operations such as altitude, heat, cold, and swamps that in some way may affect the health of the platoon.

2.11.1 Significance of Medical Intelligence. At the operational level, the objective of medical intelligence is to develop medical support that:

1. Counters the medical threat

2. Is responsive to the unique aspects of a particular theater

3. Enables the commander to conduct his operation

4. Conserves the fighting strength of friendly forces.

CHAPTER 3

Jungle Operations

3.1 INTRODUCTION

This chapter focuses on specific jungle environment characteristics, planning factors, equipment and medical considerations, and field skills for operations in the jungle. The primary users are the SEAL PL, patrol personnel, and personnel involved in planning operations involving SEALs in a jungle environment. Planners must recognize and extract information needed for use with TTP presented in chapter two and other NSW publications.

3.2 JUNGLE MISSION PLANNING

3.2.1 General. Targets should not be indiscriminately attacked. They should be part of an overall plan to degrade an entire system. Interdiction must be based on the assigned mission that directs, as a minimum, the desired results and the priorities of attack for specific systems. Based on this mission, the PL selects the specific targets and those elements against which to conduct the attack. Target selection criteria for jungle targets are the same as presented in Section 2.3.3.2.

3.2.2 Naval Special Warfare Jungle Missions. NSW missions in the jungle environment include:

1. Direct Action. NSW forces may be tasked to clandestinely attack (ambush or raid) targets in jungle areas.

2. Special Reconnaissance. NSW forces may be tasked to conduct point or area reconnaissance in jungle areas in support of intelligence collection or as an advance force reconnaissance.

3. Foreign Internal Defense (FID). NSW forces may be tasked to conduct training of special forces of a friendly nation in support of U.S. interests.

4. Unconventional Warfare (UW). NSW forces may be employed to train and support indigenous forces (guerrillas) within a jungle area.

3.2.3 Most Likely Naval Special Warfare Missions. NSW missions are normally conducted in a riverine or maritime environment. Many jungle areas border this environment and NSW forces employment is a logical military option. The most likely SEAL missions in a maritime jungle environment are:

1. Special Reconnaissance

2. Ambush

3. Raid

3.2.4 The Jungle Environment. Successful patrol operations in the jungle environment are highly dependent on having platoon members with jungle experience (see Appendix F, Tactical Lessons Learned, for a compilation of lessons learned for operations in the jungle environment). The following paragraphs briefly describe the jungle environment.

3.2.4.1 Types of Jungles. There are two basic types of jungles: Primary and Secondary. The difference is more in their history than in their appearance. The word "jungle" is synonymous with "rain forest."

1. Primary Rain Forests. Characterized by great trees and a network of vines. These are virgin tracts of jungle undisturbed by man.

2. Secondary Rain Forests. Composed of secondary growth, or vegetation that has reclaimed once cultivated or settled land. Figure 3-1 shows the global distribution of rain forests.

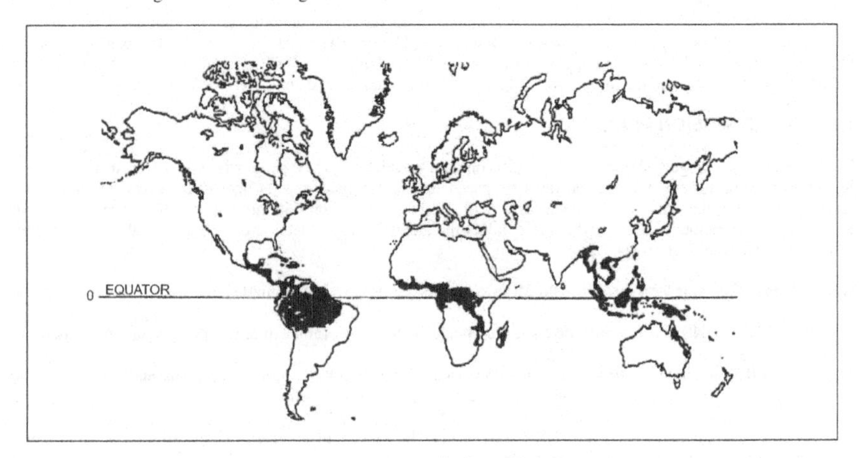

Figure 3-1. Global Distribution of Rain Forests

3.2.4.2 Categories of Jungle. There are five separate categories of jungle/rain forests, distributed in the world as depicted in Figure 3-2. Each of these rain forests may be primary or secondary. If man has cut into the rain forest, the regrowth is secondary.

1. Equatorial (tropical) Rain Forests

   a. Locations

      (1) The Amazon lowlands

      (2) The Congo lowlands (together with a coastal zone extending from Nigeria to Guinea)

      (3) Parts of Indonesia (especially Sumatra)

      (4) Several Pacific islands.

b. Climate

(1) Minimal climate variation with rain throughout the year, although occasional dry periods of one or two months can occur.

(2) High average temperatures.

c. Transition. Tropical rain forests change into subtropical rain forests on windward coasts, into monsoon forests in parts of Southeast Asia and Australia, and into mountain forests with increasing altitude.

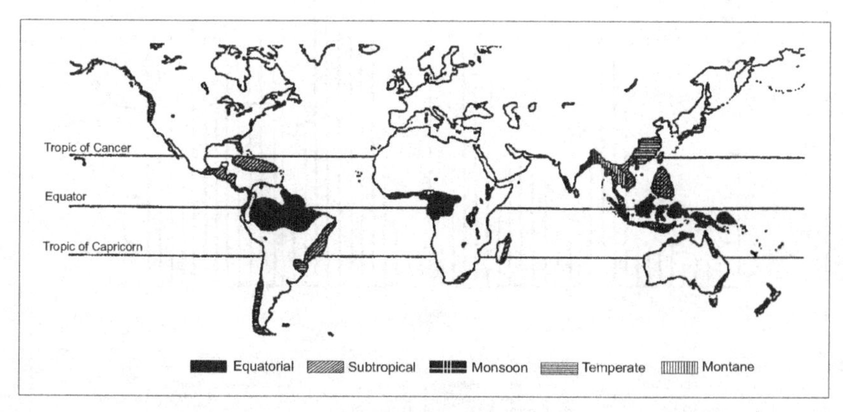

Figure 3-2. Categories of Rain Forests

2. Subtropical Rain Forests

a. Locations. From approximately 10 degrees North latitude to 10 degrees South latitude, on the following windward coasts:

(1) Central America

(2) The Caribbean

(3) The Western Ghats of India and coastal areas of Burma

(4) Vietnam

(5) The Philippines

(6) Parts of the Brazilian Coast.

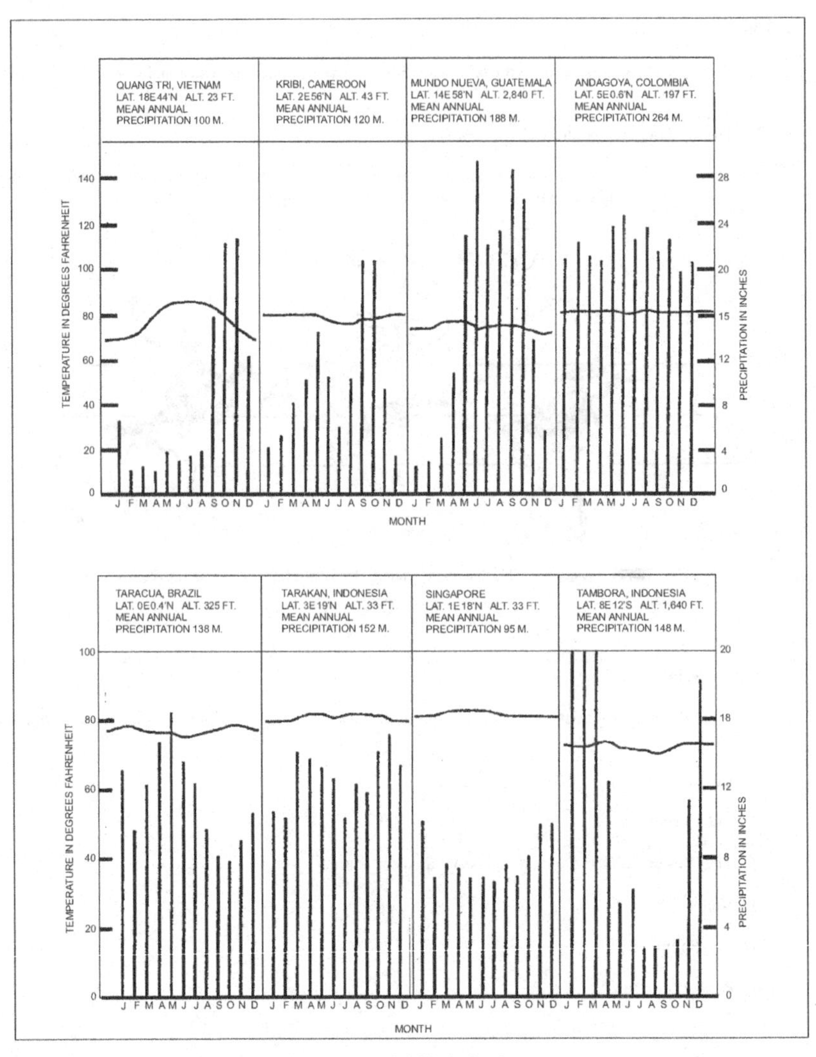

Figure 3-3. Rainfall Distribution

(7) Madagascar

(8) The Florida Everglades; small areas known as hammocks.

3. Monsoon Forests. These are also known as tropical deciduous forests.

   a. Location. These occur in regions with a high total rainfall but a definite dry season.

   b. Average temperatures. These are high, often over 77 degrees Fahrenheit, but the annual range is greater than in areas of equatorial rain forests.

   c. Undergrowth. This is likely to be denser in monsoon forest than in equatorial rain forest because of the greater openness and light penetration, and there is more "jungle" formed after clearing than in tropical rain forest.

4. Temperate Rain Forest. These are also known as temperate evergreen and laurel forests.

   a. Located in two main areas:

      (1) West Coasts between 35 and 55 degrees latitude (north & south)

      (2) East Coast between 25 and 40 degrees latitude (north & south).

   b. This type of rain forest is found in the United States.

5. Mountain Forests. These occur in the high altitude regions within or bordering tropical rain forests.

   a. Temperatures. With its lower temperatures, this forest has some similarities with the temperate rain forest.

   b. The tropical forest gives way to submontane forest at an average altitude about 3,300 feet and to mountain forest at about 4,900 feet.

3.2.4.3 Jungle Characteristics. Basic jungle characteristics are:

1. There is no winter or spring, only perpetual midsummer.

2. High rainfall is distributed more or less equally throughout the year as can be seen from Figure 3-3.

3. Vegetation consists of multiple canopies or stories of stratification with varying humidity, temperature, and light, as depicted in Figures 3-4 and 3-5.

4. Trees form the principal elements of the vegetation. Evergreen trees predominate, many with large trunk diameters (10 feet).

5. Vines and air plants are abundant.

6. Herbs, grasses, and bushes are in the undergrowth.

7. Grasses and herbs, common in the temperate woods of the United States, are rare.

8. Undergrowth consists of woody plants: seedling and sapling trees, shrubs, and young woody climbers. All of the plants grow very large. Bamboo grows from 20 to 80 feet tall and is sometimes impenetrable.

9. Riverbanks and clearings have dense, almost impenetrable growth due to the large amount of sunlight.

10. In the interior of old, undisturbed rain forests (primary jungle), it is not difficult to patrol in any direction, as it is clear of growth.

3.2.4.4 Jungle Area Descriptions

1. Swamps. Swamps are common to all low jungle areas. The two basic types are mangrove and palm.

   a. Mangrove Swamps. Coastal areas; shrub-like trees grow from three to fifteen feet; tangled root systems both above and below the water level hinder movement of boats and personnel; air and ground observation is poor; concealment is excellent.

   b. Palm Swamps. Exist in both salt and fresh water areas; movement restricted to foot; limited observation and fields of fire; concealment is excellent.

2. Savanna. Savanna is open jungle grassland where trees are scarce. The grass can be as high as fifteen feet. This thick, dense grass, combined with the heat and humidity, makes for slow and arduous patrolling.

3. Cleared Areas. Cleared areas will be places where man has made his mark.

   a. Villages and towns.

   b. Cultivated areas. Some of these may have been abandoned and be reverting to jungle.

      (1) Rice paddies. Rice paddies are flat fields that are either flooded or dry and are surrounded by six to nine foot dikes. Movement is difficult when wet and somewhat easier when dry, but concealment is always poor and cover (from ground fire only) is limited to the dikes.

      (2) Farms. Farms exist throughout the tropics. Ease of movement through farms depends on whether they are being cultivated.

      (3) Plantations. Plantations are large farms where tree crops (rubber, coconut, or fruit) are cultivated.

4. Holes and Cliffs. The dense vegetation often conceals holes and cliffs that are found in many jungle areas.

3.2.4.5 Likely Areas for SEAL Training. Riverine and maritime training opportunities are most likely to be found in the following tropical and subtropical jungle areas:

1. Southeast Asia

2. Central America

3. Caribbean

4. South America

5. Africa.

3.2.5 Jungle Planning Factors. The tropical rain forest environment is both the most difficult jungle environment and the most likely one in which SEAL operations will occur. A SEAL platoon will try to remain undetected during both ingress to and egress from the objective area. The following paragraphs discuss the primary planning considerations for the jungle environment (See the NSWMPG for a phased planning approach).

3.2.5.1 Mobility. In the tropical rain forest, both rate of advance and flexibility of maneuver are limited. See Section 3-3 for a detailed discussion of mobility in the jungle environment, time and phases of the operation, and the timing of support forces to match SEAL mobility capabilities.

3.2.5.2 Endurance. Endurance in the jungle is limited by the following factors:

1. Heat and Humidity. These can reduce endurance by 50 percent as compared to endurance in temperate areas.

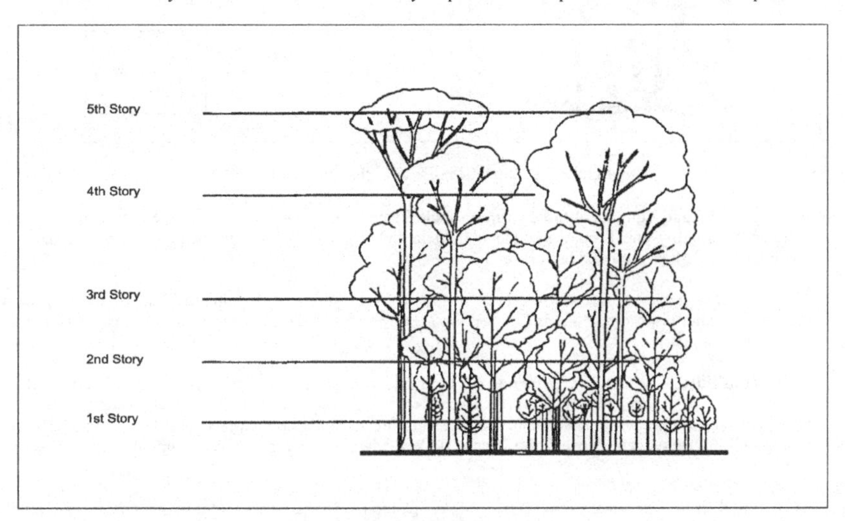

Figure 3-4. Jungle Canopies

2. Disease. This can quickly destroy the integrity of a platoon. (see Section 3.4 for preventative medicine procedures in the jungle).

3. Conditioning. Superior personal conditioning is required for the entire platoon, especially cardiovascular conditioning such as running and rapid patrolling.

4. Acclimation. Proper acclimation is required by the entire platoon upon arrival in the jungle environment. If time permits, the platoon should spend a minimum of two weeks becoming conditioned to the hot and humid jungle climate.

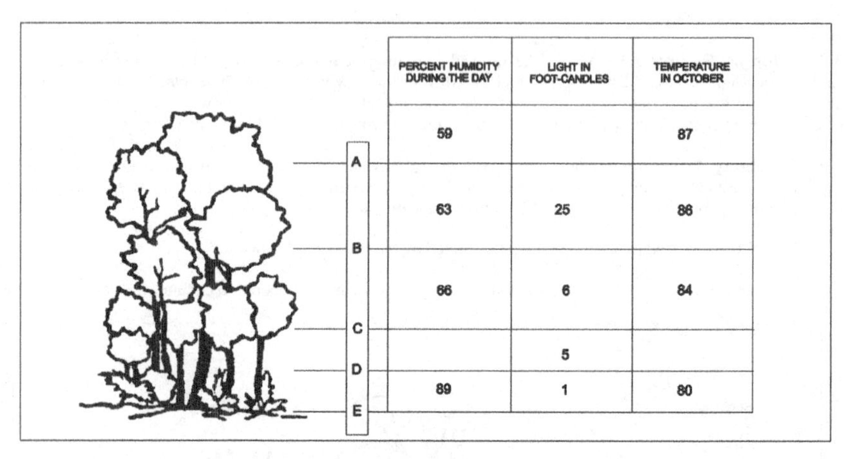

Figure 3-5. Jungle Humidity, Light, and Temperature

5. Equipment Weight. Equipment to be carried should be carefully planned to obtain the best match between endurance, mobility and firepower. "Think light to fight" is a good rule in the jungle (see the baseline equipment guidelines provided in Section 3.3.11).

6. Water. Extra water is required in the hot and humid jungle climate. Hydrate personal by drinking large amounts of water prior to operations (see Section 3.3.6 for more detail on the need for increased water in the jungle environment).

3.2.5.3 Detectability. The dense vegetation, the high tree canopy, and the frequently rugged terrain affect detectability in the jungle environment. It can be used to a platoon's advantage if it is properly equipped, the route and rate of advance are properly chosen, and proper tactics are carefully applied (see Chapter 2 for tactics). The jungle detectability factor can also aid the enemy if they have superior planning, equipment, and/or tactical advantage.

3.2.5.4 Survivability. The survivability of a SEAL platoon depends on its ability to survive the jungle environment itself, as well as to survive against jungle acclimated hostile forces that are probably familiar with the terrain. Thus, in addition to acclimation from an endurance point of view, planners should attempt to provide material that will enable the platoon to become "terrain-wise" in order or compensate for the enemy's likely superior knowledge of the jungle area (see Section 3.4 for jungle survivability techniques).

3.2.5.5 Command and Control. C^2 in jungle maritime operations is more difficult due to the thick vegetation, the canopy and the unfriendly environment for electronic equipment. Thick vegetation hampers visual communications and can block low-powered communications at high frequencies. Planners must rehearse communications under realistic conditions and provide for more equipment redundancy (see Chapter 2 Section 2.2.3 and paragraph 3.2.12 of this chapter for detailed C^3 procedures in the jungle environment).

3.2.5.6 Insertion and Extraction Planning. Carefully coordinate insertion and extraction methods, as it is important to minimize the distances SEALs move from the insertion point to the objective area and returning to the EP. Numerous insertion and extraction methods are available, but some may require logistics support and maintenance areas (see Section 2.7 for a more detailed discussion).

3.2.6 Target Planning

3.2.6.1 Target Types. Targets in the jungle are normally difficult because of natural obstacles such as rivers, cliffs, and swamps that can be used to defend enemy areas. Examples of typical SEAL targets that could be found in a jungle environment are:

1. Ammunition and supply caches
2. Guerrilla headquarters and training areas
3. Riverine areas used by the enemy for ambushes
4. Enemy river patrol boat bases
5. Chemical or munitions sites
6. Drug manufacturing sites
7. Enemy C^2 sites
8. Enemy missile sites
9. Jungle supply routes
10. Personnel targets.

3.2.6.2 Weapon Selection. Planners should compute the probability of kill of the various SEAL weapons against the specific target, and select the optimum weapon. It should be recognized, however, that a standoff weapon may be difficult to use because a clear path from launch position to target may be difficult to achieve (see weapons discipline in Chapter 2 Section 2.3.7 and Section 3.3.9 in this chapter).

3.2.6.3 Target Damage and Post-action Analysis. Targeting is complete only after damage assessment has occurred. A complete debriefing of the SEAL platoon and supporting element personnel must be conducted to:

1. Assess target damage
2. Update maps and intelligence
3. Develop new targets.

3.2.7 Threat Planning. Threat forces in the jungle are difficult to detect and track because of the heavy canopy and rough terrain normally encountered. Maximum use of intelligence sources must be requested and exploited.

3.2.7.1 Intelligence Sources

1. National Sensor Information. Information from national sensors is available through operational intelligence centers. Sources include photographic intelligence, electronic intelligence (ELINT), IR intelligence, signal intelligence (SIGINT), or HUMINT.

   a. Visual intelligence. Visual or photographic intelligence from overhead sensors is only partially useful due to the heavy vegetation and canopy normally found in jungle areas.

   b. ELINT. ELINT is most useful for determining the nature and number of units and defensive systems in the operating area.

   c. IR intelligence. IR intelligence can display the location of major population areas, base camp areas, and the location of any major equipment that has an IR signature. IR sensors can see through jungle vegetation and canopy, although attenuation is a problem.

2. SIGINT. SIGINT can be used to determine enemy movement, intentions, and degree of alert.

3. HUMINT. Timely information from friendly indigenous personnel in the jungle operating area may be possible in areas where intelligence planning has been conducted long enough. It is the most useful type of intelligence for jungle operations, given the problems in visual intelligence and the need for current local information, but it must be carefully evaluated as to the reliability of the source.

3.2.7.2 Threats to Jungle Patrols. The most likely threats that SEAL patrols will encounter in the jungle are:

1. Unfriendly indigenous forces

2. Enemy foot patrols

3. Enemy air patrols

4. Enemy river patrols

5. Booby traps

6. Defensive forces in objective areas

7. Enemy ambush.

3.2.8 Jungle Supporting Force Planning. The types of supporting forces that are likely to be required in the jungle are:

3.2.8.1 The Mobile Communications Team. The MCT must be co-located with the immediate superior in command of the patrol to ensure reliable communications.

3.2.8.2 Insertion and Extraction Forces. The patrol leader must be aware of the actions and disposition of all supporting forces throughout the operation. Support requests must be submitted as soon as possible, particularly if there is competition for assets.

1. Face-to-face planning is a must. Make every effort to meet the leadership of the supporting forces. People who know each other on a personal basis will support each other to a far greater degree than those who don't.

2. Living and working with the supporting elements is the best liaison possible, and a strong bond develops between SEAL platoons and helicopter or boat personnel.

3. If access is limited, obtain authorization for direct liaison to facilitate development of a strong inter-personal planning relationship.

3.2.8.3 Helicopter Gunship Support. Helicopter gunship support is the optimum type of CAS for jungle operations. Gunships are able to work around and, in some cases, under heavy vegetation to provide CAS.

3.2.8.4 Tactical Close Air Support. Tactical fixed wing aircraft are useful in cleared or open areas. When the target is under the canopy, the SEAL patrol must relay accurate coordinates to provide targeting. In some cases a smoke or flare may penetrate the canopy to mark a target location, however, this cannot be relied upon.

3.2.8.5 Naval Gunfire Support. NGFS is useful when operating within 10 nautical miles of a destroyer or cruiser position. Direct fire can be conducted against targets visible to the ship (or marked by a smoke or flare visible to the ship). Indirect fire can be provided on targets not visible to the ship based on provided target coordinates. SEALs should be familiar with NGFS spotting procedures.

3.2.8.6 Friendly Indigenous Force Support. Support by friendly indigenous forces is a special benefit in the jungle. It provides:

1. "Local know how"

2. Immediate reinforcement if required.

3.2.8.7 Friendly Force Support. There may be friendly conventional forces operating behind friendly front lines in a SEAL patrol area that can provide various types of support (examples: artillery, helicopter gunship). SEAL patrols should always have the location of friendly front lines plotted and committed to memory.

3.2.9 Patrol Logistic Planning

3.2.9.1 Logistic Problems. Heat, humidity, and rain in jungle areas cause the following problems:

1. Weapons rust quickly (need ample cleaning equipment and oil)

2. Battery life is shorter

3. Electrical connections corrode quickly

4. Uniforms and other clothing wear out faster.

3.2.9.2 Logistic Plans. Section 2.10 addresses logistics support and the role of NSW CSST. Logistic plans must provide support for NSW forces in the following areas:

1. Isolation planning area.

2. Isolation security personnel and food delivery.

3. Berthing area in isolation separate from planning area.

4. Rations. Meal, ready-to-eat (MRE) and others.

5. Water. People will drink more water, requiring greater water purification, storage, and transport means (more canteens or water bladders).

6. Equipment. Uniforms, boots and socks deteriorate and should be stocked for replacement.

7. Petroleum Oil Lubricants. It will be necessary to maintain and refuel outboard engines and support craft, including boats and helicopters. Oil for weapons will be used in increased quantities (10-20 percent more than normal).

8. Ammunition, Demolitions, and Pyrotechnics. These items are mission dependent. Maintain full stocks of all items in accordance with standard load lists. Coordinate this with the SEAL platoon.

9. Resupply. Air and boat resupply are the primary methods used. Air resupply provides greater flexibility, but requires advance planning.

   a. Coordinate resupply procedures used by supporting air assets.

   b. Brief staging procedures for resupply items.

   c. Ensure items for resupply are prepared and correctly packaged for the jungle environment. All items should be packed in waterproof packages. Items that deteriorate in high temperature (such as batteries) should be clearly marked as "heat sensitive material."

10. Maintenance capability. Provide a maintenance capability to support and repair all equipment.

    a. High humidity and temperature increase maintenance on all items affected by moisture and heat.

    b. Plan for a 10-20 percent increase in normal spare parts usage. Heat sensitive item usage will increase by 30-50 percent dependent on the mission requirements.

3.2.9.3 Medical Preparation. Section 2.11 addresses medical intelligence and sources of information. Plan for the following medical support for NSW personnel:

1. MEDEVAC plans and procedures

2. A designated military hospital or qualified facility, including burn and trauma units, to receive the wounded

3. Preventive medical measures against tropical diseases

4. Jungle specific medical kit supplies.

3.2.10 Tactical Route Planning. The route planning procedures provided in the NSWMPG should be used with the special considerations discussed below:

3.2.10.1 Quality of the Charts/Maps Used. Many jungle areas of the world are unreliably charted and the available charts are small scale and not useful to the small unit tactical commander. Such small-scale charts typically do not show the topographic features that are most important to the patrol, such as holes, dense vegetation areas, cleared areas, small villages and lightly populated areas, hills, cliffs, and gorges. Bear in mind also that maps of double/triple canopy areas may not show details.

1. Acquire a chart that is current and of a satisfactory scale (at least 1:50,000).

2. Chart should be marked with the latest known topographic information.

3. If U.S. maps are unavailable or seem incomplete, obtain local country maps, which may be more detailed.

4. For a dense jungle, IR photography may be the only way to find hidden characteristics under canopies.

5. Do an in-depth study of the map.

3.2.10.2 Terrain Information. Planners should analyze all features of the chart to select primary and alternate routes, route points, and route areas.

1. Ensure the platoon gets the "feel" of direction and features in the AO.

2. Locate likely navigation aids on planned routes, and train personnel in pacing and navigation skills.

3. It is far easier to follow the contour than to go against it. Take the direct route (straight line) only when in flat jungle. Straight line (azimuth only) patrolling is useless in hilly jungle.

4. Terrain features important to the jungle environment are:

   a. Swamps. Avoid if at all possible.

   b. Valleys.

   c. Ridges. Less vegetation is found at the tops of ridges and hills and animal tracks are also found on ridges, making it easier and less tiring to patrol. Making the effort to climb may save time overall.

      (1) Use ridge tops and paths selectively. Do not remain on them for a long time.

      (2) Patrolling on ridge tops must be weighed against the possibility of enemy detection or ambush.

   d. Streams and Rivers. Streams and rivers must be analyzed for width, depth, and character of current, condition of banks, bottom composition, and gradient.

      (1) Do not follow rivers, as they meander around and are bordered by thicker vegetation. The further away from the river the faster the movement.

      (2) Compute tides for rivers because tide changes may cause a river to become a stream or event mud, destroying the movement plan.

3.2.10.3 Survival Information. Obtain information on features of the operating area that, if required, could enhance the survival of the patrol, such as:

1. Vegetation and wildlife information
2. Sources of available water.

3.2.10.4 Environmental Information. Planning analysis should include:

1. Ambient light
2. Beginning of morning nautical twilight (BMNT)
3. Sunrise
4. Sunset
5. Before evening nautical twilight (BENT)
6. Moon rise
7. Moon set
8. Moon phase
9. Visibility range
10. Probability of precipitation and times of the day for precipitation
11. Times for tides.

3.2.10.5 Reduced Tactical Mobility. The baseline tactical mobility of the patrol in the jungle is eight kilometers per day (based on eight hours of patrolling in each 24-hour period). See discussion in Section 3.3.12.

3.2.10.6 Reduced Endurance. The baseline mission length (with full tactical mobility) for a patrol is 72 hours. See Sections 3.2.5.2 and 3.3.12.

3.2.10.7 Reconnaissance of the Operating Area. Obtaining complete reconnaissance information from all available resources will significantly increase the probability of mission success. This is true under all conditions, and especially when planning operations in a jungle environment.

1. Visual reconnaissance is the most effective method of providing a PL with the "eyes on" information needed for proper planning.
2. Platoon leadership visual reconnaissance by air is the most effective method.
3. OPDEC must be a factor when planning a visual reconnaissance. Conduct false flights to areas not involved in pending operations and do not set patterns of over-flying specific future target areas.

3.2.11 Jungle Tactical Communications Planning

3.2.11.1 Communications Degradation Due to Weather. The type of jungle and season of year may degrade communication capability as follows:

1. Climate. The hot and humid climate has the following effects:

   a. Battery shelf life and operating life are reduced by 20 percent. To optimize these:

      (1) Keep batteries in a cool, dry location.

      (2) Keep batteries out of direct sunlight.

   b. Radio power output and sensitivity are decreased by 20 percent.

   c. Moisture and fungus proofing are required. Normal preventive maintenance will take care of this.

2. Seasonal Propagation. Seasonal propagation will vary as follows:

   a. Rainy season (monsoon). Condition improves propagation by 10-20 percent over normal conditions.

   b. Dry season. Deteriorates propagation by 10-20 percent below normal conditions.

3.2.11.2 Communications Degradation Due to Jungle Growth. Vegetation will effect communications:

1. Transmitter efficiency is reduced by approximately 20-30 percent when in dense vegetation.

2. Raising correctly configured antennas will help HF communications by getting the antenna above the jungle canopy. See Chapter 2, Section 2.3.4 for detailed procedures.

3.2.11.3 Communications Planning Considerations. The NSWTG/TU to which the patrol reports must be aware of the cause and effect of HF radio waves' variations in the jungle environment to plan both TG/TU and SEAL communications. The most important considerations include:

1. Ensure the connectivity between the MCT, the SEAL insertion/extraction craft, and the SEALs by checking:

   a. Frequencies

   b. Times

   c. Radio equipment compatibility

   d. Correct antennas (height and direction).

2. Develop a CEOI, including a "lost communication" plan. The headquarters unit should not assume that communications are not being received by the platoon when nothing is heard from the field. The NSWTG has more transmit power than the platoon and is usually heard.

3. Utilize a repeater or relay to assure an adequate communication linkage. This could be an aircraft or a repeater on high ground.

4. Minimize communication requirements to and from the field by not requiring unnecessary contacts and using pre-arranged times.

5. Establishing communications windows.

3.2.12 Procedures for Tactical Contingencies. Contingency plans are necessary to cope with the unplanned or unexpected event. Planning should provide for communications code words that identify the tactical contingency and the action being taken. The most common events might be:

3.2.12.1 Loss of Communications. Communications may be lost because of equipment failure or temporarily loss of reception due to blockage of the propagation path by terrain. There must be a lost communications plan.

3.2.12.2 Failure to Reach Assigned Rally or Objective Points. There are many unknowns that could delay the patrol. When this occurs, there must be a back-up plan that can be executed by code word.

3.2.12.3 Medical Emergency. The nature of the jungle environment is such that a key member of the patrol could be injured (without enemy action) and unable to continue. When this occurs, a code word should be designated to indicate the action being taken and assistance required (see Section 3.4 for further discussion).

3.2.13 Insertion/Extraction Planning

3.2.13.1 Insertion. In general, keep the insertion method as simple and safe as possible, but also be prepared to be bold when the situation warrants. Some considerations are:

1. Does the method used have to blend in with the local types of airborne or surface platforms in order to arrive at the area where patrolling will commence?

2. Can the craft get close to the insertion point or will the patrol have to cross a mud flat or swampy section?

3. What will the local people be doing along the route to the insertion point? Will they alert the enemy that something unusual is happening?

4. What type of insertion platform is best for the topography of the route? Rappelling or fastroping from a helicopter into a clearing, using military or indigenous boats, or even vehicles, to get to the insertion point requires thought.

5. In jungle heat, air is less dense, which greatly decreases the lift capability of aircraft. If using an aircraft for insertion, it may not be able to carry a full load. Either reduce the load or schedule more aircraft.

6. Plan insertions just before first light. People are less aware in the early morning than in the evening, but the light factor will become a problem after sunrise.

3.2.13.2 Extraction. As in the insertion phase, ensure those responsible for extracting the platoon (i.e., boat captains and helicopter pilots) are involved in the planning.

1. Discuss primary and alternate locations and times, as well as additional methods if the primary method does not arrive.

2. Discuss communications and recognition codes.

3. Select emergency helicopter evacuation points and give them code words before going out on the patrol. If this cannot be done beforehand, be prepared to select helicopter evacuation points and pass them back from the field.

4. When planning each long term LUP and its associated area, have two nearby sites that could be reached in the event the patrol is compromised.

3.2.14 Foreign Internal Defense in the Jungle Environment

3.2.14.1. Naval Special Warfare Forces in Foreign Internal Defense. NSW forces may be called upon to conduct FID in nations with jungle environments. NSW planners and operational forces should understand the interrelationships necessary with foreign HNs. Training the forces of allied and friendly nations in UW is a SOF mission. NSW personnel should be ready to impart a wide variety of military skills for the jungle environment directly applicable to these allied and friendly forces. To accomplish this, NSW personnel are required to successfully deal with:

1. Linguistic barriers

2. Cultural inhibitions

3. Austere support situations

4. Foreign operational procedures in low intensity conflict.

3.2.14.2 Training Considerations. Planning, preparation, and procedural skills discussed in this manual are the skills taught and exercised in jungle FID.

3.2.14.3 United States Support in Foreign Internal Defense. The HN will typically need the following types of support from the NSW forces to prepare for countering both insurgency and terrorist threats.

1. Detailed tactical intelligence

2. Inexpensive, reliable, and secure communications

3. Supportable, affordable, and appropriate transportation

4. Organized and operating logistics systems

5. Means of relating positively with the civilian population

6. Medical support and training

7. Civic action and civil engineering project assistance

8. Improved development for manufacture of military goods

9. Locating non-U.S. sources of material

10. Applicable training and assistance by NSW forces.

3.2.14.4 Host Nation Personnel and Organizations

1. NSW Interrelations with HNs. To better understand the interrelationships between Naval Special Warfare and HN organizations and personnel, each country must be examined individually and each specific area identified and incorporated into the strategy best suited to achieve the U.S. goals in that region.

2. Information Sources. NSW planners and operational forces should obtain the following information when planning operations or training in foreign nations:

   a. HN military structure.

   b. Interrelationships between the HN military organization and the HN government.

   c. Determining where the bases of power lie within the nation, such as national/state police, military, etc.

   d. Respective chains of command within the HN, law enforcement, and military.

   e. Feelings of the populace regarding U.S. assistance, specifically NSW force involvement.

   f. Religious practices that may affect military operations. Since many of the jungle countries have strong religious influences, religion must be considered in determining the level of assistance and its effect.

3.2.15 Survival in Nuclear, Biological, and Chemical Warfare (under review)

3.2.16 Patrol Debrief

3.2.16.1 Debrief Schedule. Conduct the debriefing as soon after patrol completion as possible. Patrol members will likely be in a state of fatigue and under stress. Therefore, use these debriefing techniques:

1. Make the personnel being debriefed comfortable.

2. Do not make them write answers.

3. Avoid leading questions that suggest the answer.

4. Advise the PL of the desired debriefing information before the patrol departs.

3.2.16.2 Debriefing Guide

1. Terrain

   a. LZ

   b. Vegetation thickness

   c. Types and height of vegetation

   d. Map accuracy

   e. Trails used by locals

f. Streams and Rivers: crossed or followed; bridges/fords.

2. Communications

a. Problems: reception, transmission, batteries, and equipment

b. Antennas used

c. Contacts

d. Local civilians.

3. Enemy

a. Where

b. How many

c. Gender

d. Uniforms

e. Equipment

f. Weapons

g. Documents

h. Direction headed.

4. Camps

a. Location

b. Size, age, and layout of structures, buildings or defenses

c. Occupied or abandoned.

5. Food or ammunition caches

6. Communications, printing, generators, factory capabilities

7. Documents found.

3.3 JUNGLE OPERATIONS PREPARATION

3.3.1 General. This section will present information on clothing, weapons, communications, special equipment, medical requirements, and considerations for planning a jungle operation.

3.3.2 Jungle Operations Training. Training for the jungle environment is essential. Attendance at a formal jungle survival course is recommended for NSW personnel.

3.3.3 Jungle Clothing and Equipment

1. General. Give primary consideration to:

   a. Visibility. Ensure uniform and equipment blends into the environment.

   b. Operating Efficiency. Ensure all equipment and weapons are ready for instant use. Grenades, magazines, and radios must be comfortably and tactically organized for combat.

   c. Insect Protection. Use insect repellent and netting to reduce the likelihood of debilitating disease.

   d. Comfort. Test all equipment for comfort to insure optimum effectiveness. This is particularly important when wet and hot.

2. Uniforms

   a. Wear the lightweight camouflage uniform. Even when wet at night, these are difficult to see, even with NVEO.

   b. Wear loose-fitting, untailored uniforms. Tuck the blouse into the trousers when wearing load-bearing equipment.

   c. Do not starch.

   d. Wear the sleeves down.

   e. Blouse or secure trousers outside the boots.

   f. Soak the collar, cuffs of jacket sleeves, fly, waist, and bottom of trouser legs (and blousing cloth) with insect repellent (permethrin spray) and let dry. This will preclude the need to spread repellent all over the body and possibly compromise the platoon's location by smell.

   g. Ensure any non-standard items have grommets to release water.

   h. Wear an "Australian net" or bandanna around the neck.

   i. Do not wear skivvies, as they tend to chafe when wet and lead to infection. Personal preference dictates.

3. Hats. Wear a floppy hat to break up the outline of the head. Do not wear only a bandanna on the head.

4. Gloves. Wear gloves with fingers cut out. This allows easy access to the trigger. Gloves also provide protection from thorns and heat of hot weapons. Two recommendations include:

   a. Weight lifting gloves

   b. Nomex (aviator gloves).

5. Boots. Many quality boots are available, but the jungle boot is still the most reliable. Better than all-leather boots, they will dry faster. Wash the ventilating insoles with soapy water, as necessary. If possible, each member of a patrol should wear boots with the same sole pattern.

6. Personal Equipment. Plan to carry as light a load as possible to give the patrol greater endurance. The heavier the load, the greater the energy drains on personnel. The heat and humidity of the jungle aggravate this endurance reduction. Appendix E lists suggested items for the first, second, and third line of equipment.

   a. First Line. This is carried or worn on the individual at all times and consists of the following:

      (1) Essential items (e.g., watch, compass, flashlight, knife, etc.)

      (2) Immediate emergency items (e.g., field dressing, emergency food, etc.)

      (3) Survival items (e.g., fishing kit, snares, etc.)

      (4) Escape items. Easily concealed items that can pass pat down search (e.g., escape map, compass, piece of hacksaw blade, etc.)

   b. Second Line. This is carried on LBE and worn at all times.

      (1) Essential for 24 hours of efficient operating.

      (2) Water bottle, ammunition, 24 hours of rations.

   c. Third line. This is carried in the backpack or rucksack unless actually in use. It includes the remainder of mission-essential operating equipment.

7. Flotation Equipment. Always include flotation equipment when operating in the jungle. Consider the requirement as follows:

   a. Salt water provides greater buoyancy than fresh water. Compensate appropriately by requiring more buoyancy if only able to test in salt water (e.g., launching from a ship to the jungle).

3.3.4 Camouflage

3.3.4.1 Principles. Primary camouflage concerns in the jungles are divided into the three general groups:

1. Shine. Belt buckles and other metal objects reflect light. Camouflage any object that reflects light.

2. Regularity of outline. The human body, rifles, helmets, hats, and vehicles have familiar outlines and are easily identifiable. Camouflage the shape of these familiar objects.

3. Contrast with the background. The color of the uniform, equipment, and skin must blend as naturally as possible with the surroundings. Do not ruin an excellent camouflage appearance by wearing a bandoleer of ammunition over the uniform, as this creates both shine and color differences from any surroundings.

3.3.4.2 Uniforms. Wear the standard jungle camouflage uniform. Do not wear threadbare uniforms, insignia, bandoleers of linked ammunition, or anything that might shine.

3.3.4.3 Types of Camouflage

1. Camouflage "Paints." These come in a variety of containers and are effective for covering exposed skin.

   a. Team up to ensure a complete and effective job.

   b. Use the signal mirror.

   c. Use forest green, OD, or black for jungle.

   d. Alternate colors in wide bands or blotches.

   e. Don't forget the back of the ears and back of the neck.

   f. Check each other after going through water.

2. Field expedient camouflage. When prepared camouflage paint is not available, use whatever is available.

   a. Mud. Provides excellent camouflage for shiny objects such as the belt buckle, but it should not be used on the skin due to its high bacterial content and tendency to run off when sweating or during precipitation.

   b. Coloring cloth. Mud, charcoal, burnt cork, charred coffee grounds or charcoal may be used to vary the color of pieces of cloth. Do NOT use fuel, oil, or grease due to their odor.

   c. Attaching cloth. Stitch the pieces of cloth loosely overlapping, with an irregular pattern of texture, line and color. (Similar to a handmade Ghillie Suit used by snipers).

   d. Foliage. In most cases, natural foliage is preferred to artificial camouflage, but it may be difficult to secure to the body and equipment. Rubber bands cut from discarded tire tubes can be used effectively for this purpose.

Apply all camouflage in contrasting patterns. Test personnel camouflage by having them sit in the jungle, day and night, and observing them.

3.3.4.4 Preparing a Camouflaged Position. Do not leave telltale signs of presence. An area having no vegetation other than a row of evenly spaced vegetation would be a possible giveaway to the position.

1. Exercise the same care as in individual camouflage.

2. Match camouflage to the surrounding vegetation.

   a. Select a background that will virtually absorb the operator and his equipment.

   b. Remember that concealment provided by vegetation does not protect the SEAL from enemy fire.

3. Inspect the position from various observable angles.

4. Be aware of observation from the jungle canopy.

5. Enforce strict light and noise discipline.

6. Position at least one lookout while the others work.

7. Prepare an alternate position, accessible from the primary position by a covered route.

3.3.4.5 Camouflage During Movement

1. Be camouflage conscious from departure time until return.

2. Observe the terrain and vegetation, and change camouflage as necessary.

3. Ensure camouflage hasn't washed, sweated, or been wiped off.

3.3.5 Food Rations. Food rations in the jungle should be MREs. Ration loads must be as light as possible. To decrease the total weight, open the packages, remove items not desired, and tape them closed again.

3.3.6 Patrol Water Rations

3.3.6.1 Relation of Water Carried to Tactical Mobility. The baseline environment is temperate, and the baseline mobility figures are based on that, but the amount of water to be carried is always an important planning consideration.

3.3.6.2 Jungle Zone Water Planning. The amount of water to be carried per man for a baseline jungle patrol (72 hours) is 4 quarts (1.3 quarts per day per man). This can be reduced slightly if the mission is in an area where usable water can be found for periodic replenishment. In a jungle zone where water is scarce or where water supplies may be contaminated, however, the tactical commander must make a careful judgment regarding the amount of water to be carried (up to 10 quarts). Insufficient water is the first thing, short of enemy action, that can cause mission failure. Use the following guidelines for water planning:

1. Water weighs about eight pounds per gallon, or two pounds per quart.

2. Reduced water consumption reduces physical endurance.

3. Reduced physical endurance reduces tactical mobility.

3.3.7 Sleeping Equipment

3.3.7.1 Equipment. A lightweight net hammock is preferred for sleeping to keep personnel off the ground.

3.3.7.2 Protection against Insects

1. Mosquito bed nets (mosquito bars). Ensure all personnel have and use mosquito net, as it is one of the most valuable methods of protection.

    a. Make nets ready before dusk.

    b. Spray 2 percent d-Phenothrin aerosol inside bed net to kill any mosquitoes already inside.

    c. Do not let the net sag onto the body (a mosquito can bite through it).

    d. Do not touch the net during the night.

2. Head nets. In the absence of an insect repellent jacket, a head net is an alternative. Consider wearing a head net as a last-resort method to keep mosquitoes or other insects from affecting operations.

   a. Protects against mosquito bites

   b. Restricts vision of the wearer

   c. Makes sleeping very difficult.

3.3.8 Surveillance, Target Acquisition, and Night Observation Equipment. Carefully prepare critical and sensitive equipment for jungle operations. Review technical manuals to determine the best means of protecting equipment such as that listed below.

3.3.8.1 Surveillance Equipment

1. NVEO

   a. Use a Claymore bag around the neck and on the chest to carry an NVEO

   b. Carry a spare battery

   c. Waterproof NVEO seals with silicon grease

2. Imaging sensors (IR)

3. Listening devices

4. Binoculars.

3.3.8.2 Navigation Aids/Beacons

1. Radio beacons

2. IR beacons

3. Portable GPS.

3.3.8.3 Target Acquisition Equipment

1. Laser target designators

2. Smokes and flares

3. Tracer rounds.

3.3.9 Weapons and Ordnance Preparation

3.3.9.1 Weapons. The best weapons for the jungle are the M60 machinegun and M4 with M203 grenade launcher.

1. General Considerations

   a. Assigned personnel (e.g., PL, APL, and RTO at a minimum) load one full magazine with tracer rounds to identify the enemy position if taken under fire.

   b. All patrol members load the last three rounds in a magazine as tracer to alert when it is close to empty. Platoons may devise alternate methods as desired.

   c. Tape or remove all sling swivels.

   d. Check all magazines for load and function before going on operations.

2. M203 Grenade Launcher. Carry the following amounts and types of 40mm rounds:

   a. PT/RS – six High Explosive Dual Purpose (HEDP) and two anti-personnel buckshot (one in chamber)

   b. PL – four HEDP, two illumination, and two red/green star clusters.

   c. R/G - eight HEDP and two illumination.

3.3.9.2 Ordnance and Demolitions

1. Claymore mines

   a. Remove all firing wire from the plastic spool and discard the spool.

   b. Re-roll the wire in a "figure 8" fashion.

   c. Replace in its bag so the mine can be placed first and the wire can then be laid back to the firing position.

   d. Pre-connect the clacker (with circuit tester attached) to the firing wire and then stow it in the pouch.

   e. Prime at least two Claymores with 10 to 30 second fuses. Carry others with shipping plugs on blasting caps to speed later priming.

   f. Dual-prime each Claymore for both electric and non-electric firing. Pre-cut time fuse to desired times. Burn time becomes undependable the longer time fuse is exposed to moisture and humidity.

   g. Waterproof all non-electric firing systems with a gasket sealing compound like RTV, a condom, or other locally established and approved procedures.

2. Grenades

   a. Fold rigger's tape through the rings and tape the ring to the body of the grenade.

   b. Plan to have patrol members carry a mixture of grenades: smoke, CS, WP, concussion, fragmentation. In mud areas, including bunkers, fragmentation grenades are not as effective as concussion grenades.

   c. Carry violet and red smoke, as they are most visible to aircraft. WP is good for dense jungle or wet weather.

d. Camouflage grenades with black or OD spray paint.

e. Carry smoke grenades on your LBE.

f. Carry one thermite grenade per squad to destroy sensitive or classified material or enemy equipment.

g. DO NOT straighten, bend, or modify pins on grenades.

3.3.9.3 Weapons Maintenance Procedures

1. Clean weapons at least twice daily and also after crossing through mud or silty water.

2. Keep all weapons well lubricated and carry cleaning equipment on operations.

   a. For lubricant, use one or two drops of LSA (MIL-L-46000B) spread lightly over a weapon with a cloth.

   b. Use maximum lubrication on weapons to prevent rust or silt intrusion. Lubricate weapons at least once daily or after each cleaning, especially trigger assembly safety housing and slide.

3. Test firing. Test fire weapons under the same conditions anticipated on the mission. For instance if the weapon will be submerged on water crossings, test fire after submersion.

4. Keep plastic barrel caps over the muzzles when they are not in use (improvise with tape).

5. Protect ammunition from direct sunlight. If it is cool enough to hold in a bare hand, it is safe.

6. Waterproof magazines by placing them inside one or two non-lubricated condoms.

3.3.10 Communications Equipment Preparation. Communications is the entire platoon's responsibility. All personnel must know the basics in case of emergency.

3.3.10.1 General Planning

1. Radio selection. Select man-packed radios based on mission, communication distance to base camp, terrain, and type of jungle. Appendix B lists types, frequency ranges, and compatibility.

   a. The normal ratio of received to transmitted communication time is 10:2.

   b. The AN/PRC 117, a HF radio, is the preferred radio for jungle operations as it provides longer transmission range than other man-packed radios.

   c. SATCOM, which uses UHF, has a tendency to scatter in vegetation.

2. CEOI. The PL should carry a copy of the CEOI. Ensure that the RTO also carries a copy in a known location so that others may locate it in an emergency (left breast pocket, or in accordance with command SOPs). Waterproof the CEOI in a plastic bag, or laminate it.

3.3.10.2 Communications Equipment Inventory and Inspection. Inventory and inspect all radios, crypto equipment, antennas, and batteries for the following:

1. Loose knobs

2 Damaged case

3. Bent antenna

4. Cracked battery box

5. Misaligned antenna connection.

3.3.10.3 Pre-set and Verify Communications Frequencies

1. Setting. Set the frequencies in accordance with the CEOI. The best frequency ranges for operating in the jungle are in the HF range. See Appendix B.

2. Verify frequencies. Verify that the frequencies being set are the same ones being set at the key C^2 modes for the patrol. Communications are too important to rely solely on the distribution of the CEOI. Ensure that there is no error or misunderstanding.

3.3.10.4 Batteries. Lithium batteries last longer than nickel cadmium batteries and should always be used on jungle patrols.

1. Always install fresh batteries prior to departing on a jungle patrol.

2. In addition to the installed batteries, carry six spare batteries for a baseline jungle patrol (72 hours).

   a. Two batteries for the radio, two batteries for the crypto, and two batteries for a contingency.

   b. Increase the number of spare batteries by at least two if the mission involves reconnaissance communications over a longer period.

   c. Increase the spare batteries in proportion to patrol length (if more than a baseline 72-hour patrol).

3. Do not remove the batteries from the plastic wrapping until ready to use.

3.3.10.5 Handsets. Carry two spare handsets in a waterproof bag. Use padding, so they won't get crushed if the rucksack is dropped.

3.3.10.6 Pre-mission Radio Checks. Conduct these radio checks with all radio and crypto equipment in the rucksack, exactly as it would be in the field.

1. Timing

   a. Conduct checks prior to loading or connecting crypto equipment.

   b. Conduct checks again after crypto is loaded and connected.

2. Stations for Radio Checks. Conduct checks with:

   a. NSWTG/TU or immediate superior in the chain of command.

b. Other commanders or warfare commanders that may require information from the patrol (e.g., strike rescue commander, riverine patrol commander, battle group warfare commander, etc.).

c. The NSW base camp.

d. Support aircraft (insertion/extraction and CAS).

e. Support boats (NSW Boat Detachments).

f. Support ships (base ships, NGFS ships).

g. Other supporting forces.

3. Equipment handover. Conduct a face-to-face handover of compatible equipment from prior patrols to ensure communications are not the limiting factor in the patrol.

4. Communications coordination visits. Conduct visits to supporting command communications centers to ensure the use of compatible equipment. This is always a correct procedure, but is especially required (mission time permitting) when operating with commands that may not be familiar with NSW communications (examples: joint forces, naval battle groups).

3.3.10.7 Maintenance

1. Waterproof radios with silicon grease. Carry extra silicon grease in a plastic bag for field use.

2. Pack radios in waterproof containers when not in use.

3. Remove batteries.

4. Protect handsets in waterproof bags.

5. Check bags for condensation and dry them as required.

6. Insure the battery box pressure relief valve/vent on radios in plastic bags is not obstructed.

7. Clean radios after each operation.

3.3.11 Baseline Patrol Equipment

3.3.11.1 Insertion/Extraction Equipment. Insertion and extraction may require equipment for:

1. Rappelling

2. Fastroping

3. McGuire rig

4. CRRC

5. Indigenous craft.

3.3.11.2 Miscellaneous Equipment. Ensure jungle pack-out includes:

1. Lines
2. Snaplinks
3. Flotation equipment, life jackets, and CO^2 cartridges (one spare per man).

3.3.11.3 Carrying Equipment. Avoid metal on metal contact. Tape off or pad any problem.

3.3.11.4 Carrying Communications Items. Keep maps and communication and electronic equipment dry.

3.3.11.5 Obstacle Crossing Equipment

1. Mountains, Gorges, and Cliffs. Anticipate terrain features requiring use of basic mountaineering skills. Carry a snaplink and 20-foot length of tubular nylon and 20 feet of 550 cord per platoon member.
2. Rivers and Streams. Flotation equipment is required.

3.3.11.6 Baseline Weight Assignment for Patrol Positions. Appendix D provides the weights of weapons and accessories. The individual equipment and consumable weights assigned to each man for a baseline 16-man patrol (by patrol position) are listed in Figure 3-6. These are a guide for the tactical planner. The PL should recognize the relation between weights, amount of water carried (adds weight), and patrol mobility.

3.3.12 Tactical Mobility

3.3.12.1 Baseline Mobility

1. Baseline Mobility in Temperate Zones. A patrol in reasonably flat terrain in a temperate environment (defined as daily temperature range from 45 to 95 degrees Fahrenheit) should plan to cover 15 kilometers/day. This figure includes the following:

   a. Three 15-minute rest stops (one every two hours), that also serve as a "SLLS" period for alerts regarding enemy activity.

   b. One 30-minute stop for a communication window.

   c. One 30-minute delay for obstacle crossing.

3.3.12.2 Baseline Mobility in The Jungle Environment. In the jungle environment the baseline rate of advance is reduced to eight kilometers per day (assuming patrol movement for eight hours in a 24 hour period). This figure includes the following:

1. Three 15-minute rest stops
2. One 45-minute stop for a communication window
3. One 45-minute delay for obstacle crossing.

Baseline mobility in the jungle environment is eight kilometers per eight-hour daylight period, moving one eight hour period in each 24-hour period. This is a planning and training guideline only. PLs should adjust patrol speed as required for the mission.

1. Variance of Rate of Advance. The actual rate of advance is situation dependent. Unpredicted conditions can either hasten or delay the patrol. For example, usually there are approximately 12 hours of daylight, permitting longer travel in a day.

2. Rehearsal. Platoon planners should rehearse in conditions similar to patrol conditions to determine mobility.

3.3.12.3 Daily Patrol Movement. The patrol should plan to move for one eight hour period per day under normal conditions in the jungle environment. An eight-hour movement should be followed by a rest period of at least six hours. Movement for one eight hour period and one four hour period in a 24-hour period is the maximum (it can be realistically sustained for 48 hours). The PL should always be aware of conserving the energy of his men for actions at the objective, and for E&E should that become necessary. A key to tactical success is the ability of the patrol to move efficiently, but to conserve maximum energy for the most demanding phases of the mission, those in which the enemy is most likely to be encountered ("get there first with the most").

3.3.12.4 Variance of Jungle Mobility with Heavier Individual Loads. Total standard individual personnel loads in the jungle are not changed from those for temperate climates, but the weight must be distributed in different ways (typically more water). If the individual loads noted in Figure 3-6 are varied, planned patrol mobility should be reduced. Use the following rule of thumb:

1. The mobility of the patrol is governed by the mobility of the man with the heaviest load (normally the machine gunner).

2. Maximum individual load varies with mission (all three lines of equipment used).

3. Standard load for the man carrying the heaviest load (machine gunner) is 75 pounds. This means that if this is the heaviest load in the patrol, the patrol should be able to carry out the standard jungle mobility of eight kilometers in an eight-hour period.

4. For an initial maximum load of no more than 75 pounds, estimated travel distance can be increased 0.5 kilometer per eight-hour period for each two and a half pounds of load reduction for the most heavily loaded man (down to 50 pounds, which would be 13 kilometers per eight hours). If a special case occurs where the most heavily laden man carries less than 50 pounds, the mobility will be slightly increased.

5. Subtract one kilometer per eight-hour period from the baseline mobility (eight kilometers per eight hour period) for each five pounds of weight (above 75 pounds.) added to the most heavily loaded man up to 90 pounds (which would be five kilometers per eight hours.). If a special case occurs where a man is required to carry more than 90 pounds, the platoon mobility and operational effectiveness at the objective may be affected. Appendix G provides a standard load-planning sheet for use by the PL.

3.3.12.5 Optimum Time for Patrolling. SEALs can patrol in either day or night conditions in the jungle.

1. Day Conditions. Daylight patrols are faster and may be conducted, but they are not preferred.

2. Night Conditions. Nighttime patrols are significantly slower due to limited visibility but are preferred. In certain types of jungles where there is less vegetation, less canopy, and periodic cleared areas, night patrols are less impeded.

3.3.12.6 Flexibility of Maneuver. In the dense vegetation of the tropical rain forest, patrol formation options are limited. Formations that permit each member to see at least one other member should be used. If intra-patrol communications beyond visual range are in place, more flexibility in patrol type and spacing may be employed, but patrol members must always be able to reinforce each other, which is difficult when heavy vegetation limits mobility and visibility.

3.3.13 Patrol Briefings/Reports

3.3.13.1 Schedule of Recommended Pre-patrol Briefings. The following briefings should be prepared and conducted prior to the patrol:

1. Mission objective and plan
2. Intelligence
3. Route
4. Communications
5. Formations and tactics
6. Insertion/extraction
7. Special assignments.

3.3.13.2 Reports. Plan to keep detailed notes on the patrol to assist future patrols in this area and elsewhere.

3.3.14 Jungle Patrol Inspections and Rehearsals

3.3.14.1 Rehearsal. Rehearsals under jungle conditions are mandatory to test men and equipment.

3.3.14.2 Standard Operating Procedures. Rehearsal allows the platoon to refine SOPs, providing the ultimate confidence builder. Jungle skills must be honed to perfection.

3.3.14.3 Inspection and Rehearsal Schedule. The NSWTG should schedule the following:

1. Appropriate jungle rehearsal site
2. Enough rehearsal time
3. Insertion and extraction vehicles: boats, helicopters, or vehicles
4. Terrain and climate similar to expected operation area.

3.3.14.4 Mandatory Rehearsal Checklist Items

1. Preset compasses and watches
2. Conduct communication checks

3. Check extra batteries

FIRE TEAM ONE	
1. PT/Rifle Man	
Water	8 lb
Weapons	10 lb
Special equipment	2 lb
Ruck and personal items	20 lb
Ammunition and ordnance	10 lb
TOTAL BASELINE WEIGHT	50 lb
2. PL/Squad 1 Leader/Element 1 Leader	
Water	8 lb
Weapons	10 lb
Special equipment	10 lb
Rucksack and personal items	20 lb
Ammunition and ordnance	12 lb
TOTAL BASELINE WEIGHT	60 lb
3. Rifle Man/RTO	
Water	8 lb
Weapons	10 lb
Special equipment	19 lb
Rucksack and personal items	20 lb
Ammunition and ordnance	8 lb
TOTAL BASELINE WEIGHT	65 lb
4. AW Man	
Water	8 lb
Weapons	19 lb
Special equipment	0 lb
Rucksack and personal items	20 lb
Ammunition and ordnance	28 lb
TOTAL BASELINE WEIGHT	75 lb

Figure 3-6. Patrol Organization/Weight Assignments

FIRE TEAM TWO	
5. Rifle Man/GN/FT B Leader	
Water	8 lb
Weapons	10 lb
Special equipment	10 lb
Rucksack and personal items	20 lb
Ammunition and ordnance	12 lb
TOTAL BASELINE WEIGHT	60 lb
6. Rifle Man/GN/HM	
Water	8 lb
Weapons	10 lb
Special equipment	10 lb
Rucksack and personal items	20 lb
Ammunition and ordnance	12 lb
TOTAL BASELINE WEIGHT	60 lb
7. AW Man	
Water	8 lb
Weapons	19 lb
Special equipment	0 lb
Rucksack and personal items	20 lb
Ammunition and ordnance	28 lb
TOTAL BASELINE WEIGHT	75 lb
8. Rifle Man/GN	
Water	8 lb
Weapons	10 lb
Special equipment	10 lb
Rucksack and personal items	20 lb
Ammunition and ordnance	12 lb
TOTAL BASELINE WEIGHT	60 lb

Figure 3-6. Patrol Organization/Weight Assignments

FIRE TEAM THREE	
9. Rifle Man/GN/Element 3 Leader	
Water	8 lb
Weapons	10 lb
Special equipment	10 lb
Rucksack and personal items	20 lb
Ammunition and ordnance	12 lb
TOTAL BASELINE WEIGHT	60 lb
10. AW Man	
Water	8 lb
Weapons	20 lb
Special equipment	6 lb
Rucksack and personal items	20 lb
Ammunition and ordnance	21 lb
TOTAL BASELINE WEIGHT	75 lb
11. Rifle Man/GN	
Water	8 lb
Weapons	10 lb
Special equipment	10 lb
Rucksack and personal items	20 lb
Ammunition and ordnance	12 lb
TOTAL BASELINE WEIGHT	60 lb
12. Rifle Man/GN	
Water	8 lb
Weapons	10 lb
Special equipment	10 lb
Rucksack and personal items	20 lb
Ammunition and ordnance	12 lb
TOTAL BASELINE WEIGHT	60 lb

Figure 3-6. Patrol Organization/Weight Assignments

FIRE TEAM FOUR	
13. Rifle Man/APL/Squad 2 Leader/FT 4 Leader (APL)	
Water	8 lb
Weapons	10 lb
Special equipment	10 lb
Rucksack and personal items	20 lb
Ammunition and ordnance	12 lb
TOTAL BASELINE WEIGHT	60 lb
14. Rifle Man/Squad RTO	
Water	8 lb
Weapons	10 lb
Special equipment	21 lb
Rucksack and personal items	20 lb
Ammunition and ordnance	4 lb
TOTAL BASELINE WEIGHT	65 lb
15. AW Man	
Water	10 lb
Weapons	20 lb
Special equipment	4 lb
Rucksack and personal items	20 lb
Ammunition and ordnance	21 lb
TOTAL BASELINE WEIGHT	75 lb
16. RS/Rifle Man/GN	
Water	8 lb
Weapons	10 lb
Special equipment	10 lb
Rucksack and personal items	20 lb
Ammunition and ordnance	12 lb
TOTAL BASELINE WEIGHT	60 lb

Figure 3-6. Patrol Organization/Weight Assignments

4. Check navigation equipment and maps
5. Check surveillance, target acquisition, and night observation equipment
6. Check and test fire all weapons
7. Check flotation of each person and all equipment
8. Check rucksacks and LBE for proper weight and pack out
9. Check equipment noise when moving
10. Check any mission specific equipment
11. Practice SOPs emphasizing danger areas and enemy contact
12. Practice formations and tactics
13. Practice setting demolitions and mines
14. Practice setting up ambush sites
15. Practice entering and leaving boats, helicopters, or vehicles (for insertion and/or extraction, emergencies)
16. Practice rappelling and belaying
17. Practice river/stream crossing
18. Actions at the objective (in detail).

3.4. MEDICAL AND SURVIVAL TECHNIQUES

3.4.1 General. This section provides information on techniques and procedures for jungle survival, both with and separated from the patrol. Personal hygiene, first aid, and basic jungle field skills will be presented for PLs, members, and commands involved in training SEALs for jungle operations.

3.4.2 Jungle Patrol Hygiene Procedures

3.4.2.1 General Health. The hot, humid jungle climate increases the chances of contracting tropical diseases. Prevent disease with good sanitation practices and preventive medicine. Without preparation and attention to health procedures, the most physically fit person can succumb to tropical diseases and become non-operational. The preventive and treatment procedures that follow take time and platoon mobility requirements must allow for that. Failure to do so will result in decreased endurance and operational capability.

3.4.2.2 Preventive Medical Actions Prior to Deployment. Update immunizations and commence malaria prophylaxis as early as possible to observe any adverse reactions. Malaria prophylaxis is usually done two weeks prior to deployment. For platoons on alert for deployment into a possible jungle environment, this may require sustained use of malaria prophylaxis.

3.4.2.3 Medical Training on Tropical Diseases. Conduct jungle medical training on tropical diseases. Insects, dirty food, and contaminated water are the primary causes of jungle diseases.

3.4.2.4 Preventive Medicine Actions upon Arrival in Operations Zone

1. Acclimation. Allow a minimum of two weeks to acclimate to the environment.

2. Hydration. Ensure personnel drink more water than would be consumed in a temperate climate to replace the fluids lost through increased sweating. Hydrate by drinking water until urine is clear, with no yellow color. Then drink at least another quart.

3.4.2.5 Preventive Medicine Action after Deployment. Ensure personnel continue taking malaria prophylaxis; generally for eight weeks. Monitor for jungle disease. Monitor personnel for diseases possibly contracted in the jungle and keep immunizations current.

3.4.3 Diseases; Preventive Measures and Field Treatment

3.4.3.1 Jungle Waterborne Diseases. Numerous diseases and health problems come from water and heat related causes.

1. Types of diseases. Consider all water from natural sources to be contaminated. Diseases resulting from impure water include:

   a. Dysentery

   b. Typhoid

   c. Cholera

   d. Hepatitis

   e. Giardiasis

   f. Traveler's diarrhea

   g. Blood fluke infection (caused by exposing an open sore to impure water).

2. Prevention of waterborne diseases. To prevent waterborne diseases:

   a. Obtain drinking water only from approved water points.

   b. Use rainwater collected after it has been raining at least 15 to 30 minutes. This lessens the chance of impurities being washed from the jungle canopy into the water container. Even then, the water should be purified.

   c. Purify all drinking water using one of the procedures in paragraph 3.4.11.2.

   d. Do not bathe in untreated water.

   e. Keep the body fully clothed when crossing water obstacles.

3.4.3.2 Jungle Fungal and Bacterial Skin Diseases

1. Fungal and bacterial diseases. Fungi and bacteria are tiny plants that multiply fast in the hot, moist conditions of the jungle.

   a. Fungal. Sweat-soaked skin invites fungus attack (e.g., jungle jock itch, athlete's foot, etc.).

   b. Bacterial. The skin is covered with bacteria ready to invade as a secondary infection (e.g., streptococcal and staphylococcal infections).

2. Warm water immersion skin diseases. The following are common skin diseases caused by long periods of wet skin:

   a. Warm water immersion foot. This disease occurs when the feet are constantly wet from water crossings. The bottoms of the feet become white, wrinkled, and tender. Walking becomes painful.

   b. Chafing. This disease occurs when the trousers stay wet for hours, causing the crotch area to become red and painful to even the lightest touch.

3. Fungal and bacterial skin disease treatment

   a. Let skin dry

   b. Fungal: use anti-fungal cream/ointment (e.g., Loprox)

   c. Bacterial: use anti-bacterial ointment (e.g., Bactroban).

4. Prevention of fungal and bacteria skin diseases.

   a. Bathe often, and air or sun-dry the body as often as possible.

      (1) Do not sunburn; use unscented, SPF 15 sun block.

      (2) Wear a hat.

   b. Wear clean, dry, loose-fitting clothing whenever possible.

   c. Do not sleep in wet, dirty clothing.

   d. Carry one dry set of clothes just for sleeping.

      (1) Put on dirty clothing in the morning, even if wet.

      (2) This practice not only fights fungal, bacterial, and warm water immersion diseases, but also prevents chills and allows for better rest.

   e. Take off boots and massage the feet as often as possible. Dust feet, socks, and boots with foot-powder (e.g., Desenex, Tinactin, etc.) at every chance.

   f. Carry several pairs of socks and change them frequently.

3.4.4 Jungle Insects; Preventive Measures and Field Treatment. Another real danger of the jungle is insects, many of which pass on diseases or parasites.

1. Mosquitoes. Malaria/yellow fever-carrying mosquitoes are probably the most harmful of the tropical insects. Mosquitoes are most prevalent early at night and just before dawn. Malaria is more common in populated areas than in uninhabited jungle, so be especially cautious when operating around villages. The following precautions and procedures should be followed regarding mosquitoes:

   a. Apply mudpacks to mosquito bites. This offers some relief from itching.

   b. Use the prescribed prophylaxis (usually Dapsone, mefloquine HCL, or chloroquine-primaquine).

   c. Use insect repellent (but only in the manner prescribed to preclude enemy detection of the smell). 31.5 percent diethyltoluamide or DEET is considered best. Be careful, as this will damage watch crystals and plastic eyeglass lenses.

   d. Wear clothing that covers as much of the body as possible.

   e. Use nets or screens at every opportunity.

   f. Avoid the worst infested areas when possible.

2. Ticks. Ticks may be numerous, especially in grassy places. Carry out the following procedures after patrolling through grassy areas:

   a. Strip to the skin once a day or more frequently.

   b. Inspect all parts of the body for ticks, as well as leeches, bed bugs and other pests.

   c. Work as a team or group; examine each other.

   d. Brush ticks off clothing; flick them off the skin.

      (1) If ticks become attached to the body, cover them with a drop of iodine. They will let go. Heating them will also make them let go, but don't burn the skin.

      (2) Use care when removing a tick; the head may stay in and start an infection.

   e. Touch up the bite with antiseptic.

3. Fleas. Fleas are common in dry, dusty buildings. The females will burrow under toenails or into skin to lay their eggs.

   a. Remove them with a sterilized knife. Sterilize knife blade with heat from lighter or match; wipe off carbon deposits with alcohol or sterile pad.

   b. Keep the cut clean. In India and China, bubonic plague is a constant threat. Rat fleas carry this disease and discovery of dead rats usually means a plague epidemic in the rat population, which may be a prelude to an outbreak among human beings.

c. Fleas may also transmit typhus fever.

d. In many parts of the tropics, especially Malaysia and Indonesia, rats carry fleas and other parasites that in turn carry jaundice and other fevers. Keep your food in rat-proof containers or in rodent-proof caches.

4. Mites. In many tropical and temperate parts of the Far East, tiny red mites carry a type of typhus fever. These mites resemble the chiggers of the southern and southwestern United States. They live in the soil where they burrow a few inches into the ground, in tall grass, in cutover jungle, and in stream banks. Bites occur when one lies or sits on the ground. The mites emerge from the soil, crawl through your clothes and bite. Usually, the victim is unsuspecting for the bite is painless and does not itch. Mite typhus is a serious disease. To prevent being bitten, do the following:

a. Clear the ground of debris where planning to camp.

b. Burn ground off, if possible.

c. Sleep above the ground.

d. Treat clothing with insect repellent.

5. Wasps and Bees. Wasps and bees will rarely attack unless their nests are disturbed. When a nest is disturbed, leave the area and reassemble at the last RP. In case of stings, mudpacks are helpful.

a. Sweat Bees. In some areas there are tiny bees called sweat bees that may collect on exposed parts of the body during dry weather, especially if the body is sweating freely. Their presence will cause an annoying, tingling sensation. They will leave when sweating has completely stopped or they may be scraped off with the hand.

b. African (killer) bees. This is a more ferocious type of bee and is best left alone. These bees swarm and attack, inflicting stings that can be fatal. Clear the area immediately, backtrack 300 meters, and reroute movement appropriately.

6. Centipedes and Scorpions. Larger ones can inflict painful, but not fatal stings. They are normally found in dark places and under rocks or logs.

a. Shake out blankets before sleeping.

b. Check clothing and boots before putting on.

7. Spiders. Spiders are commonly found in the jungle. Their bites may be painful, but are rarely serious. Spiders rarely invade human territory. The danger from spiders normally occurs when a patrol member moves into a place where the spiders are. If you run into a spider web by accident, knock it down and back away immediately. The spider will normally avoid any large animal, or human, that encounters its web.

8. Ants. Can be dangerous to injured men lying on the ground and unable to move, or to men asleep on the ground. Avoid sleeping on the ground in the jungle. If a man is injured, rig a hammock or a bamboo platform to keep him off the ground.

9. Riceborer Moth. This small, plain-colored moth with a pair of tiny black spots on the wings is found in Southeast Asian jungle lowlands. They collect around lights in great numbers during certain seasons. The

moth should never be brushed off roughly, as the small barbed hairs of its body may be ground into the skin. The barbed hair causes a sore, much like a burn, often taking weeks to heal.

3.4.5 Jungle Parasites; Preventive Measures and Field Treatment

1. Leeches. Leeches are common in many swampy, jungle areas, particularly throughout most of the Southwest Pacific, Southeast Asia, and the Malay Peninsula. There are both land and water leeches.

   a. Problem

      (1) Although leeches are not poisonous, their bites may become infected if not cared for properly.

      (2) The small wound from a leech provides a point of entry for the germs causing tropical ulcers or "jungle sores".

   b. Prevention.

      (1) Watch for leeches on the body.

      (2) Brush them off before they have had time to bite.

      (3) Do not pull them off forcibly because part of the leech may remain in the skin.

      (4) Leeches will release themselves if touched with insect repellent, a moist piece of tobacco, the burning end of a cigarette, coal from a fire or a few drops of alcohol.

      (5) Tuck trousers securely into the boots. An alternative method is to wrap Claymore mine carrying straps, impregnated with insect repellent, around trousers and boot top. This will prevent leeches from crawling up the legs and into the crotch area.

3.4.6 Jungle Reptiles; Preventive Measures and Field Treatment

1. Snakes. Snakes are prevalent in all jungle areas, but are seldom seen when patrolling.

   a. Snakebite. If a SEAL should accidentally step on or otherwise disturb a snake, it will probably attempt to bite. The chances of this happening while traveling along trails or waterways are remote if one is alert and careful.

   b. Precautions. Use ordinary precautions. Be particularly watchful when clearing ground.

   c. Treatment. Treat all snakebites as poisonous. In the event of snakebite, follow these steps:

      (1) Remain calm, but act swiftly, and chances of survival are good. Less than one percent of properly treated snakebites are fatal. Without treatment, the fatality rate is 10 to 15 percent.

      (2) Immobilize the affected part in a position below the level of the heart.

      (3) Place a lightly constricting band 2 to 4 inches closer to the heart than the site of the bite.

      (4) Reapply the constricting band ahead of the swelling, if it moves up the arm or leg.

(5) Place the constricting band tightly enough to halt the flow of blood in surface vessels, but not so tight as to stop the pulse.

(6) Do not cut open the bite or suck out venom.

(7) Inform the HM.

(8) If possible, get part of the snake's head with 2 to 4 inches of its body attached to aid with identification. Identification assists in obtaining the proper antivenin.

(9) MEDEVAC, if required.

2. Crocodiles and Caymans. These are meat-eating reptiles that live in tropical areas.

a. Crocodiles. The marine crocodile of the Indo-Australian region is very dangerous. They are abundant along the seashores, in coral reef areas as well as in salt-water estuaries and in bays. The best course of action is to exercise great caution whenever a person is in or near water where crocodiles or alligators are evident. Do not attract them by thrashing in the water. Avoid them at all times.

b. Caymans. Caymans look like smaller versions of crocodiles. They are found in South and Central America. They are not likely to attack unless provoked.

3.4.7 Dangerous Jungle Animals; Preventive Measures and Field Treatment

1. Wild Animals. All large animals can be dangerous if cornered or suddenly startled at close quarters. This is especially true of females with young.

a. Locations/Characteristics

(1) Africa. Lions, leopards, and other flesh-eating animals abound in areas where they are protected from hunters by local laws. In areas where they are not protected, they are shy and seldom seen. When encountered, they will attempt to escape.

(2) Sumatra, Bali, Borneo, Southeast Asia, and Burma. There are tigers, leopards, elephants, and buffalo.

(3) Latin America. These jungles have the jaguar. Ordinarily, these animals will not attack a man unless they are cornered or wounded.

(4) Certain semi-domesticated jungle animals, such as the water buffalo and elephant, may appear tame, but this tameness extends only to people the animals are familiar with. Avoid them.

b. Dangers. The primary dangers from wild animals are:

(1) Bites

(2) Claw lacerations

(3) Crushing or breaking injuries related to the size and weight of the animal.

c. Field treatment. Field treatment for animal related injuries would be the same as for other types of puncture, slash, and crush or break wounds. The wound must be quickly cleaned to prevent infection.

2. Bats. Bats are prevalent and, in some cases, can cause rabies. The "vampire" bat, found in Panama, can be a danger. Since their bites require a series of rabies shots, preventive measures against such bites should be taken.

   a. Characteristics. In identifying a potential target, bats are looking for heat from exposed skin and lack of movement (victim is asleep).

      (1) The bats' approach is characterized by an extremely high pitched (inaudible to some people) series of squeaks as the bats fly above their targets.

      (2) After identifying a target they land, sometimes noisily, in low branches of nearby trees and bushes.

      (3) They then drop to the ground and crawl toward the victim.

   b. Prevention

      (1) Normally, hammocks rigged off the ground provide a slight continuous movement that bats are afraid of.

      (2) Wearing gloves, mosquito net hood, and sleeping in a full uniform, will provide a minimal heat signature to preclude attracting bats.

      (3) A roving watch moving through camp will usually cause bats to be flushed.

      (4) If a roving watch is not possible, ensure all personnel carry hammocks, hood netting, gloves, and lightweight Gortex for use during night LUPs. Sleep in pairs; one awake, one asleep.

3.4.8 Dangerous Jungle Plants; Preventive Measures and Field Treatment

1. Types of Dangerous Jungle Plants

   a. Nettles. Nettles, particularly tree nettles, are a dangerous variety of vegetation. Contact with them produces severe stinging.

   b. Ringas trees. Ringas trees are located in Malaysia. They affect some people in much the same way as poison oak (a red, itchy rash that can spread over the body if not controlled).

   c. Poison Ivy and Poison Sumac cause the same type of rash as poison oak. The danger from these poisonous plants in the jungle is similar to that in the woods of the United States eastern seaboard.

   d. Black Palm. Black palm is found in Panama. It has very sharp thorns.

   e. Rattan Palm. This palm, found in most jungle areas, also has sharp thorns.

2. Prevention. Some of the dangers associated with poisonous vegetation can be avoided by keeping sleeves down and wearing gloves.

3. Treatment.

   a. For plants causing rashes:

      (1) Do not scratch, as this will spread the rash

      (2) Apply cortisone cream liberally at least twice a day

      (3) Do not bandage unless the rash becomes raw or infected

      (3) Keep the affected area clean.

   b. For thorns:

      (1) Remove thorn

      (2) Clean wound

      (3) Use antiseptic and wrap

      (4) Change the bandage at least daily and keep the wound clean to avoid infection.

3.4.9 Patrol First Aid Procedures. The preventive measures and field treatment procedures described in the preceding paragraph are primarily the responsibility of individual patrol members. First aid consists of treatment by both the individual and the platoon HM. This section describes key first aid procedures for the jungle environment.

3.4.9.1 Infection. Infection is always possible in the jungle at the site of any broken skin. Promptly disinfect even the smallest scratch, as it can become infected much more rapidly in the jungle warmth. Failure to treat small infections can lead to blood poisoning and gangrene.

3.4.9.2 Blood Poisoning. Blood poisoning is infection of the blood stream. It can spread rapidly. It begins with an untreated infection in the skin. Symptoms are red streaks in the infected area and a hot, sensitive area around the infection. Blood poisoning must be treated immediately with strong antibiotics.

3.4.9.3 Gangrene. Gangrene is an advanced stage of infection and blood poisoning in which the tissue in the affected area begins to die. Gangrene can spread rapidly and can result in loss of a limb or death. Gangrene must be treated immediately with strong antibiotics and surgical procedures.

3.4.9.4 Heat Injuries. As the period of jungle exposure increases and the body acclimates, high temperatures and high humidity will become less of a problem. Figure 3-7 summarizes the cause, recognition, and treatment of heat injuries.

3.4.10 Procedures for Handling the Wounded and Disabled. One of the following methods should be selected depending on the severity of the wound or disability:

1. Continue the patrol. Continue the patrol if the patient is able, recognizing that his disability may slow down the patrol.

2. MEDEVAC. Call for immediate MEDEVAC. Set up a defensive perimeter and wait for MEDEVAC. The PL should decide whether or not the mission has been compromised and whether to proceed or call for extraction.

TYPE	CAUSE	SYMPTOMS	TREATMENT
Dehydration	Loss of too much water. About two thirds of the human body is water. When water is not replaced as it is lost, the body dries out.	Sluggishness and listlessness	Give the victim plenty of water
Heat Exhaustion	Loss of too much water and salt.	• Dizziness • Nausea • Cramps • Rapid, weak pulse • Cool, wet skin • Headache	• Move the victim to a cool, shaded place for rest • Loosen the clothing. • Elevate the feet to improve circulation • Give the victim cool salt water (two salt tablets dissolved in a canteen of water). Natural seawater should not be used.
Heat Cramps	Loss of too much water	Painful muscle cramps that are relieved as soon as salt is replaced.	Same as for heat exhaustion.
Heatstroke (Sunstroke)	Breakdown in the body's heat control mechanism. The most likely victims are those who are not acclimated to the jungle, or those who have recently had bad cases of diarrhea. Heatstroke can kill if not treated quickly.	• Hot, red, dry skin (most important sign) • No sweating (when sweting would be expected) • Very high temperature (105 to110 degrees) • Rapid pulse • Spots before eyes • Headache, nausea, dizziness, mental confusion • Sudden collapse	• Cool the victim immediately. This is achieved by putting him in a creek or stream, pouring canteens of water over him, fanning him, and using ice, if available. • Give him cool saltwater (prepared as stated earlier if he is conscious). • Rub his arms and legs very rapidly. • Evacuate him to medical aid as soon as possible.

Figure 3-7. Heat Injuries

3.4.11 Jungle Field Skills

3.4.11.1 Signaling. In the evasion/survival scenario, signaling to friendly aircraft can be dangerous and, in areas of multiple canopies, may be impossible for air crewman to see.

1. Use a radio if available.

2. Otherwise get to a clearing and signal with a mirror, any shiny object or an aircraft panel or signal cloth.

3. A signal fire can also be made; however, it could serve as a beacon to the enemy as well as to friendly forces. If a signal fire is required, start the fire with any dry material available and then burn green material to create smoke that will penetrate the canopy.

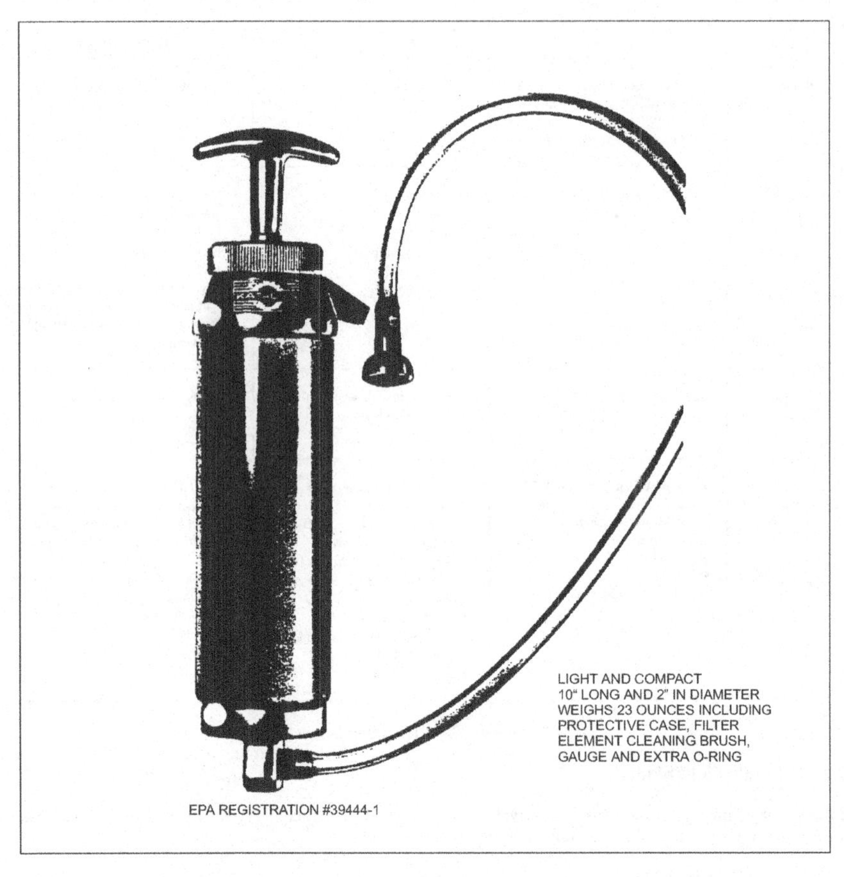

Figure 3-8. Katadyn Water Filter

3.4.11.2 Water Procurement. Water can be obtained from numerous sources; however, purity is always suspect if the water comes from existing water sources such as standing water, lakes, or streams.

1. Water Purification. Purify all suspect water before drinking by:

   a. Using the Katadyn Filter (See Figure 3-8). This is the best method of obtaining pure water, however it does not work with salt water.

   b. Boiling the water. Filter the water first to remove any large impurities (use a piece of cloth or parachute nylon). Bring the water to a full boil for at least three minutes. Allow the water to sit for three minutes to settle and cool. Pour off the top of the boiled water into a second container, leaving the bottom 0.5 inch in the first container (to avoid any sediment).

   c. Using Water Purification Tablets. Follow the instructions on the packet.

2. Obtaining water from jungle streams. Obtain nearly clear water from muddy streams or lakes using the following procedure:

   a. Dig a hole in sandy soil one to six feet from the bank.

   b. Allow water to seep in, and then wait for the mud to settle.

   c. Filter discolored or turbid water using an improvised filter of parachute nylon or a signal panel.

3. Obtaining Water from Plants. Water from plants can be used without further treatment.

   a. Coconut Palms

      (1) Use the green, unripe coconuts, about the size of grapefruit, not the mature ones.

      (2) The water in the mature ones may cause stomach cramping.

      (3) Catch drinkable sap as shown in Figure 3-9.

   b. Vines. Vines are usually good sources.

      (1) Choose a good-sized vine and cut off a three to six foot length. Make a first cut at the top.

      (2) Sharpen one end and hold a container or your mouth to the sharpened end as shown in Figure 3-10.

      (3) In general, large diameter vines yield more water than small ones.

      (4) More water is available during the day, and it will be many degrees cooler than the air temperature. Never drink from a vine that has milky, sticky, or bitter tasting sap

   c. Banana. The banana or plantain trunk can be made into an ideal source of water with a few cuts from a knife or machete as illustrated in Figure 3-11.

      (1) Cut the trunk leaving approximately three inches of it protruding from the ground.

      (2) Cut out a bowl-like reservoir in the three inches of trunk protruding above the ground. Water will immediately flow into the bowl from the roots.

(3) This water will taste bitter but if the bowl is allowed to fill and be scooped out three times, the fourth filling will be palatable and a continuing source.

(4) The same trunk can be used for periods of up to 4 days continuously. It is recommended that when the bowl is not in use, it be covered with a banana leaf to keep the insects out of the water. See Figure 3-11.

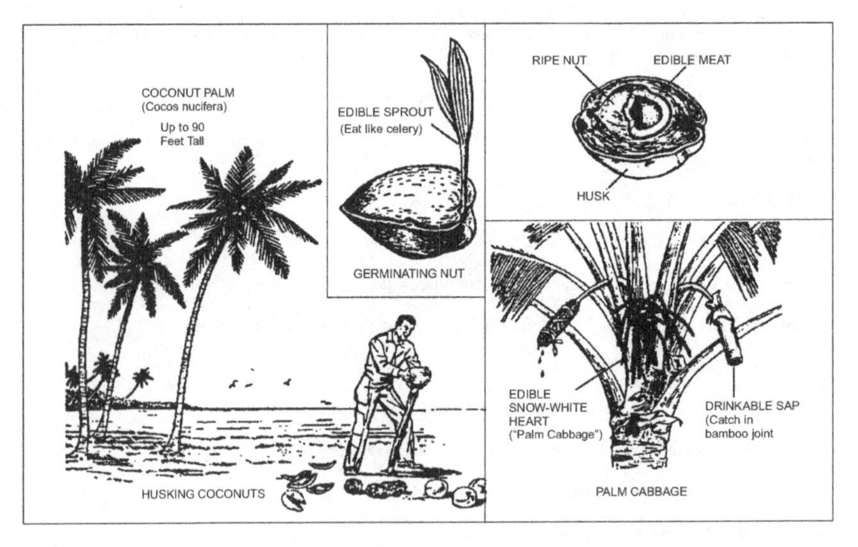

Figure 3-9. Obtaining Drinkable Sap from Coconut Palms

d. Bamboo. The stems sometimes have water in the hollow joints.

(1) Shake the stems of old, yellowish bamboo.

(2) If a gurgling sound is heard cut a notch at the base of each joint and catch the water in a container.

e. Bromeliads. These are pineapple-like plants. The up-curved leaves form natural cups that catch rainwater and dew, holding them deep in the heavy leaf base. Cut it and hold it upside down to drain. Filter the water of debris before drinking.

4. Collecting rainwater. Collect rainwater as follows:

a. Dig a hole and line it with a piece of material, like rain gear or plastic.

b. Catch water dripping from trees by wrapping a clean cloth around a sloping tree.

c. Arrange one end of the cloth to drip into a container, like a canteen cup.

Remember wildlife lives in the higher regions of the jungle canopy; so let it rain for a period of 15 to 30 minutes to ensure waste material on leaves and branches is washed away.

Figure 3-10. Obtaining Drinkable Sap from Vines

3.4.11.3 Food Procurement. Both animal and plant food are abundant in the jungle area. Streams and rivers are the best sources of animal food. Land animals come to their banks to drink and to bathe and aquatic animals live in their shallows and pools.

1. Animal food (land). Trails and roads are the normal passageways along which animals travel through tropical forests. These can include hedgehogs, porcupines, anteaters, mice, wild pigs, deer, and wild cattle. Do NOT use firearms unless it is a last resort, as this may reveal your location.

   a. Snakes should all be considered poisonous, but they can provide a food source and are delicious if not over cooked. Large monitors, iguanas, and lizards are excellent, once skinned and gutted. Bake or roast the long cylindrical tail muscle.

   b. Frogs can be poisonous

      (1) All brilliantly colored frogs should be cautiously avoided.

(2) Some frogs and toads secrete substances through the skin that have pungent odors and are often poisonous.

c. In the trees are birds, bats, squirrels, rats, and monkeys. Monkey meat is excellent, tasting somewhat like veal.

d. Larger wild animals are best left alone.

e. Insects. If desperate, many varieties of insects are available and easy to catch. Cook larger insects to kill internal parasites. Remove stinging apparatus from stinging insects before eating.

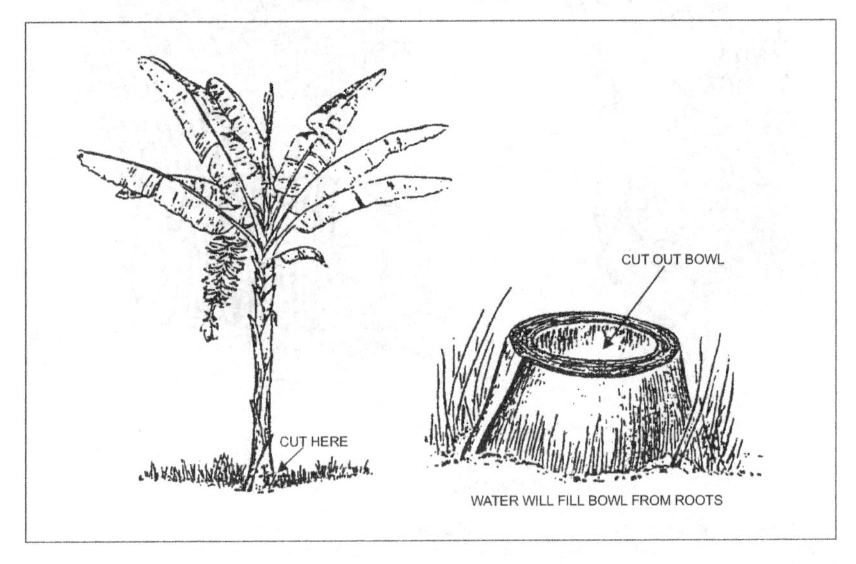

Figure 3-11. Obtaining Water from Banana Trunks

2. Animal food (water)

a. Seafood such as fish, crabs, lobsters, crayfish, and small octopods can be poked out of holes, crevices, or rock pools. Spearing is the most effective method of catching them. Sharpen a piece of green bamboo to make an effective spear. Fishhooks and a small amount of line should be carried in the jungle to catch fish that may not be within spearing distance. Bait the hooks with insects or larvae that can be caught readily in the jungle.

b. Snails and limpets cling to rocks and seaweed from the low-water mark and above.

3. Plant food. Plant food is very abundant in the jungle and is usually much easier to find than animal food. Observe the rules regarding the edible parts of plants in Figure 3-12.

a. Nuts and fruits eaten by monkeys are usually safe. Be cautious initially; sample only a small quantity and wait for any adverse effect.

b. Palms, as a group, are conspicuous, readily recognized, and usually an excellent source of food. Many species of palms yield edible fruits, flowers, terminal buds, sugary sap, or starch from the trunk.

(1) Most Western Hemisphere palms are edible or at least can be safely tested.

EDIBLE PARTS OF PLANTS	
Underground Parts	Tubers Roots and Rootstalks Bulbs
Stems and Leaves (potherbs)	Shoots and Stems Leaves Pith Bark
Flower Parts	Flowers Pollen
Fruits	Fleshy fruits (dessert and vegetable) Seeds and Grains Nuts Seed pods Pulps
Gums and Resins	
Saps	

Figure 3-12. Edible Parts of Plants

(2) Fruits of Eastern Hemisphere palms are not edible due to irritating crystals.

c. Rattan Palms (Figure 3-13) are vine-like palms that look like climbing bamboos. They are common in Pacific jungles. The edible part is the growing tip. To get the tip:

(1) Cut the vine and gradually pull it down through the jungle canopy.

(2) Removing the thorny sheath makes it easier to pull the vine.

(3) Cut off the six to eight-foot growing tip.

(4) Remove the spiny outer sheath. Cut this section in short lengths.

(5) Cook on a bed of coals.

(6) Eat the inner heart when the outer covering is well charred.

Figure 3-13. Rattan Palms Edible Parts

(7) Most rattan can be eaten uncooked, but they are not very tasty.

d. Bamboo (Figure 3-14) comes in many species. Some are climbing vines like rattan.

(1) Their stems are hollow and divided into nodes.

(2) The new shoots of this bamboo, or the buds often found along stems, can be cooked and eaten in the same way as rattan.

e. Ferns (Figure 3-15) are common in the jungle.

(1) The young fronds can be boiled as greens, although some may be too bitter.

(2) The new leaves at the top, known as fiddle heads, are the edible parts. Although covered with fuzzy hair, this can be removed by rubbing during washing.

f. Sweet Sop is a small tree with simple, oblong leaves. The fruit is:

(1) Shaped like a blunt pinecone with thick gray-green or yellow, brittle spines.

(2) Is easily split or broken when ripe, exposing numerous dark brown seeds imbedded in the cream colored, very sweet pulp.

g. Pili Nut comes in numerous species. The single, more or less triangular seeds of the nut are edible either raw or roasted.

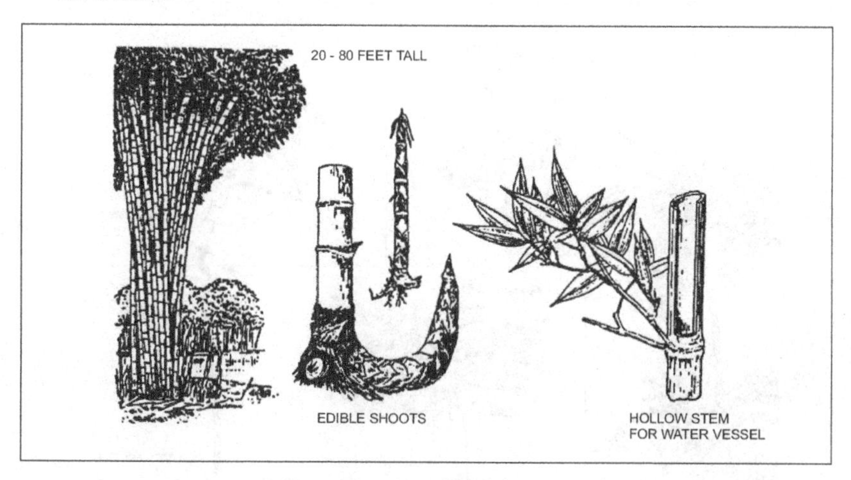

Figure 3-14. Bamboo

h. Other sources. In secondary or cutover jungle and abandoned-farmed areas, many varieties of plant foods are found: e.g., bananas, plantains, papayas, mangoes, guavas, cashews, and more.

(1) Using the edibility guidelines of Figure 3-16 and training will provide the confidence to survive in the jungle.

(2) When eating jungle plants, eat in modest quantities until the digestive system accommodates to them or stomach cramps and discomfort can result.

3.4.11.4 Fire Making

1. When and Where to Build a Fire. Whether or not to build a fire under existing patrol conditions is a difficult decision. Use these decision factors:

a. Location. Build a fire well off trails, beneath trees in a thickly vegetated area that causes the smoke to dissipate as it rises through the foliage.

b. Time. Fires are easier to disguise and will blend in better during the dawn, dusk, and during bad weather.

c. Build a fire only if it is the last resort.

(1) Use a single-person stove.

(2) Small amounts of C-4 plastic explosive will burn and can provide a heat source for cooking.

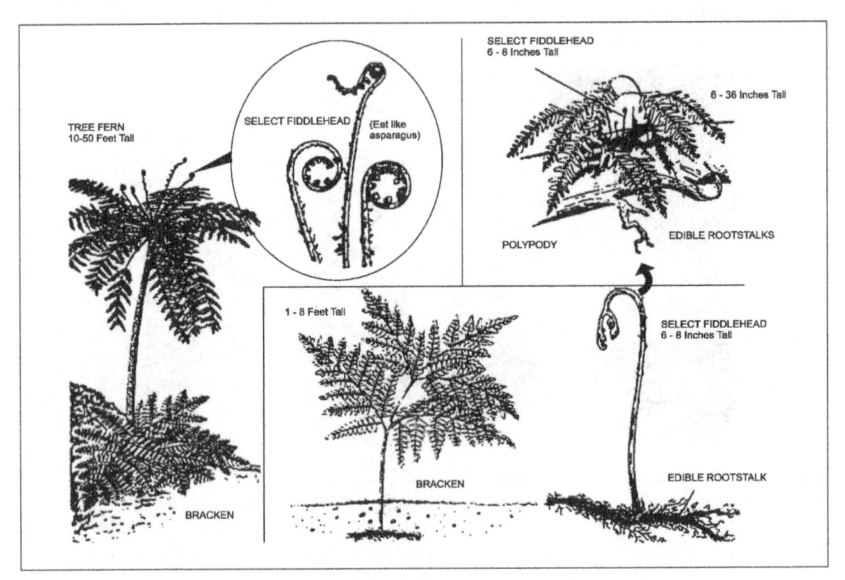

Figure 3-15. Ferns

2. Fire Starting Materials and Fuel

a. Dry Season. Dry, dead hardwoods with the bark removed and about the size of a pencil are ideal. These will burn more rapidly and create less smoke.

b. Rainy Season. It may be difficult to find dry wood. Many of the larger trees have hollow trunks, and cutting strips of the inner lining is one idea. Dry wood can also be found hanging in vine networks or lying on bushes.

(1) Don't use green bamboo as it emits dangerous fumes.

(2) If using dry bamboo, cut into each section of the sealed sections, or they will explode when they get hot.

3. Fire Construction

a. Dakota Hole Fire (Figure 3-17). This is one of the best fires for a single person.

(1) Prepare a "fireplace" by digging two holes in the ground, one for ventilation and one for the fire.

(2) Recommended holes are approximately 12 inches deep and about 12 inches apart with a side tunnel dug to connect the bases of the holes.

(3) Place dirt on a piece of cloth to be ready for immediate extinguishing to conceal the fire site in an emergency.

(4) Keep the fire small and build to keep the flames below the ground's surface.

EDIBILITY GUIDELINES – PLANTS

DO NOT EAT

- Umbrella-shaped flowering plants
- White, yellow, and red berries
- Plants that irritate the skin
- Plants with shiny leaves
- Mushrooms and fungi
- Milky sap
- Legumes
- Bulbs

Apply the edibility test to one plant at a time. Once a plant has been selected for this test, proceed as follows:

1. Crush or break part of the plant to determine the color of its sap. If clear - **Proceed.**

2. Touch the sap/juice to inner forearm or tip of tongue. If no problems, such as a rash or burning sensation, bitterness, or numbing sensation - **Proceed.**

3. Prepare plant for consumption by boiling in two changes of water. The toxic properties of many plants are water-soluble or destroyed by heat; cooking and discarding in two changes of water lessons the amount of poisonous material or removes it completely. Boiling periods should last about five minutes. **Proceed.**

4. Place about one teaspoonful of the prepared plant in the mouth for five minutes and chew, but do not swallow it. A burning, nauseating, or bitter taste is a warning of possible danger. If any of these occur, discard that plant as a food source. If no unpleasant side effects occur, swallow and wait eight hours.

5. If no ill effects occur after the eight hours (e.g., nausea, cramps, diarrhea), eat about two tablespoons and wait an additional eight hours.

6. If no ill effects occur after this period, the plant is considered edible.

Figure 3-16. Plant Edibility Guidelines

b. Trench Fire (Figure 3-18). This type of fire would be better for more personnel.

(1) Build it about 8 to 12 inches below the surface as pictured.

(2) Determine length based on how many people are involved. They should not crowd either end of the hole so that it may get an adequate amount of air to help it burn hot and eliminate smoke.

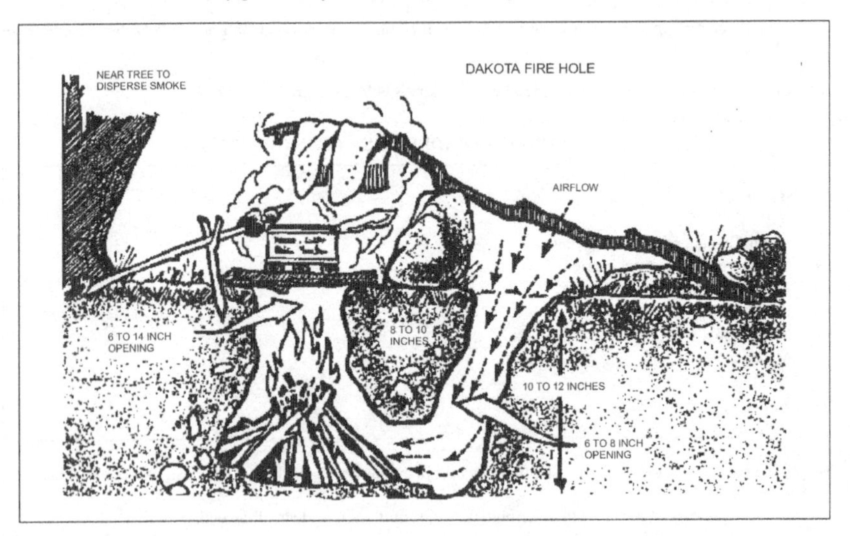

Figure 3-17. Cross Section of Dakota Fire Hole

4. Clean up after the fire. Insure all traces of fires are hidden. Cover all coals with jungle debris to make it appear similar to surroundings.

3.4.11.5 Setting up Campsites and Shelters

1. Jungle Campsite. Attempt to construct a jungle campsite that will provide maximum protection, but leave minimum signs of the site after the patrol departs. The patrol will need to construct a more complete campsite during the rainy season to avoid becoming soaked and, thus, vulnerable to jungle diseases. Use the following guidelines in building and using a campsite.

   a. Select a secure knoll or high spot, back from mosquito infected swampy areas.

   b. Scout the immediate area around the selected area for signs of people.

   c. Adjust location if necessary.

   d. Be aware of the ambient jungle noises and use these as sensors.

2. Improvised Shelter

a. Inspect the selected area for signs of insects. Avoid any obvious insect nesting areas.

b. Use a poncho in a number of ways as shown in Figure 3-19. Always keep one end of the poncho higher than the other.

c. Construct lean-to using jungle materials. Large leaves from jungle plants can be used to construct temporary cover using the lean-to method.

d. Construct something to keep personnel off the ground in the shelter. A thatch work of bamboo or other types of light brush can be used as flooring.

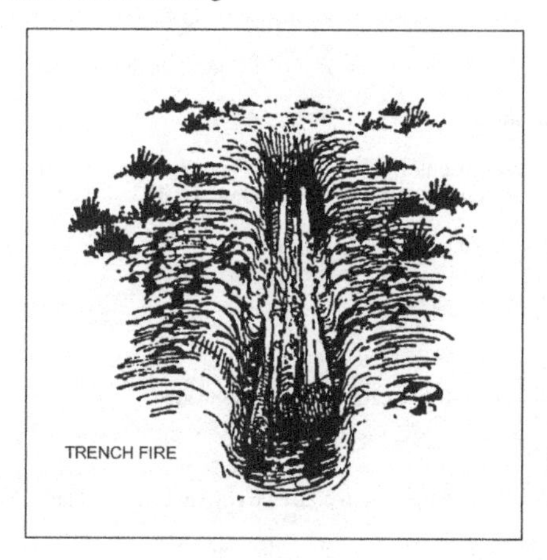

Figure 3-18. Trench Fire

3. Bed. A hammock is the best method for sleeping in the jungle and is an important item for combating insects.

   a. Carry a hammock or improvise one with the poncho.

   b. Use the hammock to stay off the ground (discourages ants, spiders, leeches, and other crawling problem insects and animals).

4. Things to Avoid:

   a. Do not sleep on the ground.

   b. Do not camp too near a stream or pond, especially during rainy season.

   c. Do not camp on game trails or near water holes.

d. Do not build a shelter under dead trees or trees with dead limbs.

3.4.11.6 Survival Equipment

1. First Line Equipment. Carry survival equipment on the body in the first line of equipment. Figure 3-20 provides suggestions for contents of a baseline survival kit.

2. Knife Skills. Each SEAL must have a basic knowledge of knife skills. Each patrol member should carry a knife, e.g., a Leatherman or Swiss Army Knife.

   a. Keep the knife sharp. Carry a whetstone and sharpen the knife as required to maintain a keen edge.

   b. Rub a small amount of lubricating or gun oil on the blade each day to keep it from rusting.

3.4.12 Counter-tracking Skills

3.4.12.1 Procedures to Avoid Leaving a Trail. It is important for all patrol members to exercise discipline in moving through the jungle so as not to leave an easily followed trail. Use the following guidelines:

1. Avoid telltale campsites. When leaving the campsite:

   a. Destroy all shelters.

   b. Bury all materials.

   c. Cover all used areas with brush to match the surrounding area.

   d. Bury all fire materials.

   e. Conduct a patrol walk through the area to police up any stray material that may have been discarded.

   f. Inspect the area to ensure it matches the surrounding area.

2. Movement through vegetation. Broken or mutilated vegetation can leave a clear path. Use the following guidelines:

   a. If possible, do not clear brush ahead of you. Move through it, returning the brush to its original place.

   b. Be careful not to break any branches or twigs for any reason.

3. Avoid footprints. Avoid soft areas. Walk on hard or rocky areas if possible.

4. Avoid dropping or leaving traceable materials. Any foreign material discarded will provide a trail. The basic rule is not to drop anything unless it is buried and covered with brush. Typical items indicating presence of humans include:

   a. Cigarettes

   b. Matches

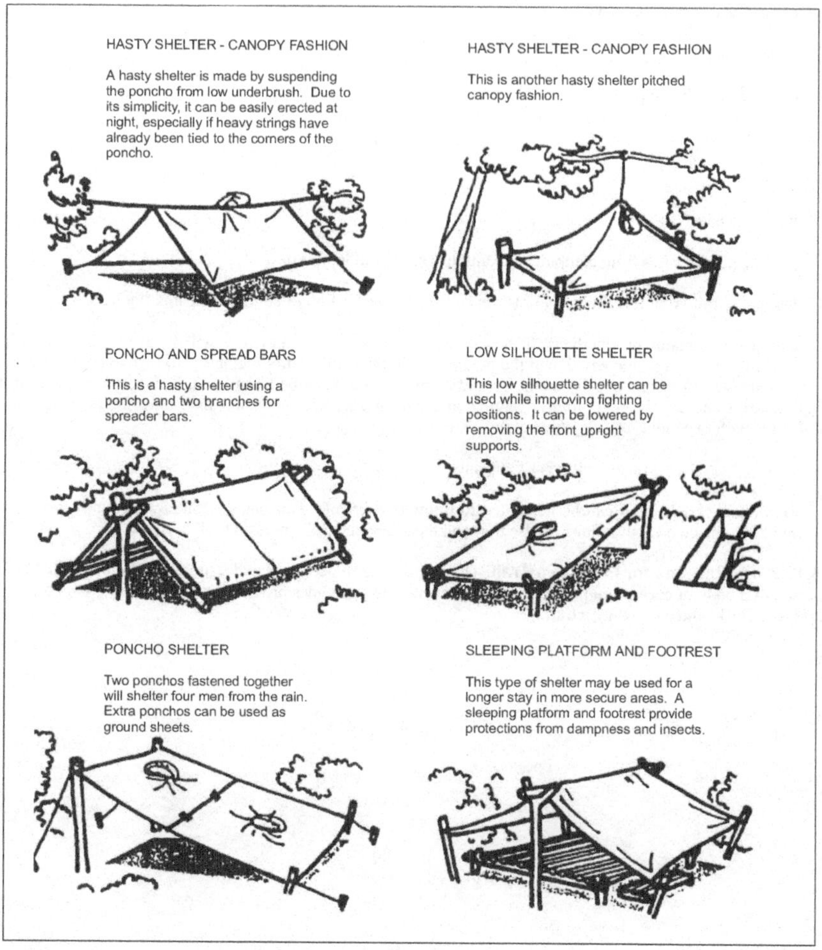

HASTY SHELTER - CANOPY FASHION

A hasty shelter is made by suspending the poncho from low underbrush. Due to its simplicity, it can be easily erected at night, especially if heavy strings have already been tied to the corners of the poncho.

HASTY SHELTER - CANOPY FASHION

This is another hasty shelter pitched canopy fashion.

PONCHO AND SPREAD BARS

This is a hasty shelter using a poncho and two branches for spreader bars.

LOW SILHOUETTE SHELTER

This low silhouette shelter can be used while improving fighting positions. It can be lowered by removing the front upright supports.

PONCHO SHELTER

Two ponchos fastened together will shelter four men from the rain. Extra ponchos can be used as ground sheets.

SLEEPING PLATFORM AND FOOTREST

This type of shelter may be used for a longer stay in more secure areas. A sleeping platform and footrest provide protections from dampness and insects.

Figure 3-19. Shelter Types

c. Gum and candy wrappers

d. Toothpicks

e. Chewing tobacco juice

f. Urine

g. Feces

h. Tissue paper

i. Wrappers or cases from equipment (examples: ammunition, food).

5. Use the terrain. Certain terrain features can be used to mask a trail or confuse potential trackers.

   a. Walk in streams or riverbeds. This will cover footprints, but mobility will be reduced, detectability (visibility) will be increased, and the possibility of injury to a patrol member will increase (jungle stream and river depths are usually unknown and holes and rocks can be hazardous). For these reasons, use the stream and riverbeds sparingly, staying in them only as necessary and then only for 50 yards or so (enough to confuse or delay trackers).

   b. Walk on rocky ground to prevent footprints.

6. Deception. A false trail can be generated to throw trackers off. This can be done by creating some of the telltale signs noted above along a false trail and then breaking off using "no trail" methods.

3.4.12.2 Procedures for Finding a Trail. The patrol will, on occasion, find it necessary to follow the trail of a person or a body of enemy troops in the jungle. Reverse the guidelines provided in the preceding paragraph. Be alert for ambush when following a trail.

<table>
<tr><th colspan="2">SURVIVAL KIT</th></tr>
<tr><th colspan="2">MINIMUM REQUIREMENTS</th></tr>
<tr><td>U.S. MONEY AND LOCAL CURRENCY</td><td>POCKET KNIFE - stainless steel, 3- to 4-inch blades</td></tr>
<tr><td>COMPASS</td><td>PISTOL - weapon of choice/issue</td></tr>
<tr><td>SIGNAL MIRROR - impact resistant plastic</td><td>550 CORD - for fishing, sewing, etc.</td></tr>
<tr><td>SNARE WIRE - stainless; for trapping</td><td>FLASHLIGHT</td></tr>
<tr><td colspan="2">MATCHES
1. Plastic or metallic container
2. Waterproof matches rolled in paraffin soaked muslin in an easily opened container like a soap container or a toothbrush case, or
3. Waterproof kitchen-type matches (cushion heads against friction).</td></tr>
<tr><td colspan="2">NEEDLES – sailmakers', surgeons', and darning; at least one each</td></tr>
<tr><td colspan="2">FISHING EQUIPMENT - stainless steel hooks with monofilament leaders (six each of the following sizes: 8, 10, 12, 14), three small flies, plastic worm, small spoon</td></tr>
<tr><td colspan="2">SIGNAL CLOTH - international orange/silver with camouflage side</td></tr>
<tr><td colspan="2">ALUMINUM FOIL - heavy duty; minimum three feet (can wrap fish hooks in this)</td></tr>
<tr><td colspan="2">MEDICAL
Bar of surgical soap
Personal medicines
Anti-malaria tablets
Water purification tablets
Small bandages (e.g., "Band-Aids")
Tropical antibiotic
Insect repellent
Lip ointment
Sun block
Aspirin</td></tr>
<tr><td colspan="2">NICE-TO-HAVE ITEMS
YARN - One foot each of red, yellow, and orange (used as bait for fishing and for bird snares).
SURGICAL TUBING - Three feet of small diameter (constrictor band, sling shot).
FLEXIBLE SAW (WIRE SAW) - 18 inch, multi-strand, stainless
3 CORNER FILE
MAGNIFYING GLASS - Map reading and fire starting.
SAFETY PINS
PENCIL FLARES
CANDLES
HAMMOCK - Light weight mesh type: dual purpose as a fish net and for sleeping</td></tr>
</table>

Figure 3-20. Jungle Survival Kit

CHAPTER 4

Desert Operations

4.1 INTRODUCTION

This chapter provides NSW personnel with guidelines for planning, preparation, TTP for operations specific to the desert environment. This is a guideline and is to be used in conjunction with the NSWMPG for planning, Chapter 2 for basic tactics, and any other NSW tactical manual with specific information for planning each phase of SEAL desert operations.

4.2 DESERT TACTICS PLANNING

4.2.1 General. Desert operations are conducted in a harsh environment of high heat, dust, bright sunlight, and extremely dry conditions. Success in the desert depends on preparation, training, using unique skills, and adjusting existing skills and techniques to meet the unique circumstances of desert operations.

4.2.2 Desert Mission Planning. Planning factors are basically the same for operations in the different environments and are presented in detain in Section 2.3.3.2.

4.2.2.1 Naval Special Warfare Desert Missions. SEAL operations involve small units. SEALs do not mass on target or carry out missions that require assault troops and/or larger units. SEAL missions in the desert environment might include:

1. Direct Action. SEALs may be tasked to clandestinely attack targets in desert areas and provide laser target designation for aircraft. SEALs do not carry out raids or DA operations that can be better dealt with by tactical aircraft, naval gunfire or other such types of assets.

2. Special Reconnaissance. SEALs may be tasked to conduct coastal, point, or area reconnaissance in desert areas in support of intelligence collection or as an advance force reconnaissance element.

3. Foreign Internal Defense. SEALs may work in an interagency activity to train, advise, and otherwise assist a desert HN's military and paramilitary forces. See paragraph 4.2.15 for additional information on FID.

4. Unconventional Warfare. SEALs and other personnel may be employed to train military and paramilitary indigenous desert nation or surrogate forces in guerrilla warfare and other direct offensive, low visibility, covert or clandestine operations, as well as the indirect activities of subversion, sabotage, collection, and E&E.

4.2.2.2 Desert Planning Factors. The planning factors for SEAL desert operations are:

1. Survivability. The requirement to carry sufficient drinking water significantly affects survivability.

2. Endurance. Endurance is reduced significantly due to the heat and the need for drinking water. This is the primary mission-limiting factor for desert operations.

3. Training. Specialized training is mandatory.

4. Mobility. When SEALs conduct maritime insertion, infiltration to coastal targets will be on foot. Mobility is therefore reduced from that of more temperate areas. Use of vehicles or aircraft are ways to increase mobility. These modes of infiltration and exfiltration are preferable for desert operations.

5. Detectability. In an open desert area, SEAL patrols are extremely detectable to air and mounted patrols.

4.2.2.3 SEAL Mission Considerations. SEALs may require the following to successfully complete desert operations:

1. Drinking water. Do not rely on resupply or caches.

2. Communications support.

3. On-call dedicated fire support: NGFS, CAS, and Artillery.

4. Mobility.

4.2.3 The Desert Environment

The following paragraphs provide a brief description of the desert environment.

4.2.3.1 Types of Deserts. The world map (See Figure 4-1) shows desert areas of the world. There are more than fifty named deserts in the world. Many of these border a body of water, either an ocean or sea. The term "desert" is applied to a variety of areas. There are salt deserts, rock deserts, and sand deserts. There are two climatic extremes for deserts: a north temperate desert and a near tropic desert. Other deserts fall in between these extremes.

4.2.3.2 Categories of Deserts. As described below, there are three generic categories of deserts:

1. Sandy/Sand Dune Deserts. Sandy/sand dune deserts are extensive flat areas covered with sand or gravel and the product of ancient deposits or modern wind erosion. "Flat" is relative in this case, as some areas may contain sand dunes over 1,000 feet high and 10-15 miles long. Other areas, however, may be totally flat for distances of 3,000 meters and beyond.

   a. Trafficability in such terrain will depend on the windward or leeward gradients of the dunes and the texture of the sand.

   b. Plant life may vary from none to scrub brush reaching over six feet high.

2. Rocky Plateau Deserts. Rocky plateau deserts have canyons interspersed by extensive flat areas with quantities of solid or broken rock at or near the surface. The canyons, or steep-walled eroded valleys, are known as wadis in the Middle East and arroyos in the United States and Mexico.

3. Mountain Deserts. Scattered ranges or areas of barren hills or mountains, separated by dry, flat basins characterize mountain deserts. High ground may rise gradually or abruptly from flat areas to a height of several thousand feet above sea level. Most of the infrequent rainfall occurs on high ground and runs off rapidly in the form of flash floods, eroding deep gullies and ravines and depositing sand and gravel around the edges of the basins. Water rapidly evaporates, leaving the land as arid as before, although there may be

short-lived vegetation. If sufficient water enters the basin to compensate for the rate of evaporation, shallow lakes may develop, such as the Salton Sea and the Dead Sea. Most of these lakes have a high saline content.

4.2.3.3 Desert Characteristics. Figure 4-1 shows that deserts comprise about 20 percent of the earth's land surface, but only about four percent of the world's population live there. Other general desert characteristics are:

1. Temperatures. Extremes of temperature are a typical characteristic of deserts. The temperature in the Arabian Desert can exceed 120 degrees Fahrenheit in the summer. In the Libyan Sahara desert, temperatures in excess of 110 degrees Fahrenheit can be expected in the summer. Hot days and cool nights are usual. Nights will be 20-40 degrees Fahrenheit colder than daytime temperatures (in some deserts, the temperature can drop as much as 60-70 degrees Fahrenheit at night). Maximum daytime temperatures in the winter will be 70-100 degrees Fahrenheit, with temperatures as low as 40 degrees Fahrenheit at night.

2. Winds. Desert winds can achieve near hurricane force. Wind chill factors therefore become a planning consideration. Dust and sand suspended in the wind can make maintenance of equipment very difficult, restrict visibility and in general, make life almost intolerable. Rapid temperature changes always accompany strong winds.

3. Sandstorms. Sandstorms can last for days. The windblown sand can be extremely painful on bare skin and can reduce visibility to the point where operations are impossible.

4. Rain. All deserts are arid, as rainfall is extremely limited. However, when rain occurs it may consist of one single violent storm in a year with high surface water runoff. Usually too much rain is received too quickly and is a liability rather than an asset. Annual evaporation rate exceeds the annual precipitation rate (less than 10 inches annually).

5. Flooding. Flooding, known as flash floods, can occur several hundred miles away from a rainstorm due to heavy collection of runoff rain in the dry streambeds. Be alert to weather reports.

6. Light. The sun and low cloud density combine to produce unusually bright and glaring light conditions during the day.

   a. Sometimes, when visibility conditions seem near perfect, distance underestimation occurs.

   b. Mirages, or heat shimmer, distort the shape of objects. Mirages are caused by heated air rising from the extremely hot desert surface. The effect of mirages is compounded with binoculars.

   c. Observation is best in the cooler parts of day: dawn and dusk.

   d. OPs should be as high above the desert floor as possible to provide the advantages of less distortion, increased distance viewing, and better depiction of shadows cast.

   e. NVEO works very well on bright desert starlight and moonlight nights.

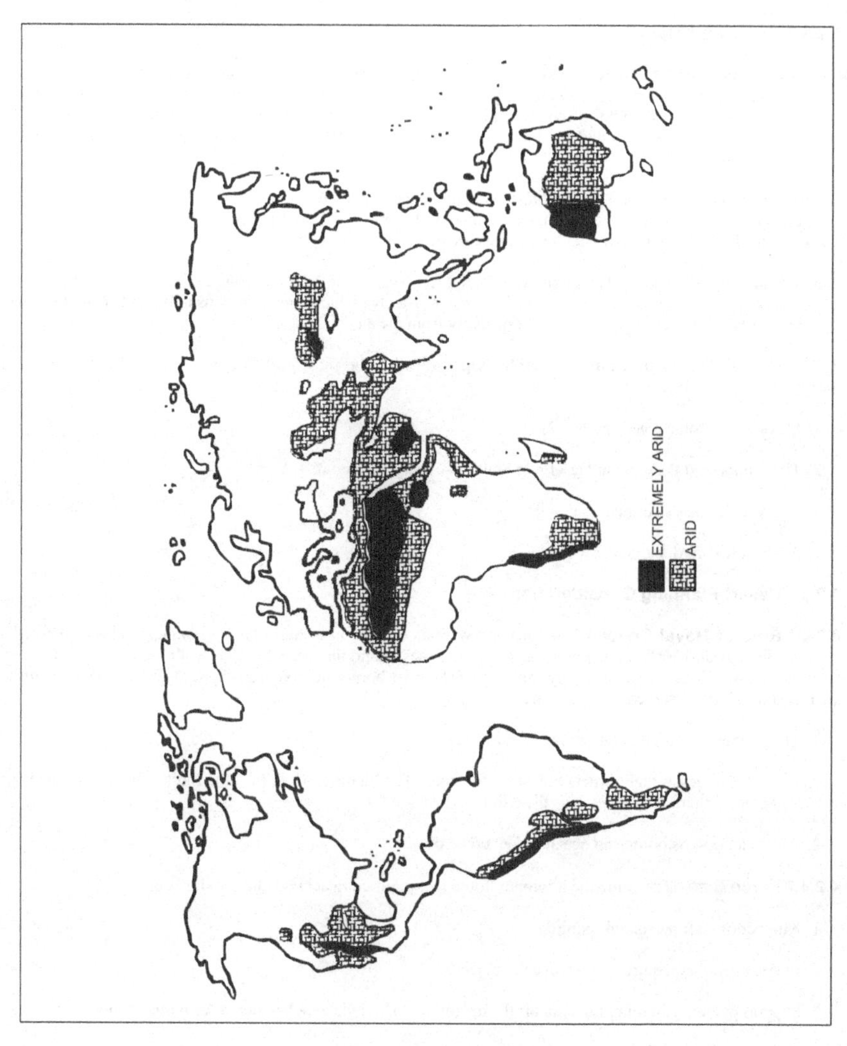

Figure 4-1. Deserts of the World

4.2.3.4 Fauna and Flora

1. Invertebrates. Invertebrates such as ground-dwelling spiders, scorpions, and centipedes, together with insects of almost every type, are found in great quantity in the desert. Drawn to man as a source of moisture or food, lice, mites, and flies can be extremely unpleasant and carry diseases such as scrub typhus and dysentery. The stings of many scorpions and the bites of centipedes or spiders can be extremely painful, though they are seldom fatal.

2. Reptiles. These are perhaps the most predominate group of desert animals. They abound in the desert, especially snakes and lizards that have adapted to the arid harsh environments. Most are harmless, but some are lethal. See Paragraph 4.4.3.7 regarding care of snakebites.

3. Vegetation. Some plants have extensive lateral root systems to take advantage of the occasional rain while others have deep roots to reach subsurface water. The available vegetation is usually inadequate to provide much shade, shelter, or concealment, especially from the air.

4.2.3.5 Useful Training Areas for SEAL Operations. The most probable desert areas NSW forces may operate in are the following:

1. Mexico's Sonoran (see Figure 4-2)

2. The Tumbe and Patagonian Deserts of South America (see Figure 4-2).

3. The Middle East's Arabian (see Figure 4-3).

4. Africa's Kalahari and Saharan (see Figure 4-3).

4.2.4 Desert Planning Considerations

4.2.4.1 Role of Naval Special Warfare. NSW units operate in a maritime environment. This is defined as missions in coastal, harbor, and riverine areas to distances inland that are consistent with the ability to insert and extract from maritime areas. Normally, maritime areas that border deserts are salt water. Examples of maritime areas that involve a desert environment are:

1. The Arabian Desert that borders the Persian Gulf

2. The African desert that borders the Atlantic Ocean, The Mediterranean Sea, the Indian Ocean, and the major rivers of northern Africa (such as the Nile)

3. American deserts bordering ocean and riverine areas.

4.2.4.2 Force Size. Determine the minimum force size required to achieve the mission's goal(s):

1. Number of staff personnel required

2. SEAL operators required to achieve the objective

3. Support personnel required to support the operation, to include attachments to/from other units.

4.2.5 Target Planning

4.2.5.1 Target Types. Targets are normally difficult to approach because of the long distance visibility that results in a high probability of detection of units moving in the desert. This makes enemy defense of targets easier. Examples of typical SEAL targets that would be found in a desert environment are:

1. Ammunition or supply caches
2. Guerrilla headquarters and training areas
3. Chemical munitions sites
4. Enemy C^2 sites
5. Enemy missile sites
6. Desert supply routes
7. Personnel targets
8. Enemy shipping in harbors and ports
9. Radar installations.

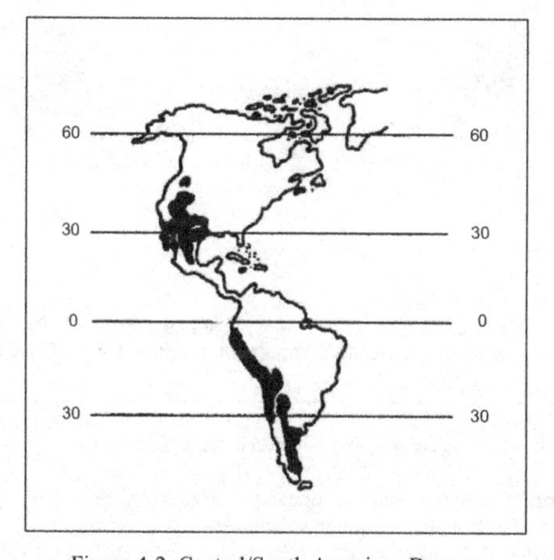

Figure 4-2. Central/South American Deserts

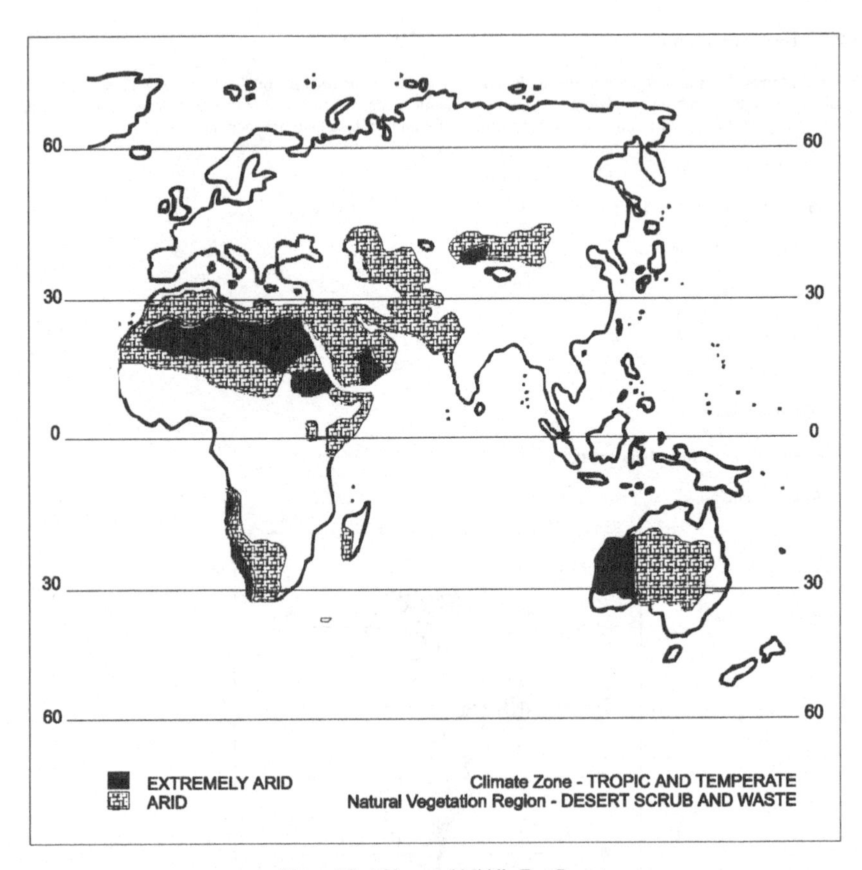

Figure 4-3. African and Middle East Deserts

4.2.5.2 Weapon Selection. Planners should compute the probability of kill of the various SEAL weapons against the type of target, and select the optimum weapon. When a standoff weapon yields the required kill probability, it should be selected. A standoff weapon is the weapon of choice in the desert environment because it may be difficult to get close to a target for a demolitions or other type of attack. Section 2.3.7 addresses weapons discipline and Section 4.3.10 addresses weapon selection.

4.2.6 Threat Planning. Detection in the desert is of particular concern because of the lack of cover and the long-range visibility. The threat of detection by air patrols is high. Maximum use of intelligence sources must be requested and exploited in order to avoid areas known to contain enemy defensive forces.

4.2.6.1 Intelligence Sources

1. National Sensor Information. Information from national sensors is available through operational intelligence centers. Sources include photographic intelligence, ELINT, IR intelligence, SIGINT, or HUMINT.
   a. Visual Intelligence. Visual or photographic intelligence from overhead sensors is the most useful type in the desert if it is timely and can assist in threat avoidance.

   b. ELINT. ELINT is useful for determining the nature and number of units and defensive systems in the operating area.

   c. SIGINT. SIGINT can be used to determine enemy movement, intentions, and degree of alert.

   d. Measure and Signature Intelligence (MASINT). Scientific and technical intelligence obtained by quantitative and qualitative analysis of data (metric, angle, spatial, wavelength, time dependence, modulation, plasma, and hydro magnetic) derived from specific technical sensors for the purpose of identifying any distinctive features associated with the target. The detected feature may be either reflected or emitted.

   e. HUMINT. HUMINT (i.e., timely information from friendly indigenous personnel in the desert operating area) can be obtained in areas where intelligence planning has been conducted for a sufficient period of time. It is the most useful type of intelligence for up to date local information in the operating area, especially when corroborated by MASINT. HUMINT must be carefully evaluated with regard to the reliability of the source.

4.2.6.2 Threats to Desert Patrols. Desert warfare is ideal for motorized and armored forces, long range standoff weapons, NVEO, and combat air support. Enemy forces in the desert will likely consist of armored vehicles with CAS.

1. Cover and Concealment. There is little cover and concealment in the desert. Once compromised, a force on foot is easily located.

2. Reaction Forces. These usually consist of armored vehicles or aircraft.

3. Engaging the Enemy. A SEAL patrol on foot cannot carry enough weapons and ammunition to successfully engage in a firefight with an armored force, airmobile force, or combat helicopters.

4. Escape. It is very difficult to escape on foot from mobile enemy forces in the objective area.

4.2.7 Desert Supporting Force Planning. The types of supporting forces that are likely to be required in the desert environment include MCT, insertion/extraction forces, desert vehicles, tactical CAS, NGFS, friendly indigenous forces, and other friendly forces.

4.2.7.1 Mobile Communications Team. The MCT will be required to maintain reliable communications with the patrol from the command center.

4.2.7.2 Insertion/Extraction Forces. Involve the supporting forces (pilots, coxswains, drivers, etc.) in the early planning so they can determine if the plan falls within the capability of their assets and personnel.

4.2.7.3 Desert Vehicles. Where possible, plan to use desert-capable vehicles for all desert patrols.

4.2.7.4 Tactical Close Air Support. Tactical fixed-wing aircraft are useful in cleared or open areas. SEAL patrols should plan to have CAS available if they are detected and require cover for extraction.

4.2.7.5 Naval Gunfire Support. NGFS is useful when SEALs are operating within 10 nautical miles of a destroyer or cruiser position. Direct fire can be conducted against targets visible to the ship (or marked by a smoke or flare visible to the ship). Indirect fire can be provided on targets not visible to the ship based on target coordinates provided by the SEAL patrol. SEALs should be familiar with NGFS and spotting procedures. NGFS maybe necessary to protect an EP.

4.2.7.6 Friendly Indigenous Forces. Support by friendly indigenous forces is of special benefit in the desert.

1. Friendly indigenous forces provide "local know how."

2. They can also provide immediate reinforcement, if required.

4.2.7.7 Friendly Force Support. There may be friendly conventional forces operating near front lines in a SEAL AO that can provide various types of support (examples: artillery, helicopter gunships). SEAL patrols should always have the location of friendly front lines plotted and committed to memory.

4.2.7.8 Supporting Force Requirements. Consider the following in supporting unit planning:

1. Location of assets. Are the assets available in the AO or will they have to be imported?

2. Sufficiency of assets. Are there enough assets and back-ups, to handle the mission requirements?

3. Support facilities. Are there adequate facilities available to the supporting forces during the operation? These include:

   a. Messing

   b. Berthing

   c. Equipment storage, hangar space, pier space, hazardous material storage (including demolitions and ammunitions), and, vehicle lifting equipment (cranes, etc.).

   d. Refueling and watering facilities

   e. Communications

   f. Electronics.

4. Base Posturing. Proper base posturing is mandatory. Effective base perimeters must be established to protect against sappers and terrorist raids. See the NSW Ashore and Afloat Basing Manual for specifics.

4.2.8 Desert Patrol Size

4.2.8.1 Patrol Size. SEALs are specialists in maritime SO. The three operational components of a SEAL desert platoon consist of:

1. FT. Four men
2. Squad. Eight men or two FTs
3. Platoon. 16 men (two squads or four FTs).

4.2.8.2 Patrol Types. Although a full platoon of 16 men may be used on some missions, the PL may tailor the size of the patrol to fit the mission tasking. An eight-man squad would generally be the maximum size for most desert missions.

4.2.9 Patrol Augmentation Requirements

4.2.9.1 Augmentation Types. Various types of personnel can augment the patrol to meet the mission requirements. These persons must be physically conditioned to the desert environment to the same degree as the SEALs or their presence may endanger the mission. Examples of possible augmentations are:

1. Language Capability. Most SEAL missions in the desert will be rapid and language augmentation may not be necessary. However, language capability will be required at the base for logistics, and for obtaining information on local conditions from the local people.
2. Military Intelligence Capability.
3. Psychological Warfare Capability.
4. Civil Affairs Capability.
5. Indigenous Guides. A guide, if required, should also speak English and be able to serve as an interpreter.

4.2.10 Logistic Planning. Logistical support is always a challenge, and an arid environment burdens all types-supply, communications, and maintenance. Section 2.10 presents details on logistics support and the role of CSST.

4.2.10.1 Logistic Problems. The desert terrain and environmental features require extra organizational inventory and may require special equipment acquisition. Heat, extreme dryness and blowing sand in desert areas cause the following problems:

1. Water supplies carried are critical, as there are few sources of obtaining water in the desert.
2. Weapons become gritty and sandy quickly (need proper cleaning and dry lubrication supplies).
3. Battery life is shortened.
4. Uniforms and other clothing wear out faster.

4.2.10.2 Logistic Plans. Logistic plans must provide support for NSW forces in the following areas:

1. Isolation Support:
   a. Isolation planning area
   b. Isolation security personnel and food delivery
   c. Berthing area in isolation separate from planning area.
2. Rations. MRE. Note that MREs may have sodium contents to high for desert missions. Test different types of rations under local conditions.
3. Inventories. Maintain full stocks of all items in accordance with standard load lists. Particular attention must be afforded to high usage mission dependent items. These may include:
   a. Water purification pumps and chemicals
   b. Canteens, water bladders, and other water storage systems
   c. Insect and rodent control items, including mosquito netting and scarves for covering the face and neck
   d. Appropriate camouflage that conforms to the color scheme and patterns of the region
   e. Goggles w/sunglass lenses and sunglasses.
   f. Dry batteries for various pieces of electronic equipment.
   g. Medical items for problems unique to the desert environment and region of operations (see Section 4.4).
   h. Equipment, uniforms, boots, and socks.
   i. Ammunition, demolitions, and pyrotechnics.
4. Resupply
   a. Establish, as soon as possible, the means of shipping replacement items to the support base.
   b. Create a high/low reorder system that replenishes the items when they reach certain predetermined low levels.
   c. Air and boat resupply are the primary methods used. Air resupply provides greater flexibility, but requires advance planning.
   d. Coordinate resupply procedures used by supporting air assets.
   e. Brief SEAL platoon staging procedures for resupply items.

f. Ensure items for resupply are prepared and correctly packaged. All items should be packed in dust/sand proof packages. Items that deteriorate in high temperature (such as batteries) should be clearly marked as "heat sensitive material."

5. Maintenance Capability. Provide a maintenance capability to support and repair all equipment.

a. High temperature and blowing sand will increase maintenance on all items affected by heat and grit.

b. Plan for a 10-20 percent increase in normal spare parts usage. Heat sensitive item usage will increase by 30-50 percent, depending on the mission requirements.

4.2.10.3 Medical Preparation. Medical unit requirements for desert operations are essentially the same as for temperate climates. When planning for medical support, increased vehicle evacuation time must be considered due to increased dispersion and large areas over which operations are conducted. This problem can be further complicated if the enemy does not recognize the protection of the Red Cross, thereby inhibiting air evacuation within the range of enemy air defense weapons. The importance of trained HM is critical to overcoming this. The incidence of illness from heat injuries and diseases are higher than in temperate climates. Fevers, diarrhea, and vomiting, for example, cause loss of water and salt, which can culminate in heat illnesses. Cold weather injuries can also occur during a desert winter.

1. Medical Essential Elements of Intelligence. Proper medical preparations require comprehensive and detailed medical intelligence. To develop medical intelligence, information is gathered, evaluated, and analyzed. See Section 2.11 for specifics on medical intelligence.

2. Medical Plans. Plan for the following medical support for NSW personnel:

a. MEDEVAC. Procedures are likely to encounter unique problems due to the desert terrain.

(1) There may be little in the way of natural protection for the wounded while awaiting evacuation.

(2) Wounded management problems and the effects of shock may be compounded by the lack of water and the rapid onset of dehydration.

(3) There is generally little in the way of terrain masking for an incoming MEDEVAC platform.

b. A designated military hospital or other qualified facility to receive the wounded. This facility should have burn and trauma units.

c. Preventive medical measures against diseases.

d. Desert specific medical kit supplies.

3. Medical Training. Provide training for medical problems unique to the given desert region of operations. This should include:

a. Hygiene considerations

b. Preventive medical measures against heat disorders to include sunburns

c. First aid

d. Common diseases and heat illnesses (symptoms and treatment)

e. Insect or other type bites (prevention and treatment)

f. Dangerous animal life or vegetation.

4. Medical Personnel. Medical personnel support, to include staff doctor and corpsmen augmentation, may be required.

5. Medical Equipment. Medical equipment for desert operations that should be considered above the standard load-out:

a. Individual first aid kits should include eyewash and lip balm (Chapstick).

b. Sunblock should be provided to all individuals (SPF 15, or greater).

c. Special immunizations should be administered prior to entering the region.

4.2.11 Tactical Route Planning. Route planning should be conducted in accordance with the NSWMPG to identify potential enemy threats during the operation. Special considerations for the desert environment include:

4.2.11.1 Quality of Charts/Maps. Many of the desert areas of the world are unreliably mapped, and the available maps are small scale and not useful to the small unit tactical commander. Such small-scale maps typically do not show the topographic features that are most important to the patrol. The patrol planner should acquire an up to date map of a satisfactory scale (at least 1:50,000), preferably marked with the latest known topographic information.

4.2.11.2 Terrain Information. Current and detailed intelligence terrain studies are mandatory. These must identify and accurately locate water sources, steep ridges, etc., that may be deciding factors in routing to the target area. This information is critical in:

1. Determining mobility in the difficult desert environment

2. Evaluating options and locations for insertion and extraction

3. Developing an E&R plan.

4. Planners should analyze all features of the chart to select primary and alternate routes, route points, and route areas. Terrain features important to the desert environment are:

a. Oasis

b. Valleys

c. Wadis

d. Rivers (width, depth, character of current, condition of banks, bottom composition, and gradient)

e. Areas with shifting sands and large dunes

f. Valleys or gorges subject to flash flooding

g. Caravan routes and desert roads

h. Populated areas

i. Enemy Location

j. Other water locations.

4.2.11.3 Survival Information. (see Section 4.4) Obtain information on features of the operating area that may enhance the survival of the patrol, such as:

1. Vegetation and wildlife information.

2. Sources of available water. For dense, remote deserts, request IR photography. This may be the best way to find currently available sources of water in the desert.

4.2.11.4 Environmental Information. Planners should provide analysis to include:

1. Ambient light
2. BMNT
3. Sunrise
4. Sunset
5. BENT
6. Moonrise
7. Moonset
8. Moon phase
9. Visibility range
10. Probability of precipitation
11. Times for tides
12. Daytime temperatures
13. Nighttime temperatures.

4.2.11.5 Reconnoitering the Area of Operation. Plans for reconnaissance of the AO should be carefully made. Routine reconnaissance should be conducted on a continuing basis so the timing of a mission is not cued to the enemy by a reconnaissance vehicle. Reconnaissance in the desert is best carried out by aircraft. Desert operational forces must be provided as much information about their mission as available. If the opportunity to reconnoiter the AO is available, key members of the operation must be afforded the opportunity to participate.

1. It is incumbent upon the staff planners to investigate the feasibility of providing aerial reconnaissance to the operational personnel.

2. Operational personnel should participate in the reconnaissance. Every possible means of documenting what is seen through cameras, videos, and recordings must be accessed.

3. Whenever it is impossible to conduct a reconnaissance, determine if recent available overhead photography is available.

4.2.11.6 Reduced Tactical Mobility. The rate of movement over the desert floor will depend on the following:

1. Type of travel: foot or by vehicle

2. Requirement for noise discipline

3. Distance from insertion point to the ORP

4. Condition of the terrain along the route of approach

5. Soil type and condition

6. Number and difficulty of obstacles to be traversed

7. Physical condition of the platoon and attachment personnel.

The baseline tactical mobility of the patrol on foot in the desert is four kilometers per day (based on an eight-hour patrol in each 24-hour period).

The baseline tactical mobility of the patrol in vehicles in the desert is 40 kilometers per day (based on an eight-hour patrol in each 24-hour period, proceeding at a suitable speed for the situation).

4.2.11.7 Infiltration Route. Select avenues of approach to the target or ORP based on a detailed reconnaissance of the area of operation. Consider no portion of the desert terrain impassable. The enemy will defend the most approachable directions the heaviest so, in reality, the most difficult approach may be the safest.

1. Give strong consideration to difficult approaches during any visual reconnaissance or during map and photo study.

2. Plan and memorize the route before leaving, so that if required, the platoon can follow it quickly and accurately without reference to the map.

4.2.12 Tactical Communications Planning

4.2.12.1 Communications Planning Considerations. The NSWTG/TU to which the patrol reports must familiarize itself with the local effects of HF radio waves in the desert environment to plan both TG/TU and SEAL communications (See the SEAL Communication Handbook.) The most important considerations include:

1. Connectivity. Ensure the connectivity between the MCT, the SEAL insertion/extraction craft, and the SEALs by checking:

    a. Frequencies

    b. Times

    c. Radio equipment compatibility

    d. Correct antennas: height and direction.

2. CEOI. Develop a CEOI, including a "lost communication" plan.

    a. The NSWTG/TU should not assume, when nothing is heard from the field, that communications are not being received by the platoon. The NSWTG has greater transmit power than the platoon and is usually heard.

    b. Use a repeater or relay to assure communication linkage. This could be aircraft or a high ground mounted repeater.

    c. Minimize communication requirements to and from the field by not requiring unnecessary contacts.

    d. Use pre-arranged communication windows.

4.2.12.2 Weather Related Communications Degradation. The type of desert and the season of the year may degrade communication capability, summarized as follows:

1. Climate. The hot and dry climate has the following effects:

    a. Battery shelf life and operating life are reduced by 20 percent.

        (1) Keep batteries in a cool, dry location if possible.

        (2) Keep batteries out of the direct sunlight.

    b. Radio power output and sensitivity are decreased by 20 percent.

    c. Sand and dust proofing is required.

    d. Normal preventive maintenance schedules should be modified to provide for inspection for sand or grit and their removal.

2. Seasonal Propagation. Seasonal propagation will vary as follows:

    a. Winter. Cooler, slightly moister conditions improve propagation by 10 percent over normal conditions.

    b. Summer season. Dry seasons deteriorate propagation by 10-20 percent below normal conditions.

4.2.13 Desert Procedures for Tactical Contingency. Tactical contingency plans to cope with the unplanned or unexpected event should provide for communications code words that identify the tactical contingency and the action being taken. The most common of such events in desert operations are:

1. Loss of Communications. Communications may be lost because of equipment failure or temporarily lost due to some geographic feature that blocks the propagation path. In this case, there must be a lost communications plan.

2. Failure to Reach Assigned Rally or Objective Points. This is to be expected in the desert environment where there are many unknowns that could delay the patrol. When this occurs, there must be a back-up plan that can be executed by code word.

3. Medical Emergency. Regardless of a SEAL's superior conditioning, the nature of the desert environment is such that a key member of the patrol could be injured (without enemy action) and be unable to continue. Code words should be designated to indicate the action being taken and assistance required.

4.2.14 Insertion/Extraction Planning

4.2.14.1 Insertion/Extraction Vehicles. Determine what type of insertion platform is best given the desert topography along the route. This may include:

1. SDV

2. Special Boats

3. Helicopters (may involve fastroping or rappelling)

4. Fixed wing aircraft

5. High Mobility Multipurpose Wheeled Vehicles. These vehicles can also be used for infiltration, actions at the objective, and exfiltration.

4.2.14.2 Insertion/Extraction Considerations

1. Does the method used have to blend in with the local type of assets, such as passenger cars, camels, motorbikes, etc., in order to arrive unnoticed at the area where the patrolling to the objective will commence?

2. Can a surface craft get in close to the drop point or will it be necessary for the patrol to move over a shallow estuary or reef?

3. What will indigenous personnel (e.g., nomads or coastal fishermen) be doing in the vicinity of the insertion point?

4. Will indigenous personnel alert the enemy that the patrol has been inserted or that something unusual is happening? There tends to be less awareness in the early morning hours or during the middle of the day when the heat demands that individuals minimize movement. Level of awareness not withstanding, the light factor becomes a greater problem after sunrise.

5. Can the platform selected carry the patrol and its equipment?

a. Aircraft do not carry as much payload in hot, less dense air.

b. Heavier wheeled and track vehicles bog down in sand easier than lighter vehicles.

6. Determine how the vehicle will approach the objective area and where the primary and secondary insertion/EPs will be designated.

7. Plan false insertions or extractions for the patrol as a means of OPDEC.

4.2.15 Foreign Internal Defense in the Desert

4.2.15.1 Naval Special Warfare Forces in Foreign Internal Defense. NSW forces can be called upon to conduct FID in HNs that contain a desert environment. These operations may also involve participation in Low Intensity Conflicts in the desert. NSW planners and operational forces should understand the interrelationships necessary with foreign HNs. Training the forces of allied and friendly nations in UW is a mission of SOF. NSW personnel should be ready to impart a wide variety of military skills for the desert environment directly applicable to these allied and friendly forces. To accomplish this, NSW personnel are required to work through:

1. Language barriers.

2. Cultural inhibitions. The importance of respecting a HN's cultural mores cannot be overemphasized.

3. Austere support situations.

4.2.15.2 Training considerations. The planning, preparation, and procedural skills discussed in this manual are the skills that will be taught and exercised in desert FID.

4.2.15.3 United States Support in Foreign Internal Defense. The HN will typically need the following types of support from the NSW Forces to prepare for countering both insurgency and terrorist threats:

1. Applicable training and assistance for resistance fighters (very likely in most desert operations)

2. Detailed tactical intelligence

3. Inexpensive, reliable, and secure communications

4. Supportable, affordable, and appropriate transportation

5. Organized and operating logistics systems

6. Means of relating positively with the civilian population

7. Medical support and training

8. Civic action and civil engineering project assistance

9. Improved development for manufacture of military goods

10. Locating of non-U.S. sources of material

4.2.15.4 Host Nation Personnel and Organizations

1. NSW Interrelations with HNs. In order to better understand the interrelationships between NSW forces and HN organizations and personnel, each country must be examined individually and each specific area of interest must be identified and incorporated into the strategy to best achieve the operational goals in that region.

2. Information Sources. NSW planners and operational forces should obtain information in the following areas when operating or training in foreign nations:

   a. HN military structure.

   b. Interrelationships between the HN military organization and the HN government.

   c. Where the bases of power lie within the nation such as national/state police, military, religious leaders, etc.

   d. Respective chains of command within the HN law enforcement and military.

   e. Feelings of the populace regarding U.S. assistance, specifically NSW involvement.

   f. Religious practices that may affect military operations. Since many of the desert countries have strong religious influences, religion must be considered in determining the level of assistance and its effect.

4.2.16 Survival in Nuclear, Biological, and Chemical Warfare (under review)

4.3. DESERT PATROL PREPARATION

4.3.1 General. This section will present information on clothing, personal equipment, organizational equipment, medical aspects, and planning considerations in preparing for desert operations.

4.3.2 Desert Training. SEALs should not be assigned to a desert mission unless they have had training in desert operations, preferably training keyed to the specific desert area.

1. Planned operations are preferred.

2. Time sensitive operations may be carried out, but the likelihood of success of such missions in the demanding desert environment (without adequate area-specific training) is significantly diminished.

4.3.3 Desert Clothing

4.3.3.1 Desert Uniforms. Because most desert regions experience extremes of temperatures, the clothing worn must keep a SEAL warm at night or when temperatures drop low. It must also prevent loss of too much sweat through evaporation and provide protection from the sun's harsh rays during the hot days.

1. General. Clothing should be loose and layered. The standard issue desert uniform satisfactorily meets those requirements when properly worn.

2. Uniform Trousers and Top. Wear the full uniform, sleeves down, to provide total protection over the trunk, arms and legs. This prevents severe sunburn and dehydration (sweat will not evaporate as rapidly). Full

coverage of the body also provides protection from insects. The clothing should be arranged to allow layering with undergarments and outer shells to compensate for the great temperature ranges.

a. Lightweight desert camouflage uniforms are the best.

b. Clothing helps ration perspiration by absorbing it and cooling the body through evaporation.

3. Boots. These should provide good support and protection from the heat, sand, and rough terrain. They should be lightweight and not too tight to allow for swelling of the feet from the heat. Danner brand and other cold weather type boots are too insulated and would become uncomfortably hot in the desert heat and sun. Use good quality desert boots.

a. Jungle boots allow sand to penetrate through vents.

b. Leather combat boots wear out quickly, especially in rocky terrain. Use saddle soap or natural oils to keep leather boots from cracking.

c. Boots without steel plates in the sole are cooler.

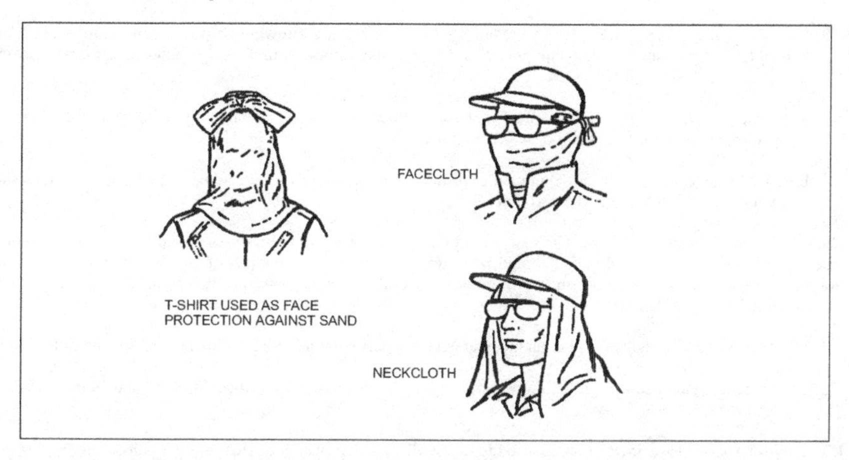

Figure 4-4. Head Protection

4. Headgear. A hat must be worn. One that covers the back of the neck and provides shading for the ears and face is preferred. It should allow an air space above the scalp. The desert nomads use a turban for increased protection against the heat. These are good and can be improvised as shown in Figure 4-4.

5. Gloves. Gloves provide protection from both hot weapons and irritating vegetation and rocks. Use weight lifting gloves or any preferred gloves with trigger finger cut off to allow trigger access.

4.3.3.2 Coloring. When conducting operations in the desert, consideration must be given to camouflage. The AO will have a general pattern or color and will dictate what camouflage type will be best. The standard desert camouflage issue uniform is sufficient for most regions.

1. Jungle Uniforms. Wearing jungle camouflage or green utility uniforms may appear adequate for conduct of operations at night in the desert, but plan for the possibility of remaining in the target or AO during daylight hours for unplanned reasons. Therefore, the uniform worn should blend best in both day and night situations.

2. Visibility. At night, green or solid dark colored uniforms stand out as black against lighter desert backgrounds.

4.3.4 Personal Equipment. See Appendix E for a list of recommended first, second, and third line equipment.

4.3.4.1 Weatherproofing. All equipment must be protected from water, sand, heat, light, and dust.

1. Waterproofing. This will normally involve making the equipment temporarily waterproof during the insertion (and extraction) if movement is over water.

2. Dust proofing. Protect equipment with cloth covers, seals, etc. Blowing dust and sand from wind and helicopter blades can get into equipment and create malfunctions in radios, weapons, cameras, vision devices, demolition accessories, etc.

   a. Plastic coverings will keep out dust and sand. However, such protection must be removed prior to operating electronic equipment or it will overheat.

   b. Oil or grease lubrication on equipment will allow dust and sand to accumulate. Dry graphite lubricant should be used in any desert environment

4.3.4.2 Weight Considerations. The more weight added to a person, the less mobile he will be and the more water he will consume. An individual should never carry more than 80 pounds of equipment in the desert to include his weapon, ordnance, uniform, food, and water. (Remember: water weighs two pounds per quart or eight pounds per gallon). See Paragraph 4.3.7 for a more complete discussion of water requirements.

1. Carefully plan all weights for platoon members, paying particular attention to amounts of ammunition.

2. Ensure all weight is equitably distributed among platoon members. Appendices B, D, E, and G are provided to assist in this effort.

4.3.4.3 Heat Considerations. Increased body heat will increase sweating. Heat will be gained not only from the direct rays of the sun but also by reflective heat from the desert sand, conductive heat from sitting on rocks or desert sand, and hot desert winds. The temperature can be 30 degrees cooler a foot above the ground than on the sand. Rule of thumb: Add 30 degrees Fahrenheit to air temperature to estimate sand temperature.

4.3.5 Camouflage Considerations. Individual camouflage is personal concealment a SEAL uses in combat to surprise and deceive the enemy. All patrol members must know how to use the ground, surroundings, and other aids for effective concealment.

4.3.5.1 Personal Camouflage. The desert camouflage uniform is mandatory in desert operations. Use only desert color schemes for operations in the desert.

4.3.5.2 Equipment Camouflage. Dark green web gear and backpacks are highly visible and reveal a concealed SEAL. All equipment (weapons, rucksacks, web gear, and LBE) must be individually camouflaged. Use tan or beige paint for most equipment. Tape is another option and is best for rifles. Use spray paint for ease. Use a six-foot by six-foot desert camouflage net (cut from a larger equipment camouflage net) for personal concealment and shade. This should be a standard item carried on patrol.

4.3.5.3 Shadows. Another problem encountered in the desert is the awareness of an individual's shadow. Observation from overhead can easily detect shadows not visible from an observer on the ground.

4.3.5.4 Light. Be wary of light discipline in the desert. The glow of a red lens flashlight alone can give away a platoon's position.

4.3.6 Food

4.3.6.1 Meal, Ready-to-eat. Although these are the preferred food on patrol, do the following before operations:

1. Open MRE package and remove any unnecessary contents to reduce weight and bulk.

2. Reseal MRE package with tape.

4.3.6.2 Improvised Foods. The best food source for quick energy on desert operations may be those improvised by the individual.

1. Do not use items with too much sodium.

2. Use nutritious meal replacement bars of complex carbohydrates.

4.3.6.3 Food Consumption. If SEALs expend more calories than they take in, they will be more prone to heat illnesses. Since men may lose their desire for food in hot climates, they must be encouraged to eat. The heavier meal of the day should be eaten in the cooler hours. Never eat food if there is no water available. Food requires water to digest and eating actually increases dehydration. Rule of thumb: no water-no food.

4.3.7 Water

4.3.7.1 Water Consumption. Water is a physiological as well as a psychological need. The human body requires a given amount of water for a certain level of activity. On patrol a SEAL cannot carry all the water he needs. In order to compensate, do the following:

1. Super-Hydration/Over Drink System. This is a method of ensuring adequate water intake. Thirst by itself is not a good indicator of how much water must be consumed. Relying on this indicator alone, most individuals will consume only 2/3 of their daily water needs. The average man can hold approximately two quarts of water. Do NOT try to reduce water intake to test the lower limits of your body's daily requirements!

   a. Begin to drink water at least 6 hours prior to mission departure. Drink at least one quart per hour for the six hours prior to patrol commencement.

   b. Continue to drink water throughout the insertion phase. Drink water from non-personal canteens placed in the insertion craft.

2. Hydration Test. Prior to mission departure check for proper water consumption with a urine check.

   a. The color of urine gives a rough indicator of hydration status. Dark urine indicates water deficiency. Clear, uncolored urine is correct.

   b. Inability to urinate reflects an emergency situation showing advanced stages of dehydration. Immediate water intake and medical attention is required.

3. Water Temperature. Do not consume ice-cold water (or beverages). The temperature confuses the body's thermostat resulting in less sweating and subsequent overheating.

SALT REQUIREMENTS	
Two methods of making up a solution of 0.1 percent salt in drinking water are described below. Table salt is used throughout, and the container should be vigorously stirred or shaken after adding the salt	
Addition of Table Salt to Water	
Table Salt	**Amount of Water**
2 ten-grain crushed salt tablets or ¼ teaspoon	1-quart canteen
4 ten-grain crushed salt tablets	2-quart canteen
1¼ level mess kit spoons	5 gallon can
9 level mess kit spoons or 3/10 pound	36 gallon lister bag
1 pound	100-gallon tank
1 level canteen cup	250 gallons (in water trailer)
ADDITION OF SATURATED SALT SOLUTION [1]	
Amount of Solution	**Amount of Water**
½ canteen cup (quart size)	1-quart canteen
1 canteen cup (2-quart size)	2-quart canteen
1 mess kit spoonful	1 gallon [2]
2/3 canteen cup	36-gallon lister bag
4 canteen cups	250 gallons (in water trailer)
Notes 1. To prepare the solution (approx 26 percent) dissolve 9 level mess kit spoons in 2/3-canteen cup of water. 2. Follow logical progression i.e., 3 spoonfuls to 3 gallons, 12 spoonfuls to 12 gallons, etc.	

Figure 4-5. Salt Requirements

4. Salt and Electrolytes. In addition to water, salt and electrolytes are lost in sweat. The daily diet contains adequate amounts of salt; however, when acclimating, additional salt is necessary. When the body needs

salt, salty water does not taste salty. Use .1 percent saltwater during acclimation, and as necessary on patrol. See Figure 4-5 for salt requirements.

a. Salt tablets. Never swallow whole salt tablets, as they will pass through the body and will not be dissolved. Crush and dissolve the salt in the water until the water tastes slightly salty.

b. Alternating drinks. Sip salt water alternately with pure water.

5. Salt/Electrolytes During Acclimation. Drink .1 percent salt water in equal amounts with pure water. Drink an equal amount of fluid containing an electrolyte supplement. Drink at least 9 quarts a day (more in climates where daytime temperature exceeds 100 degrees Fahrenheit).

6. Salt/Electrolytes on Patrol. Follow the same guidelines used for acclimation.

a. Reduction of salt/electrolyte intake will be required if diet is salty.

b. Indicators of too much salt are headaches and tendencies to dehydrate. Adjustment of salt intake is a personal responsibility, and the correct amount will vary among individuals.

c. Indicators of insufficient salt are headaches, lightheadedness, and premature exhaustion.

7. Smoking. Smoking or any use of tobacco increases the desire for water and should be avoided.

4.3.7.2 Carrying Water

1. Ensure canteen tops are NBC compatible (have gas mask tube access).

2. Use standard issue 2-quart canteens or 2-liter Nalgene type rectangular, plastic bottles. Improvise with 2-liter plastic soft drink containers. Wrap these with riggers' tape.

3. Drink the water from the rucksack first; then drink water from the canteens on the web belt. Small Nalgene four-ounce bottles should be carried on the person as a first line water supply for E&R. Four bottles are recommended, two in each trouser pocket.

4.3.7.3 Other Beverages

1. Carbonated Beverages. Do not substitute carbonated beverages for water. They produce a full feeling in the stomach that defeats the super hydration/over drink system.

2. Alcoholic Beverages. NEVER drink alcoholic beverages when water is in short supply because alcohol requires extra water for the body to process it.

4.3.8 Sleeping. Since the primary time for movement is during the night, sleeping (if required) will occur during lay up in the day. In the most severe desert climates, it may be too hot to sleep during the day. If this is the case, the PL should plan sleeping periods at dawn and dusk, with movement at night. Daytime would be spent resting in camouflaged, shaded positions.

1. Camouflage Net. Plan to utilize this to provide both shade and concealment.

2. Temperature. Temperatures change from night to day and from season to season. Consider this during mission planning.

3. Sleeping bag and ground pad are highly recommended.

4.3.9 Surveillance, Target Acquisition, and Night Observation Equipment. Electronic devices such as cameras, NVEO, beacons, laser target designators, or sensors may be required in accordance with the mission plan. These items should be tested to ensure they are in proper working order prior to launching on the operation.

1. Assign patrol members the responsibility for preparing and carrying designated items.

2. Conduct required training for this equipment.

4.3.10 Weapons and Ordnance

4.3.10.1 Weapons Selection. The patrol must carefully select the best weapon or weapons combination to satisfy the mission requirement.

1. M4 Assault Rifle. This is the standard rifle and has the following characteristics:

   a. Reduced reliability in the desert or sand environment.

   b. Maximum effective range of approximately 500 yards.

   c. Lighter weapon and ammunition load-out weight.

   d. Compatible with the M203 grenade launcher (but adds weight).

2. M14 Assault Rifle

   a. More reliable in desert/sand environment.

   b. Maximum effective range of approximately 660 yards.

   c. Heavier weapon and ammo load-out weight.

   d. Greater knock down power than M4.

   e. Ammunition compatibility with Mk 43 Mod 0.

3. The M203 Grenade Launcher is not compatible with the M14 Assault Rifle. A patrol must include a M79 Grenade Launcher as a separate weapon to provide that capability.

4.3.10.2 Test Firing. Test fire weapons under the same conditions that will be encountered on the patrol. For example, if the insertion has the patrol wading through heavily salinated water along a desert shoreline to reach a beach or boat, ensure there is no weapon malfunctioning from immersion in salt water. There may not be another opportunity to test fire while on patrol.

4.3.10.3 Maintenance. Petroleum lubrication on weapons will cause dust and sand to adhere to the surfaces of the weapon parts, creating the potential of a jam.

1. Dry graphite is the recommended lubricant.

2. Weapons systems should be cleaned at least twice daily.

3. Keep plastic barrel caps or tape over the muzzles to prevent sand and grit from getting into the barrels.

4.3.10.4 Ordnance and Demolitions

1. Keep ammunition and other ordnance items away from direct sunlight. A rule of thumb is that if they can be held in bare hands, they are safe.

2. Conduct a thorough target analysis. Determine what combination of ordnance and its placement will produce maximum damage with minimum explosive weight.

   a. In the desert, extra demolition weight translates into sweat, water consumption, and reduced endurance.

   b. Appendix D provides a summary of weapons and ordnance weights.

4.3.11 Communications. Communications are adversely effected by desert conditions. These conditions include:

1. Dust

2. Sand

3. Mirages

4. Heat

5. Great Distances

6. Wave Propagation

7. Difficulty in rigging raised field-expedient antennas

8. High winds.

Communications should be planned for an absolute minimum level unless tactical requirements dictate otherwise.

4.3.11.1 Equipment Selection. Equipment carried depends on the situation with regards to:

1. Intra-squad communications requirements

2. Electronic transmissions to support assets

3. Enemy DF capabilities

4. C^2 communication requirements

5. Use the radio type that best suits the mission. For example, short range, near line-of-sight desert terrain calls for a VHF radio (PRC 113). See Appendix B for man pack radios available for desert operations.

4.3.11.2 Ultra High Frequency Satellite Communications. UHF SATCOM is the preferred type of communication for the desert. It is less detectable than HF. Generally, there should be a primary and back-up communications capability. The back-up can be mechanical signals vice electronic transmissions to reduce communications equipment weight.

4.3.11.3 Very High Frequency. The desert environment is likely to degrade the transmission range of radios, particularly VHF fitted with encryption equipment. This degradation is more likely to occur in the hottest part of the day, approximately from 1200 to 1700. Changing to non-encrypted communications will increase range.

4.3.11.4 Antennas. Electrical grounds are poor in desert terrain because the surface of the desert floor lacks moisture. Poor grounding reduces radio communications ranges. Whip antennas lose one-fifth to one-third of their normal range. Complete antenna systems such as horizontal dipole antennas and vertical antennas with counterpoises will be required.

1. General

   a. Keep all radios cool and clean.

   b. Keep radios in shade when possible.

   c. Keep radios covered with a damp towel. Do not block air vents.

   d. Ensure sufficient batteries are available to last 150 percent of the expected mission duration time.

   e. Pour water or urinate on ground wires to increase/improve grounding characteristics.

4.3.12 Patrol Preparation

4.3.12.1 Acclimation. Getting used to the heat requires time. The acclimation period differs for different people. In general, better physical fitness equates to faster acclimation. See Figure 4-6.

1. Normal Period. Allow a minimum of 14 days for SEALs to acclimate (21 days is preferable), with progressive degrees of heat exposure and physical exertion.

2. Heat Effects. Acclimation increases heat resistance, but it does not reduce the fatiguing effect of heat. Work has to be done slowly and efficiently in the heat of the day. Heat can be so severe during the day that men will find it difficult to even sleep.

3. Alternative Hours. Make every effort to work during cooler evening and early morning hours.

4.3.12.2 Preparation

1. Ensure personnel receive a minimum of six hours of rest per day.

2. Give 15 to 20 minutes of rest per hour if doing heavy work.

3. Ensure all personnel drink water containing electrolytes at least hourly.

4. Drink at least one quart after meals.

5. Stay in the shade as much as possible; if in the sun, keep full uniform on with sleeves rolled down. Do not wear T-shirts or shorts in the sun.

6. Be alert for heat injuries.

4.3.13 Baseline Patrol

4.3.13.1 Duration of a Foot Patrol. The baseline desert mission is 36 hours. The patrol should be able to cover a total of 6 kilometers. (3 kilometers to the objective and 3 kilometers back to the EP). 8 to 10 quarts of water will be required depending on daytime temperatures. It entails the following phases:

1. Night One

    a. Last light insertion

    b. Infiltration to target.

2. Next Day

    a. Laying up/conducting reconnaissance.

3. Night Two

    a. Mission accomplishment

    b. Exfiltration

    c. Extraction.

4.3.13.2 Duration of a Vehicle Patrol. The baseline vehicle patrol is 60 hours. The patrol should be capable of covering 100 kilometers. (50 kilometers in and 50 kilometers out to the EP). Each man should maintain 8 to 10 quarts in his personal equipment. The remaining water should be carried in the vehicles. Drink water from the vehicles first. With vehicles, each man will be able to carry more equipment and weapons/ordnance.

4.3.13.3 Baseline Mobility. Mobility and duration in the desert depend on water and its weight. For reference purposes, baseline mobility is:

1. Foot patrol: 4 kilometers per day (based on an 8-hour patrol in 24 hours).

2. Vehicle patrol: 40 kilometers per day (based on an 8-hour patrol in 24 hours).

4.3.13.4 Patrol Rest Stops. The figures in paragraph 4.3.13.3 includes the following rest stops:

1. Three 15-minute rest stops (one every two hours) also serve as a "SLLS" period for alerts regarding enemy activity.

2. One 45-minute stop is for a communications window.

3. One 45-minute delay is for obstacle crossing.

4.3.13.5 Baseline Weight Assignment for Patrol Positions. Appendix D provides the weights of weapons and accessories. The individual equipment and consumable weights assigned to each man for a baseline 16-man patrol (by patrol position) are listed in Figure 4-7 at the end of this section. The table is based on patrol use of the M4 rifle. Weapon and ordnance weight will increase if the M14 rifle is used. The primary reason for using the M14 rifle is the increased range that may be necessary for some missions. These weights are a guide for the tactical planner. The PL should recognize the relation between weight, amount of water carried (adds weight), and patrol mobility.

PERIODS OF ACCLIMATION		
Moderate Conditions: Desert air temperature below 105 °F		
	Hours of Work	
	MORNING	AFTERNOON
First day	1	1
Second day	1½	1½
Third day	2	2
Fourth day	3	3
Fifth day	Regular duty	
Severe Conditions: Desert air temperature above 105 °F		
	MORNING	AFTERNOON
First day	1	1
Second day	1½	1½
Third day	2	2
Fourth day	2½	2½
Fifth day	3	3
Sixth day	Regular duty	

Figure 4-6. Guide for Increasing Length of Work Periods

4.3.14 Rehearsals. Every effort should be made for operational forces to rehearse. Time permitting, rehearsals should be conducted in terrain and climatic situations similar to that expected in the AO. This will aid in acclimating men to the expected environment of the actual operation.

4.3.14.1 Rehearsal Planning. Identify an appropriate rehearsal area early in the planning process.

1. Arrange transportation to and from the rehearsal area.

2. Establish OPSEC early.

3. Be objective in determining the conditions under which the rehearsal will be run.

4. Involve the supporting forces in the rehearsal planning.

5. Ensure the communication plan, the C^2 arrangements, and the evaluation criteria are realistic for the rehearsal.

4.3.14.2 Rehearsal Advantages

1. Operational Forces

   a. Effects on personnel can better be anticipated.

   b. Negative operational performance of equipment can be discovered and compensated for before the actual operation.

2. Supporting Forces

   a. Supporting force personnel will better understand their role.

   b. Supporting force equipment will be tested to determine its effectiveness and reliability prior to the mission.

FIRE TEAM ONE	
1. PT/Rifle Man	
Water	20 lb
Weapons	7 lb
Special equipment	2 lb
Rucksack and personal items	11 lb
Ammunition and ordnance	10 lb
TOTAL BASELINE WEIGHT	50 lb
2. PL/Squad 1 Leader/Element 1 Leader (PL)	
Water	20 lb
Weapons	7 lb
Special equipment	10 lb
Rucksack and personal items	13 lb
Ammunition and ordnance	10 lb
TOTAL BASELINE WEIGHT	60 lb
3. Rifle Man/RTO	
Water	20 lb
Weapons	7 lb
Special equipment	19 lb
Rucksack and personal items	12 lb
Ammunition and ordnance	7 lb
TOTAL BASELINE WEIGHT	65 lb
4. AW Man	
Water	20 lb
Weapons	19 lb
Special equipment	0 lb
Rucksack and personal items	8 lb
Ammunition and ordnance	28 lb
TOTAL BASELINE WEIGHT	75 lb

Figure 4-7. Patrol Organization/Weight Assignments

FIRE TEAM TWO	
5. Rifle Man/GN/FT B Leader	
Water	20 lb
Weapons	10 lb
Special equipment	0 lb
Rucksack and personal items	15 lb
Ammunition and ordnance	15 lb
TOTAL BASELINE WEIGHT	60 lb
6. Rifle Man/GN/HM	
Water	20 lb
Weapons	7 lb
Special equipment	8 lb
Rucksack and personal items	15 lb
Ammunition and ordnance	10 lb
TOTAL BASELINE WEIGHT	60 lb
7. AW Man	
Water	20 lb
Weapons	19 lb
Special equipment	0 lb
Rucksack and personal items	8 lb
Ammunition and ordnance	28 lb
TOTAL BASELINE WEIGHT	75 lb
8. Rifle Man/GN	
Water	20 lb
Weapons	10 lb
Special equipment	0 lb
Rucksack and personal items	15 lb
Ammunition and ordnance	15 lb
TOTAL BASELINE WEIGHT	60 lb

Figure 4-7. Patrol Organization/Weight Assignments

FIRE TEAM THREE	
9. Rifle Man/GN/Element 3 Leader	
Water	20 lb
Weapons	10 lb
Special equipment	0 lb
Rucksack and personal items	15 lb
Ammunition and ordnance	15 lb
TOTAL BASELINE WEIGHT	60 lb
10. AW Man	
Water	20 lb
Weapons	20 lb
Special equipment	6 lb
Rucksack and personal items	8 lb
Ammunition and ordnance	21 lb
TOTAL BASELINE WEIGHT	75 lb
11. Rifle Man/GN	
Water	20 lb
Weapons	10 lb
Special equipment	0 lb
Rucksack and personal items	15 lb
Ammunition and ordnance	15 lb
TOTAL BASELINE WEIGHT	60 lb
12. Rifle Man/GN	
Water	20 lb
Weapons	10 lb
Special equipment	0 lb
Rucksack and personal items	15 lb
Ammunition and ordnance	15 lb
TOTAL BASELINE WEIGHT	60 lb

Figure 4-7. Patrol Organization/Weight Assignments

FIRE TEAM FOUR	
13. Rifle Man/APL/Squad 2 Leader/FT 4 Leader (APL)	
Water	20 lb
Weapons	7 lb
Special equipment	10 lb
Rucksack and personal items	13 lb
Ammunition and ordnance	10 lb
TOTAL BASELINE WEIGHT	60 lb
14. Rifle Man/Squad RTO	
Water	20 lb
Weapons	7 lb
Special equipment	19 lb
Rucksack and personal items	12 lb
Ammunition and ordnance	7 lb
TOTAL BASELINE WEIGHT	65 lb
15. AW Man	
Water	20 lb
Weapons	19 lb
Special equipment	0 lb
Rucksack and personal items	8 lb
Ammunition and ordnance	28 lb
TOTAL BASELINE WEIGHT	75 lb
16. RS/Rifle Man/GN (RS)	
Water	20 lb
Weapons	7 lb
Special equipment	10 lb
Rucksack and personal items	13 lb
Ammunition and ordnance	10 lb
TOTAL BASELINE WEIGHT	60 lb

Figure 4-7. Patrol Organization/Weight Assignments

4.4 MEDICAL AND SURVIVAL TECHNIQUES

4.4.1 General. The desert is one of the most difficult environments in which to survive. If a SEAL is placed in a survival situation, he will probably be both surviving and evading capture. There will be major concerns to contend with, but if prepared, the chances for success increase dramatically. The major enemies, besides the opposition, are: FATIGUE, PAIN, HEAT, COLD, HUNGER, and THIRST.

4.4.2 Survival Preparation. Desert areas are characterized by scanty rainfall. Extremes of temperature are as characteristic of deserts as lack of rain and great distances. Summer daytime heat in any desert of the world will make an individual's sweat glands work at capacity. Each SEAL will need drinking water and electrolyte to maintain that production. Most important in any desert is the extremely long time between opportunities to hydrate unless water is carried. Water is the biggest problem when surviving in the desert.

4.4.2.1 Travel and Movement. First, if possible, always wait until nightfall or early morning to travel. This not only conserves body fluids and helps prevent fatigue, but also makes it more difficult for the enemy to find tracks or detect movement in the area. Move towards a coast, a known route of travel, a water source, or an inhabited area. Things that will aid desert movement include:

1. Along a coast, wetting your clothes in the sea can conserve perspiration.
2. Follow the easiest route possible. Avoid loose sand and rough terrain. Follow trails unless there is too much risk of encountering the enemy or unfriendly indigenous inhabitants.
3. Care for the feet. Use good quality boots. Keep feet dry. Tape or pad any areas that may be subject to blisters. Avoid blisters! They inhibit mobility and are subject to infection.
4. Avoid crossing sand barefooted as it can burn the feet.
5. The absence of terrain features makes estimating distance difficult.
6. In hot deserts, reserve all strenuous activities for the cooler portions of the day or for night. Do not hurry! One can survive longer on less water if perspiration is kept down.
7. Take routes not expected by the enemy in order to reduce the risks of capture or compromise.
8. Minimize tracks and avoid leaving signs for the enemy to pick up movements and location.

4.4.2.2 Essential Clothing and Equipment. In hot deserts, one needs clothing for protection against sunburn, heat, sand, and insects, so don't discard any of it.

1. Solar Protection
   a. The head, the back of the neck, and other unprotected areas such as arms, must be kept covered during the day. Wear a cloth neckpiece to protect the back of the neck from the sun. Wear all clothing loosely. (To make a desert "burnoose," see Figure 4-8.)
   b. Open clothing only when well shaded. Reflected sunlight can cause sunburn.
   c. Keep heat out of the body by keeping clothes on. It may feel more comfortable without a shirt or pants, because sweat evaporates fast. But then more water is needed. Moreover, bare skin invites sunburn that

can be a painful problem. By keeping the operational uniform on, some degree of camouflage will be maintained to prevent detection by the enemy.

d. Light-colored clothing turns away the heat of the sun better than dark clothes.

e. Clothing also keeps out the hot desert air.

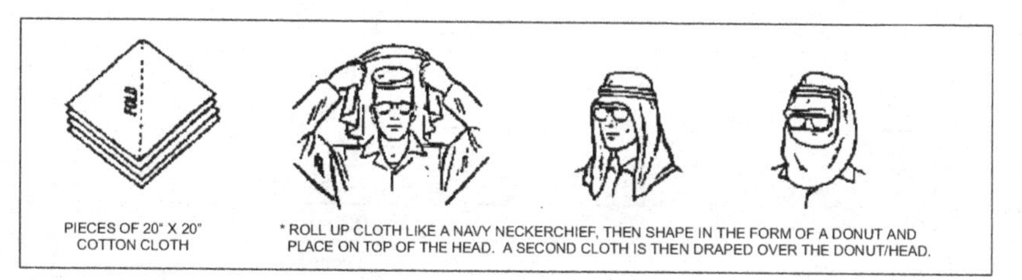

Figure 4-8. Improvised Burnoose or Arab Headdress

2. Foot Protection. Keep sand and insects out of the shoes and socks, even if frequent stops are necessary to clean them out.

3. Thermal Protection. Be prepared for cold weather at night. The temperatures can drop 60 to 70 degrees Fahrenheit after the sun goes down and cold can be as incapacitating as the heat during operations.

4. Desert Survival Gear. Figure 4-9 provides a reasonable equipment kit for surviving and evading in the desert.

4.4.3 Medical and First Aid

4.4.3.1 Health and Hygiene

1. Personal Hygiene. Proper standards of personal hygiene must be maintained as a deterrent to disease.

   a. Cleaning the areas of the body that sweat heavily is especially important.

   b. Underwear should be changed frequently and foot powder used often.

2. Health. Check for signs of injury, no matter how slight, as the dirt of the desert, or insects, may cause infection of minor cuts and scratches. Using small quantities of disinfectant in washing water can reduce the chance of infection. Minor sickness can have serious effects in the desert. Prickly heat and diarrhea, for example, can upset the sweating mechanism and increase water loss, increasing the possibility of heat illnesses. The buddy system can help ensure that prompt attention is given to these problems before they incapacitate individuals.

DESERT SURVIVAL EQUIPMENT KIT	
Knife	40 lb test fishing line
Compass	Snare wire
Plastic signal mirror	25 ft of 550 cord
Personal first aid kit	Tweezers
Minimum 4 qt of water	Small Desert Survival Manual
Flop hat (desert camouflage)	Comfortable, good hiking boots
White laundry bag for snakes	Sunglasses
"Steel sack" garbage bags	Sunscreen/sunblock
Small flashlight/batteries	Signal panel
3 ft of plastic tubing	Collapsible container
6 x 6 ft square cammie netting	Bottle of iodine tablets
Safety pins (6)	Small sponge
60 x 60 in sheet of clear plastic	4 oz Nalgene bottles (water filled)

Figure 4-9. Desert Survival Equipment

3. Diseases. Diseases found in the desert include:

   a. Plague

   b. Typhus

   c. Malaria

   d. Dengue fever

   e. Dysentery

   f. Cholera

   g. Typhoid.

Vaccines or prophylactic measures can prevent some of these. High levels of field hygiene and sanitation are necessary to prevent disease, especially where there are no vaccines or prophylactic measures.

4.4.3.2 Fungus Infections and Prickly Heat. The excessive sweating common in hot climates can aggravate prickly heat and some forms of fungal infections of the skin. The higher the humidity, the greater the possibility of their occurrence. Although most deserts are not humid, there are exceptions, and these diseases are likely in humid conditions.

TYPE	CAUSE	SYMPTOMS	TREATMENT
Dehydration	Loss of too much water. About two-thirds of the human body is water. When water is not replaced as it is lost, the body becomes dried out	Sluggishness and listlessness.	Give the victim plenty of water.
Heat Exhaustion	Loss of too much water and salt.	• Dizziness • Nausea • Cramps • Rapid, weak pulse • Cool, wet skin • Headache.	• Move the victim to a cool, shaded place for rest • Loosen the clothing • Elevate the feet to improve circulation • Give the victim cool salt water (two salt tablets dissolved in a canteen of water); natural seawater should not be used.
Heat Cramps	Loss of too much water	• Cramps that are relieved as soon as salt is replaced	• Same as for heat exhaustion
Heatstroke	Breakdown in the body's heat control mechanism. The most likely victims are those who are not acclimated to the desert, or those who have recently had bad cases of diarrhea. Heatstroke can kill if not treated quickly.	• Hot, red, dry skin (most important sign) • No sweating (when sweating would be expected) • Very high temperature (105 to 110 degrees) • Rapid pulse • Spots before eyes • Headache, nausea, dizziness, mental confusion • Sudden collapse	• Cool the victim immediately. This is achieved by putting him in a creek or stream; pouring canteens of water over him; fanning him; and using ice, if available • Give him cool saltwater (prepared as stated earlier if he is conscious) • Rub his arms and legs very rapidly • Evacuate him to medical aid as soon as possible

Figure 4-10. Heat Injuries

4.4.3.3 Heat Illnesses: Symptoms and Treatment. There are three levels of heat illness: heat cramps, heat exhaustion, and heat stroke. (See Figure 4-10 for a summary of heat illness symptoms and treatment.) To avoid these illnesses, SEALs must be physically fit, thoroughly acclimated, and sufficiently hydrated with water and necessary salts. Use the Buddy System! Each individual should keep an eye on his buddy to recognize heat stress symptoms quickly. Thirst is not an adequate warning.

1. Heat Exhaustion. The patient is first flushed, then pale, and sweats heavily. His skin is moist and cool. He may become delirious or unconscious.

   Treatment. Treat the patient by placing him in the shade, flat on his back. Give him salt dissolved in water.

2. Heat Cramps. The first warning of heat cramps usually is cramps in leg or belly muscles.

Treatment. Move to shaded area and give salt in any form to balance loss. Administer normal saline 500-1,000 cc intravenously in acute cases. 0.1 percent salt solution in cool water orally will afford both relief and continued protection.

3. Heat Stroke. Heat stroke may come on suddenly and can be fatal. It is necessary to recognize heat stress symptoms quickly. The face is red and the skin is hot and dry. All sweating stops. Headaches are severe and the pulse is fast and strong. Unconsciousness may result.

   Treatment. Cool the patient. Loosen his clothing; lay him down flat, but off the ground and in the shade. Cool him by saturating clothes with water and by fanning. Do not give stimulants. When suffering from heat stroke, the most dangerous of the heat illnesses, there is a tendency to creep away from teammates and attempt to hide in a shady and secluded spot.

4.4.3.4 Dehydration. Thirst is not an adequate warning of dehydration because the sensation may not be felt until there is a body deficit of one to two quarts of water. Force fluids daily. Be on constant lookout for fatigue and dehydration. See Figure 4-11.

1. Partial dehydration. This type affects alertness, lowers morale, and degrades proficiency.

2. Loss of efficiency. A loss of 2½ percent of body weight by sweating (about two quarts) results in a 25 percent efficiency loss. A loss of fluid equivalent to 15 percent of body weight is usually fatal.

4.4.3.5 Sunburn/Exposure Problems and Prevention

1. Radiant Light. The sun's rays either direct or bounced off the ground, affect the skin and can also produce eyestrain and temporarily impaired vision. Overexposure will cause sunburn. It is characterized by:

   a. Reddened skin

   b. Blistering of the skin

   c. Dehydration

   d. Feeling cold since the blood comes to the skin's surface. Remember that the sun is as dangerous on cloudy days as on sunny days.

2. Prevention. Use of sunblock ointments will help prevent the overexposure and damage to the skin. When exposing the skin to sunlight, remember:

   a. Use of Sunscreens (SPF 15 or greater) or sunblock lotions is important.

   b. Sunburn ointments are not designed to give complete protection against excessive exposure.

   c. Excessive sunbathing or exposure to desert sun can lead to serious problems or even fatalities.

4.4.3.6 Respiratory Diseases and Cold Weather Injuries. SEALs may tend to stay in thin clothing until too late in the desert day and become susceptible to chills, so respiratory infections may be common. Personnel should gradually add layers of clothing at night (such as sweaters), and gradually remove them in the morning. Where the danger of cold weather injury exists in the desert, commanders must guard against attempts by inexperienced personnel to discard cold weather clothing during the heat of the day.

1% - 5% of Body Weight	6% - 10% of Body Weight	11% - 20% of Body Weight
Thirst Vague discomfort Economy of movement Anorexia (no appetite) Flushed skin Impatience Sleepiness Increased pulse rate Increased rectal temperature Nausea	Dizziness Headache Dyspnea (labored breathing) Tingling in limbs Decreased blood volume Increased blood concentration Absence of salvation Cyanosis (blue skin) Indistinct speech Inability to walk	Delirium Spasticity Swollen tongue Inability to swallow Deafness Dim vision Shriveled skin Painful urination Numbness of the skin Anuria (decreased or deficient urination)

Figure 4-11. Dehydration Symptoms

4.4.3.7 Snakes. As a general rule, all snakes can be dangerous. Bites even from non-poisonous snakes can become infected.

1. Snakes seek cool areas under rocks, trees, and shrubs.

2. Prior to sitting or resting in these areas, always check for snakes.

3. Check before putting on boots and clothing in the mornings.

4. In the event of snakebite, follow these steps:

   a. Remain calm, but act quickly.

   b. Immobilize the affected part in a position even or slightly below the level of the heart.

   c. Place a one-inch wide constricting band two - four inches up on the heart side of the bite. Bind the limb as tightly as a sprained ankle, taking care not to constrict blood flow.

   d. Do not attempt to cut open the bite or suck out the venom.

   e. Seek medical help. If possible, take the snake's head to aid with identification and ensure administration of proper antivenin.

4.4.3.8 Medical Equipment Kit. Figure 4-12 provides a list of those items that should be included in an individual's basic medical kit. Most items are available through Naval Supply channels:

1. Be medically self-sufficient. Whatever is carried is better than what will be found there.

2. Carry plenty of Imodium or a reasonable substitute to combat diarrhea. This is particularly important since diarrhea can further dehydrate the body. Contact medical personnel as soon as possible.

MEDICAL EQUIPMENT KIT			
Item	**Quantity**	**Item**	**Quantity**
Nylon bag, zipper closure	1	Cannula, trachea (#5 nylon)	1
Ringers lactated injection 500 ml	1	Restricting band, tubing (penrose)	1
7.5" x 8"inch battle dressing	1	Needle and catheter unit 18 ga	1
Battle dressing, small 4" x 7"	1	Tourniquet, web with buckle	2
Large battle dressing 11 3/4"	1	Airway, oropharyngeal	1
Tape, adhesive 1"	1	Needle catheter, unit 16	2
Providone-iodine impregnated pad	9	Needle, catheter unit 14	1
IV Injections set (tubing)	1	Sytyres #3	2
Surgical knife blade #10, package of 6	1	Sterile strips (package)	2
Surgical knife blade #11, package of 6	1	Triangular bandage 37" x 37" x 37"	2

Figure 4-12. Medical Equipment Kit

4.4.3.9 Miscellaneous Medical Considerations

1. Eye and Ear Protection. In some desert areas as much as 80 percent of the ambient light is reflected back at the individual. Eye and ear protection is a must for protection from constant bright sunlight and blowing sand.

    a. Sunglasses with UV protection are necessary. Wear goggles with sunglass lenses if possible. Even though the glare does not seem painful, the very high light intensities of the desert will cause a decrease in one's night vision. If sunglasses are not available, the next best help is to make slit goggles as shown in Figure 4-13.

    b. Another method is to shade the eyes from above with a hat or burnoose that has cloth down the sides of the face.

    c. Smear charcoal or dirt on the bridge of the nose and below the eyes to cut down on the glare.

    d. Have an eye wash solution available for the patrol to cleanse the eyes, avoiding sand and dust irritation and injury.

4.4.4 Signaling in the Desert

4.4.4.1 Improvised Signals

1. Make flares from tin cans filled with sand soaked with gasoline (light it with care). Add oil or pieces of rubber to make dense smoke for daytime signal.

2. Burn gasoline or use other bright flame at night.

3. Dig trenches to form signals.

4. Line up rocks that will throw a shadow that can be easily seen by aircraft overhead.

5. If brush can be found in the area, gather it into piles and have it ready to light when a search aircraft is heard or sighted.

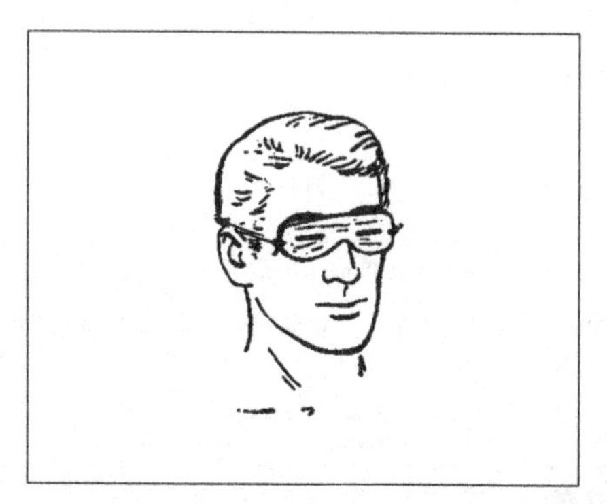

Figure 4-13. Improvised Slit Goggles

4.4.4.2 Prepared Signals

1. Smoke flares and smoke grenades are the best for use in the daytime. Flares and bright flames are difficult to see in the bright daylight.

2. A signal mirror is a very good signaling method. Practice using the signal mirror (should be part of survival kit) or use a brightly polished metal object as a substitute. At all times, keep in mind that signals used to achieve pick-up by friendly forces can also be used by the enemy to gain knowledge of one's location to aid their attempts at capturing evaders or even ambushing friendly rescue forces.

4.4.5 Water Procurement

4.4.5.1 Effects of Water Loss. In the desert, a person's life will depend on his water supply. (Figure 4-14 shows how long one can survive on specific amounts of water at various temperatures.) The normal temperature of a human body is 98.6 degrees Fahrenheit. Any variation from that reduces one's efficiency. An increase of six-eight degrees for any extended period is fatal! When a human sweats, his body loses water, and he must replace the loss by drinking water. Otherwise he will pay for the loss in reduced efficiency and perhaps death. If a man is working in temperatures of 110 degrees Fahrenheit, he can do only about half as much work as he could normally do; one can walk only half as far as could be accomplished in cooler temperatures with plenty of water.

4.4.5.2 Minimizing Water Loss. The only way to conserve water is to ration sweat. As discussed in paragraph 4.4.2.2, protect against excessive evaporation of perspiration (and many annoying insects) by keeping fully clothed. Clothing helps control perspiration by not letting the perspiration evaporate so fast that some of its cooling effect is missed. While an individual may feel cooler without a shirt, he will perspire more (and likely will sunburn). Water rationing recommendations:

1. When drinking water, drink in small sips.

2. If the supply is critical, use water to moisten the lips only.

3. Keep small pebbles in the mouth or chew grass as a means of relieving thirst feelings.

4. Prevent water loss by breathing through the nose.

5. Limit talking.

6. Keep in the shade as much as possible during the day.

7. Do not lie directly on the ground. Improvise methods, if possible, to get a few inches off the ground. The temperature can be 30 degrees cooler a foot above the ground, and that difference can save a lot of sweat.

8. Slow motion is better than speed in hot deserts. If one must move about in the heat, he will last longer on less water by taking it easy. Don't fight the desert - it is impossible to win!

9. Drink water only as needed. Remember that the enemy knows the importance of water also and will use his knowledge of water sources for locating survivors in the desert.

4.4.5.3 Sources of Desert Water

1. Solar Still. A solar still can be made from a sheet of clear plastic stretched over a hole in the ground. Whatever moisture is in the soil plus that from plant parts (fleshy stems and leaves) when they are used as a supplementary source, will be extracted and collected by this emergency device. Obviously, where the soil is extremely dry and no fleshy plants are available, little, if any, water can be obtained from the still. However, in such situations, the still can be used to purify polluted water such as body wastes. The parts for the still are a piece of plastic film about six feet square, a water collector-container or any waterproof material from which a collector-container can be fashioned, and a piece of plastic tubing about 1/4 inch in diameter and four - six feet long. The tubing is not essential, but makes the still easier to use. See Figure 4-15. Certain precautions must be taken.

   a. If polluted water is used, make sure that none is spilled near the rim of the hole where the plastic touches the soil and that none comes in contact with the container, or there is a chance that this freshly distilled water will be contaminated.

   b. Do not disturb the plastic sheet during daylight "working hours" unless it is absolutely necessary.

   c. If a plastic drinking tube is not available, raise the plastic sheet and remove the container as few times as possible during daylight hours.

   d. It takes from 30 minutes to an hour for the air in the still to become re-saturated and the collection of water to begin again after the plastic has been disturbed.

2. Use of Plants

   a. Certain types of vegetation only grow with roots in or near surface water. Insect or animal scratchings around plants are good signs that moisture is present.

b. If a solar still is set up near these, the efficiency will increase because of the amount of moisture in the soil at these points.

c. If plants are used to get water out of the pulp, discard the pulp after chewing or sucking out the moisture. Barrel cactus is a good source for this type of moisture retrieval. Cut off the top and mash or chew the pulp in this cactus.

No Walking At All	Maximum daily temperature (°F) in shade	TOTAL AVAILABLE WATER PER MAN, U.S. QUARTS					
		0	1 Qt	2 Qt	4 Qt	10 Qt	20 Qt
		DAYS OF EXPECTED SURVIVAL					
	120	2	2	2	2.5	3	4.5
	110	3	3	3.5	4	5	7
	100	5	5.5	6	7	9.5	13.5
	90	7	8	9	10.5	15	23
	80	9	10	11	13	19	29
	70	10	11	12	14	20.5	32
	60	10	11	12	14	21	32
	50	10	11	12	14.5	21	32

Walking At Night Until Exhausted And Resting Thereafter	Maximum daily temperature (°F) in shade	TOTAL AVAILABLE WATER PER MAN, U.S. QUARTS					
		0	1 Qt	2 Qt	4 Qt	10 Qt	20 Qt
		DAYS OF EXPECTED SURVIVAL					
	120	1	2	2	2.5	3	
	110	2	2	2.5	3	3.5	
	100	3	3.5	3.5	4.5	5.5	
	90	5	5.5	5.5	6.5	8	
	80	7	7.5	8	9.5	11.5	
	70	7.5	8	9	10.5	13.5	
	60	8	8.5	9	11	14	
	50	8	8.5	9	11	14	

Figure 4-14. Water/Temperature/Time of Survival Chart

3. Miscellaneous Sources. Carry one Katadyn pump per squad to filter what water is found. **TREAT OR FILTER ALL WATER OBTAINED IN THE DESERT!**

a. Rainwater can be caught in a tarpaulin, poncho, raft, or plastic that funnels to a container.

b. Condensation often forms on the surface of objects and plants in more humid regions such as found along the coastlines of some deserts (the area SEALs are more likely to operate in). Cool stones, collected from below the hot surface of the desert, if placed on a waterproof tarp may cause enough dew to collect for a refreshing drink. Exposed metal surfaces like aircraft wings or tin cans are the best dew condensers. They should be clean of dust or grease to get the best-flavored water. If an individual expects a good drink, he must collect the dewdrops very soon after sunrise. When near the coast with salt water, one can construct a beach well or wells. When the tide is low, dig ten feet above the high tide mark until water seeps in.

(Several holes can be dug). Allow it to settle for one hour. Normally the top two inches is drinkable without too much saline content.

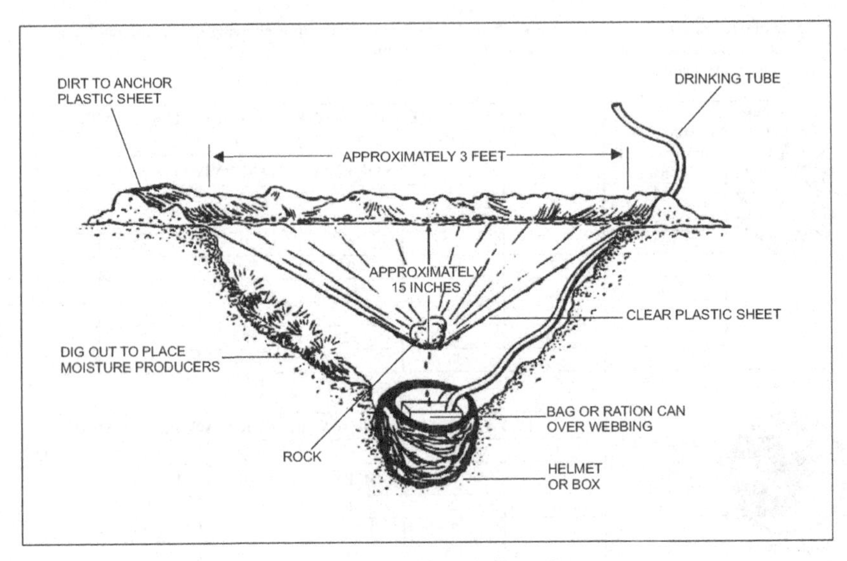

Figure 4-15. Improvised Solar Still

4. Water Indicators

   a. Animal tracks often converge towards isolated water holes.

   b. Populated areas are always near water. Often the presence of herds of goats, sheep, or cattle will indicate water sources are very near.

   c. Parrots and pigeons always nest near potable water. Birds fly towards water in early evening and mornings. Watch the flight of birds, particularly at sunset and dawn. Birds circle water holes in desert areas.

   d. Bees remain in close contact with water sources.

   e. Check the base of cliffs or rock formations, the base of sand dunes on the leeward side and dry riverbeds for likely places to locate water or moisture.

5. Unique Geographic Water Sources.

a. In Northern Mongolia: All roads lead to water. It is possible to tell one is going in the right direction when the trail he is on joins another. The "arrow" formed by two connecting trails points toward the water. In desert and near-desert regions, wells are gathering places for native peoples as well as stopping places for caravans. Campfire ashes, animal droppings, and generally disturbed surface show that others have camped there. Such indications will also indicate that a well is not far off. Paths leading from the camping area should lead to the well.

b. In the Libyan Sahara: Doughnut-shaped mounds of camel dung often surround the wells. Unless the small mound ring is recognized one could easily miss the well.

c. In the Arabian Desert: Near the Persian Gulf and the Red Sea and in the Libyan Sahara near the Mediterranean Sea, the air is quite moist. This moisture condenses on cool objects. Often condensed moisture or dew will be heavy enough to drip from metal awnings or roofs on cool mornings. In Arabia, this morning dew and even fog extends inland several miles (the likely location of SEAL missions). Occasionally fog occurs as much as 200 miles from the Persian Gulf.

d. In the Sahara type deserts, the Saxual tree has a sponge-like bark that contains water. Mash, squeeze, or chew the bark to extract moisture.

4.4.6 Food Procurement. Eat nothing during the first 24 hours, and do not eat unless you have water. It is important to ration food from the beginning.

4.4.6.1 Animals. In most deserts, animals are scarce. Their presence depends on water and vegetation, and true deserts offer little of those.

1. Look for animals at waterholes, in grassy canyons, low-lying areas, dry riverbeds where there is a greater chance of moisture, or under rocks and in bushes.

2. Animals are most commonly seen at dusk or in the early morning.

3. The smaller animals are the best and most reliable sources of food (rabbits, mice, ground squirrels, coyotes, etc.). Find their burrows and try to snare them with a loop snare, trap, or deadfall when they come out at dusk or dawn. If the animal is warm-blooded, boil it until the meat falls off the bones. The blood is drinkable in small quantities and it can replace salt content. However it should not be consumed unless water is available.

4. Look for land snails on rocks and bushes.

5. Lizards may be trapped or snared. Skin them prior to cooking and eating.

6. If birds are caught, be sure to skin them rather than plucking them. Boil them in water. Use the water as stew afterwards.

4.4.6.2 Plants. Usually, where there is water, plants can be found. Look for some soft part that is edible. Try all soft parts above the ground such as flowers, fruits, seeds, young shoots, and bark.

1. During certain seasons, some grass seeds or bean bushes may be found. All grasses are edible, but the ones found in the Sahara or Gobi are neither palatable nor nutritious.

2. Try any plant found. Tasting a plant is not fatal, even if the plant is poisonous.

3. The fruit of the prickly pear cactus is edible and in fact is often used to make jelly by desert natives. There is also some moisture extract to be gained from this fruit.

4.4.6.3 Insects. Many insects provide an edible food and energy source.

1. Ants can be eaten but first the heads should be removed.

2. Bees provide some nourishment in addition to perhaps indicating that a source of water is near. Remove their head, stinger, wings and legs before consuming them.

3. To eat crickets and grasshoppers first remove legs, heads and wings.

4. Scorpions can be eaten, but remove the tail section that contains the stinger and the poison pouch.

5. Leave centipedes alone entirely.

4.4.7 Shelter and Fire Making

4.4.7.1 Emergency Shelters. Shelter from the midday sun is advisable.

1. Manmade Shelters. A shelter will be necessary, mostly to protect from sun and heat. Make a shelter of a parachute, as shown in Figure 4-16 (or from other sources of protection such as poncho liner, camouflage material, etc.). The layers of cloth separated by airspace of several inches make a cooler shelter than a single thickness. The parachute can also be placed across a trench dug in the sand.

2. Natural Shelter. Normally limited to the shade of cliffs or the lee side of hills. If one camps or travels in a desert canyon or dry riverbed, he should be prepared to make a quick exit. Cloudbursts, although rare, do cause sudden and violent floods that sweep along a dry valley in a wall of roaring water.

4.4.7.2 Fire Making. In most deserts the fuel for making fires is extremely rare.

1. Wherever one finds plant growth, use all twigs, leaves, stems, and underground roots for burning.

2. Stems of palm leaves and similar wood serve as fuel near oases.

3. Dried camel dung is the standard fuel where woody fibers are lacking. In the Gobi, dried heifer dung is the preferred fuel. Heifer dung has a symmetrical shape, in contrast to the broad irregular pattern of cow dung. It burns with a hot blue flame, in contrast to the smoky yellow flame of cow dung, sheep droppings, etc.

4.4.8 Other Considerations. There are other factors that may come to play in survival in the deserts.

4.4.8.1 Local Inhabitants. In normal times desert people are hospitable and will provide food and water to survivors. During war, the natives of desert areas may be hostile. This hostility may result not from the personal feelings of the natives but from a promise of a government hostile to friendly forces to pay for prisoners. Under such conditions, one must make every effort to avoid contact with natives. Remember, the natives of most Middle Eastern deserts have a very strict code which they rigidly adhere to:

1. Duty toward God

2. Protection of tent neighbors

3. Attention to the laws of hospitality

4. Punishments are harsh. When working with the nomadic desert tribes in the Middle East, remember to abide by their customs and avoid conflicts.

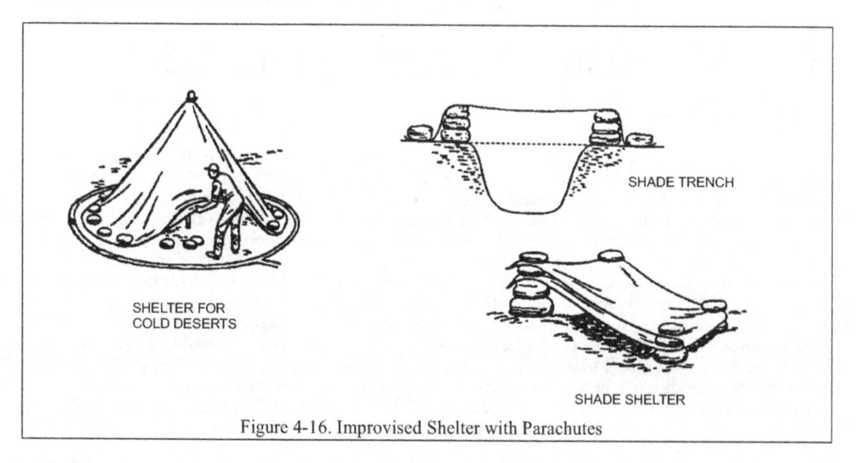

Figure 4-16. Improvised Shelter with Parachutes

4.4.8.2 Travel Routes

1. Coastal Routes. If it is not possible to follow a trail in the Saharan or Arabian deserts the best bet is to head for the coast, providing its location is known and the distance is within range of the surviving SEALs' strength and water supply.

2. Sandstorms. If wind and dust impede progress or shut out visibility:

   a. Stop traveling.

   b. Mark the direction of travel by some means such as a row of stones, or other markers.

   c. Lie down and sleep out the storm with the back to the wind.

   d. Cover the face with cloth.

   e. Don't worry about being buried by a sandstorm. One won't have to sleep out a storm more than a few days at the most. Remember sandstorms are not blizzards. Traveling in zero visibility and losing the way is the real danger.

CHAPTER 5

Mountain and Arctic Operations

5.1 INTRODUCTION

SEAL operations in arctic, subarctic, or extreme cold-weather regions are some of the most difficult tasks required of a NSW operator. Successful mission prosecution will be significantly affected by geographic and environmental factors. Morale, health, medical problems, and equipment failure rates will be affected by the environment and have a direct bearing on combat effectiveness. Planning and tactical considerations relative to target interdiction will often become secondary to problems that the environment inflicts on personnel and equipment. The education of NSW staff and operational personnel regarding mountain and arctic operations is critical for clarifying mission success criteria against the environmental constraints. Failure to identify limiting factors and to plan accordingly will substantially increase the probability of mission abort, compromise, or unnecessary casualties. Considerable planning, training, and environmental conditioning (acclimatization) in geographic areas that represent real target scenarios are essential. The ability to successfully conduct cold-weather missions is directly proportional to the amount of training and experience acquired in a similar environment. The information presented herein will serve as a background resource regarding mountain and arctic operations. This information is presented in a manner to facilitate a realistic understanding of the operational environment. SEALs must identify problems and exercise initiative to create solutions unique to each mission.

5.1.1 Naval Special Warfare Operations. NSW personnel are trained to conduct clandestine operations as delineated by NWP 3-05 (Naval Special Warfare). Many of these missions may be prosecuted in arctic or subarctic regions. When conducting operations in northern latitudes, however, the enemy may present less of a threat than the operating environment. Climate, water temperatures, and terrain will dramatically affect operational performance throughout each phase of the mission.

5.1.1.1 Mountain and Arctic Impact. Cold weather operations require that personnel develop heightened levels of stamina, determination, and initiative. The subarctic regions are characterized by the widest fluctuations of temperature on earth. Prolonged exposure to cold will aggravate or intensify difficulties that are inherent in NSW missions. The physical terrain presents radically different problems by season to mobility, cover and concealment, communication, camouflage, and equipment. This environment greatly reduces the margin for error. The effects of cold and precipitation, combined with attempts to negotiate varying terrain features, require SEALs to closely scrutinize personnel selection, mission planning, equipment, and training to increase the probability of success.

5.1.2 Target Considerations. NSW Group One (NSWG-1) and NSW Group Two (NSWG-2) are tasked to conduct arctic and subarctic operations in support of their respective theaters of operation. Many of the baseline requirements necessary to conduct these missions are similar. Environmental training, acclimatization, equipment application, and unit SOPs are necessary to achieve a general winter warfare capability. Winter warfare skills that apply to one AO may be applicable in another, but there are constraints (geographic and target related) that are peculiar to each AO. The requisite skills and methods necessary to engage targets in the Northern Pacific will vary from those required in Northern Europe and may have to be acquired and rehearsed separately. Joint Intelligence Centers in Europe, the Atlantic, and the Pacific provide target folders that highlight the relevant characteristics of each AO. Target analysis will assist in planning and identify mission and training requirements.

5.1.2.1 Naval Special Warfare Group One. The geographic region in which NSWG-l elements will conduct mountain and arctic warfare operations lies between 45 and 60 degrees north latitude. Siberian and Asiatic weather patterns combined with oceanographic factors create an extremely diverse environment within the Northern Pacific AO. The time of year will dramatically influence tactics. Light data, temperatures, wind chill, visibility, water and land access routes, and terrain are subject to seasonal change. Failure to recognize the impact of changing conditions will directly affect the operational unit.

The location of the target, enemy capabilities, and geographic and environmental factors will influence the types of missions that NSWG-1 can successfully conduct in the Northern Pacific. In some instances, winter conditions will favor the offensive unit. Hardened terrain and long periods of darkness will enhance mobility and concealment. Conversely, weather patterns are more severe and will limit insertion options and affect survivability. Coastal approaches will be impeded by sea-state and ice. High winds will affect parachute operations. Summer operations will be affected by long periods of daylight, increased foliage, and boggy terrain. The enemy order of battle may include a combination of early warning radar sites, EW and ground control intercepts, coastal surveillance, airborne reconnaissance, and hydroacoustic capabilities.

5.1.2.2 Naval Special Warfare Group Two. The geographic region in which NSWG-2 targets are located is NATO's northern flank, predominately in Europe between 65 and 70 degrees north latitude. The seasonal variation of daylight ranges from virtually no sunlight in winter to no sunset during summer. Winter conditions in the continental interior are severe, being affected by the Siberian high and Icelandic low. The Gulf Stream and North Atlantic Drift prevent ice formation and allow access through the Barents Sea. Terrain features vary from plains in the southern interior to north central mountain ranges. The extreme north is predominately barren tundra. Permafrost is prevalent and contributes to swampy conditions during the spring and summer. Frozen rivers that typically thaw from the south to the north also contribute to boggy ground conditions in the interior. Operational units will be faced with rapidly changing weather patterns and varying terrain. Access routes may be easier to negotiate in winter due to darkness, snow cover, and hardened terrain features that enhance surface movement.

The locations of the target areas provide access from land or sea. Barren tundra in the north will enhance mobility but provide little concealment. Insertion locations will affect route selection and mobility in areas where mountains and fjords are prevalent. Border patrols and frontier troops are stationed along the western border. For many of the eastern approaches, the enemy order of battle will be similar to that which will be encountered by NSWG-1 forces. Additionally, transits by sea or land to remote targets will be vulnerable to detection by indigenous fishermen, hunters, and local inhabitants.

5.1.2.3 Wet/Cold Conditions. Temperature is a major factor in determining wet/cold and dry/cold characteristics. Wet/cold conditions occur when temperatures range from 14 to 32 degrees Fahrenheit. Variations in day and night temperatures cause alternate freezing and thawing. Wet snow and rain may accompany these conditions. Clothing should consist of a waterproof or water-repellent and wind resistant outer layer with an inner layer of sufficient insulation to provide protection in moderately cold weather. Waterproof footwear is required.

5.1.2.4 Dry/Cold Conditions. Dry/cold conditions occur when average temperatures are lower than 14 degrees Fahrenheit. The ground is typically frozen and the snow is dry at these temperatures. Lower temperatures and higher winds increase the need for physical protection. The insulation characteristics for clothing should carry a rating to -65 degrees Fahrenheit. A water-repellent and wind resistant outer layer must protect inner layers of insulation. Insertion methods, regardless of surface temperatures, will typically require personnel to recover from being wet (and therefore physically vulnerable) and stabilize core temperatures with dry insulation before proceeding on their mission.

5.1.3 Naval Special Warfare Mission Planning. Mission planning procedures for mountain and arctic operations will follow the same planning process as other types of NSW missions. Consideration for time windows, communications, insertion and extraction methods, contingency requirements, supporting forces, intelligence, and logistics will be a part of the mission planning process. Environmental and topographical conditions will be the major factors that affect planning criteria. Mission planners who are unfamiliar with arctic operations may prove to be a liability to the operational unit. Failure to recognize the impact of arctic and subarctic conditions on personnel and equipment capabilities will undermine the ability to conduct successful operations. It is imperative that NSW mountain and arctic experts be consulted during the planning phase.

5.1.3.1 Time Considerations. Time critical events affect the success of all NSW missions. Mountain and arctic operations require that each phase of the mission be measured against the effects of the environment. Weather patterns, topography, and sea state will dramatically affect the insertion, infiltration, time-on-target, exfiltration, and extraction. The equipment load-out will require modification to ensure survivability. The increased weight of equipment will accelerate operator fatigue, limit choice of patrol routes, and decrease operational radius. Movement over difficult terrain and high seas will increase the time required to reach the target. Finally, time of year will affect movement. Low light conditions in winter will allow movement throughout the majority of the day, although this advantage should be weighed against the more severe conditions that may be encountered. Arctic operations during summer months may appear more feasible from a standpoint of physical survivability, but clandestine movement may be limited due to long periods of daylight. Mobility will be impeded by flooding and boggy ground conditions. Enemy patrol and reconnaissance efforts may become more effective as weather conditions permit.

5.1.3.2 Intelligence. Rapidly changing environmental conditions, and enemy order of battle require that timely intelligence be provided to the NSWTG/TU. Fleet intelligence centers, aerial reconnaissance platforms, satellites, and HUMINT sources can provide information for planners regarding a specified AO. SEALs are extremely vulnerable from the time they leave the insertion platform until the extraction is complete. Communication with higher headquarters is difficult in arctic and subarctic regions. The NSWTG/TU must be suitably located to act as a relay between higher headquarters and the NSW element. A communication system must be established and maintained to ensure that essential information regarding weather patterns, enemy activity, and contingency operations is current.

5.1.3.3 Communications. NSW forces employ several types of communication equipment, dependent on the operational environment, supporting forces, and higher headquarter. Communications requirements in northern latitudes present special problems. Satellite and HF communications capabilities and limitations should be considered and appropriate back up measures taken. Enemy DF capabilities may preclude the use of certain types of equipment, requiring modification of reporting procedures. Cold weather in arctic areas affects communication equipment by reducing the efficiency of certain components, making communication particularly difficult. Ionospheric disturbances and the aurora effect are examples of atmospheric interference. Snow, ice, or rain may cause static. Radio wave propagation charts must be consulted to identify the minimum and maximum frequencies to achieve optimum communications at scheduled hours on designated days. Frequency planning will identify alternate frequencies that should be used at specified times of the day and night to enhance HF communication. Communications and electronic operating instructions, brevity codes, and reporting procedures must be promulgated for all operational elements. Communications compatibility with supporting forces (CAS, NGFS) and frequency change times are essential for coordinating concurrent missions between separate elements. The location of higher headquarters (NSWTG/TU) will affect communication linkage. A plan must be developed to ensure that mission objectives and contingencies can be achieved should the SEAL element lose its communications capability.

5.1.3.4 Insert/Extract Methods. NSW forces routinely use air, surface, and subsurface platforms during the insertion and extraction phases of an operation. The objective is to conduct the above activities in a clandestine manner that will not cause the enemy to alter or heighten its defensive posture. The use of indigenous insertion and extraction platforms may enhance the operation. Locations, times, and masking of intent are basic elements for consideration. Arctic operations may involve extreme cold and increased sea-states. Planners should be aware of time and distance parameters. Cold weather, wind, and poor visibility will increase the time required to accomplish even the most routine tasks. Heightened sea-states will impede the progress and navigation of CRRC. The operational radius, transit times, and effects of environment and sea-states on personnel and equipment should be closely studied to identify realistic capabilities. Failure to do so may contribute to mission failure or compromise. Terrain features will affect mobility over land. Snow, ice, glaciers, swamps, mountains, and vegetation will influence route selection and must be considered when planning insertion/extraction locations and methods.

5.1.3.5 Logistics. After identifying the various mission requirements, staff planners will coordinate times and locations for assembly of personnel and service assets. Fleet asset positioning, staging areas, ordnance transport regulations, etc., must be preplanned. Careful selection of mission-specific equipment for NSW operations is critical to arctic operations, as resupply may not be possible after departure to the AO. The NSW element is limited to those supplies that can be man-packed or are sled-portable. This directly impacts on the time the element can remain self-supported in the field. The AO, enemy activity, and reliability of resistance groups will determine the feasibility of resupply, cache, or parachute drops.

5.1.3.6 Supporting Forces. Operational planners need to be aware of the overall support structure. Coordination is required to support the SEALs in areas that are outside the primary mission. These forces include: MEDEVAC, NGFS, CAS, CSAR, and the E&R structure within the AO. Adherence to time schedules and cognizance of fleet asset locations during all phases of the operation will have a direct bearing on the mission. Communication and intelligence links should be established to ensure compatibility between planning elements, support forces, and operational units. NSW operations generally require fleet support during the mission. It is vital that fleet personnel tasked to support the mission be well versed in SO tactics and requirements to facilitate timely and efficient deployment of NSW personnel. To ensure continuity, personnel who support rehearsals and FMP training should be used during the actual operation.

5.1.3.7 Operation Security. OPSEC consists of those actions that are necessary and appropriate to deny the enemy information concerning planned, ongoing, and completed operations. OPSEC must be introduced early in the operational planning sequence and continue through every phase of the operation. Loading and transportation of equipment and personnel should be disguised and precautions taken during satellite over flight periods. Mountain and arctic operations may occur at great distances from homeport and staging bases. A credible cover story must be developed to supplement OPSEC efforts.

5.1.3.8 Operation Deception. When appropriate, the cognizant Fleet Information Warfare Center should be consulted during the planning phase. The advance force commander can employ OPDEC measures to cause the enemy to realign its forces or prematurely disclose its order of battle. Fleet or NSW forces, as a separate mission, may conduct OPDEC measures. Preserving the clandestine nature of mountain and arctic missions is essential for success.

5.1.4 Training. The complex nature of NSW operations requires a wide spectrum of capabilities to ensure mission success. Mountain and arctic missions require NSW personnel to develop specialized skills and confidence in their ability to successfully complete the mission. Each individual must learn to prevail through hardship, work as a team member, and achieve the objective regardless of circumstance. Mountain and arctic operations test the individual and unit to the utmost. Toughness, resourcefulness, and initiative are qualities required to operate in mountain and arctic AOs. Training objectives must be designed to ensure that each individual can conduct the mission in all weather, climate, and terrain conditions associated with mountain and arctic warfare. NSW skills and attitudes acquired in temperate areas will not suffice in the arctic environment without proper conditioning (mental and physical) and training. Self-discipline is the most important individual attribute. The individual must move beyond personal survival if mission success is to be achieved.

5.1.4.1 Leadership. Aggressive leadership is essential to overcome the challenges of the environment. Individuals must recognize that there are two threats: the enemy and the cold. A positive attitude is essential to resist the insidious effects of the cold. Tasks become more difficult and require more time to accomplish. Acceptance of these tenets will reduce frustration and foster a positive attitude. Proper knowledge, equipment, and training will contribute to proficiency and mission success. Cold weather operations require additional initiative and forcefulness of character. Organization and preparation are essential in developing habits that will sustain personnel through periods of physical hardship. The arctic environment will not tolerate omissions in judgement and skill.

5.1.4.2 Cold Weather Personnel. The advantages of specialization in cold weather operations by specific NSW organizations are threefold. First, individuals understand what to expect regarding their professional development, and therefore take an active interest. Secondly, they train primarily in this specialty that enhances familiarization, proficiency, and confidence regarding cold weather operations. Thirdly, they deploy to areas that facilitate follow-on training, thus improving their proficiency. Continuing this training concept throughout the year ensures the development of a true operational capability. A small and dedicated group of mountain and arctic experts (four men) will generally out-perform a platoon size force (sixteen men) if even a small percentage of the individuals in the larger force are unskilled or unacclimated.

5.1.4.3 Training Objectives. Mountain and arctic operations are complex and difficult. The use of mountain and arctic equipment requires the development of new skills and techniques. To walk on snowshoes, one day of instruction is generally sufficient. However, several days use of snowshoes during normal training will rapidly increase proficiency. In a period of two weeks an operator can be taught enough ski techniques to enable him as an individual to negotiate flat or rolling terrain with greater speed than if he were on foot or snowshoes, but he will not yet be able to operate effectively as a combat skier within a platoon. Approximately eight weeks of intensive training are needed in order to become a military skier capable of operating proficiently in any type of terrain.

The environment and terrain may vary dramatically from one target to another. Research and development is an ongoing process regarding arctic maritime operations. Exposure limits and the effects of sea-state conditions on personnel, equipment, and time/distance factors can be determined through the development of realistic training. The following section outlines areas in which training is required to develop a realistic mountain and arctic capability. Specialization is necessary to develop expertise in some skill areas, but all personnel must understand the principles of survival, mobility, and the application of equipment to meet the general mission objectives listed below.

1. Acclimatization

    a. Altitude

    b. Environment.

2. Survival Techniques

    a. Navigation

    b. Shelter construction

    c. Fire making

    d. Survival diet/hunting and fishing techniques

e. Signaling principles.

3. Medical Factors

a. Nutrition requirements

b. Cold weather injuries.

4. Equipment Familiarization

a. Clothing (cold/wet, cold/dry)

b. Departmental equipment (cold weather problems)

c. Mountain and arctic personal equipment.

5. Mobility Techniques

a. Mountaineering/alpine

b. Skiing: cross country, downhill, and use of pulk (flat bottom sled pulled by one or two skiers and used for transporting equipment over snow covered terrain)

c. Rock movement

d. Snow movement (snowshoe, glissade)

e. Ice movement/rope work/ice rescue procedures.

6. Water Operations

a. Small boat transits

b. Navigation (northern latitude - over the horizon)

c. Beach landing

d. Cliff assault

e. Cold weather diving.

7. Tactics (SOP)

a. Patrolling formations with and without pulks

b. Camouflaging

c. LUP and deception trail procedures

d. OPs

e. Cold weather weapons employment

f. Cold weather demolitions

g. Cold weather communications

h. Offensive/defensive skills

i. Combining movement techniques (snowshoes, skies, roped up, etc.) with fighting tactics

j. Parachuting

k. Cache procedures

l. Fishhook procedures for LUP and breaks

m. Jump-off procedures

n. Ice crossing and rescue procedures

5.1.4.4 Training Standards and Realism. Selection of suitable training areas ensures that climate and terrain conditions likely to be encountered during subsequent cold weather operations are experienced during training. Operational readiness evaluations that include the development of tasks, conditions, and standards should be established to ensure that mission specific criteria is achieved. Training and FMPs, which meet those conditions likely to be encountered during actual operations, will better serve NSW personnel than overly structured exercises that preclude realism.

5.2 ENVIRONMENTAL IMPACT

5.2.1 General. The purpose of this section is to provide environmental information to NSW personnel conducting operations in arctic and subarctic regions. The information addresses factors that impact on individual survival and successful mission prosecution. Areas of concern include meteorological, topographical (terrain and vegetation), and mountain hazard information. NSW personnel must be familiar with the characteristics of the environment in which operations may occur. Failure to observe nature's patterns, obstacles, and warning signs will degrade the capabilities of the individual and unit. Medical and dietary considerations will take on added significance. See Section 2.11 and Appendix N for medical intelligence sources. NSW personnel should understand areas of physical susceptibility that can be minimized or overcome by proper preventative measures.

5.2.2 Arctic and Subarctic Regions. Arctic regions raise issues of military importance by virtue of their strategic location. Methods of warfare in any theater are characterized by the geographic nature of the theater and by the distribution, frequency, intensity, and duration of environmental factors. In no other military theater of the world do these factors result in such extremes. Many combinations of environmental conditions affect the vast regions included within the Arctic. The severe climate, lack of cover, and limited mobility would seem to render it unsuitable for major combat operations. Swamps, rivers, and lakes can be an asset to movement while frozen, and a liability in the summer. On the ice pack of the Arctic Ocean, temperatures are influenced through the year by the movement of air from other regions. The temperature on the ice pack remains fairly constant and quickly returns to temperatures appropriate to the season with the passing of weather fronts. Coastal temperatures throughout the subarctic closely resemble those of the icecap. The coldest months of the year are normally February and March. During the summer, relatively high temperatures may occur along coastal areas due to offshore winds. If the wind direction reverses, an immediate and often drastic temperature drop may occur.

Military operations have been conducted in the subarctic in all seasons. The following paragraphs will discuss the characteristics of the subarctic environment, especially during the winter season. It is the winter season and its freeze-up in autumn and break-up in the spring that produces the majority of problems in the subarctic. The combination of forest and deep snow found in northern areas poses particular problems in subarctic mountain warfare. Operations in summer are normally conducted in the same manner as those in temperate regions.

1. Forests of spruce, alpine fir, birch, and pine intermingle with rivers, streams, lakes, and swamps to dominate the subarctic landscape. The forest decreases the prospect of obtaining detailed information by ground or air reconnaissance. LUPs can be constructed and well concealed. Artillery and CAS is difficult, and all vehicle movement in thick forest is generally limited to roads. The major differences between arctic and subarctic regions are vegetation features, precipitation, and wide temperature ranges.

2. Wind along the subarctic coastline can be strong and gusty in winter and cause extremely rough seas, create dangerous wind-chill exposure problems, and seriously reduce visibility. Winds are greater and storms are more severe in the southern boundaries of the subarctic than along the polar edges.

3. Cloud cover is most widespread in summer and autumn, and least widespread in winter and spring. Cloud heights are lower in arctic and subarctic regions than in more temperate zones. During the summer, extremely dense fog is common along coasts and can seriously reduce visibility. Fog conditions may actually aid insertion and extraction, although navigation will be more difficult.

4. Since it is impossible to determine what time of year NSW personnel may be committed to operations in these climates, an extensive study of yearly climatic variations and characteristics of this potential AO is necessary. Climatic variations and the physical terrain present radically different problems by season to movement, cover and concealment, camouflage, and choice of equipment for operators. NSW targets are not located in areas representative of the extremes found at interior Siberian stations, but temperature changes between summer and winter, although less severe along coastal areas (30 degrees average between warm and cold seasons), are still among the harshest in the world.

5.2.3 Meteorological Factors. Among the requirements necessary to conduct successful operations in the arctic and subarctic is a thorough understanding of the environment. Weather and terrain conditions will have a profound affect on even the most ordinary tasks. Those variations require operators to have specific knowledge about the natural obstacles that will be encountered. Further, individuals must develop skills and tactics in order to move beyond mere survival and successfully use the environment to their advantage.

5.2.3.1 Atmosphere. There are four primary characteristics of the atmosphere. First, the atmosphere is divided into six layers. Secondly, the troposphere is that part of the atmosphere that extends from the earth's surface to approximately 40,000 feet. Thirdly, the troposphere is the only part of the atmosphere that contains water vapor. Finally, the troposphere is the only part of the atmosphere in which clouds form.

5.2.3.2 Pressure. Pressure is a characteristic of the atmosphere; it decreases with an increase in elevation. A barometer is used to measure changes in atmospheric pressure. A drop in barometric reading means a low-pressure system is in the area and an increase means a high-pressure system. An altimeter is generally used for measuring increases and decreases in elevation, but it can also be used for reading barometric pressure. The air is composed of several different parcels of air that may be extremely large. These are known as large air masses and will usually have different pressure readings and characteristics.

1. High pressure. This is an air mass with a higher pressure reading than the air surrounding it. Its characteristics are as follows:

a. Airflow is clockwise (Northern Hemisphere; counterclockwise in Southern Hemisphere)

b. It is associated with clear skies

c. Generally winds will be mild in the area of high pressure.

2. Low pressure. This is an air mass with a lower pressure reading than the surrounding air. Its characteristics are as follows:

a. Airflow is counterclockwise (Northern Hemisphere; clockwise in Southern Hemisphere)

b. It is associated with bad weather.

3. Weather patterns. Air masses forming over oceans are termed "Maritime", and are usually wet or have a concentration of moisture. Those forming over land are called "Continental", and are normally dry. These air masses will form over either tropical (warm) or polar (cold) oceans or landmasses. Air masses can be categorized into the following weather patterns.

a. Maritime polar. Wet and cold.

b. Maritime tropical. Wet and warm.

c. Continental polar. Dry and cold.

d. Continental tropical. Dry and warm.

4. Pressure and winds. These have a dominant influence on our weather for the following reasons:

a. High pressure usually brings good stable weather. The air from a high-pressure system is trying to gradually flow out to equalize its pressure with the surrounding air.

b. Low pressure brings unstable weather, usually bad. As the low begins to build vertically, the system becomes unstable, continues to build, and causes turbulence. This process creates bad weather.

c. Pressure differences cause winds, which flow from higher to lower pressure.

5.2.3.3 Formation of Wind and its Patterns. Air at the earth's surface near the equator is heated by the sun and rises. As it rises, it expands, circulating toward the polar regions. It cools as it rises because there is less pressure at higher elevations. As it moves, it cools and returns to earth at different areas of the world. It then flows back towards the equator. This creates different wind patterns that are affected by regional temperature, climate, and hemisphere. The earth's rotation plays a key role because it causes the winds to flow at an angle instead of in a straight line. Air masses generally move in a west to east pattern. Air in a high-pressure area will move to a low-pressure area to equalize the pressure. This can cause a shift in wind patterns.

1. In high mountains, the ridges and passes are seldom calm, but strong winds in protected valleys are rare. Normally, wind speed increases with altitude since the earth's frictional drag is strongest near the ground. This effect is intensified by mountainous terrain. Winds are accelerated when they converge through mountain passes and canyons. Because of the funneling effects, the wind may blast with great force on an exposed mountainside or summit. Usually, the local wind direction is controlled by topography. Stronger winds are experienced in the winter along the coast than in the interior at that time. These winds are the

result of the turbulence developed from the contrasting temperature of the warmer maritime air streams reacting with the inland cold caused by lack of solar heating in winter.

2. The force exerted by wind quadruples each time the wind speed doubles. Therefore, wind blowing at 40 knots pushes four times harder than a twenty-knot wind. With increasing wind strength, gusts become stronger and may be fifty percent higher than the average wind speed. The speed of the winds accompanying local storms is less than that of winds with traveling storms. Local winds assume an erratic pattern due to sunshine and varying degrees of intensity on uneven terrain. During the winter season, or at high altitudes, personnel must be constantly aware of the wind chill factor and associated cold weather injuries. Examples of wind patterns in the Northern Hemisphere include:

   a. Trade Winds. These are prevailing, almost continuous winds blowing from the northeast (southeast in the Southern Hemisphere), caused by southward (northward in the Southern Hemisphere) moving air. The trade winds are in the subtropical regions (about 10 to 30 degrees latitude) and influence tropical storms as they travel from east to west.

   b. Prevailing westerlies. These are caused by a dominant west to east motion of the atmosphere, centered over the mid-latitude regions (about 30 to 60 degrees latitude) and characterized by high level winds that are predominantly from the west. These winds will also generally drive mid-latitude storms from west to east.

   c. Polar easterlies. These winds will dominate low-level air circulation north or south of about 60 degrees latitude. Few major storm systems will develop in the smaller Arctic region, but its chief influence is its contribution to the development of mid-latitude storms by forcing air masses into the prevailing westerlies at lower latitudes.

5.2.3.4 Physical Aspects. Radiation from the sun is the earth's primary source of heat. The energy that heats the earth also warms the air. Heat is transferred to the air in three ways:

1. Reradiation and radiation. Reradiation is how the sun heats 85 percent of the air. The solar energy radiates into the earth and is then reradiated back out into the atmosphere. Direct solar rays radiate only 15 percent of the air. The reradiation is normally the closer layer to the earth while the radiated air is farther out.

2. Direct contact. Air being warmed by direct contact at the earth's surface and rising.

3. Convection. As an air mass is heated and rises, the atmosphere traps and circulates the heat, inducing the air to move laterally. This makes the earth much warmer than it would be from pure solar radiation alone.

5.2.3.5 Humidity. This is the amount of water vapor in the air. All air has some water vapor, and therefore, some percentage of humidity that will range from 1 to 100 percent. It is the condensation of vapor that causes most weather phenomena, e.g., clouds, rain, snow, etc. Warm air holds more moisture (higher humidity) than cold air. Air holding the maximum amount (100 percent) of moisture at that temperature is saturated. Air generally holds only a portion of the water vapor it could hold at a particular temperature. This percentage is called relative humidity. As warm air rises and cools it loses its ability to hold moisture and the moisture condenses into clouds. Droplets or ice crystals form and enlarge in rising air masses, eventually dropping out of the clouds as precipitation. The two essential elements for precipitation are a moist air mass and vertical lifting.

5.2.3.6 Precipitation. Arctic regions generally have little precipitation (snowfall and rainfall). At the northern limits of the North Atlantic, the precipitation is less than eight inches annually, an amount corresponding roughly to that found in the semi-arid parts of the Western United States. Near the entrance to Franz Jose Fjord in Eastern

Greenland, the precipitation levels are similar to the amount found in Death Valley, California. Precipitation is developed from warm ocean currents; thus amounts up to 40 inches are found on the southern coast of Greenland and up to 60 inches in the Aleutians/Northern Japan region. Where the moisture falls as snow, approximately 10 inches of snow equals one inch of rainfall. The precipitation rate is largely a result of the continuous flow or circulation of warm waters by the Gulf Stream, that flows out of the tropics of the Caribbean, along the coast of the United States, and across to the North Sea. The Japanese current flows from the vicinity of the Philippines, east of Japan, and across the Aleutian chain.

1. The subarctic precipitation characteristics of Eurasia are similar to Alaska. During the spring season, the operator is likely to encounter several types of precipitation that can seriously endanger the mission. Assuming an operator can remain dry, the cold winters will not preclude effective operations. Winter storms are frontal in nature and much of the terrain will be windswept rocky waste interspersed with deep wind-crusted snowfields that can seriously impede movement. From October to May, precipitation is in the form of snow. Summer months are the wettest.

2. Break-up/Freeze-up. The mud is the dominant seasonal factor in operations during the spring and autumn. With the spring thaw and break-up come flooding and ice jams. Routes become almost unusable and severely restrict operations. This condition is also present during the autumn freeze-up. During both the spring thaw and the autumn freeze-up, it normally takes four to six weeks before conditions are sufficiently dry or frozen for normal operations. In areas where permafrost is a factor this condition may continue through the summer months.

5.2.3.7 Fronts. The circulation of warm and cold air can determine weather patterns. These characteristics, affected by wind and geographic conditions (mountains), are known as fronts.

1. Cold Fronts. This occurs when cold air is advancing into and under warm air. The cold front has a steeply slanted, unstable boundary. It is characterized by rapid movement and a cloud progression of cirrus to cirrocumulus, followed by cumulus to cumulonimbus. The winds in the area of a front will help you to determine the type moving into the area. Winds signaling the approaching cold front generally are from the south or southeast, and will sharply change direction to the southwest and then west or northwest as the cold front begins to pass.

2. Warm Front. Warm air advances into and over cold air. It has a shallow front edge and moves slowly. The cloud progression is cirrus to cirrostratus to altostratus, followed by stratus to nimbostratus. This front can carry heavy precipitation. It is slow moving and the weather conditions last longer than a cold front. The winds signaling the approach of a warm front are normally from the northeast, east, or southeast, and their speeds increase. The winds will change direction to the south as the warm front passes.

3. Occluded Front. This involves a rapidly moving cold front that overtakes a warm front. The air between the fronts is forced aloft and will combine its precipitation with that of whichever air mass is over the observer. An occluded front will usually cause turbulence and precipitation because of the proximity of the two air masses and their very different pressures.

5.2.3.8 Clouds. The size and shape of clouds and their altitudes define cloud types. The two major families of clouds are cumulus and stratus. It is possible to tell the difference between local and traveling storms and to estimate their probable occurrence. Clouds are formed by rising air. This air rises from convection effects (e.g., temperature, reradiation), turbulence due to surface irregularities (e.g., desert sands or mountains), orographic uplift (e.g., air rising over a mountain), or general ascent of large air masses or “thermals”.

1. Cloud Types. There are three main types of clouds. Their names can be combined, or prefixes can be added to further identify their characteristics and elevation in the sky.

   a. Cirrus. Thin, feathery, high clouds.

   b. Stratus. Featureless, layer-type cloud, usually hard to distinguish in the mountains.

   c. Cumulus. Lumpy, vertically formed clouds (cotton balls).

2. Cloud Height. Clouds are also classified by altitude. Understanding the cloud type will assist the operator in making weather predictions.

   a. Cirrus clouds are composed of ice crystals and may transpire from 19,000 to 35,000 feet. They may be detached white clouds with a fibrous structure or extensive thin veils. Cirrus clouds can provide warning of bad weather hundreds of miles away and up to one day in advance of an approaching front.

   b. Altostratus clouds are the middle dark clouds that form a continuous uniform sheet at elevations from 7,500 to 19,000 feet. They may cover the sky completely, and may develop from the descending and thickening cirrus clouds or the merging of high cumulus clouds. Altostratus clouds should be observed for signs of bad weather. When these thicken, especially if preceded by cirrus clouds, precipitation within five to ten hours is likely.

   c. Cumulus clouds look like large patches of cotton or giant heads of cauliflower. They form at any height from 1,500 to 12,000 feet. The base is flat, but the top varies in size and height. Cumulus clouds are brilliant white in direct sunlight but are dark on the shaded side. Altocumulus clouds are similar in form but smaller. They appear in uniform layers or cloud decks from 7,500 to 22,000 feet. Cumulus clouds should be watched for possible growth leading to cumulonimbus clouds.

   d. Cumulonimbus (thunderhead) clouds are overgrown and darkened cumulus clouds that are likely to yield precipitation. A single cumulonimbus cloud mass may extend from a base at 2,500 feet to a height of 35,000 feet. The tops of cumulonimbus clouds are made up of ice crystals and have a fibrous texture. They are typically associated with cold fronts and can occur at anytime of the year.

   e. Stratus clouds are similar to altostratus, but much lower. They may develop from a fog layer in which the bottom portion has evaporated or may develop where upslope winds are blowing. They often are associated with moisture and precipitation, and are usually seen below 3,000 feet except where they are forced up over a mountain or lie in high valleys. If they reach the ground, they become fog. Nimbostratus clouds yield steady rain.

   f. Cloud caps often form above pinnacles and peaks. As with all clouds, they indicate moisture. Cloud caps should always be watched for changes. If they descend, they often bring winds with them. If they grow and descend, bad weather can be expected.

5.2.3.9 Arctic Whiteout. Whiteout occurs when there is an unbroken snow cover and a uniformly overcast sky so that light reflected from the sky is approximately equal to that from the snow surface. The presence of ice crystals aggravates the condition. Blowing snow and rotor downwash can also cause whiteouts. This is an extremely dangerous condition and any movement should be heavily weighed against the total disorientation it creates.

5.2.3.10 Grayout. This condition occurs over a snow-covered surface during twilight or when the sun is close to the horizon. There is an overall grayness to the surroundings, the sky is overcast with dense clouds, and there is an absence of shadows with a resulting loss of depth perception. This phenomenon is similar to whiteouts, but a horizon is usually distinguishable.

5.2.4 Field Expedient Weather Analysis Forecasting. Predicting weather is often an educated guess or can be a conclusion based on meteorological evidence. The use of a portable aneroid barometer, altimeter, thermometer, wind meter, or hygrometer may assist in making local weather predictions. Predictions based on past results may or may not be accurate. Natives of an area may supply weather lore that is accurate. An individual remaining in one mountain region for several weeks in any season can add indications for that area based on his experience. As an example, shower-type precipitation is often produced by cumulus or cumulonimbus clouds and often produces sudden wind gusts and temperature fluctuations.

1. Buy's Ballot Law. In the Northern Hemisphere, winds around a low-pressure area blow counterclockwise. This leads to Buy's Ballot Law, that states: In the Northern Hemisphere, if you stand with your back to the wind, the low-pressure area is to your left. This is important because it allows you to determine whether you are likely to catch the full impact of the bad weather or not. Lows that pass south generally bring much more prolonged bad weather and precipitation than those that pass to the north. This method of forecasting should be repeated over a 24-hour period for an accurate analysis. Buy's Ballot Law can be used in all areas of the Northern Hemisphere.

2. Instruments. Steadily falling barometric pressure usually indicates an approaching storm. Steadily rising pressure usually indicates a clearing. This can be measured with either a barometer or altimeter.

3. Contrail lines. A basic way of analyzing a low pressure is to note the contrail lines from aircraft. If they don't dissipate within two hours it indicates a presence of low pressure. This usually occurs 24 hours prior to a frontal system. Contrail lines can also form cirrus clouds in a low-pressure system by causing "almost saturated" air to trigger the formation and spreading of a deck of cirrus clouds (these are called aerodynamic contrails).

4. Cirrus clouds. These can precede a storm by 24 hours or more. A ring around the moon or sun indicates the presence of water vapor. It appears in the form of a thin cirrostratus cloud. The larger the ring, the more water vapor present and the closer the storm.

5. Lenticulars. These are optical (lens-shaped), stratus/cumulus clouds called "mountain wave clouds", because of their appearance like that of an ocean wave. They are good indicators of moisture in the air and high-speed winds aloft. When preceding a cold front, the winds and clouds will lower.

6. Changing Weather. A change in the weather may be indicated by a marked shift in pressure, wind velocity, or wind direction, a change in temperature that is abnormal, or a change in the moisture content of the air within a 12-hour period.

7. Traveling Storms. The following indicates the approach of a traveling storm:

   a. A thin veil of cirrus clouds spreads over the sky, thickening and lowering until altostratus clouds are formed. The same trend is shown at night when a halo forms around the moon and then darkens until only the glow of the moon is visible. When there is no moon, cirrus clouds only dim the stars but altostratus clouds completely hide them.

   b. Low clouds that have been persistent on lower slopes begin to rise at the time upper clouds appear.

c. Various layers of clouds move in at different heights and become abundant.

d. Lens-shaped clouds accompanying strong winds lose their streamlined shape and other cloud types appear in increasing amounts.

e. A change in the direction of the wind is accompanied by a rapid rise in temperature not caused by solar radiation. This may also indicate a warm, damp period.

f. A light green haze is observed shortly after sunrise in mountain regions above the timberline.

8. Local Disturbances. Indications of local thunderstorm showers or stormy weather are:

a. An increase in size and rapid thickening of scattered cumulus clouds during the afternoon.

b. The approach of a line of large cumulus or cumulonimbus clouds with an advance guard of altocumulus clouds. At night, increasing lightning windward of the prevailing wind gives the same warning.

c. Massive cumulus clouds hanging over a ridge or summit.

9. Strong Winds. Indications from strong winds seen at a distance include:

a. Plumes of blowing snow from the crests of ridges and peaks or ragged shreds of cloud moving rapidly.

b. Persistent lens-shaped clouds, or a band of clouds over high peaks, ridges, or downwind from them.

c. A turbulent and ragged banner cloud that hangs to the lee of a peak.

10. Fair Weather. Fair weather may be associated with the following:

a. A cloudless sky and shallow fog, or layers of smoke or haze at valley bottoms in early morning.

b. A cloudless sky that is blue down to the horizon or down to a point where a level haze layer forms a secondary horizon.

c. Conditions under which small cumulus clouds appear in the forenoon and rise or vanish during the day.

d. Clear skies except for a low cloud deck that does not rise or thicken during the day.

e. A red sky at sunset.

f. A decreasing halo around the sun or moon.

g. Dew on the ground in the morning.

h. Small snowflakes or ice crystals, which indicate that the clouds are thin and fair weather may exist at higher altitudes.

11. Bad weather. Bad weather is indicated by the following:

a. A gradual lowering of the clouds. This may be the arrival or formation of a new lower stratum of clouds. It can also indicate the formation of a thunderhead.

b. A red sky in the morning.

c. An increasing halo around the sun or moon.

d. An increase in humidity and temperature.

e. Cirrus (mares' tails) clouds.

5.2.5 Mountain Barriers. The following information outlines the effect that mountains will have on wind and precipitation.

1. Wind and Mountains. To overcome a mountain barrier, winds will accelerate on the side that the wind flows from (windward side). This gives the wind upward momentum. This momentum is further increased because there is less friction higher in the atmosphere. Wind velocity is the greatest as the wind overcomes the ridge crest and it decelerates as it descends the far slope (leeward side).

2. Air Masses with Precipitation. If air masses have precipitation, they will drop precipitation on the windward side, enough to cover the mountain. Windward sides usually have more precipitation than the leeward sides. Mountain ranges will have more precipitation because air masses are forced higher to overcome the barrier. This causes them to cool and drop more precipitation than normal. At a canyon floor, winds usually are deflected to align with the canyon's direction. The amount of deflection depends on the steepness and length of the canyon. Wind flow over small-scale obstacles will be much more complicated. Each obstacle can affect wind differently.

5.2.5.1 Environment. Mountain weather is erratic. Hurricane-force winds and gentle breezes may occur short distances apart. The weather in exposed places contrasts sharply with the weather in sheltered areas. Weather changes in a single day can be so variable that in the same locality one may experience hot sun and cool shade, high winds and calm, gusts of rain or snow, and then intense sunlight again. This results from the life cycle of a local storm or from the movement of traveling storms. The effects of storms are modified by the following local influences that dominate summer storms:

1. Variation in altitude

2. Differences in exposure

3. Distortion of storm movements and the normal winds by irregular mountain topography.

5.2.5.2 Temperature. The temperature normally falls from three to five degrees Fahrenheit for every 1,000-feet gain in altitude. This is called the Adiabatic Lapse Rate. In an atmosphere containing water vapor, the temperature drops one degree Fahrenheit for each 300-foot rise in altitude. In dry air, it drops one degree Fahrenheit for each 150-feet rise in altitude.

1. Temperature Inversion. On clear and calm nights, valley temperatures will be colder than ridgetop temperatures. This is caused by convection and gravity. As the warm air rises, the heavier and denser air wedges its way underneath and settles on the valley floor. On cold, clear, calm mornings higher temperatures may often be encountered as altitude is gained. This reversal of the normal situation is called

temperature inversion. Temperature inversion occurs when air is cooled by ice, snow, and heat loss from radiation and settles into valleys and low areas.

2. Chinook Winds. These winds force air currents up and over mountaintops. The air is warmer at the lower elevations and cools as it gains altitude due to a decrease in pressure. On the reverse side of the mountain the temperature increases due to heat released by the water vapor in the cloudy air and an increase in pressure. The temperature may be warmer on the leeward slope at the same elevation than on the windward slope.

5.2.5.3 Thunderstorms. Although individual thunderstorms are normally local and of short duration, they can be part of a large weather system that may hinder mountain operations. In the alpine zone above the timberline, thunderstorms may be accompanied by freezing precipitation and sudden gusty winds. Ridges and peaks are focal points for concentrated and dangerous electrical activity.

1. Local thunderstorms develop from rising air columns, a result of intense heat from the sun on a small area. They occur most often in the middle or late afternoon. Scattered, fair weather cumulus clouds continue to grow larger and reach a vertical depth of several thousand feet, and may quickly turn into thunderstorms.

2. Thunderstorms that occur at night or in the early morning are associated with major changes in weather. This results in a long period of foul weather before clearing on the high summits. Thunderstorms occurring at these times may also be part of a storm line. These are followed by a prolonged period of cool dry weather.

5.2.5.4 Lightning. Many casualties have been reported due to lightning during mountain operations that involve an increased risk of being struck. Mountain climbers are often found on prominent peaks and exposed ridges that are subject to lightning strikes and lesser discharges. The reason for this is that the ridges help produce the vertical updrafts and rain-cloud conditions that generate lightning. There are precautions that can be taken by the climber.

1. The best way to avoid lightning in the mountains is to stay off exposed peaks or ridges and stay out of an unprotected flat expanse during an electrical storm. Avoid being under prominent or isolated trees. If an individual is caught in an exposed place and has time before the storm reaches him he should get as far down the mountain and away from the exposed ridges as possible. The middle of a ridge is preferable to the end of a ridge.

2. If lightning strikes seem imminent or they are striking nearby, an individual should seek protection from direct strikes and ground currents. A flat shelf, slope, or slightly raised area dominated by a nearby high point gives protection from lightning strikes. If there is any choice, a spot on dry, clean rock is preferred over damp or lichen-covered rock. A scree slope is good. The climber should tie himself down if a severe shock would cause him to fall.

3. Lightning kills by passing through the upper torso and the heart causing cardiac arrest. Insulation of the hands and upper body from the ground by keeping them elevated away from the ground helps deny electricity a path through the upper body and the heart. Avoiding upper body proximity to the ground and keeping objects attached to the upper body away from the ground increase the chances for survival in a near strike.

5.2.5.5 Fog. Several types of fog are encountered in areas of extreme or arctic cold. They are predominately ice fogs and are classified according to their source. When moisture is released into the air from any source at low temperatures, ice crystals are formed. These hang in the air when little or no wind is present. Perspiration and moisture cause human-animal fog from the breath. Overflowing streams causes water fog. Water products of combustion will cause town fogs, vehicle fogs, weapon fogs, or aircraft fogs. Normally these fogs appear when the temperature drops below -25 to -30 degrees Fahrenheit. Fog will form around weapons being fired as a result of the moisture released by the combustion of the propellant charge. These fogs affect concealment, visibility, and the operation of aircraft and other equipment. Observed fire of AW, air defense weapons, and wire guided missiles is hampered by fogs that form from moisture released by combustion of propellants resulting in an identifying signature. Inversion effects cause fogs at warmer temperatures (25 to 40 degrees Fahrenheit) as warm rains melt snow during break-up periods.

On windward slopes, persistent fog, cloudiness, and precipitation often continue for days. They are caused mainly by the local barrier effect of the mountains on prevailing winds. Any cloudbank appears as a fog from within. Fog limits visibility and causes whiteout conditions that hamper operations by increasing the chance of accidents. It does, however, aid surprise attacks. When traveling without landmarks, it is necessary to use a compass, altimeter, and map to maintain direction. If fog and precipitation occur at the same time, extra clothing is needed for protection against cold and wetness.

5.2.5.6 Traveling Storms. The most severe weather conditions, storms with strong winds and heavy precipitation, are due to widespread atmospheric disturbances that usually travel easterly. If a traveling storm is encountered in the alpine zone during winter, all the equipment and skill of the individual will be tested against low temperatures, high winds, and blinding snow.

1. Traveling storms result from the interaction of cold and warm air. The center of the storm is a moving low-pressure area where cyclonic winds are the strongest. Extending from this storm center is a warm front that marks the advancing thrust of warm air and the cold front that precedes onrushing cold and gusty winds. The sequence of weather events, with the approach and passing of a traveling storm, depends on the state of the storm's development, and whether the location of its center is to the north or south of a given mountain area. Scattered cirrus clouds usually merge into a continuous sheet that thickens and lowers gradually until it becomes altostratus. At high levels, this cloud layer seems to settle. A stratus deck may form overhead at lower altitudes.

2. A storm passing to the north may bring warm temperatures with southerly winds. It may be partly clearing before colder air with thundershowers or stormy conditions move in from the northwest. However, local cloudiness often obscures frontal passages in the mountains. The storm may go so far to the north that only the cold front phenomena of heavy clouds, squalls, thundershowers, and colder weather are experienced. The same storm passing to the south would be accompanied by a gradual wind shift from northeasterly to northwesterly with a steady temperature fall and continuous precipitation. After colder weather moves in, the clearing at high altitude is usually slower than the onset of cloudiness, and storm conditions may last several days longer than in the lowlands.

3. Rapidly changing weather conditions often create glaze. Glaze, commonly known as freezing rain, occurs most frequently during spring and autumn when surface air temperatures are between 23 to 32 degrees Fahrenheit. Rime is a coating of ice deposited on trees and other objects by freezing fog. Accumulated rime can reach to ½ centimeter in 24 hours, and can coat boats, weapons, and equipment and impair their function. Hoarfrost is another form of precipitation of operational importance. It is caused by direct condensation of atmospheric water vapor on solid objects. Again, this can coat weapons, radios, and equipment and jeopardize proper functioning.

5.2.6 Hydrographic and Coastal Conditions. Threat and environmental analysis may require that SEALs approach targets from the sea. The above-mentioned weather features will continue to play a role during planning and actual conduct of the mission. Hydrographic conditions require careful analysis to ensure that the unit reaches the AO following insertion at sea. In planning and conducting beach landings or extractions, two important areas must be considered, tides and surf timing. Tides must be calculated during the planning phase to determine the optimum times to land/launch, i.e., amount of beach to cross, and surf timing is critical to a successful beach landing or launch. The coxswain must be well rehearsed in the practice of determining the low sets with greatest interval to effect a safe launch. Although the timing is not as critical in landing, the coxswain must still be well rehearsed in riding the back of a wave into the beach without broaching the CRRC. The following factors will affect the operational unit during the transit and coastal landing phases of the mission.

1. Water Temperature. The conduct of clandestine maritime operations in the arctic and subarctic is difficult and hazardous. Exposure to these waters should be limited if possible. The combined effects of water, air, and wind will have a debilitating affect on personnel unless properly protected. Depending on the nature of the operation, a wet-suit, dry-suit, or exposure-suit will be required.

2. Tides and Currents. Tidal ranges in the Arctic are slight, particularly in areas of a straight coastline. In bays, deeply indented coastlines, and other restricted areas, the tidal range may be considerable, in some regions as much as 20 feet. Tidal conditions in the Arctic have not been completely and progressively studied. Furthermore, since wind and ice conditions exercise such a pronounced influence, predictions are not as reliable as they are in other regions of the world. In areas where the tidal range is great, more caution must be exercised. Local wind or ice conditions may result in unpredictable tides. In areas of tidal variation, the beach may be covered with ice floes left as a result of a preceding tide. This condition should normally be expected in any ice area. In the marginal ice areas, it is highly desirable that data on currents be collected so the movement of ice may be evaluated. The flowing together of warm and cold currents gives rise to a vapor condition or fog.

3. Surf and Swell. Sea-state conditions will profoundly affect the transit and coastal landing phases of the mission. Surrounding weather conditions will directly influence the sea-state. Conditions must be monitored closely prior to launch. Failure to obtain and apply current intelligence will result in the inability to negotiate the transit. CRRC and/or Zodiacs are routinely employed for transit purposes. Weight factors (personnel and equipment) and sea-state will directly influence the operational radius. Distance to the landing site will have to be shortened and transit time lengthened prior to launch as the sea-state increases. Surf and coastal formations will determine the landing capability and procedures required at the shoreline to avoid the loss of equipment and undue exposure of personnel to the water.

4. Ice. The formation of ice will affect the transit and landing. Navigation should ensure that floating ice is avoided at sea. As the coastal approach is reached, the degree of ice formation at the objective will affect the location of the landing where the NSW unit must beach their boat(s).

5. Fog. Coastal areas are subject to dense, low-lying fog and cloud cover. This condition will severely restrict visibility and increase the navigation problem.

5.2.7 Terrain Factors. Terrain features play a major role in route selection, equipment requirements, and the requisite training necessary to negotiate the AO. Rates of movement and the operational radius will be affected. NSW personnel should be familiar with the types and composition of terrain that will be negotiated. Features that will be discussed include: mountains, rock, ice, snow, glaciers, and forests. Planning considerations that are affected include time windows, specialized equipment, training, load-bearing factors, contingency planning, casualty evacuation, and route selection. A SEAL, recognizing the influence of weather and climate on survival, must understand the terrain he will negotiate during the transit to the target.

5.2.7.1 Mountains. Mountains are defined as landforms more than 1,500 feet above the surrounding plain and characterized by steep slopes. Slopes are commonly from 40 to 45 degrees. Cliffs and precipices may be vertical or overhanging. Mountains may consist of an isolated peak, single ridges, glaciers, snowfields, compartments, or complex ranges, extending for long distances. Most of the terrain is an obstruction to movement. Mountains generally favor the defense, but attacks can succeed by using detailed planning, rehearsals, and surprise. There are usually few roads in the mountains. Most are easily defended, since they follow the easiest avenues of travel in the valleys and through passes. Ridgelines often have good trails, since they form the next easiest routes, but trails are generally observable. Detailed maps show roads and many of the trails. Terrain intelligence, photograph interpretation, and local residents may supply additional information. Terrain familiarization will determine the most feasible routes for cross-country movement when there are no roads or trails. Intelligence should include topographic or photographic map coverage and detailed weather data for the AO. When planning mountain operations it may be necessary to use additional information on size, location, and characteristics of landforms. This includes drainage, types of rock and soil, and amount and distribution of vegetation.

5.2.7.2 Formation Mechanics. Most mountain ranges are the result of stresses in the earth's interior. In order to relieve these stresses, thick sections of the crust slowly bend (fold) or fault (break). The resultant surface relief caused by folding and faulting is then altered by the processes of erosion. The following are examples of mountain types.

1. Fault Block Mountains. Fault Block Mountains are bordered on one or more sides by faults, dividing the crust into up and down moved or tilted blocks. The Tetons, Jackson Hole of Wyoming, and the Sierra Nevada of California are examples.

2. Folded Mountains. Folded mountains, such as the Appalachians, have numerous faults but the principal structures are large-scale folds that are modified by erosion.

3. Dome Mountains. Dome mountains are usually the result of the upward movement of magma and the supplement folding of the rock layers overhead. Erosion may strip away the overlying layers, exposing the central igneous core. Examples are the Ozark Mountains of Arkansas and Missouri and Stone Mountain, Georgia.

4. Volcanic Mountains. Volcanic mountains are built by magma coming to the surface from below. The mountains of the Hawaiian Islands and Indonesia are examples of this type.

5.2.7.3 Weathering and Erosion. Once mountains have been created, the forces of nature begin a relentless task of tearing them down. As soon as land is raised above the sea and exposed to wind and running water, the forces of weathering and erosion begin to act. Weathering, both mechanical and chemical, breaks the rocks into smaller pieces without moving the pieces very far. Erosion then transports the pieces to another location by gravity, wind, water, or ice. The most important type of weathering in mountainous regions is called frost wedging. This is the result of moisture in the rocks and crevices freezing and thawing repeatedly. The resulting expansion and contraction wedges off angular flakes and blocks of rock that fall down the slope and are accumulated as talus and scree. Scree and talus resulting from wedging action are generally poor for climbing due to their instability, although scree may be descended rapidly using certain techniques.

5.2.7.4 Rock Types. Rocks can be classified by their method of origin. There are three basic types of rock classifications:

1. Igneous. These are "fire-formed" rocks that have solidified from a hot liquid melt, called magma. Igneous rocks rise from a depth in the earth as a molten magma. If the magma cools and solidifies before reaching the surface, the igneous rocks are termed intrusive. Intrusive rocks, such as granite, cool slowly and result in a dense crystalline structure that forms a tough, hard rock, generally excellent for climbing. Intrusive rocks normally have a great many small cracks and fissures that may be used for hand and foot holds. The cores of most major mountain ranges in the world are granite, and in general the older the mountains, the more granite rock has been exposed at the surface and the better the climbing. When magma rises to close to the

earth's surface, the molten rock may either flow onto the surface and cool or be ejected in a volcanic explosion as ash and lava. These igneous rocks are termed extrusive, as they cool and solidify in the atmosphere. Extrusive rocks that are ejected by volcanic action have very little strength or cohesion and are very difficult for climbing. Extrusive rocks that cool more slowly, such as basalt, can be almost as good for climbing as granite.

2. Sedimentary. Rocks that are deposited by the action of water, wind, and/or ice, or chemically precipitated from water. Sandstone, shale, and coal are sedimentary rocks usually deposited by rivers and oceans, whereas limestone is precipitated from seawater.

3. Metamorphic. These "changed" rocks were originally igneous or sedimentary rocks that, due to temperature and/or pressure within the earth, have been altered physically and/or chemically. Examples of metamorphic rocks are slate from shale, marble from limestone, and gneiss from granite.

5.2.7.5 Rock and Ice Formations. Rock and ice will take on specific characteristics as the environment shapes and changes their development.

1. Bedrock. Unbroken solid rock, overlaid in most places by rock fragments or soil.

2. Chimney. A vertical crack in a rock or ice face that will admit the body. Many types of chimneys are encountered from clean splits in the rock or ice to right-angled corners containing cracks.

3. Cornice. An overhanging formation of snow, usually on the leeward side of a ridge or at the top of a gully.

4. Crevasse. A deep horizontal fissure or crack in thick ice or glaciers.

5. Fissure. A narrow opening in rock, snow, or ice produced by splitting a narrow crack.

6. Knoll. A small hill.

7. Outcrop. An occurrence of a different rock or mineral in a larger rock or rock face. It is usually harder than the parent rock.

8. Peak. The top of a mountain.

9. Ravine. A deep cleft in a mountainside or in the floor of a valley. Normally quite narrow with steep sides.

10. Rock-fall. The accumulation of rocks that have fallen from a cliff or slope. Occurs on all steep slopes and particularly in gullies or chutes.

11. Slab. A relatively smooth portion of rock at an angle, usually of 30 to 75 degrees from the horizontal. Also, the main cohesive body of an avalanche.

12. Wall. A rock face inclined between 75 to 90 degrees from the horizontal.

5.2.7.6 Arctic Soil. Talik is sandy subsoil that does not contain enough moisture to freeze. Ice lenses are deposits of pure ice that may form in either talik or permafrost. Permafrost is perennially frozen material.

5.2.7.7 Snow. The deep winter snow affects mobility in many ways. It impedes movement cross-country and on roads. It blankets terrain features, hiding obstacles such as stumps, rocks, ditches, small streams, fallen trees, minefields, and other artificial obstacles. Snow cover acts as a thermal cover retarding the freezing and thawing of the underlying ground and ice. If snow cover arrives before the ground freezes, the underlying surface will remain soft and trafficability will be poor. When the snow melts, it saturates the ground and makes it impassable. Traction on compacted snow is better during extreme cold weather. Snow crust will occasionally bear the weight of an individual or small group on foot.

5.2.7.8 Forests. Coniferous trees are characteristic of northern forests that are composed of a few simple tree families. Spruce, fir, larch (tamarack), and pine are the chief conifers and are combined with such broad leaf species as aspen, birch, beech, maple, alder, and willow. The more severe and rugged the terrain and weather, and the higher the elevation, the poorer the forest formation, to the point of the absence of any trees at all.

1. Jeffrey Pine. This large tree usually grows at between 6,000 and 8,000 feet elevation. Jeffrey pines can grow up to six feet in diameter and up to 200 feet in height. The needles group in groups of three and are about four to eight inches in length. The cones are approximately seven inches long and five inches wide, with prickly scales pointing down. The bark is reddish brown with large scales pealing off mature trees. The bark smells like vanilla or pineapple.

2. Lodgepole Pine. This tall straight tree usually grows at between 8,000 and 10,000 feet of elevation. They can grow up to four feet in diameter and up to 150 feet in height. Their needles are short and stubby, grow in groups of two, and are approximately two inches in length. The bark is covered with small gray scales. The prickly cones are about two inches long and one inch in diameter.

3. Western White Pine. These trees are found at elevations of between 9,000 and 11,000 feet. They grow up to five feet in diameter and up to 175 feet in height. The skinny needles grow in groups of five and are about two inches long. The cones are three to six inches long and one inch in diameter. The slightly curved cones have no scales. The bark of a young tree is quite similar to Lodgepole Pine, but the bark of a mature tree is rough and broken into small blocks.

4. White Fir. Found at up to 7,500 feet elevation. Barrel-shaped cones are borne erect on the uppermost boughs. It has up to three-inch long gray-green needles that grow individually beside, and above the branch. Needles grow opposing each other, facing slightly upward in a V-position.

5. Western Juniper. A low evergreen shrub from 12-30 feet in height, growing low and spreading upright. The bark is reddish brown and shreds. The tiny scale-like leaves are straight, sharply tipped, ridged, and nearly at right angles to the branches. The blue berry is bitter but edible. Junipers often grow on rocky slopes in semi-arid areas.

6. Mountain Alders. These small trees are found between 4,500 and 8,500 feet elevation. The leaf margins are fine teeth, superimposed on rather coarse saw teeth. The little cone-like catkins grow in bunches and dangle from the tree year round. It is one of the most reliable indicators of water since it is usually confined to areas where water is always present.

7. Willow. Most varieties of willows have long, narrow, and smooth or toothed leaves, or oblong lance-like leaves with short stems.

8. Quaking Aspen. A member of the willow family. Aspen is usually found between 6,000 and 9,000 feet elevation. The leaves are bright green and heart shaped. The bark is smooth and whitish in color. They are normally the first tree species to emerge after a forest fire or avalanche, so they may mark the location of a past fire or avalanche.

9. Black Cottonwood. Found between 5,000 and 9,000 feet elevations, black cottonwood is also a member of the willow family and indicates water in the area. The leaves are bright green and spear shaped. The bark is gray and rough.

10. Mountain Mahogany. Mountain mahogany grows in thickets and is found between 5,000 and 10,000 feet elevations. It is a small crooked tree 12 to 20 feet in height. It has smooth, small, round leaves. It is also called "iron wood" because it is very hard and dense.

5.2.8 Mountain Hazards. The following are examples of natural hazards that should be considered when planning mountain operations.

5.2.8.1 Rockfall. This is probably the most common hazard encountered by the military mountaineer. The structure and composition of a rock area should be studied. Rock that has been subjected to severe weathering will be more prone to rockfall. "Soft" rock that is found in sedimentary formations is more prone to rockfall than is solid, glaciated granite. Stratified rock such as slate and shale can be extremely loose and unstable. Indicators of rockfall must be learned and observed in the field. Fresh debris at the bottom of a cliff, scree cones at the bottom of gullies, and scars on snowfields below rock pitches are all excellent indicators of rockfall. It is also important to know at what time rockfall is most likely to occur. Generally, rockfall occurs early in the day on east and south faces as the sun first warms them and occurs in late afternoon on west and north faces. There is, however, not a 100 percent correct rule to be followed. The NSW mountaineer must use all his knowledge and experience to minimize the danger involved. Areas where rockfall is likely should be avoided in favor of ridges.

5.2.8.2 Lightning. The danger from lightning is greater on rock than on snow or ice. During a thunderstorm the following steps should be taken:

1. Avoid summits and ridges

2. Stay away from prominent objects

3. Avoid gullies filled with water

4. Avoid overhangs and recesses

5. Avoid cracks in wet rocks

6. Squat on dry ground or rucksack.

5.2.8.3 Avalanches. There are two main causes of avalanches. The weight of large amounts of accumulated snow and steep slopes that exceed the cohesive forces within the snowpack, or between the snowpack and ground. These two elements combined can produce an avalanche:

1. Terrain. Ground surface conditions have considerable effect upon snow. A broken, serrated, or boulder-strewn surface provides a good anchor for a snowpack. In other conditions, this surface can provide dangerous stress concentrations in the snowpack. Slides breaking off at ground level do occur but are unlikely. Smooth, even slopes of bare earth, solid rock, or grass favor massive ground-level avalanches, typical of the high alpine zone. Contours of a mountain influence the avalanche; terraces, talus, basins, and outcrops are effective barriers. Barriers either divert the moving snow or give it room to spread out and loose momentum. Gullies collect and channelize the descending snow, making favorable slide paths that must be avoided. Ridges lying parallel to the slide path are normally secure. The convex portion of the slope is more likely to avalanche because the snow layers settling upon it are placed under tension. Avalanches usually

fracture at the sharpest point on the curve, increase to full speed instantly, and pulverize rapidly. Snow is under compression on the concave portion of the slope. The terrain features and grade are both important. Slides are not always likely on steeper slopes. Vertical faces do not hold enough snow to avalanche. Grades below 30 degrees are of less concern for dry snow slides unless an avalanche is induced by unusual circumstances. Grades between 30 and 45 degrees are most critical for dry slab avalanches. Snow does not normally accumulate in large quantities on steeper slopes. Slush-flows often occur on shallow slopes, especially where the ground is impermeable in the spring. The most dangerous slopes are those that have a convex portion above the treeline where tensile stresses can develop and have a concave portion below where the falling snow can accelerate. The upper portion is usually the starting zone. The dimensions of the slope (length and width) determine the size of the snow slide and amount of destruction.

2. Slope Steepness. Avalanches can be produced on slopes as gentle as 15 degrees. The majority of avalanches occur on slopes between 25 and 60 degrees. Slopes above 60 degrees are generally too steep to build up significant amounts of snow. Any slope above 30 degrees should be evaluated for stability before movement over snow. A field-expedient inclinometer may be constructed from two equal-length ski poles. One pole is marked at exact center. To determine if a slope is greater or less than 30 degrees, the unmarked pole is mated to the mid-point of the marked pole at a 90 degree angle. By placing the ski pole tip of the marked pole on the ground, a determination can be made by observing the full-length pole as it is oriented up the fall line.

   a. Slope Profile. Dangerous slab avalanches are more likely to occur on convex slopes, but may also occur on concave slopes. Concave slopes are usually more stable than convex slopes and, therefore, safer. However, not only do avalanches occur on concave slopes, but also they may be triggered from the flat ground below the slope.

   b. Slope Aspect. Snow on north-facing slopes is more likely to slide in midwinter. South-facing slopes are more dangerous in spring and on warm sunny days. Slopes on the windward side are more stable than leeward slopes.

   c. Vegetation. Vegetation, except grass, has a restraining effect on avalanches. The existence of heavy cover indicates that slides in that location are rare. It is a mistake, however, to consider all forested areas as safe. Scattered timber is not a deterrent. Slopes where the timber has been destroyed by fire are potentially good for snowslides. Climax avalanches in heavy snow years often destroy forested areas that have grown up since the last major slide.

   d. Exposure. Slopes facing the sun favor avalanches produced by thawing. Loose snow avalanches are more common on slopes opposite the sun. Cornices form along ridges and crests that lie at right angles to the prevailing wind. A cornice can suddenly release and start an avalanche on the slope below. Lee slopes are the most probable locations for overloads of wind-driven snow, but snow transported to the wind-beaten slopes, and that which remains, is packed and stabilized.

3. Climate and Weather. In addition to the terrain factors, climate and weather are the other basic elements for the avalanche phenomenon. Storms that deposit up to one inch of new snow each hour are common. About 80 percent of all avalanches occur as a direct result of the additional load deposited during a snowstorm. The remaining 20 percent are delayed action avalanches that occur for no obvious reason.

   a. Temperature. Temperature fluctuates widely and rapidly in the mountains. Prolonged spells of extremely low temperatures occur and there are occasional intrusions of warm air masses, usually in connection with a storm. Rainfall may occur in the coastal zones and create avalanche conditions. Temperature greatly affects the cohesion of snow. A fall in temperature retards the settlement of the slab and increases the chances of forming a hoarfrost layer under the slab on which it can glide.

b. Wind. Wind action during storms in the mountains is strong, and its influence on snow is the most important of all the contributing factors. It transports snow from one exposure to another during storms and fair weather, thus promoting overloads on certain slopes. It also modifies the size and shape of snow particles.

4. Types of Avalanches. Avalanches may be classified according to the type of snow involved, manner of release, or size. Classification according to the type of snow involved is normally used, and all slides are divided into two general groups: loose-snow avalanches and slab avalanches.

a. Loose-snow Avalanches. An avalanche of loose snow always starts on the surface from a point or a narrow sector. From the starting point it grows like a fan, expanding in width and depth. The speed and nature of its development depend on whether the snow is dry, damp, or wet.

b. Dry Loose-snow Avalanches. These are composed of loose snow, possibly drifted but not wind-packed. They normally start at a point of origin and travel at high speed on a gradually widening path, increasing in size as they descend. A dry loose-snow avalanche is always shallow at the start and volume depends on the snow it can pick up during its run. Thus, a dry-snow avalanche of dangerous size can only occur on a long slidepath, or from a large accumulation zone that funnels into a constructed outrun. Heavy snowfall at low temperatures produces the phenomenon of the "wild snow" avalanche, a form-less mass pouring down the mountainside. They are actually avalanches of mixed air and snow. Windblast, a side effect of large, high-speed avalanches, is powerful enough to damage structures and endanger life outside the actual avalanche path.

c. Damp and Wet-slide/Loose-snow Avalanches. These resemble avalanches of dry snow with the same point of origin and gradually become wider. Their mass is many times greater than that of a dry avalanche, and is more destructive. Being heavier and stickier, they develop more friction and travel at a slower rate. The principal hazard of damp and wet-snow avalanches is to fixed installations. Such avalanches travel at comparatively low speed, causing them to stop suddenly when they lose momentum and to pile up in towering masses. This is in contrast to the dry-slide that tends to spread out like the splash of a wave. Damp and wet-slides solidify immediately on release from the pressure of motion.

d. Packed-snow Slab Avalanches. Wind-packed snow called windslab or snowslab causes more deaths than other avalanche types and is equal to the wet-slide avalanche as a destroyer of property. Hard-slab is usually the result of wind action on snow picked up from the surface. Soft-slab is usually the result of wind action on falling snow. Windslab avalanches behave entirely different from loose snow. They have the ability to retain an unstable character for days, weeks, or even months that leads to the unexpected release of delayed-action avalanches. These are often triggered from minor causes such as a skier cutting a slope or from no observable cause at all.

(1) The windslab avalanche combines great mass with high speed to produce maximum energy. It may originate either at the surface or through the collapse of a stratum deep within the snowpack. It starts on a wide front with penetration in depth. The entire slab field, which includes the top, sides, and bottom, will release at almost the same time. An angular fracture line (instead of a point) roughly following the contour always marks the place where the slab has broken away from the snowpack. In a packed-snow avalanche, the main body of the slide reaches its maximum speed within seconds. Speeds of 60 miles per hour are common. It exerts full destructive power from the place where is

starts, whereas a loose-snow avalanche does not attain its greatest momentum until near the end of its run.

(2) Due to the characteristic delayed release action as stated above, the slab avalanche is the most dangerous of all types. A series of slab avalanches may stabilize conditions only locally, leaving an adjacent slab as lethal as an unexploded shell. If found on the surface, windslab has a dull, chalky, nonreflecting appearance and has a hollow sound underfoot. If the slab is hard, it often settles with a crunching sound that an experienced mountaineer recognizes as a danger signal.

e. Climax Avalanche. The climax avalanche is a special combination type. Its distinguishing characteristic is that it contains a large proportion of old snow and is caused by conditions that have developed over time (at least one month and possibly an entire season). Climax avalanches seldom occur and require an unusual combination of favorable factors. Whenever they occur, the penetration of a climax fracture is always in great depth, usually to the ground. They travel farther and spread out wider than ordinary avalanches on the same slide path. During heavy snow years these avalanches may travel further than the normal slide path, and destroy forests and structures.

f. Cornices. A cornice is a snow formation related to the slab. They build up on the lee side of crests and ridges that lie at or near right angles to the wind. Occasionally they are straight-walled, but their characteristic shape is that of a breaking wave. The obvious hazard from cornices is due to the fractures of the overhang from simple overloading or weakening due to temperature, rain, or sun erosion. These falling cornices are large enough to be dangerous by themselves. They may also release avalanches on the slopes below.

5. Triggers. A loose-snow slide usually occurs during or immediately after a storm before the new snow has had a chance to form into a strong layer. As internal changes occur, an avalanche may release without apparent cause. Some triggers are more obvious.

a. Overloading. Added weight from a storm is the most frequent cause of avalanches. New snow accumulates until the weight overcomes the strength at some depth. A combination of failures may occur such as collapse of a weak layer and the spread of a tension fracture. The overloading may occur just in the starting zone from wind deposition.

b. Localized impacts. These include the impact of an avalanche or cornice falling from above and the cutting action of a skier moving on the slope. These localized impacts are effective, depending on where they are applied and the degree to which stresses have built up in the snow.

c. Explosives. Ski areas and highway departments use explosives to control critical avalanche slopes. The objective is to keep the snow layers from accumulating to great depth in starter zones and to test the slopes for stability.

6. Hazard Forecasting. The accurate prediction of an avalanche occurrence either in time or location is impossible. The actual avalanche occurrence is governed by a set of variables that cannot be reduced to a simple mathematical formula. The experienced mountaineer can usually recognize the development of a hazardous situation in time to avoid the danger area. Terrain and climate are the two basic causes of avalanches. Terrain is a constant; climate is the variable. Systematic avalanche studies have identified several factors that contribute to the avalanche hazard. These factors, which are subdivisions of climate and weather, are as follows:

a. Depth and Condition of the Base. A 24-inch depth is sufficient to cover ground obstructions and to provide a smooth sliding base. Greater depths destroy such major natural barriers as terraces, gullies, outcrops, and clumps of small trees. If one of the lower layers consists of highly faceted crystals, especially depth hoarfrost, the slope is dangerous because it may easily fail on that layer. The presence of a layer of depth hoarfrost or a thick layer of granular or sugar snow can be detected by an experienced observer by probing. Snow pits are commonly dug to observe the layering.

b. Old Snow Surface. A loose snow surface promotes good cohesion with a fresh fall but allows deeper penetration of any avalanche that starts. A crusted or wind-packed surface means poor cohesion with the new snow but may restrict the avalanche to the new layer.

c. Types of New Snow. Dry snow has little cohesion except for the mechanical interlocking of snowflakes. Thus, loose snow avalanches can easily occur. Wet snowflakes are cohesive and undergo rapid settlement to form a layer. When slightly damp, a strong slab may readily form under the action of wind.

d. Snow Density. Snow density is an indication of its strength and is commonly measured in snow pits. The density of each layer normally increases throughout the lifetime of a snow cover although the layer of snow just over the ground can decrease.

e. Snowfall Intensity. When the snow piles up at the rate of one inch or more per hour, the pack is growing faster than stabilizing forces (such as settlement) can take care of it. If vision is restricted to 100-200 meters, it is snowing greater than one inch per hour. This sudden increase in load may fracture a slab beneath and result in a slide.

f. Precipitation Intensity. Based on experience, it is concluded that with a continuous precipitation intensity of 3/32 inch of water or more per hour and with wind action at effective levels, the avalanche hazard becomes critical when the total water precipitation reaches one inch. This is one of the newest methods used in avalanche forecasting. It requires interpretation by trained weather station personnel with special equipment.

g. Settlement. Settlement of snow is continuous. With one exception, it is always a stabilizing factor. The exception is shrinkage of a loose snow layer away from a slab thus depriving it of support. In new snow, a settlement ratio less than 15 percent indicates that little consolidation is taking place; above 30 percent, stabilization is proceeding rapidly. Over a long period, ordinary snow layers shrink up to 90 percent, but slab layers may shrink no more than 60 percent. Abnormally low shrinkage in a layer indicates that a slab is forming.

h. Wind Action. Wind action is an important contributing factor. It overloads certain slopes at the expense of the others. It grinds snow crystals to simpler forms and it constructs stable crust and brittle slab, often side by side. Warm wind (the downslope Chinook of North America, and the Foehn of Europe) is a more effective thawing agent than sunlight. An average velocity of 15 knots is the minimum effective level for wind action in building avalanche hazards.

i. Temperatures. Air temperature determines the type of snow that falls. At temperatures below 15 degrees Fahrenheit the new snow is unstable and does not settle and form a strong layer. The settlement is rapid above 28 degrees Fahrenheit. Sudden temperature changes can induce dangerous thermal stresses in a slab.

7. Avalanche Conditions. Avalanches obey mechanical laws that can be identified and evaluated by trained personnel with special equipment. General indicators of avalanche prone areas include:

a. Evidence of previous avalanches

b. Steep slopes between 30 and 45 degrees

c. Severe changes in temperature

d. Heavy snowfall

e. Lee slopes

f. Cornice ridges

g. Snow plumes and high winds

h. Visible fracture lines in the snow

i. Audible settling of the snowpack

j. Underlying smooth slopes (grass, etc.)

k. Slushy "spring" snow.

8. Use of Avalanche Transceivers (beacons)

a. Transmits and receives on one or two frequencies (2.275 kHz or 457 kHz).

b. Dual frequency has less range than single frequency models.

c. The frequency can only be directionally found in the range that the beacon is transmitting.

d. Frequency is not one commonly used for voice data transmissions.

e. Power is delivered by two AA batteries that should be replaced if the red battery check light flashes less than five times. Always carry spares.

f. Temperature range is –5 to 100 degrees Fahrenheit.

g. Move quartz watches one foot away.

9. Locating Victims

a. Start search for victims at last seen position.

b. Spread searchers 30 yards apart or less.

c. Stop every 15 to 20 paces and listen carefully.

d. Turn beacon slowly 360 degrees until signal volume is the loudest.

e. Turn beacon volume down one increment at a time.

f. Keep volume as low as possible to hear clearly.

g. Probe and dig at strongest signal on lowest volume setting.

h. After locating victim, turn off victim's beacon before continuing search.

10. If you are a victim:

a. Don't panic

b. Try to stay on the surface

c. Discard equipment like skies, poles, and rucksack

d. Make swimming motions keeping head uphill

e. Kick off the ground if possible

f. Hold your breath and keep your mouth shut

g. Position hands and arms to protect your head and form an air pocket

h. Many victims die of suffocation from snow in their mouth and nose

i. Try breaking through immediately after you stop because the snow will set up quickly

j. Don't waste air by yelling unless you hear rescuers above you

k. Don't struggle, conserve oxygen and energy, and try to relax.

11. Avalanche Rescue. Speed is critical. A search leader must be identified and personnel trained in this type of search. The rescuers must determine if the area is safe to be in and then work quickly and together. Most victims are carried to the place with the greatest deposit of snow. If the avalanche made a turn, victims usually will be deposited on the outside of the turn. The beacon search should begin where the victim was last seen and work downhill. If multiple victims are buried and one contact is made, leave two men to dig at that spot and keep searching for the other victims.

12. Safety rules

a. Never travel alone.

b. Only one man is exposed to the threat at a time (everyone else watches him).

c. Stay out of run out areas

d. Always wear avalanche transponders (beacons).

e. Everyone carries snow shovel, probes, and beacons..

f. Do not assume the slope is safe after the man has crossed.

g. Avoid traveling below or on top of cornices. Conduct a through route study during mission planning phase.

h. Don't assume tree covered slopes are safe.

i. Plan an egress route if an avalanche cuts lose above you.

j. Gather local avalanche history information when possible.

k. If you must cross a potential avalanche slope:

(1) Unsnap rucksack waste band and loosen shoulder straps.

(2) Remove ski pole wrist straps, ice axe wrist loops, and ski straps.

(3) Close up clothing (zip up to keep snow out).

(4) Check transceivers to be certain they are working.

5.2.9 Glaciers. Glaciers are the world's greatest earthmovers. These rivers of ice flow under the influence of gravity and constantly restructure the mountains around them. They often provide the best route in otherwise impassable terrain. Glaciers can be very small or large enough to cover an entire continent. They may move from several inches to several hundred feet a year, and when the melt rate exceeds the rate of downward movement, a glacier can actually retreat back up the valley. Movement is usually caused by gravity, by basal slippage over the bedrock, or internal flow in the ice.

5.2.9.1 Formation of Glacial Ice. Glaciers can occur in any mountainous terrain where the annual snowfall exceeds snowmelt, and the terrain permits deep deposits of snow. Once the snow depths exceed ten feet the weight of the underlying snow is compacted into firn (generally old snow that has survived for more than one full year) snow (ten feet of snow makes 1 foot of firn snow). This recrystallizes into firn ice and then into glacial ice. The region of the glacier where the snow accumulation exceeds the melt year-round is called the accumulation zone. Further down the glacier, where melt exceeds accumulation, there is an area called the ablation zone. The (estimated) boundary between the two zones is called the firn line.

5.2.9.2 Types of Glaciers. Glaciers are differentiated by the following characteristics:

1. Cirque Glacier. Does not advance beyond the bowl in which it is formed.

2. Hanging Glacier. A glacier that is forced out of its origin area and over a cliff or precipice so that it breaks off at the snout and tumbles down as an ice avalanche.

3. Valley Glacier. A glacier which advances down a valley. (Glaciers form U-shaped valleys, while rivers form V-shaped valleys.)

4. Piedmont Glacier. A glacier that moves out of its valley origins onto an open plain or into the sea creating a fan-like pattern.

5.2.9.3 Glacial Features. The most common features of a glacier are moraines. These are piles of rock and other debris that have either fallen onto the glacier or have been pried loose by the glacier as it moves along. Among the many glacial features are four basic types of moraines.

1. Lateral Moraine. This is the rock debris along the valley wall.

2. Medial Moraine. This is formed when two glaciers come together and the lateral moraines are forced out into the middle of the glacier.

3. Terminal Moraine. This is the rock debris at the snout of the glacier that has been dredged up and pushed down the valley. Many times a glacial stream will also flow out from under the terminal moraine.

4. Ground Moraine. If the glacier stops and then begins to recede, the rocks and debris under the glacier can be exposed. This is called ground moraine. Generally, the outside of a moraine wall is stable, but the side facing the glacier can be steep and loose and should be avoided. A moraine may present the only path, but it can be awkward and tiring due to the jumbled rocks.

5. Crevasses. These are the result of irregularities in the bedrock under the glacier or stresses in the ice. They are called transverse, longitudinal, or lateral depending on their orientation to the direction of the glacier's movement. Crevasses can be 200 feet deep, but are normally between 40 to 100 feet in depth. Crevasses are hindrances in traveling across glaciers because they may have to be bypassed. They can be hazardous when blowing snow conceals them by forming a snowbridge across the top. These bridges must be tested before crossing and can give way without warning. In summer the ice is less covered with snow and most crevasses are exposed, but they can still be hazardous at night or during periods of reduced visibility.

6. Bergschrund. A bergschrund is a small crevasse. This is the separation of the glacial ice at the point where the glacier transitions from the new snow or ice on the steep mountainsides to the valley floor. The bergschrund can be very high and even overhanging and can create a serious obstacle to movement when attempting to move from the valley floor onto the mountainsides.

7. Seracs. These are ice walls and towers that have been forced upward due to pressure within the glacier. They are unstable and can unexpectedly fall. This is another hazard and route planning should give them a wide berth when possible.

8. Icefalls. These result from the flow of the glacier down a steep ridge. The ridge forces the ice up into a jumbled mass as the glacier flows over the rock. These present formidable obstacles and should be avoided during movement.

9. Nunatak. This is bedrock that protrudes up through the glacier creating an obstacle to glacial movement. This can create pressures in the ice and help to bring about some of the hazards mentioned earlier like crevasses, seracs, and icefalls.

10. Rock and Ice Avalanches. These can occur when seracs and blocks of ice come loose and cascade onto the glacier. These are dangerous when traveling near the valley walls or under a hanging glacier. Detour around these areas when possible.

11. Water Hazards. This is a moulin or glacier mill. Melt-water from the glacier has found a crevasse and is moving to the ground below the ice. Glacial streams above this point may also be hazardous. They are very cold, can be deep, and create a hazard to ground movement. In summer, glacial thaw water can form in troughs, freeze at night, and form glacial swamps. These should also be approached with care.

5.2.10 Coastal Terrain. The arctic contains coastal landforms that vary from extensive coastal plains to rugged mountains cut by deep fjords. The advance and retreat of the great ice sheets have left the marks of glaciation. Existing glaciers project into the sea or approach very close to it. Frozen ground is a phenomenon common to the arctic. The term frozen ground includes permafrost, the active layer of permafrost, dry frozen ground, and other variations. Frozen ground is influenced by constant factors including such features as snow cover, vegetation, hydrography, and ground heat conductivity.

5.2.10.1 Glacial Effects. Glaciers that reach the sea may provide the best approach to an inland objective, but getting a force safely on top of a glacier can be difficult and dangerous. The surface may also turn out to be too heavily crevassed for safe use. Crossing glaciers requires special training and equipment. In over-glacier operations units should be lightly equipped so as to allow maximum mobility. Movement may require special footwear. Thorough and continuous route reconnaissance must be made regardless of the proximity of the enemy.

1. In many flat coastal plains and other poorly drained areas, lakes and ponds have been formed as the result of glacial action. In summer when the ice has gone, they constitute an obstacle to movement inland. They become breeding grounds for swarms of mosquitoes and black flies and hinder the direction and control of forces. By contrast, in winter, frozen lakes may be used as airstrips, provided the ice has become thick enough to bear the weight of aircraft.

2. In certain parts of the arctic raised beaches commonly occur. They exist as a series of flat topped gravel ridges, left by rising land surfaces, some of which have water or ice-filled marshes behind them. These beaches, resembling giant gravel terraces parallel to the shoreline, may extend for several hundreds yards inland. For light aircraft, they make good landing strips. Glacial action has left many coastal areas encumbered with great quantities of rocks and boulders.

5.2.10.2 Coastal Ice. Low temperature is no guarantee of ice thickness or stability. Since water bodies cool and warm at a slower rate than the air, ice formation, even in temperatures of -50 degrees Fahrenheit, can take weeks to several months before the surface is safe for travel. A good rule of thumb from Eskimo military personnel: if the option exists to move over land rather than over ice, do so. If ice movement is unavoidable, stay close to shore or move to shore at the earliest opportunity. For practical purposes there are four types of ice likely to be encountered in the field:

1. Blue Ice. It is by far the best type for travel. Normally the color is light blue or green. In some cases, where water depth is less than three feet, the ice will be clear and the bottom visible. Cracks may be visible, but not a sign of weakness if they run in the same direction as the current. These cracks are caused by ice contraction in extreme cold.

2. Candle Ice. Candle ice is normally encountered in the spring, but can occur anytime that surface water finds its way through the ice by melting from the surface to the bottom. This type of ice appears as a series of icicles or candles. Because the horizontal strength of the ice is weak, there is no cohesive strength. Even when thick, this ice is very dangerous to cross.

3. Rotten Ice. Rotten ice may be encountered at any time. It can be caused by a thaw or by incomplete freezing. Rotten ice often indicates the presence of contamination. Generally it is dull and chalky in color, and is very brittle. Rotten ice has no strength and should be avoided.

4. Unsupported Ice. This occurs when there is a space between the ice and water. It is normally found in areas where the water table has fallen due to tidal action. Unsupported ice can be detected by cutting a hole into the ice. If the water rises less than 3/4 of the way up the side or does not rise at all, then the ice is unsupported. This ice is dangerous and should not be crossed.

5.2.11 Human Factors. Cold weather will be the primary adversary faced by NSW personnel during mountain and arctic operations. Historically, cold weather has produced more casualties during combat than conflict against the enemy. NSW operations are generally conducted without direct support. Personnel must understand the debilitating affects of cold weather. Chances for MEDEVAC during clandestine operations are remote. The best trained and disciplined individuals will be able to prevent cold weather casualties. Acclimatization, preventive measures, and cold-weather injuries are discussed below.

5.2.11.1 Acclimatization: Symptoms and Adjustments. NSW personnel deployed to high mountainous elevations require a period of acclimatization before undertaking operations. The expectation that a freshly deployed, non-acclimated unit can go immediately into action is unrealistic and could be disastrous. Even the physically fit individual experiences physiological and psychological degradation when thrust into high elevations. Time must be allocated to allow for acclimatization, conditioning, and training. Training in mountains of low or medium elevation (4,500 to 7,500 feet) does not require special conditioning or acclimatization procedures, but some individuals will have some impairment of operating efficiency at these low altitudes. Above 7,500 feet (high elevation), most non-acclimated operators can be expected to display some altitude effects.

A person is said to be acclimated to high elevations when he can effectively perform physically and mentally. The acclimatization process begins immediately on arrival at higher elevation. If the change in elevation is large and abrupt, a percentage of personnel may suffer the symptoms of acute mountain sickness (AMS). Disappearance of the symptoms (from four to seven days) does not indicate complete acclimatization. The process of adjustment continues for weeks or months. The behavioral effects will vary depending on the person. Some personnel may experience a deterioration of memory, inability for sustained concentration, and decreased vigilance. Others may become irritable, experience impairment of night vision, or substantial loss of physical strength and breath due to low blood-oxygen saturation levels.

Judgement and self-evaluation are impaired in the same manner as a person who is intoxicated. During the first few days at high altitude, leaders have extreme difficulty in maintaining a coordinated operational unit. The roughness of the terrain, and the harshness and variability of the weather add to the problems of non-acclimated personnel. Although strong motivation may succeed in overcoming some of the physical handicaps imposed by the environment, the total impact still results in errors in judgment.

1. Somatic Factors. Somatic factors include age, gender, general health, sleep quality and amount, and body characteristics. Somatic factors are typically among the first factors considered when selecting individuals for a specific task. Poor health and a lack of quality sleep can significantly degrade individual performance. Finally, somatotype, or body characteristics, such as surface to volume ratio and the amount of insulating fat, can have profound effects on physiological and metabolic processes, and ultimately on performance. This research has led to the development of biomedical intervention techniques designed to optimize human performance capabilities under extreme cold environmental conditions. The procedures developed include hyperhydration, carbohydrate loading, and physiological acclimation to name the most promising.

2. Psychological Factors. Psychological factors include fear, attitude, and motivation. Excessive fear can paralyze an individual, while positive attitudes and motivation can result in human endurance or accomplishment beyond physical explanation.

3. Activity Factors. Activity factors include intensity, rhythm, position, and duration. Intensity refers to the general difficulty of the activity; this factor relates to the weight of the load that is carried or the degree of concentration required to perform a task. Rhythm refers to the pace of speed of the activity. Position refers to the physical position of the body during the performance of the task. Duration refers to the length of time the activity is continued. Activity factors define the work that is performed and determine the requirements and products of the metabolic processes (i.e., oxygen uptake, calories used, heat generated).

4. Environmental Factors. These include noise, vibration, temperature, altitude, and depth underwater. Pollution, toxic substances, biological agents, and radiation may also be involved.

5. Modifying Factors. Modifying factors include training, equipment, acclimation, and conditioning. Some of these factors, such as equipment (e.g., clothing), are completely external to the organism. Others, such as training and conditioning, begin external to the organism, but are fundamentally internal events, both mental and physical. Similarly, acclimation is a physiological response to external conditions (e.g., the organism being exposed to a hot or cold environment), but the changes are so subtle and occur so gradually that the individual might not perceive them. Modifying factors, in the forms of thermal protection, mobility, equipment, and appropriately designed man-machine interface features are the principal means by which human performance can be enhanced under extreme environmental conditions.

6. Thermoregulation. In many NSW mission scenarios it is the temperature of an environment that acts as a limiting factor to operational effectiveness. Thermoregulation is the means by which body temperature is maintained despite exposure to high or low temperatures during operations in extreme environments. The following paragraphs summarize the process of thermoregulation.

   a. Human Thermostat. The human "thermostat" is set at an internal temperature of about 98.6 degrees Fahrenheit with individual differences of ±0.5 degrees Fahrenheit (37 degrees Celsius ± 0.3 degrees). The corresponding skin temperature is approximately 91.4 degrees Fahrenheit (33 degrees Celsius). The human thermoregulation system has evolved to maintain these optimum operating temperatures. There are three primary means by which optimum body temperature is maintained: through shifts in blood volume and heat transfer near the surface of the skin; through heat loss by evaporation of sweat; and by the generation of heat through physical movement. Thus, as ambient temperature decreases, the body attempts to reduce the rate of heat loss by reducing the flow of blood to the cooling skin. This is accomplished by reducing the inside diameters of the peripheral blood vessels (vasoconstriction). When skin temperature drops below 86 degrees Fahrenheit (30 degrees Celsius), involuntary shivering is activated to generate metabolic heat. Voluntary movements also serve to increase metabolic heat generation.

   b. Ambient Temperature Increases. When ambient temperature increases, the heat exchange process is reversed. The peripheral blood vessels dilate and heart rate increases to facilitate the dumping of body heat into the cooler air at the surface of the skin. Sweating occurs at a skin temperature of about 94 degrees Fahrenheit (34.5 degrees Celsius) to permit the further transfer of heat to the environment through evaporation. Under some conditions, it becomes impossible for optimum core temperature to be maintained by the thermoregulation system. Maintenance of optimal core temperature may be impossible when ambient temperatures are higher than internal (i.e., core) temperatures, when heavy work is required (with associated heat production), or when heat transfer is impaired by clothing or equipment. As a result, core temperature rises. Variations in skin and core temperature beyond narrow limits can significantly degrade human mental and physical performance and ultimately result in serious injury and death.

5.2.11.2 Personal Hygiene and Sanitation. The principles of personal hygiene and sanitation that govern operations on low terrain also apply in the mountains. Leaders must ensure that personal habits of hygiene are not neglected. Standards must be maintained as a deterrent to disease, and as reinforcement to discipline and morale.

1. Personal Hygiene. This is especially important in the high mountains during periods of cold weather. In freezing weather the operator may neglect washing due to the cold temperatures and scarcity of water. This can result in skin infections and vermin infestation. If bathing is difficult for any extended period, the operator should examine and clean his skin often. This helps reduce skin infections.

a. Crotch should be ventilated in the evening and use baby wipes instead of toilet paper. Use powder to keep dry.

b. Beards and shaving both have positive aspects. If daily shaving is preferred, shave at night to allow facial oils to reestablish during sleep and then apply additional protective screens prior to exposure. Beards may conceal frostbite, but noses and ears provide early indication of frostbite before bearded areas are endangered. If a beard is worn, proper hygiene is required to prevent bacteria or lice build up. Commercially available wipes with alcohol offer some advantages, but alcohol may induce frostbite if exposure occurs before alcohol evaporates. Use all types of cleaning pads or wipes with caution, avoiding water-based creams and lotions, to prevent frostbite. The non water-based creams can be used for shaving in lieu of soap. Chapstick may be used on lips, nose, and eyelids. Topical steroid ointments should be carried for rashes. The teeth must also be cleaned regularly to avoid diseases of the teeth and gums. Underwear should be changed at least twice a week. When operating in areas where resupply is not possible, a complete change of clothing should be carried. If laundering of clothing is difficult, clothes should be shaken and air-dried. Sleeping bags should be cleaned and aired as the situation allows.

c. Whenever changing socks, operators should examine their feet for wrinkles, cracks, blisters, and discoloration. Keep nails trimmed but not too short. Long nails wear out socks and short nails do not provide proper support for the end of the toes. Medical attention should be sought for any possible problems.

2. Sanitation. In rocky or frozen ground, it is difficult to dig latrines. If latrines are constructed, they should be located downwind from the position and buried after use. In tactical situations, the operator may dig "catholes" in a designated downwind location. Since water freezes, it can be covered with snow and ice or pushed down a crevasse. In rocky areas above the timberline, waste may be covered with stones.

5.2.11.3 Water Consumption. Even when water is plentiful, thirst should be satisfied in moderation. Quickly drinking a large volume of water may actually slow the individual down or promote severe cramping. A basic rule is to drink little but often, 16 ounces for every pound lost. Liquids lost through respiration, perspiration, and urination must be replaced if the operator is to operate efficiently. Maintenance of liquid balance is a major problem in mountain operations. The sense of thirst may be dulled by high elevations despite the greater threat of dehydration. Hyperventilation and the cool, dry atmosphere bring about a three- to four-fold increase in water loss by the lungs. Hard work and overheating increase the perspiration rate. The operator must make an effort to drink liquids even when he does not feel thirsty. A man's body weight is 60 to 70 percent liquids. A small decrease in water volume will produce a large decrease in performance.

1. Water uses

   a. Digestion

   b. Absorption of nutrients

   c. Excretion of wastes

   d. Blood circulation

   e. Body temperature.

2. Statistics on decreased water volume

a. 1 percent decrease = thirsty

b. 2 percent decrease = uncomfortable and efficiency decreases by 25 percent

c. 3 percent decrease = dry mouth, low urine output, decrease blood volume

d. 4 percent decrease = confusion and decreased physical performance

e. 5 percent decrease = headache

f. 20 percent decrease = death.

3. How water is lost

a. Urine, 16 ounces

b. Respiration, ten ounces (increase by four in cold) up to two liters a day

c. Sweating and evaporation, 24 ounces

d. Stools, 16 ounces

e. Injuries (blood loss is equal to a 3:1 ratio of water loss)

f. Average daily water loss is 1,000-2,300 milliliters and as high as 2,000 milliliters per hour with strenuous activity.

5.2.11.4 Factors Affecting Cold Injury. Cold injuries may be divided into two types: freezing and non-freezing. The freezing type is known as frostbite. The non-freezing type includes hypothermia, dehydration, trench foot, and immersion foot. Cold injuries result from impaired circulation and the action of ice formation and cold on the tissues of the body. Temperature alone is not a reliable guide as to whether a cold injury can occur. Low temperatures are needed for cold injuries to occur, but freezing temperatures are not. Significant injuries may occur when the temperature falls below 50 degrees Fahrenheit. Wind speed can accelerate body heat loss under both wet and cold conditions. Many other factors in various combinations determine if cold injuries will occur. These factors include humidity, wind speed, exposure time, activity, type and condition of clothing, and other factors that include:

1. Previous cold injuries. Creates a greater risk of subsequent cold injuries, but usually in a different area.

2. Race. Blacks are more susceptible to cold mountain injuries than Caucasians.

3. Geographic Origin. Personnel from warmer climates are more susceptible to cold injury.

4. Ambient Temperature. The temperature of the air (or water) surrounding the body is critical to heat regulation. The body uses more heat to maintain the temperature of the skin when the temperature of the surrounding air is 37 degrees Fahrenheit than when it is 50 degrees Fahrenheit.

5. Wind Chill Factor. Personnel should be familiar with the wind chill factor. When the forecast gives a figure that falls within the increased danger zone or beyond, caution must be taken to minimize cold injury. The equivalent wind chill temperature is especially important when the ambient temperature is 30 degrees

Fahrenheit or less. Tissue can freeze if exposed for a prolonged period and frequent warming of the part is not practiced. The greater the wind chill, the faster freezing of tissue will occur. See Appendix I for the wind-chill table.

6. Type of Mission. In operations requiring prolonged immobility, and long hours of exposure to low temperatures, the absence of re-warming increases the number of cold injuries.

7. Terrain. Minimal cover and wet conditions increase the potential for cold injury.

8. Clothing. Clothing for cold weather should be worn loose (to trap air) and in layers (to conserve body heat). Clothing should be clean since prolonged wear reduces its air-trapping abilities and clogs air spaces with dirt and body oils. Wet clothing loses insulation value therefore accumulation of perspiration should be prevented. Appropriate measures should be taken when a change in weather or activity alters the amount of clothing needed to prevent a cold injury and overheating.

9. Moisture. Water conducts heat more rapidly than air. When the skin or clothing becomes damp or wet the risk of cold injury is increased.

10. Dehydration. The most overlooked factor causing cold injuries is dehydration. Individuals must retain their body fluids. In cold weather the human body needs special care, and the consumption of water is important to retain proper hydration.

11. Age. Age is not a significant factor within the usual age range of SEAL operators.

12. Fatigue. Mental weariness may cause apathy and lead to neglect of duties vital to survival.

13. Concomitant Injury. Injuries resulting in shock or blood loss reduce blood flow to extremities and may cause the injured individual to be susceptible to cold injury.

14. Discipline, training, and experience. Well-trained and disciplined operators suffer less than others from the cold.

15. Nutrition. Good nutrition is essential for providing the body with fuel to produce heat in cold weather. The number of calories consumed must increase as the temperature becomes colder. If adequately clothed and protected, personnel in cold climates do not require more than normally provided rations of 3,500 to 4,500 calories a day.

16. Activity. Excess activity results in loss of large amounts of body heat by perspiration. This loss of body heat combined with the loss of insulation value provided by the clothing (due to the contamination of clothing with perspiration) can make an individual susceptible to cold injury.

5.2.11.5 Principles of Body Heat. Body heat may be lost through five different mechanisms:

1. Radiation. The movement of heat rays from a warm object to a colder object is radiation. An uncovered head can radiate one half of the body heat on a 40 degrees Fahrenheit day. At 50 degrees Fahrenheit, up to 75 percent of the total body heat output can be lost through radiation. Personnel must keep all extremities covered to retain heat.

2. Conduction. Touching a cold object (bare rock or ice), sitting in snow, being rained on, and cold-wet clothing all cause rapid heat loss through conduction. Air is a poor conductor. Water is an excellent

conductor, as is snow, rocks, and any other cold object. Insulate and place layers with air space between body and the cold conductors.

3. Convection. The movement of air is called convection. The body continuously warms the air next to the skin by radiating a temperature almost equal to that of the skin. This layer of warm air must be controlled by the use of clothing that insulates, allows water vapor (perspiration) to pass through it, and is windproof.

   a. An eight miles per hour wind removes four times as much heat as a wind of four miles per hour.

   b. A temperature of 15 degrees Fahrenheit with no wind is safe, a temperature of 15 degrees Fahrenheit with 20-25 mile per hour winds is life threatening.

   c. Surface to volume ratio. Smaller ratio equals greater loss, greater ratio equals lesser loss.

   d. Clothing is a key in prevention.

   e. Convection is also a major heat loss factor in water. Swimming at a rapid pace multiplies heat loss. Only swim in cold water if there is no possibility of rescue or if unintentionally submersed in cold water. Otherwise stay still and calm.

4. Evaporation. The evaporation of perspiration causes heat loss. Wet clothing can cause heat loss by conduction. Proper ventilation reduces perspiration and keeps clothes dry. Dressing in layers allows the operator to remove or add clothing as needed.

5. Respiration. Respiration also cools the body. Breathing cold air at high altitudes during a strenuous climb can significantly affect body temperature. Heat escapes when warm air is exhaled. In higher elevations, breathing is deeper and more rapid to compensate for lower quantity of oxygen in the atmosphere. As much as 2,300 calories may be lost in a day. Placing a wool scarf or mask over the mouth and nose warms inhaled air and helps keep the body warm.

5.2.11.6 Cold-weather Injuries. Some of the most common cold-weather injuries are described as follows:

5.2.11.6.1 Shock. Shock is the depressed state of vital organs due to the cardiovascular (heart) system not providing enough blood. Shock should be assumed in all injuries and treated accordingly. Even minor injuries can produce shock due to cold, pain, fear, and loss of blood.

5.2.11.6.2 Dehydration. A man's body weight is 60 percent water. A small decrease in water volume will produce a large decrease in performance. Dehydration is the loss of body fluids to the point that it prevents or slows normal body functions. Dehydration precedes all cold-weather injuries and is a major symptom in AMS. It contributes to poor performance in all physical activities even more so than lack of food. Cold weather requirements for water are no different than in the desert. They may exceed desert requirements because of the increased difficulty in moving with extra clothing and through the snow. At high altitudes, the air is dry. Combined with a rapid rate of breathing, as much as two quarts of liquid may be lost each day. The operator needs four to six quarts of water each day to prevent dehydration when performing physical labor in a cold or mountainous environment. Alcohol is not considered an adequate liquid replacement as it contributes to dehydration. Coffee and tea cause excessive urination and should be avoided in large quantities. Factors contributing to dehydration include:

1. The thirst mechanism does not function properly in cold weather.

2. Water is often inconvenient to obtain and purify.

3. There is a lack of moisture in the air in cold climates and at high altitudes.

4. Cold causes frequent urination.

5.2.11.6.3 Hypothermia. Hypothermia is the lowering of the body core temperature at a rate faster than the body can produce heat. Hypothermia may be caused by exposure or by sudden wetting of the body. Even on moderate days from 40 to 50 degrees Fahrenheit with little precipitation, if heat loss exceeds heat gain and the condition of the operator is allowed to deteriorate, hypothermia can result.

1. Factors contributing to hypothermia include:

   a. Dehydration

   b. Poor nutrition

   c. Diarrhea

   d. Decreased physical activity

   e. Immersion in water

   f. High winds

   g. Inadequate protective clothing.

2. Stages of Hypothermia

   a. Mild Hypothermia

      (1) 95-98 degrees core temperature. Sensation of chilliness, skin numbness; minor impairment in muscular performance, particularly in fine movements with the hands, shivering begins.

      (2) 93-95 degrees. Personnel become more uncoordinated, weak, stumbling, slow paced, mild confusion, and apathy.

      (3) 90-93 degrees. Personnel become grossly uncoordinated, frequently stumble and fall, unable to use hands, display mental sluggishness with slow thought and speech.

   b. Severe Hypothermia

      (1) 86-90 degrees. Cessation of shivering, severely uncoordinated with stiffness and inability to walk or stand, incoherent, confusion, irrationality, fruity acetone breath, and urine soaked clothing. 50-80 percent mortality from re-warming even in a hospital setting.

      (2) 82-86 degrees. Severe muscular rigidity, semi-consciousness, dilation of pupils, no heartbeat or respiration.

(3) Below 82 degrees Fahrenheit. Unconsciousness; eventually death due to cessation of heart action at temperatures approximating 68 degrees Fahrenheit.

3. Hypothermia Treatment

a. Mild Hypothermia

(1) Apply first aid to wounds before treating hypothermia.

(2) Prevent further heat loss.

(3) Remove person from the elements.

(4) Remove wet clothing, put on dry clothes and use sleeping bags.

(5) Use a heat source, heat packs, or another body (avoid heat injury). Place heat source around neck, axilla and groin areas. Re-warm core first.

(6) Warm liquids only to conscious and alert victim. No alcohol or caffeine.

(7) If improved, drive on! If condition deteriorates, MEDEVAC!

(8) Keep person quiet and treat gently. Rough treatment can cause atria fibrillation (low output of blood volume) or ventricular fibrillation (no output at all).

b. Severe Hypothermia (below 90 degrees Fahrenheit)

(1) MEDEVAC victim as soon as possible.

(2) Do not re-warm victim in the field.

(3) Do not give anything by mouth to victim.

(4) Avoid giving medications due to low core temperature. The liver works slower and metabolic functions decrease that causes slower drug absorption and longer lasting effects from the medications. This in turn makes it easy to overdose the victim.

(5) Understand that low core temperature causes the lungs to retain higher levels of CO^2. When giving oxygen you can cause a hypoxic state.

(6) Life threatening injury! Treat it as such.

(7) No one should be considered cold and dead until he has been warm and dead!

4. Prevention

a. Clothing. Use proper layering of clothes that fit loosely and provide a wind barrier that is highly water-resistant.

b. Hydration. Maintaining proper hydration is key to good health in any environment, especially a winter one.

5.2.11.6.4 Immersion or Trench Foot. This is damage to the circulatory and nervous systems of the feet that occurs from prolonged exposure to cold and wet at above freezing temperatures. This can happen whether wearing boots or not. Personnel may not feel uncomfortable until the injury has already begun.

1. History. The name comes from WWI, when men spent weeks with their feet in flooded trenches. The British had over 115,000 casualties due to trench foot and frostbite. During WWI, Americans had 60,000 injuries. Only 15 percent ever returned to combat. Identical injuries among victims of sea warfare as a result of being in cramped, flooded life rafts led to the name "immersion foot."

2. Factors contributing to trench foot include:

   a. Getting feet wet

   b. Failure to change socks when wet or not having dry socks

   c. Improper hygiene.

3. Mechanism of Injury

   a. Many days or weeks of wet feet in near freezing weather

   b. Actual immersion in water is not necessary; wet socks and boots will do the same damage

   c. Decrease in blood circulation due to the cold temperatures

   d. Contact with cold water.

4. Diagnosis

   a. Numbness after 7-10 days

   b. Electric-like shock increases exhaustion due to inability to sleep.

5. Grades

   a. Minimal: red skin and slight sensory change

   b. Moderate: edema, redness, blebs, bleeding, and irreversible nerve damage

   c. Severe: edema, blebs, massive bleeding, and gangrene.

6. Treatment

   a. Prevention is the best key to treatment

   b. Clean and dry feet

c. Elevate

d. Keep victim warm

e. Severe cases require amputation in a hospital setting.

7. Prevention

a. Spare socks (powder helps)

b. Remove socks and boots and massage feet to increase circulation

c. These two points brought the injury rate from immersion foot from 85 percent to 5 percent for the British.

5.2.11.6.5 Blisters. When a hotspot is first noticed and prior to the formation of a blister, cover the hotspot with tape (over the area and beyond it). Use benzoine tincture to help the tape adhere to and toughen the skin. Once a blister has formed, cover it with a dressing large enough to fit over the blister, and then tape it. Never drain blisters unless surrounded by redness or draining pus that indicates infection. If this occurs, drain the blister from the side with a sterile needle. After cleaning with soap and water, gently press out the fluid leaving the skin intact. Make a doughnut of moleskin to go around the blister and apply to the skin. For toe blisters, wrap the entire toe with adhesive tape over the moleskin. (Toenails should be trimmed straight across the top, leaving a 90-degree angle on the sides. This provides an arch so that the corners do not irritate the skin).

5.2.11.6.6 Frostbite. Frostbite is the freezing or crystallization of living tissues. Exposure time can be instantaneous or within minutes. The extremities are usually the first to be affected. Heat is lost faster than it can be replaced by blood circulation or from direct exposure to extreme cold or high winds. Damp hands and feet may freeze quickly since moisture conducts heat away from the body and destroys the insulating value of clothing. Heat loss is compounded with intense cold and inactivity. With proper clothing and equipment, properly maintained and used, frostbite can be prevented. The extent of frostbite depends on temperature and duration of exposure. Superficial frostbite involves only the skin. The layer immediately below the surface usually appears white to grayish while the surface is hard and the underlying tissue is soft. Deep frostbite extends beyond the first layer of skin and may include the bone. Discoloration continues from gray to black and the texture becomes hard as the tissue freezes deeper.

1. Factors contributing to frostbite include:

a. Dehydration

b. Below-freezing temperatures or wind chill

c. Skin contact with supercooled metals or liquids

d. Use of caffeine, tobacco, or alcohol

e. Neglect

f. Tight-fitting or constrictive clothing

g. Immobility

h. Hypothermia

i. Plastic boots versus leather boots (findings from the Mt. McKinley study).

2. Factors about Frostbite

a. Blacks are at risk three to six times more than whites.

b. Whites with type O blood are at greater risk than whites with type A or B.

c. Smokers are at increased risk (nicotine is a vasoconstrictor).

d. Previous cold injury increases risk, but usually to a different area.

e. Greater altitude increases odds of injury, but not from colder temperature. The hypoxic state hinders vasodilatation of the extremities.

3. Mechanism of Tissue Injury

a. Ice crystals form between cells, extracting water from the cells

b. Ice crystals form in cells and rupture cells

c. Obstruction of blood flow by tight-fitting constricting clothing or injury.

4. Diagnosis of Frostbite

a. Starts with burning pain that goes away if unattended.

b. Skin turns from red to yellow, pale to purple.

c. Tissue is hard and firm with a waxy appearance.

d. Absent sensation, particularly if greater than one inch in diameter.

e. Ears and nose often noticed by someone else first.

f. Feet are usually first due to conduction from ground, followed by nose, ears, and fingers.

5. Treatment of Frostbite

a. Superficial Frostbite

(1) Treatable in the field

(2) Re-warm at body temperature; keep clean, dry, and warm

(3) Prevent further cold contact

(4) Blisters, edema, and pain are common after thawing.

b. Deep Frostbite

(1) MEDEVAC if possible

(2) Do not re-warm in the field.

(3) Protect frozen parts and treat victim as a litter patient if possible

(4) Re-warm at hospital

(5) In the worst case have the victim walk on the frozen injury, but be certain not to re-warm the injury for this will cause severe tissue damage.

6. Prevention of Frostbite

a. Avoid all tight fitting clothing and gear

b. Avoid vasoconstrictors such as alcohol, nicotine, and caffeine

c. Dry clothing and footwear increases health of the tissue

d. Lip balm for nose and ears helps protect tissue

e. Anti-perspiration can help keep feet from sweating

f. Common sense and the ability to know your limitations are important.

5.2.11.6.7 Constipation. Constipation is the infrequent or difficult passage of stools. Factors contributing to constipation include:

1. Lack of fluids

2. Improper nutrition

3. Failure to defecate as required.

5.2.11.6.8 Carbon Monoxide Poisoning. This is the replacement of oxygen in the blood with carbon monoxide. A contributing factor is inhalation of fumes from burning fuel such as fires, stoves, heaters, and running engines without proper ventilation.

5.2.11.6.9 Snow Blindness. Snow blindness is sunburn of the cornea of the eye due to exposure to ultraviolet radiation. The condition is called photophthalmia.

1. Factors contributing to snow blindness include:

a. Reflection of sunlight from all directions off the snow, ice, and water

b. Ultraviolet rays that cause vision problems, even on cloudy days.

2. Diagnosis. Symptoms may not develop until 8-12 hours after exposure

a. Gritty, sandy feeling

b. Headache

c. Blurred vision with a pink cast to the vision

d. Leads to ulceration in worst cases and can cause permanent damage to the eye.

3. Treatment

a. Heals of its own accord in a few days

b. Relieve pain with cold compress

c. Keep in a dark environment and cover both eyes with patches

d. Ophthalmic ointment containing an anti-inflammatory steroid may shorten heal time

e. Pain medications may be given in severe cases

f. Avoid anesthetics as these may cause greater damage to the eyes.

5.2.11.6.10 Sunburn and Wind Chapping. Personnel can get sunburned even though the temperature of the air is below freezing. On snow, ice, and water, the sun's rays are reflected with more intensity from all angles and irritate the lips, nostrils, and eyelids. A sunscreen applied to the parts of the face exposed to reflected light will give effective protection. Soap or shaving lotions with a high alcoholic content should not be used since they remove natural oils that protect the skin from the sun. If the skin blisters, apply a disinfectant and cover, since the blistered area, especially lips may become badly infected. Chapping due to cold and wind is rarely serious. Any greasy substance (except those that have a water base that can freeze on the skin) can be used for treatment. Personnel will experience sunburn problems especially at higher altitudes in the spring of the year, as the days get longer and the sun reflects off the snow increasing its intensity. Personnel who become sunburned must ensure they drink plenty of water to promote healing and replace fluid loss from damaged skin.

1. Risk Groups

a. Red hair personnel first, blondes second, brown hair third, and black last.

b. Europeans, Mediterraneans, Indians, and Blacks respectively.

c. Children are at greater risk than adults.

d. People on medications are at a very high risk.

e. An increase in altitude will increase the ultraviolet strength of the exposure.

2. Treatment amounts to cold wet dressing soaked in boric acid (one teaspoon per quart of water).

3. Prevention involves the use of sunscreen with sun block. Lip and nose balm are helpful.

5.2.11.6.11 Chilblains. This is a nonfreezing injury that is uncomfortable, but causes no impairment.

1. Location is on the cheeks, nose, ears, and knees.
2. Causes are repeated exposure of bare skin to wet windy conditions at temperatures of 60 degrees to near freezing.
3. Appearance is white, red with an itchy sensation. Often cracks and becomes rough to touch.
4. Treatment is to use moisturizing ointment on injured site. Protective clothing will help.

5.2.11.7 Heat Injuries. Heat injuries, associated with hot weather, can occur during sunny winter days if proper planning is not exercised. Planning, periodic inspections of personnel clothing and equipment, a balance of water and food intake, and rest can preclude most heat injuries.

5.2.11.7.1 Sunburn. Sunburn is the burning of exposed skin surfaces by ultraviolet radiation. Factors contributing to sunburn include:

1. Fair skin
2. Failure to acclimate to direct sunlight
3. Exposure to intense ultraviolet rays for extended periods.

5.2.11.7.2 Heat Cramps. An accumulation of lactic acid in the muscles and a loss of salt through perspiration cause heat cramps. A contributing factor is strenuous exertion that causes the body to heat up and produce heavy perspiration.

5.2.11.7.3 Heat Exhaustion. Heat exhaustion may occur when an individual exerts himself in a hot environment and overheats. The blood vessels in the skin become so dilated that the blood flow to the brain and other organs is reduced. Factors contributing to heat exhaustion include:

1. Strenuous activity in hot areas
2. Non-acclimated personnel
3. Inadequate diet
4. Not enough water or rest.

5.2.11.7.4 Heat Stroke. Heat stroke is a life-threatening situation caused by overexposure to the sun. The body is so depleted of liquids that its internal cooling mechanisms fail to function. Factors contributing to heat stroke include:

1. Prolonged exposure to direct sunlight
2. Overexertion.

5.2.11.8 Acute Mountain Sickness. AMS is a temporary illness that may affect both the non-acclimated beginner and experienced operator. The reported incidence of AMS varies from 8-100 percent of exposed individuals. Personnel are subject to AMS at altitudes as low as 8,000 feet. Incidence and severity increases with altitude, or when quickly transported to high altitudes. Disability and ineffectiveness occurs in 50 to 80 percent of the individuals rapidly brought to altitudes above 10,000 ft. At lower altitudes, or where ascent to altitude is gradual, most personnel can complete assignments with moderate effectiveness and little discomfort. Symptoms include severe headache, lassitude (weakness/exhaustion), irritability, nausea, vomiting, anorexia (chronic lack of appetite), indigestion, flatus (gas), constipation, and sleep disturbances. Dehydration may intensify the symptoms of AMS, so even the best provisions are ineffective if water needs are not met. Dehydration at high altitudes results from the combination of low environmental humidity and a number of physiological and behavioral factors. Hypoxia induces fluid shifts, diuresis (increased or excessive excretion of urine) and hyperventilation, and personnel decrease fluid intake due to blunted thirst from the cold, hypoxia, and nausea. The onset of symptoms begins 6-12 hours after ascent, and peaks in intensity in 24-48 hours, resolving in 3-7 days as acclimatization takes place. A small number of individuals may have symptoms for longer periods. The highest incidence occurs with sea-level residents who fly into the high mountain area, because flying is the fastest means of ascent and allows no acclimatization. The recommended profile is to take one day of rest at 8,000 feet (2,500 meters) and one additional day at rest for every additional 2,000 feet (600 meters). It is advisable to sleep as low as possible. This is reflected in the mountaineer's axiom: "climb high, sleep low." Acetazolamide 250 milligrams every six hours for 48 hours is effective in preventing or reducing AMS. Side effects such as paresthesia (abnormal skin sensations) are common. It is contraindicated for those with sulfa allergies. High altitude cerebral edema may occur in individuals who ascend rapidly to altitude although it has a low incidence of occurrence. High altitude cerebal edema is potentially fatal if left untreated. Symptoms can include severe headache, nausea, vomiting, and extreme lassitude. Left untreated, general symptoms that will manifest include visual change, anesthesia, paresthesia, rigidity, hemiparesis, clonus, pathological reflexes, hyperreflexia, bladder and bowel dysfunction, hallucinations, and seizures.

5.2.12 Diet. Success in cold weather operations depends on proper equipment and nutrition. Because the combined effects of cold and altitude modify eating habits, precautions must be taken. If possible, at least one hot meal should be eaten each day. Fatigue from physical activities and/or loss of sleep may cause individuals to become disinterested in eating properly. Decreased food consumption due to the perceived effect of preparation and the unpleasant taste of cold rations usually result in nutritional deficits. Although hot food is not a necessity, it does increase morale and a sense of warmth and well being. Because of the increased loss of water, drinking fluids is a critical component of the cold weather diet. A conscientious effort must be made to drink often and as much as possible (> 5-6 liters/day). Most of the weight loss in the first 2-4 days is a result of water loss (dehydration) and decreased food intake due to the loss of appetite. Provide for group meals when possible. People will eat more when meals are consumed socially. Schedule breaks for meals and snacks but don't allow snack foods to substitute for meals. Snacks should augment or supplement meals.

5.2.12.1 Nutritional Acclimatization. The following elements are characteristic of nutritional acclimatization in cold or mountain operations:

1. Loss of weight (2-7 pounds.) during the first days at elevation.

2. Loss of appetite associated with symptoms of mountain sickness.

3. Loss of weight usually stops with acclimatization (if adequate water is consumed).

4. An increase in carbohydrate intake.

5. At progressively higher elevations the tolerance of fatty foods may decrease in some individuals.

5.2.12.2 Dietary Components. Energy is produced from all three major food components: protein, fats, and carbohydrates. All are required to maintain a well-functioning body. Each may provide energy, bulk, vitamins, and minerals. The recommended proportion of each food component varies depending on the physical exercise conditions (intensity, duration, and work-rest ratio), environmental conditions, type of training/acclimation, and pre-mission diet.

1. Protein. Proteins consist of amino acid units linked together in a unique order. When consumed, proteins are broken down into individual amino acids that are absorbed through the intestine into the blood. Once in the blood, amino acids are delivered to cells and: (1) used to make new proteins (enzymes, muscle, and other body tissues) or; (2) converted into glucose and used for 5-15 percent of the energy during exercise. The energy value of proteins is four kilocalories/gram of weight. Primary protein sources with high amounts of the 29 primary amino acids (eight amino acids are considered to be essential) include: egg whites, milk, cheese, legumes (beans), poultry, fish, and meats, but other foods (vegetables, cereals) also contain proteins in varying and lesser amounts.

   Protein should comprise no more than 30 percent of the total daily calories. The amino acid tyrosine may be beneficial in reducing the stress of high altitude exposure.

   The recommended daily protein requirement for the average 176-pound (80 kilogram) active SEAL is 1.0-1.5 grams/kilogram body weight or 80-120 grams (2.8-4.2 ounces/day). The lower value (2.8 ounces) is recommended for SEALs during cold missions to avoid increased urination and loss of water associated with eliminating ammonia (a protein waste product). To provide a more constant availability of amino acids to the body, protein intake should be distributed over an entire day. If any amino acid is low, the effectiveness of the others is reduced. If a protein source is included in every meal or snack, it may improve the fat burning process, providing improved energy and endurance.

2. Fats (and oils). Fats are the most concentrated forms of food energy (nine kilocalories/gram). For a healthy diet, only 25-30 percent of the total daily caloric intake should come from fats, however, during some missions or for logistical reasons, rations may contain more than 30 percent fats. Main sources of fats are meats, nuts, oils, butter, eggs, whole milk, cheese, and fats added to pre-packaged foods during the manufacturing process. Avoid all saturated and hydrogenated or partially hydrogenated fats/oils.

3. Carbohydrates. Oxygen tension is increased by a high carbohydrate diet. Carbohydrates are also an important source of energy for muscle and brain tissue. In the form of glucose and other sugars, carbohydrates are involved in all energy-producing pathways in the body's cells. Carbohydrates contain four kilocalories/gram of weight. Carbohydrates are required for replacing/maintaining energy stores (glycogen) in both the muscles and liver. If the diet contains inadequate amounts of carbohydrates, then these stores become depleted and the body resorts to fat for energy. Fat is readily available and in large supply, but is converted slowly into energy and requires five percent more oxygen than carbohydrate adding a burden to the already overtaxed oxygen economy. Carbohydrate is the major energy source for all physical activities that require more than 65 percent of your maximum aerobic capacity. Carbohydrates are necessary to prevent low blood sugar (glucose) and the associated loss of mental performance (glucose is the only fuel the brain can use). If more carbohydrate is eaten than can be stored as glycogen (stored form of glucose in the muscles and liver), then the excess is converted to fat and stored in the adipose (fat) tissue. For the active athlete (SEAL), carbohydrates should comprise 40 percent of the total daily calories. High sources of carbohydrates are foods such as pasta, rice, breads, sweet potatoes, unrefined grains, vegetables, fruits, and juices. High carbohydrate diets have been recommended as a "non-pharmacological" method to reduce the symptoms associated with AMS.

5.2.12.3 Vitamins. Vitamins are classified into two groups on the basis of their ability to be stored in fat or water. Overdoses can occur with fat-soluble vitamins such as vitamin A (because of the storage in fat tissue). Taken as Beta-Carotene (which is water-soluble), no overdose occurs. The fat-soluble vitamins include vitamins A, D, E, and K. Vitamin C has been reported to facilitate heat acclimatization. Multiple B vitamins have been reported to lessen fatigue during work in the heat, and may help with sleep and stress. Vitamins A, E, and C are anti-oxidant and may be desirable at high altitude to help deal with the free radicals from oxygen which occur in the physiology when stressed by high altitudes and high exercise levels. The water-soluble vitamins include vitamins B and C that are found in cereals, vegetables, fruits, and meats. A well-balanced diet provides all of the required vitamins. Since water-soluble vitamins are not stored for the long term, a proper diet throughout the day ensures adequate levels of these vitamins. If an improper and unbalanced diet is anticipated, then vitamin supplements (not to exceed 100 percent recommended daily allowance) may be considered during lengthy (weeks-months) deployments or missions.

5.2.12.4 Minerals. Mineral elements can be divided into two groups: those needed in amounts of several hundred milligrams (350-1200 mg) such as calcium, phosphorous, and magnesium; and trace elements needed in amounts of only a few milligrams a day such as iodine, iron, and zinc. These mineral requirements will be contained in a balanced diet of meats, vegetables, grains, and fruits.

Salt (sodium). Without the proper amount of salt (sodium, chloride, potassium, etc.) the body cannot function properly or retain the correct amount of water. Sustained heavy sweating rates increase the loss of sodium and a number of other nutrients including chloride, potassium, calcium, magnesium, iron, and nitrogen. Eating adequate food salted a bit more than usual provides ample sodium. Sodium supplementation is unnecessary. Salt is amply provided in most foods. Excess salt consumption can place an added burden on water requirements, and can decrease sweating.

5.2.12.5 Metabolism. Eating a balanced diet provides the energy needed to conduct daily activities and to maintain the internal body processes. Normal metabolism ensures the proper use of vitamins and minerals. Since swimming, snowshoeing, cross-country skiing, and climbing may involve high-energy requirements, a balanced diet with carbohydrates is a necessity. The efficiency of the body to work above the basal metabolism varies from 20 to 40 percent, depending on the individual. Over 50 percent of caloric content of food is released as heat energy and is not available for muscle energy. Heat is also a by-product of energy metabolism during physical exertion. Metabolic heat is dispersed through evaporation of sweat, convection, and conduction. During inactivity in cold weather, resting metabolism may not provide enough heat. This will alert the "internal thermostat" and initiate heat production through shivering muscles. Shivering requires extra energy from carbohydrate stores and can consume at least 440 calories per hour for a 200-pound man.

5.2.12.6 Dietary Requirements. An adequate diet is absolutely essential for survival in cold weather operations. The body is not generally heated by external sources. The body heats itself, and the energy source for this heat is food consumption. When units deploy to a cold weather environment, individuals must increase their caloric intake significantly. If the operational window is short (up to four days) and the activity minimal, personnel may be able to subsist on 1,500 to 2,500 calories each day. Lower temperature levels, greater activity, or lengthy operational windows (> 3-4 days) will require daily sustenance levels of 3,500-4,500 calories. Heavier work may require daily diets of 5,500-6,000 calories, but may be very difficult to consume unless augmented with fats.

1. With an ascent to high altitudes, the physiology begins the acclimatization process. The cardiovascular system labors to provide the needed oxygen to the body. Large meals require the digestive system to work harder than usual to assimilate food. Since blood is diverted from the lungs to the digestive system, indigestion, shortness of breath, cramps, and illness may result. Therefore, light meals that are high in carbohydrates are best while acclimatizing to higher elevations. Personnel should eat moderately and rest

before and after strenuous physical activity. Since fats and protein are harder to digest, they should be primarily consumed at night at a reasonable interval before resting, but not just before resting. A diet high in carbohydrates requires eating or drinking frequently throughout the day. A balance of fats and proteins, early in the morning and at night, before and after physical activity, ensures that the metabolism will be serviced. Carbohydrate intake at lunch, beginning in mid-morning and continuing to mid-afternoon, is the key to maintaining energy levels.

2. Meals should be light, easy to digest, and able to be eaten either hot or cold. Commercial outfitters provide off-the-shelf dehydrated products that may be used to supplement the military diet. The Naval Health Research Center tested the acceptability and convenience of these dehydrated meals during Arctic tundra exercises. Results indicate a range of individual preferences and acceptability of these meals by SEAL operators. The Natick (US Army) Laboratory, Food Technology Division, has developed several high calorie, low bulk, and low weight food bars as part of the new lightweight rations. These compact items can be added to standard meals or used separately as emergency rations during E&R. Personnel should consider supplementing the diet with energy bars, jerky, cheese, nuts, dried fruits, and powdered juices. Bouillon cubes dissolved in hot water can replace salt, warm bodies, stimulate the appetite, and stimulate digestion. Hot beverages such as soups, powdered milk, cocoa, hot chocolate, and cider should also be considered. Since coffee and tea, colas, and other caffeine products are diuretic (increase urination), they should not be relied on for hydrating or rehydrating the body. Warm meals should be provided when possible. At higher elevations, the cooking time may be doubled. A larger cooking pot (one-half gallon) may be a more efficient tool than a coffeepot when preparing meals for a group.

5.2.12.7 Liquids. Even the best provisions are ineffective if water needs are not attended too. Inadequate hydration may decrease the body's ability to adjust to cold stress. The body loses liquid at an exceptional rate in cold weather conditions. The heavy exertion of movement on foot in the snow, preparation of lay-ups and defenses, etc., extracts its toll in sweat and loss of moisture in the breath. Exposure to cold can cause a reduction in the sense of thirst and consequently reduce water consumption. Details of water consumption are presented in section 5.2.11.3. Water is usually available from lakes, streams, or by melting ice or snow. Water obtained from lakes and streams needs to be boiled and/or sanitized. Whenever possible, water should be obtained from running streams or from a lake. If a hole is cut in the ice to obtain water, the hole should be clearly marked. If no free water is available, ice or snow must be melted. Ice produces more water in less time than snow. When melting snow, begin with a small amount of water and snow at the bottom of the pot and add snow as the snow in the pot melts. Boiling or using water-sterilizing tablets should purify the resulting water.

5.2.12.8 Rations. The proper intake of essential calories depends on the total ration consumed during the course of a 24-hour period. An individual in a cold environment needs more calories than normal to meet the increased energy requirements needed for:

1. Heating the breathed air
2. Humidifying the breathed air
3. Loss of body heat due to the cold (through convection and conduction)
4. Extra work of moving about in snow wearing heavy cold-weather clothing
5. Extra work of transporting the weight of the pack, radios, weapons, rounds, tent, etc.

5.2.12.8.1 Meal, Ready-to-eat. The MRE (1,223 calories) rations may freeze in sub-freezing temperatures. Individuals may carry the liquid-containing food items inside their field jackets. Most SEAL operators consider these rations unacceptable for cold weather exercises.

5.2.12.8.2 The Long-range Patrol Ration. The Long Range Patrol Ration (LRPR) (1,131 calories) is a lightweight nourishing ration that can be easily prepared in hot water. Four meals a day are necessary to maintain proper caloric intake. Eating LRPRs without rehydration can aggravate a dehydrated condition and constipation. LRPRs are no longer readily available through the military supply system and have been replaced with the Ration Cold Weather (RCW). Both MREs and LRPRs can be modified prior to entering the field. Individual meal packages can be augmented with supplemental foods placed into plastic bags and resealed.

5.2.12.8.3 Ration, Cold Weather. The newly developed RCW provides up to the 4,500 Kcals calculated for an individual to live and fight in the cold during a long-term conflict. This ration is available in 3,300 Kcal units with optional supplemental packets of 1,200 Kcals. Natick Laboratories is planning to develop a new modular approach to military rations that will allow the individual to assemble rations for specific mission requirements. The RCW has decreased weight and space requirements. This decreases the bulk that personnel must carry into the field and the amount of trash that must be disposed of or carried out. It provides two hot meals and easy-to-eat high-energy snacks. The ingredients for numerous hot drinks are also included in this ration. No liquid-containing food items are contained in this ration. The freeze-dried items can be eaten dry, but should always be reconstituted with hot water to prevent dehydration.

5.3. MOUNTAIN AND ARCTIC CLOTHING AND EQUIPMENT

5.3.1 General. NSW personnel must be familiar with all equipment at their disposal to successfully conduct mountain and arctic warfare missions. Survival and operating requirements will provide the foundation for equipment selection. Failure to anticipate changing environmental conditions and improper equipment selection will degrade the operational capability of the individual and unit. This section will discuss the factors that determine clothing and equipment selection. Clothing factors include: principles of heat loss, weight, transportability, camouflage, comfort, etc. The performance of communications equipment, weapons, and optics will be degraded in cold weather. Problems and preventive measures will be highlighted. Staff planners and operational personnel should closely evaluate the benefits and drawbacks of each type of equipment. Additional training may be required to establish operational proficiency with generic and specialized equipment.

5.3.2 Clothing and Equipment Factors. Several factors influence the design and use of mountain and arctic clothing and equipment. This section will discuss considerations for proper selection and use.

5.3.2.1 Heat Production. The body produces heat in the following manner:

1. Voluntary. This is a product of muscular activity associated with work.

2. Involuntary. Dilation or constriction of the blood vessels as a physiological reaction to cold (i.e., shivering).

5.3.2.2 Heat Loss. Heat loss occurs through radiation, conduction, convection, evaporation, and expiration. Details of these processes are covered in section 5.2.11.5. The following steps should be taken to minimize heat loss from:

1. Radiation: Trap heat in layers of clothing.

2. Conduction: Avoid direct contact with cold objects.

3. Convection: Wear protective windproof clothing and hat.

4. Evaporation: Avoid excessive sweating.

5. Expiration: Wrap protective clothing over the mouth.

5.3.2.3 Clothing Design Principles. Clothing is designed to meet the following criteria:

1. Insulate. Insulation material reduces the amount of heat lost to the environment. By increasing and decreasing the amount of insulation, personnel can regulate the amount of body heat retained. Cold weather clothing systems consist of several layers.

2. Layer. Several layers of clothing provide more insulation and flexibility than one heavy garment. Layering creates dead air space that traps and retains air warmed by the body. The more dead air space, the greater the insulating value. Layers can be added as it gets colder or removed as temperatures rise or the workload increases.

3. Ventilate. Ventilation helps maintain a comfortable body temperature by allowing the individual to rid himself of excess heat and moisture. Ventilation may be achieved by opening clothing, using zip vents, or removing layers to regulate body temperature.

5.3.2.4 Heat Retention. Four principles should be considered to ensure that a clothing system provides optimum warmth for the individual. Remembering and applying the principles associated with the acronym COLD will assist in heat retention.

1. Keep Clothing **C**lean. Clothing keeps the individual warm by trapping air against the body and in the pores of the clothing itself. Additionally, clothing pores transport moisture (sweat) from the body through the fabric (wicking) regulating heat build up. If the clothing is dirty and the pores become clogged, the above functions will be reduced.

2. Avoid **O**verheating. Select the amount of clothing required to remain comfortable. This will change as a result of workload or temperature. Let the environment cool the body rather than perspiration. As the workload increases, unbutton or remove layers of clothing. When activity slows down, add clothing as required.

3. Wear Clothing Loose and **L**ayered. Cold weather clothing should be fitted to accommodate layers and freedom of movement. The insulation value achieved by the layering technique will be reduced if clothing is tight or restrictive.

4. Keep Clothing **D**ry. Clothing that is inside the protective shell and close to the body should be kept dry. Wet clothing will conduct heat away from the body. Snow or water should be brushed from the outer layer of clothing to prevent moisture from soaking through the material.

5.3.2.5 Weight. Clothing and equipment should be carefully considered against the requirements of the mission. Environmental conditions, terrain, time factors, and the mission will determine the type and amount of equipment selected. Sustainability and achievement of the objective are primary considerations. Weight and stowage requirements require careful analysis for each phase of the mission.

5.3.2.6 Transportability. Terrain and distance to the objective will determine the amount of equipment that can be carried and the means of transportation. NSW operations are generally unsupported following insertion. Equipment must be resupplied, cached, man-carried, or transported by sled or pulk.

5.3.2.7 Standardization. NSW personnel should wear clothing and stow equipment in a similar manner (common to all). This approach will enhance familiarization between squad and platoon members. Standardization is vital during emergency actions, enemy contact, and low light conditions.

5.3.2.8 Clothing Requirements. Careful consideration should be given to the requisition of cold weather clothing. Rapidly changing technology has created state-of-the-art designs. The design configuration must meet the requirements delineated by personnel and mission. Clothing should be:

1. Purposeful (fulfill desired characteristics)

2. Protective (wind/waterproof)

3. Insulating (provide warmth)

4. Camouflaged

5 Buoyant (water operations)

6. Versatile (support layering principle, ventilate, provide-zips, velcro, pockets, etc.)

7. Durable (all weather/terrain, tear resistant, non-freezing, and water-resistant).

5.3.3 Personal Equipment. The proper use of personal clothing and equipment is essential. Survival, comfort, and tactical application require an understanding of clothing types and the preparation of personal and unit equipment. The cold weather environment is a deadly adversary. It is the responsibility of NSW personnel to use cold weather equipment and survival skills to their full advantage. Personnel should be constantly aware of the changing environmental conditions and monitor their heat production to maintain a comfortable body temperature. A clothing system helps to control body temperature by regulating the following layers:

1. Vapor Transmission Layer. This layer draws moisture away from the body and keeps it dry.

2. Insulating Layer. This layer holds and maintains a warm layer of air around the body. The insulating layer may be comprised of several layers of clothing.

3. Protective Layer. This layer protects the insulating layer from the dirt and wetness of the outside environment. It should breathe and be wind and water-resistant.

5.3.3.1 Individual Clothing and Equipment. The current clothing system worn by NSW personnel is based on layering techniques. Temperature and activity determine the appropriate number of layers to be worn. Design, material, and principles of wear determine the location of the article of clothing in relation to the body. Outer clothing, boots, and bivy sacks should be waterproofed. The following items compromise a typical clothing system.

1. Thermal Underwear. Comprised of various materials (e.g., polypropylene or silk layer) which wicks body moisture away from the skin. This layer is designed to keep the insulating layers dry.

2. Insulating Layer. This layer is comprised of natural fiber (e.g., wool) or synthetic fiber (e.g., capilene). Synthetic fibers are lighter and dry easier. The insulating layer may consist of pullover material, trousers, and/or a sweater. Suspenders should be used instead of a belt, as they will not reduce circulation. The insulating layer must be protected from the outside environment and remain dry. Insulating layers may be removed during periods of activity or added when the individual becomes inactive.

3. Protective Layer. This layer is worn over the insulating layer(s) and serves to retain heat and protect the individual from wind and precipitation. Synthetic material should allow moisture from the inside to pass outward while preventing water from passing inward.

4. Headwear. Head coverings are constructed of synthetics or wool. A balaclava or head cover may be used. Since the majority of body heat is lost through the head at low temperatures, the individual must wear the appropriate level of protection. The head is the first part of the body to uncover when overheated and the first part to cover when cold. Head coverings may be rolled up to cover only the top of the head or rolled down to cover the ears, back of the head, and neck. Some are designed with a face piece with holes for the eyes and mouth.

5. Hands. Hands and fingers are extremities and can quickly become numb. It is essential to keep the hands and fingers functional in order to perform most tasks and efficiently operate weapons. Gloves or mittens may be used. Mittens are preferred because they provide greater warmth to the fingers, but dexterity is lost. Gloves allow the freedom to climb, tie knots, and shoot. Liners may be used with either gloves or mittens. The outer shell will protect the liner and hands from moisture. When using mittens, a trigger finger design will allow for weapons firing. Important in keeping the hands warm is protecting the wrist from cold. This is accomplished by the use of a wristlet composed of wool with a thumbhole that partially covers the hand and is worn midway up the forearm.

6. Feet. Feet are prone to the same problems as hands. Keeping the legs warm will help prevent toes from freezing during inactivity. Additionally, the moisture created from the feet will contribute to frostbite unless proper precautions are taken. At least two pairs of socks should be worn. The socks will provide warmth and blister protection. A good sock is dense enough to prevent abrasion in high compression areas, densely and uniformly woven, and incorporates a wicker material. As the inner layer becomes wet, it can be replaced with a dry pair or rotated with the outer layer. Vapor barrier socks may be worn between the inner layer (polypropylene, silk) and the outer layer (wool). Foot powder will reduce perspiration and increase comfort.

7. Boots. Boots must be fitted properly. NSW personnel will encounter different types of terrain. The terrain and intended method of transit will determine the appropriate type of boot to be used. Boots fitted for cross-country skiing may not be suited for climbing. Support, traction, and adaptation to snow shoes and skis should be considered. The foot lengthens, widens, and swells during transit as a result of terrain and load bearing. Both feet may not be the same size. The appropriate socks and pack (load) should be worn when fitting boots. A leather boot worn in the northern maritime environment should have high uppers (5 ½ to 7 ½ inches) to protect the ankle in rough terrain, thick rubber soles for good traction, mid-sole for insulation shock absorption, and semi-rigid shank to prevent slip on small holds. The boot should also have open wide tops for ease of fitting when frozen or wet, minimum seams to decrease leaks, and a gusseted tongue to keep water out. The plastic boots are waterproofed to keep the inner insulating boot dry. The inner boot can be removed and comes in different insulating properties. Plastic boots need to fit well from the start because of their non-molding, rigid shape.

8. Gaiters. Gaiters prevent water from streams, foliage, and snow from leaking into the top of the boot. They are constructed of synthetic material and wrap around the boot and lower leg. Presently, gaiters have been designed which fit completely over the boot.

9. Over-whites. Over-whites are designed to camouflage the arctic warrior. The top and bottom may be worn separately or together as terrain features dictate. The most efficient over-whites are constructed of nylon, which absorbs less water than cotton, weighs less when wet, and dries quickly.

10. Eye Protection. Sunlight and snow can degrade vision and damage the eyes. Appropriate glacier glasses and/or goggles are required by mountain and arctic personnel to prevent injury. Jet ski goggles with a good seal are excellent for boat transits.

11. Sleeping System. The sleeping bag with bivy sack operates by the same principles as the clothing system. The bag provides insulating value and the sack provides protection against wind and precipitation. The variable temperature ratings for different sleeping systems must be matched to the appropriate temperature ranges of the environment. Ground pads are essential in preventing heat loss.

12. Tent. The requirement for shelter is absolute during arctic operations. Clothing and training will not ensure survivability without erecting or improvising a shelter during severe weather conditions. Tents should be easy to pitch in darkness, well camouflaged, and provide multiple exits and ventilation. Poles and sleeves should be color-coded. High winds require that tent anchors (stakes, ice screws, or deadman) be used.

13. Stoves. The ability to produce heat quickly and reliably is essential. Heat will warm rations, melt snow for water, dry clothes, and provide reflective heat in shelters. The stove should be multi-fuel burning. The burner jet must be clean and the pump seal should fit securely.

14. Water Containers. Containers that can be thawed should be selected. Water bottle insulators should also be used. This will allow the container to be carried on the outside of the pack. Always replenish the water supply when possible. One container should be carried close to the body.

15. Rucksacks. Packs should allow you to carry weight close to your body and center the load over your hips and legs with minimum carrying capacity of 5,200 cubic inches (approximately 3 cubic feet).

5.3.3.2 Cold Weather Survival Tips. The following steps will increase the comfort and operational efficiency of mountain and arctic personnel:

1. Cold Fingers. When wearing gloves or mittens, the fingers may still get cold. Pull the fingers out of the individual compartments and bunch them together into a fist. Swinging the arms will also increase circulation.

2. Sleeping Bag. Branches and/or a mat will help insulate the bottom. A headcover should be worn while sleeping to prevent heat loss. Use the flap to cover the zipper and prevent freezing.

3. Food Prior to Sleep. Consuming hot liquids or snacks high in carbohydrates prior to sleeping will provide fuel for heat.

4. Cold Feet. Wear extra clothing on the legs, exercise in place, or swing the legs to conserve heat and increase circulation to the feet.

5. Waterproofing. Commercial spray should be applied to seams of old or worn clothes to preserve watertight integrity.

6. Storing Clothes. Clothes removed prior to sleep can be stored inside the sleeping bag, provided they are clean and dry. This will increase insulation and keep the clothes warm. Boot liners can be placed at the foot of the sleeping bag.

7. Water Bottle. The bottle should be approximately 2/3 full to prevent freezing. Prior to sleeping, fill the bottle with hot water and place inside the sleeping bag. This will help warm the bag.

5.3.3.3 Load Bearing Equipment. Appendix F lists suggested equipment for first, second, and third line equipment. NSW personnel are responsible for maintaining and carrying a great deal of equipment during mountain and arctic operations. The design of clothing, LBE, and rucksack should allow easy access during planned and emergency operations. Items frequently needed or changed should be located in the outer pockets or flap of the rucksack. Modifications may be required and the rucksack should have a separate component (bolt pack) that can be removed and carried should the pack be ditched. One member of a unit may carry an item (e.g., tent) that is shared by two or more individuals. Individuals should stow equipment in an accessible manner for easy location by other members of the unit. Equipment organization is described as follows.

1. First Line. This equipment is used primarily for survival. Survival items should be stored in the pockets of the outer garment. It should include:

    a. Uniforms

    b. Boots

    c. E&E equipment

    d. Rations.

2. Second Line. These items should be carried by LBE, in a vest, belt gear, or bolt pack. Second line equipment includes weapons, ammunition, and enough supplies to survive should the third line (rucksack) be abandoned.

3. Third Line. This equipment is generally stored in the rucksack.

    a. Sleeping bag

    b. Ground pad

    c. Over-head cover

    d. Food, fuel, and stove

    e. Mission essential equipment.

4. Rucksack arrangement

    a. Bolt pack is secured to the rucksack for quick access during IAD's.

    b. Bolt pack has twenty-four hours of food, medical, stove, cooking gear, and extra socks ready for quick escape from the LUP.

c. Pack all heavy equipment low and over the hips for skiing.

d. Extra equipment and clothes are water-proofed with a plastic liner.

e. Pack with stuff sacks to compartmentize equipment.

f. Keep the pack as streamlined as possible.

5.3.3.4 Water Operations. The conduct of clandestine maritime operations in the subarctic is arguably the most difficult of modern warfare concepts. The critically dangerous environment one encounters when leaving the water for land, or water for an ice-shelf, is virtually unresearched. NSW personnel will operate in every type of environmental condition ranging from cold/wet to cold/dry. Some mission tasks may be accomplished with the use of wetsuits. Other operations will require the use of exposure or drysuits with protective undergarments.

1. The CRRC is a likely surface-insertion vessel for many operations. In subfreezing conditions, rubber contracts and becomes less pliable. Boats with frames or floorboards may perform better in extreme cold conditions. Ice can also freeze fittings and make inflation difficult. A de-icer that is not harmful to rubber compounds should be used on all valves and metal parts. Outboard-motor engines can and must be winterized and tested prior to cold-weather operations.

2. Diving Clothing. Diving operations may require exposure to the water for two to three hours. During the first five to ten minutes the body will be painfully cold, especially around the mouth, hands, and crotch. After exertion, the body will become quite warm but the extremities will tend to cool down. When swimming on the surface, a diver is affected slightly by atmospheric conditions. It is vital that in zero and subzero temperatures a diver changes into warm clothing immediately on leaving the water. This may be achieved by wearing a drysuit with sufficient warm clothing beneath. Individuals should be aware of the effect of condensation or water within the suit. Damp underclothes will soon become cold. Diving equipment should be kept to a minimum. Extra weight will be required to establish proper trim.

   a. Wetsuits may only perform adequately for a period as short as 1½ hours. The major disadvantage to the wetsuit is that it offers no protection in subfreezing air temperatures, so operations must be launched and terminated at a point where divers can be easily warmed and dried. Additionally, the wetsuit is not practical for dive operations that involve carrying first and second line equipment. Land activities are difficult to accomplish wearing a wetsuit and typically lead to overheating.

   b. Drysuits have been developed and tested by the Naval Experimental Diving Unit, with positive results for submerged operations of up to six hours. The Navy-developed Passive Diver Thermal Protection System, while excellent for diving operations, has proven undesirable as a transition garment for limited land patrolling. Additionally, extremities tend to get very cold and leakage may occur. The fit of swim fins will be tight due to the extra socks or booties, and will tend to constrict blood flow to the feet. Divers require larger-than-normal fins and straps for cold-water operations. The VIKING/NOKIA drysuit will provide the greatest level of thermal protection for passive operators (SDV transits). Care must be taken to avoid puncturing or tearing the material. The zipper and valves must be kept free of sand or dirt. Personnel should be familiarized with the Overboard Discharge System to ensure comfort and safety during submerged operations. Drysuits may be used for surface transits. Undergarments should be donned using the layering principle in a manner that will not restrict movement.

   c. Splash suits are designed to provide thermal protection from sea spray, winds, and rain during surface transit. If immersed the suit will remain dry initially, but water will soon begin to "seep" through the gortex material. Waterproof seals should be properly and tightly fitted to avoid flooding. Undergarments should be worn in the same manner as with dry suits. Splash suits are not intended for surface or subsurface swimming.

d. The head is a major source of heat loss. Whether personnel are operating on the surface or submerged, adequate protection is required since a substantial amount of blood will continue to circulate through the head (unlike hands and feet) during prolonged exposure. A separate wet or drysuit hood may be required to minimize heat loss.

e. The hands will lose dexterity rapidly when exposed to cold water. Digital manipulation is critical to the operator during routine diving operations, emergency actions, and weapon employment. The systems recommended to keep the hands warm will diminish dexterity. Training will be required to overcome the loss of dexterity. The following configurations are recommended:

(1) A high wrist, lobster claw mitt. It is designed with three fingers and a 6mm thickness.

(2) A five-finger neoprene glove may be worn under the lobster claw mitt for additional protection.

(3) A thinsulate glove worn under a neoprene mitt or windsurfing glove.

f. The layer principle may be applied to foot protection. Polypropylene and wool socks worn under thinsulate "booties" will increase protection. Operators should avoid restrictive layering that will result in cramping.

3. Effects of Cold on Diving Equipment. In sub-zero temperatures the mechanisms of compressed air, oxygen, and mixed gas may freeze, especially when an individual has remained on or near the surface for any length of time. If diving equipment should freeze, it must be thoroughly thawed before any pressure is put on connections or hoses. This can be achieved by immersing the equipment in water and must be accomplished before the swimmer dons the equipment to enter the water. Diving equipment most at risk from the cold includes:

a. Cuffs and neck seals of drysuits

b. Gloves and hoods

c. Mouthpiece valves

d. Relief valves

e. Bypass and main bottle valves

f. Draeger and open circuit demand valves.

5.3.4 Mountaineering Equipment. Movement over mountainous terrain normally requires the aid of special equipment. Movement up steep cliffs, across mountain streams, and deep chasms also involves using special techniques to assist the climber. Specific equipment has been designed to aid the operator in overcoming obstacles such as rock, snow, and ice areas quickly and safely. Mobility across mountainous and arctic terrain is essential if the tactical objective is to be achieved. NSW personnel use a variety of equipment to assist movement over various terrain features. Familiarity and proficiency with each type of equipment is necessary for safe and proper use. This section will discuss the equipment that assists mobility in the mountain and arctic environment.

5.3.4.1 Rope. Climbing rope used by NSW operators is a kernmantle design and constructed of nylon. A waterproof (silicon impregnated) kernmantle should be used if possible. Standard lengths are 120, 135, 150, and 165 feet. Since ropes should not be cut, desired lengths should be determined in advance and carried on the mission. Rope diameter will vary with intended use. The standard diameter for most uses is 11 millimeters. Seven millimeters is adequate for hauling and utility ropes. Non-water proofed kernmantle or nylon rope must be rinsed within 48 hours of exposure to salt water. If unable to rinse, it should be retired from any lead climber role. Nylon rope is used for a variety of purposes and situations. Operators may use it for climbing or hauling heavy loads. It may easily serve as the lifeline for an operational unit.

1. Rope terms

   a. Bight. The bight in a line is a "U" formed by one length of rope.

   b. Loop. A complete circle formed by a length of rope.

   c. Running End. The moving end used to tie a knot.

   d. Standing End. The stationary or non-working end; also called bitter end.

2. Types

   a. Static ropes have a minimum stretch of 1.5 to 10 percent and maximum durability. These are used for rappelling, river and stream crossing, hauling, Jumar climbing, etc.

   b. Dynamic ropes have a maximum stretch of 10 to 40 percent and maximum energy dissipation. These are used for lead climbing and rappelling.

   c. Laid rope

      (1) "Green line" or "gold line"

      (2) Right or left hand lay

      (3) Standard length is 120 feet

      (4) Static breaking strength is 4,500 pounds

      (5) Not waterproof.

   d. Kernmantle

      (1) A two-part rope. The core is "kern" and is 80 percent of the strength. The sheath is the "mantle" and is the remaining 20 percent of the total strength of the rope. The mantle's purpose is abrasion resistance and ease of use. The kern's purpose is strength.

      (2) Standard lengths are 120, 135, 150, and 165 feet.

      (3) The standard diameter for most uses is 11 millimeters. For stirrups, utility ropes, and hauling lines; seven millimeter is adequate.

(4) Static breaking strength (established by static tests).

11mm	4,500 lb
9mm	3,000 lb
8mm	2,200 lb
7mm	1,700 lb
6mm	1,200 lb

(5) These ropes can come waterproofed or non-waterproofed.

3. Knots. Some of the basic knots the operator needs to be familiar with are listed below. See Army Training Circular 90-6-1 (Military Mountaineering) for complete details and graphics of mountaineering knots.

a. Overhand knot secures a loose rope after a knot. When tied with a double line, it can be used to form a bight in a line.

b. Water knot can be used to tie one-inch tubular nylon together or to form a bight in one-inch tubular nylon.

c. Double figure eight secures round cord around a tree or a rock or the end of a line to a sit harness.

d. Bowline secures a line around a tree or pole and ties someone into the middle of a rope. It forms a loop to haul someone up or down without a harness.

e. Double fisherman's knot joins two lines of the same size together.

f. Clove hitch ties into anchors and can adjust line length without being untied.

g. Munter hitch is used to rappel or belay. Also used to lower a heavy load under control.

h. Prusik knot is used to ascend or descend a climbing line. Also used as a break to hold a line.

4. The following rules should be applied to protect the rope:

a. The rope should not be stepped on, or dragged on the ground unnecessarily because small particles of dirt will be ground between the strands and slowly cut them.

b. The rope should never come in contact with sharp edges of any type. Nylon rope is easily cut, particularly when under tension. If a rope must be used around an edge that could cut it, the edge must be padded.

c. Keep the rope dry. If it should become wet, hang it in large loops above the ground and allow it to dry. A rope should never be dried with an open flame.

d. Never leave a rope knotted or tightly stretched for longer than necessary.

e. Never allow one rope to rub continually against another. Allowing nylon rope-on-rope contact is extremely dangerous since the heat produced by the friction will melt the nylon.

f. The rope should be inspected prior to use for frayed or cut spots, mildew or rot, or defects in construction (new rope).

g. Climbing ropes should never be spliced since the handling characteristics will not be acceptable at the point of the splicing.

h. Mark all climbing ropes at their midpoints to facilitate the establishment of retrievable rappels and three-man party climbs.

i. The rope should not be marked with paints or allowed to come in contact with oils or petroleum products since these will weaken it.

j. The ends of a new rope or cut ends should be whipped with thread and fused with a flame. Nylon thread used in parachute repair may be used for this purpose.

k. The rope should never be subjected to high heat or flame since this can weaken it.

l. Nylon ropes can be washed with a mild soap and water solution, after which they should be rinsed thoroughly.

m. When not in use, ropes should be coiled and hung on wooden pegs rather than on nails or other metal objects. They should be stored in a cool place out of direct sunlight.

n. When in areas of loose rock, the rope must be inspected frequently for cuts and abrasions.

o. Ultraviolet radiation (sunlight) will deteriorate the rope.

p. Ropes must have a "rope history" logbook that denotes each use of the rope. It should list the type of rope use, like leader falls, dynamic rappels, number of rappels and type of equipment load put on the line. It should have the manufactured date and the date the rope was put in service. This book should be stored with the rope and give the user the complete history to determine if the rope is safe for use.

5.3.4.2 Slings. A sling is a length of cordage formed into a loop and tied off with a water knot. Slings can be made in various lengths according to their intended use. Standard lengths are six feet (quick draw), four and a half feet (shoulder length), and nine feet (double shoulder length). Slings can be used for runners on a roped party climb, improvised rappel seats, and many other uses. Nylon tubular webbing can also be used to make slings. Slings made of webbing are preferred, as they are much lighter and less bulky. Bar tacking is preferred over tying knots (one inch tubular nylon is preferred) so the sling can move freely through carabiners. Inspect rope, knots, and bar tacks before and after each use and replace if necessary.

5.3.4.3 Snaplinks (carabiners). A snaplink is used to attach a rope to a harness, as a friction device when rappelling, to attach a belayer to an anchor, to connect a rope to an intermediate protection point, to connect a climbing aid to points of direct support, and to assemble a friction brake system. Snaplinks are metal (aluminum or alloy) loops with a hinged, spring-loaded gate on one side. Snaplinks come in many sizes and shapes to suit different needs. Some snaplinks have a locking mechanism with a threaded sleeve on the gate. The sleeve tightens over the gate-opening end to hold the gate closed. The following information applies to all snaplinks:

1. The weakest part of a snaplink is the gate.

2. Locking pins should be checked to ensure they are not loose.

3. The metal should be checked for any cracks, grooves, flaws, butts, or rust.

4. The gate should snap securely from open to a closed position with no gap between the locking pin and notch.

5. If an engraver is used to mark snaplinks, it should be applied only to the gate, never to the load-bearing side.

5.3.4.3.1 D-shaped Snaplinks. This snaplink is stronger than the oval type since its shape permits the largest part of the load to be applied to the solid side opposite the gate, but it is not as universally adaptable for other applications as the oval snaplink. D-shaped snaplinks are made of metal, aluminum, alloys, and are of many sizes and thickness, all of which are available with or without locking gates. Their strength and durability vary, which must be considered before use.

5.3.4.3.2 Oval Snaplinks. These are used for most applications of ascending and descending, and form a basis to build on. They are made of metal, aluminum, and alloys, and are of many sizes and thickness. Both sides of an oval snaplink bear the strain equally under load weight. There are many modified ovals and most are available with or without the locking gate.

5.3.4.3.3 Inspection and Care. Snaplinks should be inspected before, during, and after use. The metal should be checked for cracks, grooves, burrs, rust, and flaws. The gate should open and close freely without binding. There should be no lateral movement when the gate is open. The gate-spring action should spring shut when released. The locking slot should have a slant or notch so that the gate remains shut under the impact of a climber's fall. The gate pins should not work their way out of their holes and should not be shorter than their holes. If there is a locking mechanism, it should be inspected to ensure that threads are not stripped and that the sleeve locks the gate tightly. If burrs or rough areas are identified, the snaplink should be filed smooth. Rust should be removed with steel wool and oil or solvent. The spring should be oiled as needed. The snaplink should be boiled in water for 20 to 30 seconds to remove cleaning agents since solvents cause dirt to cling to the snaplink and rub off on ropes. It is better to use a dry graphite-based lubricant on snaplinks since it does not retain dirt.

5.3.4.4 Pitons. A piton is a metal spike that is hammered into a crack in a rock to provide a secure anchor for a rope that may be attached by a snaplink. Pitons are made of either malleable or chrome-molybdenum steel and alloys. There are two types of pitons: blades that hold when wedged into tight-fitting cracks, and angles that hold from wedging and blade compression. Pitons are described by their thickness, design, and length.

1. Vertical Pitons. The blade and eye are aligned. Vertical pitons are used in flush vertical cracks. Their supporting strength ranges from 1,000 to 2,000 pounds.

2. Horizontal Pitons. The eye of the piton is at right angles to the blade. Horizontal pitons are used in flush horizontal cracks and in offset or open-book type vertical or horizontal cracks. They are recommended for use in place of vertical pitons in vertical cracks since the torque on the eye tends to wedge it into place, giving it more holding power than the vertical piton under the same circumstances. Its supporting strength is 1,000 to 2,000 pounds.

3. Wafer Pitons. These are used in shallow, flush cracks. They have little holding power and their weakest point is the rings provided for the snaplinks. Their maximum strength is 600 pounds.

4. Knifeblade Pitons. These are used in direct-aid climbing. They are small and fit into thin, shallow cracks. They have a tapered blade that provides optimum holding power.

5. Realized Ultimate Reality Piton. These are hatchet-shaped pitons about 1-inch square. They are also designed to fit into thin shallow cracks.

6. Angle Pitons. These are used in wide cracks that are flush or offset. The supporting strength is 2,000 pounds. Maximum strength is attained only when the legs of the piton are in contact with the sides of the crack.

7. Bong Pitons. These are angle pitons that are more than 1-1/2 inches wide. Bongs are commonly made of steel or aluminum alloy and usually contain holes to reduce weight and accommodate snaplinks. They have a high holding power and require less hammering than standard pitons.

8. Skyhook (Cliffhanger). This is a small hook that clings to tiny nubbins, ledges, or flakes. It requires constant tension and is used in a downward pull direction.

5.3.4.4.1 Inspection and Care. By carefully retrieving pitons, they may be used many times. Pitons must be placed so that they do not bend backwards. They should be driven until only the eye protrudes, which is the point where they are secure and provide the needed protection. Choosing the proper size and shape to fit the specific crack is key for emplacement and ease in retrieving. Pitons should be inspected before, during, and after use. The ends should be watched for "mushrooming" caused by continued pounding. The blades should be straight and edges tapered. After pounding pitons back into shape, the metal eventually loses its temper and shape. Rusted and chipped places should be filed and rubbed smooth. A light coat of oil preserves pitons during storage. On a climb, pitons should be kept dry since oils make them slippery, and oils can rub off onto other equipment.

5.3.4.5 Piton Hammer. A piton hammer consists of the following: a flat, steel head; a handle made of wood, metal, or fiberglass protected with tape; a blunt pick on the opposite side of the hammer, and; a safety lanyard of nylon cord, webbing, or leather. The hammer is about 10 inches long and weighs 12 to 25 ounces. The lanyard secures the hammer to the climber's body. A piton hammer is used to drive pitons, remove pitons, clean cracks, and pry objects loose.

5.3.4.4.1 Inspection and Care. The hammer should be inspected for serviceability to include the head, pick, handle, shaft, and lanyard. It should be free of burrs, cracks, and rust.

5.3.4.6 Chocks. Chocks are easier to emplace than pitons. In its simplest form, a chock is a stone that has become wedged in a crack. When a sling can be threaded behind one of these stones, it makes an excellent anchor and eliminates the need for placing a piton. In addition to natural chockstones, artificial ones are available. Chocks can be made from hexagonal machine nuts. Chocks are directional in their holding ability. They hold well when pulled in one direction, but tend to pop out easily when force is applied in the opposite direction. Chocks come in a variety of sizes and shapes and are designed to fit at least two sizes of cracks.

1. Wire Stoppers. The smallest chocks are wire loops of cable called "stoppers". These come in different sizes and shapes to accommodate various holes and crack widths. Wire chocks are easier to emplace and retrieve than standard chocks, but they can be levered out by rope action. The wire loop is stronger than a rope or webbing loop and is ready for use in even the most difficult placements. The uniform taper makes them excellent for stacking.

2. Copperhead. These are soft metal devices placed into shallow grooves or piton holes. The softhead bites well and resists the rotation problem that can cause harder metals to dislodge. They are lightweight, versatile, and made in various lengths and diameter.

3. Spring-activated Camming Devices. These devices provide convenient, reliable placement in cracks where standard chocks are not practical (e.g., parallel side cracks, flaring cracks, or cracks under roofs). These are placed quickly and easily, which saves time and effort, but they can be hard to remove. These devices should not be placed any deeper than needed to achieve good placement. They can be extended with a runner or

sling, but should not be placed where the shaft is forced against an edge because a falling climber could snap the shaft.

4. Chock Picks. These are used to retrieve chocks that cannot otherwise be removed from cracks. They can also be used to clean out cracks and holes. Since they are made from thin metal, they can cause injury to a climber if landed on. They can be purchased or made from pitons, thin metal, or other objects. If a chock/nut is stuck, work it carefully with a pick so the force is applied directly to the nut and not the cable juncture.

5. Bolts. Nail-like shafts made from metal that are drilled into the rock to provide support. There are two types: contraction bolts that are squeezed together when driven into a rock, and; expansion bolts that press around a surrounding sleeve to form a snug fit in a rock. Bolts usually require drilling a hole into the rock, which is time consuming and exhausting to emplace. Once emplaced, bolts are the most secure protection for an omni-directional pull. Bolts should be used only when chocks and pitons cannot be emplaced. A bolt is hammered only when it is a nail or self-driving type. The hanger (for snaplink attachment) and nut are placed on the bolt. The bolt is then inserted and driven into the hole. A hand drill must be carried in addition to a piton hammer.

5.3.4.6.1 Inspection and Care. Chocks and related hardware should be inspected before, during, and after use. All metal surfaces should be smooth and free of rust, corrosion, dirt, and moisture. Burrs, chips, and rough spots should be filed smooth and wire-brushed or rubbed clean with steel wool. Items that are cracked or warped indicate excessive wear and should be discarded. Moving parts should move freely; lubricate them if needed. Wires should be smooth with no broken strands. When not in use, chocks should be stored in a cool, dry area.

5.3.4.7 Pulleys. A pulley is vital to vertical hauling and crevasse rescue. A pulley should be small, lightweight, strong, and should accommodate the largest diameter of rope being carried. Snaplinks can also be used as pulleys.

5.3.4.7.1 Inspection and Care. All metal parts should be smooth and free of rust and dirt. The pulley wheel should rotate freely. Lubricate and wipe clean to prevent soiling ropes. When not in use, pulleys should be stored in a cool, dry area. When the wheel is difficult to rotate or the metal is warped or bent, the pulley should be replaced.

5.3.4.8 Ascenders. This is a single-direction clamping device for climbing up 8.8 to 12-millimeter rope. Ascenders are used to climb vertical and near-vertical faced ice and rock. Ropes (prusik) and stirrups can be used for ascent in direct-aid climbing. Mechanical ascenders are an efficient mechanical prusiking device.

5.3.4.8.1 Inspection and Care. The cam and safety should be free of excessive lateral movement and provide smooth, non-sticking action. The external portions should be rigid and free from cuts, nicks, and gouges. Mechanical ascenders should be kept clean and dry. When not in use they should be stored in a cool, dry area.

5.3.4.9 Descenders. Descenders are a mechanical device that is secured to a rope and allows an individual to lower himself and equipment. Snaplinks and their variations are the primary means of descending. Figure eights, pear-shaped carabiners, and other descenders can also be used.

5.3.4.10 Mechanical Belays. There are occasions when a mechanical belay is advantageous. A mechanical belay offers an easier system of statically catching a falling climber. All of the devices mentioned below should be used with kernmantle rope since there is less friction. Nylon-laid ropes can break or become damaged if used in a mechanical belay device due to increased friction.

1. Sticht Plate. A bight of the climbing rope is placed through the slot in the plate and secured into a snaplink. The belayer moves the free end of the rope sideways, forcing the plate against the snaplink, snugging the

rope tight. Sticht plates come with a combination of notches for belaying with different size ropes or with a double rope technique. They can also be used for rappelling.

2. Figure Eight. The figure eight device contains two rings of different diameters. A bight of rope is inserted through one ring and fed over the other ring in a horseshoe manner. It is placed in the narrow collar of the figure eight. Figure eights come in a variety of sizes and can also be used in rappelling.

3. Snaplinks. These can also be used for a mechanical brake system. This is a friction brake system similar to that used in rappelling. The carabiner brake system, although complex to set up, has the advantage of not requiring any special equipment other than snaplinks.

5.3.4.10.1 Inspection and Care. Since mechanical belay devices are made of metal and alloys, they should be inspected before, during, and after use. Devices should be inspected for cracks, chips, grooving, rust, moving parts, and excessive wear.

5.3.4.11 Harnesses. Harnesses are prefabricated nylon webbing used as seat, chest, and combination apparatuses. They are used for ascending and descending vertical and near-vertical slopes. There are many sizes, shapes, and designs available. Improvised seat and chest harnesses can be made using rope or webbing and knots. Prefabricated (stitched) harnesses are more comfortable and helpful in direct-aid climbing. Advantages include less risk of injury from the rope in a fall, reduced risk of suffocating while suspended after a fall, and they allow the climber to conduct free prusiks or overhang aid climbs with less strain.

5.3.4.11.1 Inspection and Care. The harness should be inspected before, during, and after use for dirt, tears, rips, fraying, grease spots, and excessive wear. Seam parts should be inspected to ensure adequate binding and sewing without ripping. Metal parts should be free of rust, burrs, nicks, chips, and cracks. Plastic should not be cracked, chipped, grooved, or broken. The harness should be machine-washed (gentle cycle in warm water) and air-dried. The metal parts should be oiled.

5.3.4.12 Ice Axe. The ice axe is the most important item of equipment for the mountaineer operating on snow and ice. The versatility of the axe lends itself to balance, stepcutting, probing, self-arrest, belays, anchors, direct-aid climbing, and ascending and descending snow and ice covered routes. The ice axe should be covered with a rubber-tip protector and stowed when not needed.

1. The ice axe is composed of a pick and an adz. The curved pick is used for hooking action and digging power. The angle varies from 65 to 70 degrees for snow and 55 to 65 degrees for ice. The teeth on the pick may extend one-fourth, one-half, or the full length of the pick. The adz (chopping portion of the head) is used primarily in stepcutting and self-arrest. It may have a curved angle or it may be flat along its length, and straight or rounded from side to side. The head is made of either stainless steel with a riveted chrome-molybdenum steel pick or a one-piece chrome-molybdenum steelhead. The head usually has a hole directly above the shaft for attaching a snaplink or wrist loop. The ice axe shaft may be made of wood, hollow aluminum, or fiberglass. The spike at the bottom of the shaft can be used for self-belays or self-arrests.

2. Ice axe safety

   a. Never use the axe as a walking cane.

   b. Carry the axe with the spike forward and down.

   c. Maintain safe distances between climbers.

d. Watch the position of the axe at all times.

e. Lanyard the axe to the prominent hand.

f. When the axe is on a pack, do not throw the pack or bend over with people near-by.

g. The ice axe should always be held in the uphill hand.

3. Methods of using the axe while on the move

a. Driving the spike into the ground with the pick in the aft position (the cane fashion) is commonly used to assist movement on slopes up to 40 degrees.

b. Driving the spike into ground with the head of the axe turned to the side (the stake position) affords more area for a handhold. The stake position is commonly used as a precursor to self-belay. This method is most useful on slopes above 40 degrees.

c. The cross body position is used with the spike angled off to one side while traversing a slope steeper than 40 degrees. You are not driving the shaft deep into the snow, but merely stabbing it into the slope as you move. You are using the cross body position as a quick balance assist as you move.

5.3.4.12.1 Inspection and Care. Never use a grinder to sharpen the axe because it will fatigue the metal. Use files to sharpen and clean off dirt, sand, and rust.

5.3.4.13 Ice Screws. Ice screws provide artificial protection for climbers and equipment in operations involving steep ascents. They are screwed into ice formations.

5.3.4.14 Ice Pitons. Ice pitons are used to establish anchor points for climbers and equipment when conducting operations on snow and ice.

5.3.4.15 Wire Snow Anchors. The wired snow anchor (or fluke) provides security for climbers and equipment in operations involving steep ascents by burying the snow anchor deep in the snow.

5.3.4.16 Snow Saw. The special tooth design of the snow saw cuts easily into frozen snow and ice. The blade is a rigid aluminum alloy of high strength about 1/8-inch thick and 15 inches long with a pointed end to facilitate entry on the forward stroke. It is used in step cutting, shelter construction, or removing frozen obstacles.

5.3.4.17 Snow Shovel. The snow shovel is made of a special lightweight aluminum alloy and is used for snow removal. The handle should be telescopic or folding. The shovel should have a flat or rounded bottom. It is used for avalanche rescue, shelter construction, step cutting, and removing obstacles.

5.3.4.18 Crampons. Crampons are devices worn on boots to provide firm footing on ice or extremely hard snow. They consist of light metal frames with 10 to 12 points protruding from the bottom. The crampons are attached to the boot with straps, and the buckles must be outboard to prevent the locking of one crampon to the other while walking. The operator must constantly monitor his foot placement so as not to inflict serious puncture wounds to himself or platoon members.

5.3.4.19 Mobility. Ground conditions during winter months can make movement difficult unless special equipment is employed. NSW personnel can take advantage of these conditions by using snowshoes, skis, snowmobiles, pulks, or by skijoring. The equipment that is used requires specialized training to attain proficiency and minimize injury. Proficiency with equipment and skills will greatly enhance the capability of the unit. The operational radius will increase, weight will be transported more efficiently with less fatigue, and less time will be required to achieve the objective. Planners should be aware of limiting factors that will affect the operating unit. Terrain, ground conditions, weight (fatigue), requirements for trailbreaking, and hours of darkness will determine the rate of movement. The enemy threat will determine the appropriate method of movement and may preclude the use of mechanized support.

1. Snowshoes. Snowshoes prevent post-holing and assist traveling in deep soft snow. The crampons attached to the bottoms of modern snowshoes provide excellent traction and will allow the operator to negotiate steep slopes. Snowshoes are relatively easy to use and require minimal training to achieve proficiency. They can be used to pull heavier loads, require little maintenance, and are easily carried. Snow depth, brush, and friction will contribute to operator fatigue.

2. Skis. Skis permit an increased rate of movement and greater maneuverability. Proficiency at skiing requires concentrated training. Cross-country skiing will decrease the time required to reach the objective. Skis will generally be staged prior to the assault. The poles may be used without skis to give personnel greater stability when walking over difficult terrain. Skis require maintenance, are easily broken, and may not be able to be locked out of a submarine due to their length.

3. Waxing. Wax is used on the bottom of the skis to make them glide forward while preventing back slip. Variations in temperature and snow conditions require waxes of varying properties. Temperature range and snow condition for each type of wax are usually marked on the container. As the weather warms and the snow becomes wetter, apply softer wax. Conversely, the colder the weather becomes, the harder the wax should be. If possible wax skis in a heated shelter, because wax sticks better to a warm surface. During movement, carry wax in an inside pocket to keep it warm. Make sure the ski bottom is dry and free of old wax, particularly wax different from that to be applied. If necessary, wax can be applied directly over a different wax, if the new wax is softer than the one already on the ski. Wax can be removed using any hard-edged tool such as a paint scraper or pocketknife, but be careful not to cut into the base of the ski.

   Wax comes in three different types for different purposes. First the base wax seals the entire bottom of the ski. The glider wax gives the ski its glide and should be applied tip to tail in three thin layers. This wax should be colder than ambient air temperature and/or snow conditions. Kicker wax is the secret of backcountry skiing. This wax sticks to the snow and releases after pressure is released. Always apply the hardiest wax first and the warmer waxes second. Smooth the wax by rubbing it with the heel of the hand or a cork. If a fire or stove is available, heat the wax to make it easier to smooth out. Do not put the skis on the snow until the wax has cooled to air temperature. Test skis for proper waxing before starting on a long march. If the wax is not gripping the snow correctly, reapply more kicker wax or extend the kicker wax area under foot. Lastly move down one warmer wax color.

4. Skins. Skins are used for long climb ascents, pulling pulks, and patrolling in areas of rolling terrain and changing weather conditions.

5. Pulk. These sleds allow NSW units to carry operational equipment greater distances with less fatigue. A four or six foot pulk may be selected for use. Personnel using skis must become proficient in techniques for moving pulks across varying terrain.

6. Snowmobiles. Snowmobiles have been used predominately for administrative purposes. The advantages of speed and load bearing should be weighed against the threat of enemy detection and compromise.

7. Skijoring. This method of transportation is achieved by using a vehicle (snowmobile, etc.) to pull NSW personnel wearing cross-country skis. The limiting factors for employment are similar to those of a snowmobile. Additionally, terrain and the availability of trails must be considered.

5.3.5 Communications. Reliable communications are difficult to establish and maintain in northern latitudes. Extreme cold produces different coefficients of expansion for plastic and metal parts. Atmospheric disturbances impede transmissions, and batteries freeze, lose power, and become useless. The following are problems peculiar to communications in the mountain and arctic environment:

1. Aurora. The aurora is a high glow in the atmosphere produced by the effects of solar flaring. The particles are attracted to the polar-regions, so aurora displays are most common in high latitudes. The aurora is an electron density in the ionosphere. The density affects radio waves through "black-out" or anomalous propagation (uneven transmission of energy). HF communications are most affected.

2. Snow Conditions. Drifting snow under high-wind conditions causes static electricity. This does not cause a loss of carrier wave, but can create a "noise" problem. Ice on the antenna may cause an increase in radiation resistance and weaken signal transmission. FM radio communication becomes spotty during snowy periods. During clear, cold weather the rated range is usually exceeded.

3. Lodestone. These deposits are scattered throughout the arctic. Lodestone is a magnetic rock that behaves like a natural magnet. If the magnetic field of the lodestone is strong enough it can bend the carrier wave of a VHF radio to the point where the radio cannot send or receive.

4. HF Systems. The high range spectrum can be a fairly dependable method of communicating at high latitudes if the operator is knowledgeable of the propagation difficulties. In northern areas, the HF band is subject to interference from magnetic storms, ionospheric disturbances, and the Aurora Borealis. As the greatest aurora disturbances occur between 60 and 70 degree north latitude, a major portion of Alaska, Canada, Scandinavia, and northern Russia are subject to this phenomena. As a general rule, HF communications are the most sporadic and unpredictable during early morning and late evening hours. In severe atmospheric disturbances, complete HF "black-out" is also common. Ways to improve HF communications are:

   a. Get a good operation frequency from the MCT.

   b. Know the direction and time of operation of the receiving station.

   c. Establish a ground. Arctic conditions with snow, ice, and frozen ground make it difficult to ground equipment.

   d. Build a proper antenna (type, length, and orientation).

      (1) Near vertical incident skywave, the footprint can be 200-250 feet.

      (2) Ten-foot whip for use on water.

      (3) Inverted "V."

      (4) Whip with a counterpoise. The counterpoise is necessary to reduce the amount of power lost to the earth. This is particularly important for HF and vertically oriented antennas. The radial design is usually a good compromise between performance, portability, and the time needed to install the system.

5 Continuous Wave (CW). Potential enemies in the northern latitudes have sophisticated EW capabilities. CW signals have the potential to override a frequency to the extent that even a weak signal can be heard. For long-range HF operations in a jamming environment, CW is the most effective method of transmission.

6. SATCOMs. At extreme northern latitudes, the antenna take off angle may be as little as 10 degrees. This factor will often limit the transmission capability because a higher angle is required to overcome mountainous terrain. This is the most reliable long-range communications.

7. Frequency Planning. Frequency propagation charts should be carefully consulted prior to operations in northern latitudes. The appropriate frequency will depend on the time of day.

8. The mountain terrain distorts line of sight communication. The patrol needs to plan movement to high ground to coincide with communications checks.

5.3.5.1 Radios and Electronic Equipment. The following list outlines problems and recommendations regarding radios and electronic equipment use during cold weather operations:

1. Metal, plastic, and rubber parts become brittle in conditions of extreme cold, therefore, equipment should be handled carefully to prevent breakage.

2. Equipment must be kept clean, dry, and warm.

3. Vapor barriers, such as cellophane, will protect microphones.

4. Jacks and plugs must be kept free of condensation.

5. All lubricants should be capable of retaining their characteristics at temperatures as low as -65 degrees Fahrenheit. In all cases, no lubricant is better than an inadequate one.

6. Warm equipment should be cooled slowly. When warm equipment contacts extreme cold air, plastic and glass may shatter.

7. Cold or frozen equipment should be gradually warmed to reduce the effects of sweating.

8. Do not place the equipment directly on snow.

9. Do not warm the equipment with direct heat.

10. Plan and prepare for the communication window while inside a tent (select frequencies, cut antenna length, prepare equipment, etc.).

5.3.5.2 Batteries. Lithium and nickel-cadmium are the primary batteries used for NSW communication equipment. The following are recommendations for cold weather use.

1. Battery life will be reduced, so bring spares.

2. Lithium batteries have a longer life (72-96 hours at 9/1-transmit/receive ratio) at lower temperatures (below 5 degrees Celsius) than nickel-cadmium batteries.

3. Keep batteries clean and dry.

4. Maintain a usage log.

5. Do not place batteries on snow.

6. Nickel-cadmium batteries must be kept warm.

    a. Place battery pack close to the body when transmitting.

    b. Place batteries in the sleeping bag when sleeping.

    c. Rotate batteries every eight hours with spares.

5.3.5.3 Antennas. The proper use of antennas is critical to successful communications at northern latitudes. Construction of antenna sites using special techniques may be required. Three limitations will affect antennas.

1. Precipitation Static. Charged particles of snow or glacial silt, when driven against antennas, can produce a high-pitched crackle that can "black-out" frequencies. Although it is impossible to eliminate static, shellacking or taping exposed metal surfaces can reduce interference.

2. Antenna Sites. The inverted "V" antenna needs height and length to achieve its maximum gain. Antenna locations can often be easily spotted against an all-white background.

3. Ground Sites. During winter it is extremely difficult to find locations for antennas. In addition to high electrical resistance, frozen ground will resist anchor stakes. During conditions of extreme cold, metal becomes brittle and coaxial sockets and metal masts will break easily.

5.3.5.3.1 Antenna Applications. The following considerations should be applied when constructing an antenna site:

1. Apply a thin coat of silicone lubricant to antenna and transmission lines at temperatures below 40 degrees Fahrenheit to maintain pliability.

2. Do not overstress plastic antenna components in extremely cold weather.

3. In cold weather, allow more time to set up the antenna.

4. Use 100-pound monofilament fishing line and a 4-6 ounce weight when constructing an antenna in trees. The antenna may be suspended with 550 cord.

5. If trees are not available, use inverted ski poles that are secured in the snow.

6. A longwire antenna is cut (length = 984/frequency) and suspended from a height of twelve inches at the radio to three inches at the end. Offset antenna 10 degrees from the intended line of communication.

7. A wire counterpoise placed under the antenna will improve communication during dry snow conditions.

8. Remove snow and ice from the antenna.

9. If the communications gear is operable, but the communicator is unable to receive or transmit, check the following possible errors or problems:

a. Antenna design is inappropriate

b. Frequency for skywave is incorrect

c. Intervening terrain is obstructing transmission

d. Location of one or more end circuits is incorrect

e. Antenna is not aligned properly

f. Strong aurora activity

g. Jamming is in effect

h. Precipitation static is overriding transmission or reception.

5.3.5.4 Handsets. Cold causes the metal and plastic handset parts to become brittle. Care should be taken not to force connections. Keep the handset in a plastic bag and carry extras.

5.3.5.5 Tactical Communications. Radio communications can be performed more easily during a halt than while on the move, especially for satellite and burst communications. General guidelines are:

1. Make communications just before the patrol moves out to reduce the enemy's ability to DF your location.
2. Ensure that all patrol members are ready to move prior to making communications. This ensures that the patrol can move immediately upon completion or if compromised.
3. The last communications of the day should be made from a security halt since you will occupy the position for only a short time. It should not be conducted from your LUP.
4. Give your position to headquarters from a known point at your last communications window of the day.
5. It may be necessary to keep radios on all night if needed in the morning. Also, if an emergency occurs, fast support will be needed.
6. Communications is everyone's responsibility.

5.3.6 Weapons and Ordnance. Cold weather and moisture will adversely affect the performance of weapons and ordnance. Special techniques and maintenance are required to ensure proper operation during arctic and maritime operations. This section will outline the problems associated with cold weather. NSW personnel must be familiar with the cold weather characteristics of weapons and ordnance to ensure proper care and effective employment.

1. In extreme cold-weather conditions weapons are subject to sluggish operation. During cold-weather operations, weapons must be cleaned with a dry cleaning solvent to remove all lubricants and rust prevention compounds. In very cold temperatures, no oil is better than heavy oil or too much oil. Often overlooked is the grease around firing pin springs inside bolts. Failure to remove causes light strikes and failures to fire. Following cleaning, the operator should sparingly apply a cold-weather lubricant to moving parts.

   a. LSA lubrication for temperatures ranging from 32 to 65 degrees Fahrenheit.

b. Graphite lubrication for temperatures ranging from 32 to -65 degrees Fahrenheit.

c. LAW lubrication for temperatures ranging from -35 to -65 degrees Fahrenheit.

2. Cold weather will increase the rate of breakage and malfunction. These problems can be attributed to the cold, although snow and ice in the weapon may also contribute. The hardened metal parts in AWs are particularly susceptible. When a weapon is fired at sub zero temperatures, breakage can occur within the first few rounds. Short bursts will gradually warm the barrel to normal operating temperature. Ice and snow in the weapon can block the muzzle and freeze moving parts. Every effort should be made to ensure a weapon remains free of ice and snow. Muzzle caps should be used. Condensation forms on all weapons when they are taken from the cold into a heated shelter. This type of condensation is referred to as "sweating." All moisture must be wiped off the weapon before it is refrozen or the condensation will freeze, especially in the internal parts, and malfunctions will occur. Bring extra sears, firing pins, and extractors for 30 percent of the weapons. Weapons should be left outside under a tent flap during cold-weather operations.

3. A problem with visibility close to the ground occurs when the weapon is fired in sub-zero temperatures. As the round leaves the weapon, the water vapor in the air is crystallized, creating minute ice particles that produce ice fog. This fog will hang over the weapon and follow the path of the trajectory. This can obscure the gunner's vision and, if the air is still, the ice fog will disclose the shooter's position.

4. Conditions of extreme cold have less effect on direct-fire weapons than most other types, but there is a noticeable effect on muzzle velocity and accuracy. A determination of this effect and the necessary adjustments must be made prior to firing to ensure accuracy. As a general rule you can figure sighted point of impact will drop one minute at 100 meters (one inch at 100 meters) for every 20 degrees Fahrenheit drop in temperature, and the same for every 20 percent increase in humidity.

5. Metal, plastic, and rubber will become brittle and more susceptible to damage as the temperature drops. Equipment (e.g., blasting caps) should be checked prior to use.

   a. Steel. The impact strength of steel will decrease as the temperature falls.

   b. Volume and Dimensional Changes. A decrease in the volume or dimension of solid materials will occur due to contraction. Close tolerance parts, such as bearings, will be affected. Liquids will expand when frozen.

   c. Chemicals and Liquids. Chemical reactions will occur at slower rates. Fuses will burn more slowly and batteries will lose their charge. Lubricants and liquids will become thick or freeze.

6. Ammunition should be kept at the same temperature as the weapon and should be carried in bandoleers, with additional ammunition carried in the pockets of the outer parka. Magazines must be cleaned of all oil and preservatives and checked frequently. All ice and condensation must be removed from magazines and belt.

7. Point detonating munitions are less effective in deep snow due to the insulating effect of the snow in absorbing blast and fragmentation. Frequently the snow will cause duds by providing inadequate impact to fuse detonation.

8. Gloves or mittens will decrease the chance of hands freezing to weapons and ordnance.

9. Demolition charges should be constructed prior to entering the field. Material subjected to cold temperature will lose pliability and become brittle.

10. Weapons must be waterproofed prior to water operations. Once a weapon is wet, adequate drying/de-icing in the field is unlikely. Ordnance and weapons should be thoroughly tested under similar environmental conditions prior to insertion.

11. Specialized items such as radio-firing devices, starlight scopes, beacons, and other electronic aids will have a shortened battery life.

5.3.6.1 Handguns. The 9-millimeter and .45 caliber pistols are affected by condensation since they are more likely to be brought inside the tent. Freezing will occur around the slide and magazine. The weapon should be checked frequently to remove ice. Breakage of the extractor or firing pin may occur. In these conditions the Smith & Wesson .38 caliber revolver is very reliable.

5.3.6.2 Assault Rifles. Parts most subject to breakage are sears, firing pins, operating rods, and those parts subject to the forces of recoil. The buffer group's ability to absorb shock will be decreased. In cold weather the M14 7.62 millimeter rifle is very reliable.

5.3.6.3 Machine Guns. Machine guns have a high rate of malfunction and breakage in extreme cold-weather conditions. Condensation will cause a freezing of parts. Freezing and hardening of the buffer will often cause violent recoil, thereby increasing the cyclic rate of fire. When this happens, breakage occurs. A "short" recoil may be encountered early in firing and require immediate-action drills to cock the weapon. Increased spare parts (e.g., sears and bolts) should be included with all weapons. If very light cold-weather lubricants are not available, machine guns will have fewer malfunctions if fired at a slow rate of fire. Hot barrels should not be placed on the snow.

5.3.6.4 Grenade Launchers (40millimeter). The plastic hand guard will crack and fall off when frozen. Taping is necessary to prevent this.

5.3.6.5 Rocket Launchers. Propellant burn rate is affected by cold. The propellant, because of its slower burning qualities in the cold, can be dangerous to fire. Operators should wear a mask for protection from slow burns. The back-blast danger area must be tripled. The range of the weapon is reduced and the operator must fire high in order to overcome the effect of the slower burning propellant. Dud rates are significantly higher in cold weather due to the shock absorbing effect of snow.

5.3.6.6 Indirect-fire Weapons. Because of the smothering effect of deep snow, mortar impact rounds are less effective in the north during much of the year. There are also problems with land-proximity fuses using electrolyte, as the cold causes them to freeze and malfunction. Deep snow and soft terrain can reduce the effect of impact by as much as 80 percent. Low temperatures will also cause malfunction of illumination rounds and devices due to freezing of the parachute and its components. Effective firing ranges are also reduced; and corrections are difficult to implement. Mortar base plates will slide on snow. Breathing on sights will cause fogging.

5.3.6.7 Hand-grenades. A cold hand-grenade will freeze to a damp glove. Personnel cannot throw a grenade as far as normal in cold-weather clothing. It is difficult to pull the pin and maintain firm control with arctic mittens or gloves on. Deep snow greatly reduces blast effects. Sticks may be attached to the grenade to prevent it from sinking into the snow on impact.

5.3.6.8 Optics. Optical devices will be adversely affected by cold weather. Equipment that requires battery power will become inoperable unless the operator can keep the battery warm (close to the body or under the arms using extension wires and clips). Condensation will occur on the eyepieces from the operator's breath or by rapid exposure to changing temperature. Liquids in the lensatic compass will thicken unless the compass is kept close to the body. This will contribute to inaccurate readings.

5.4 MOUNTAIN AND ARCTIC TECHNIQUES

5.4.1 General. This section will provide the NSW operator with an overview of the various techniques required when operating in the mountain and arctic environment. Operational requirements will dictate the type of equipment to be used and the method of employment. Proficiency as an arctic operator requires discipline and training. The information contained herein is provided for the purpose of familiarization. Unit SOPs will determine the manner and circumstances in which these techniques are applied. General guidelines should be followed, but techniques and tactics will change as a result of individual imagination and the need to persevere against the elements and the enemy. In order to conduct successful missions, NSW personnel must be comfortable and confident in the environment.

5.4.2 Eight Steps to Mountain and Arctic Survival. The AO and seasonal conditions will determine the appropriate survival equipment and tactics. Changing weather patterns and threat scenarios require that the NSW operator be familiar with several techniques. The use of common sense and appropriate equipment may mean the difference between life and death. NSW personnel will carry survival equipment in the pockets of their clothing or hip pouch (first line). Operational planning should ensure that the survival equipment is appropriate for the intended AO. Factors essential to individual survival include signaling devices, food, heat, and shelter.

5.4.2.1 Recognition. The operator needs to recognize that a survival situation exists before it becomes clear that his life is in danger. The man in charge is responsible for controlling the seriousness of the situation with a positive attitude and well thought out plan of action.

5.4.2.2 Inventory. Inventory all assets that are essential to make it to the EP (compass, map, radios, etc.). Individuals need to ask themselves: What can I salvage and what can I make (for shelter, for gathering water, etc.)? Keep everything that you possibly can with you to better your situation.

5.4.2.3 Survival Shelters. Man can survive for an amazing length of time without food or water, but without protection from the elements, particularly in a harsh environment, he will survive only a short time. Therefore, sufficient knowledge in shelter construction is essential. Any type of shelter, whether it is a permanent building, tentage, or a survival shelter must meet four basic criteria to be safe and effective. The criteria are:

1. Protection from the Elements. The shelter must protect the individual from rain, snow, wind, sun, etc.

2. Heat Retention. It must have some type of insulation to retain heat to prevent fuel and energy waste.

3. Ventilation. Ventilation is important if burning fuel for heat. Ventilation will prevent the accumulation of carbon monoxide. Ventilation will reduce carbon dioxide given off when breathing.

4. Drying Facility. Some type of drying facility is necessary in any environment where wet clothes are a possibility.

5.4.2.3.1 Natural Shelters. Natural shelters are usually the best since they take less time and materials to make into an adequate shelter and are normally much easier to camouflage. The following may be made into natural shelters with some modification:

1. Caves or Rock Overhangs. Can be modified by building walls of rocks, logs, or branches across the open sides.

2. Hollow Logs. Can be cleaned or dug out, and also modified with ponchos, tarps, or parachutes hung across the openings.

3. The following should be considered when using natural shelters:

   a. Animals

   b. Lack of Ventilation

   c. Gas Pockets

   d. Instability.

5.4.2.3.2 Man-made Shelters. Many configurations of man-made shelters may be used. Only your imagination and materials at hand limit them. Snow is the most suitable material for building natural winter shelters because it provides the best insulation, natural camouflage, and there is usually an abundance of it. The following are man-made shelters that can be used for survival in the winter:

1. Snow Wall. A snow wall is used to shelter one or two men from the wind. If made correctly, it will protect the individual from precipitation. The basic principles for construction include:

   a. Wind. Determine the wind direction.

   b. Compacting. Once the wind direction has been determined, construct a wall of compacted snow to shield from the wind.

   c. Dimensions. The wall should be constructed a minimum of three feet high, and as long as the body.

   d. Half Horseshoe. The wall should be constructed in a half horseshoe fashion for extra protection around the head.

   e. Poncho. The long axis of the poncho can be attached to the top of the snow wall, and then angled down to the ground to add more protection. Skis, ski poles, branches, etc., can be used for stability.

2. Snow Trench. A trench can be built into a drift or wherever there is more than three feet of consolidated snow. The trench is the most efficient snow structure in terms of the least amount of time to build and moving of snow. A snow trench provides shelter for one or two men.

   a. Trench. A trench should be approximately three feet wide and six to seven feet long, and depending on snow depth, five to six feet deep.

   b. Construction. One man digs the trench, and the other man digs the entrance. They both dig toward each other leaving a three-foot snow bridge at the entrance of the trench.

   c. Sleeping Bench. Sleeping benches are dug into both sides in a concave fashion. They must be deep enough to accommodate the body.

   d. Cold Hole. A cold hole is dug the length of the trench approximately twelve inches deep. It is constructed to let the cold air settle at a level below the sleeping bench to keep occupants warmer.

e. Framework. Skis, ski poles, or branches are laid across the short axis of the trench and covered with a poncho or other available material. The poncho can be covered with eight inches of snow to add extra insulation.

f. Insulation. If no sleeping mat is available, line sleeping benches with branches.

3. Snow Cave. A snow cave is used to shelter from one to 16 men in areas where deep and compacted snow is available. The basic principles for construction include:

a. The tunnel entrance should give access to the lowest level of the chamber. This is the bottom of the pit where cooking is done and equipment is stored. It should be 45 degrees from the downwind side.

b. The sitting space and sleeping areas should be on a higher level than the highest point of the tunnel entrance so that it will be in an area of the cave where warmer air is trapped.

c. The roof is arched for strength and to prevent drops of water that form on the inside from falling on the floor. The roof must be at least 18 inches thick and include a ventilation hole.

d. A large cave is usually warmer and more practical to construct and maintain than several small caves.

4. Poncho Shelter. This is one of the easiest shelters to construct. It should be one of the first types of shelters considered when planning a short stay in any one place. The one-man poncho shelter is constructed as follows:

a. Find the center of the poncho by folding it in half along its long axis.

b. Suspend the center points of two ends using cords, sticks, etc.

c. Stake the corners down; ensuring the peak of the shelter is high enough to allow the sides to be staked out far enough to make room for a man inside.

d. Ensure the hood is tied off to prevent water from entering and dripping inside.

5. Brushwood Bivy. This type of shelter is easily constructed in an area where willows of similar length and slender supple bushes grow. A poncho or similar piece of material is required for cover.

a. Find or clear an area with two parallel rows of willows at least four feet long and approximately one and a half to two feet apart.

b. Bend and tie the willows together to form several hoops that will form the framework of the shelter.

c. Cover the frame with a poncho, tarp, etc.

d. The shelter may then be insulated with dirt, moss, leaves, brush, snow, tundra, pine boughs, etc.

e. Block one end permanently and hang a poncho or other material over the other end to form a door.

6. Lean-to. A lean-to is an easily built shelter constructed of materials from heavily forested areas. It does not offer a great degree of protection from the elements during inclement weather.

a. Select a site where two trees at least three to four inches in diameter are located far enough apart so a man can lie down between them. Two sturdy poles may also be used by inserting them into the ground the proper distance apart.

b. Find a pole to support the roof. It should be at least three to four inches in diameter and long enough to reach between the two trees. It should be tied horizontally between the two trees, approximately five feet off the ground.

c. Find several long poles to use as stringers. The length will depend on the size of the shelter to be constructed.

(1) Pole Dimensions. They should be at least two inches in diameter and seven to eight feet long for a one-or two-man shelter.

(2) Placement. One stringer is required every one and a half foot along the length of the lean-to. These are laid from the horizontal roof support to the ground on one side.

(3) Roof. These stringers may be tied or just laid on the roof support depending on the material at hand. The framework may be covered with a poncho, canvas, etc.

(4) Ends. Ends may be blocked to provide extra protection.

7. A-frame Shelter. An A-frame shelter may be constructed using the same techniques as the lean-to. Connect both of them together into one shelter.

8. Fallen Tree Bivouac. Fallen tree bivouacs are easily constructed in areas where evergreen trees grow.

a. Underside branches are cut away to make a hollow area underneath.

b. The remaining branches form the top and sides of the shelter.

c. The cut branches may then be interwoven into the top and sides for extra protection, or used on the ground for insulation.

9. Tepee. Tepees are generally more difficult to construct than the other shelters due to the amount of material needed. They are often smoky if a fire is built inside.

a. Select six to ten sturdy poles at least two to three inches in diameter and 12 to 14 feet long to construct the frame.

b. Tie all poles together about two feet from the top, and spread the butt ends into a circle large enough to accommodate men and equipment. This cone-shaped frame may then be covered with any number of materials including:

(1) Boughs

(2) Brush

(3) Bark

(4) Cardboard

(5) Plastic

(6) Canvas

(7) Ponchos

(8) Animal skins.

c. Create a vent near the top for smoke to escape. Make the door small enough to be sealed with a pack or similar object.

10. Quincy Hut. A man-made shelter that normally takes 90 minutes to 3 hours (depending on size of shelter) to construct. Snow is shoveled and formed into a tight packed pile about six feet high for two people. Sticks 10 to 12 inches are stabbed into the packed pile of snow to ensure that the shelter does not collapse while digging out the shelter. Leave the snow to set. This happens quickly in -20 degrees Fahrenheit and below (30 minutes). Once the pile of snow is set, the procedure for digging a snow cave is used to burrow into the snow pile until the 10 to 12 inch sticks are exposed from the inside of the shelter.

11. Tree and Pit Bivouac. This is located under large evergreen trees. There is less snow around the base of the tree than on the open ground beside the tree, creating a pit. Dig a trench towards the trunk and enlarge the natural pit. Leave at lease six inches of snow on the ground for insulation. Cut five to eight poles about six feet long, push them into the pit, and secure them to the tree at a 45 degree angle. Cover the poles with poncho, boughs, or other material and cover with snow.

12. Reflector Walls. A reflector wall is used to retain heat that is generally lost to the surrounding air. To build a reflector wall, stack rocks or logs two to three feet high and three to four feet long. Construct this wall one and a half to two feet from the fire, directly opposite the shelter. Secure the wall with stakes.

5.4.2.4 Signals/Communication. The methods of communication include:

1. Visual. Visual signals are usually the best because they will pinpoint your location and be seen at greater distances under good weather conditions. Enemy proximity will determine the appropriate method for visual signaling. Methods include:

   a. Fires

   b. Smoke

   c. Pyrotechnics

   d. Mirrors

   e. Strobe light

   f. Flashlight

   g. Signaling panels

h. Sea markers (dye)

i. Clothing

j. Arrangement or alteration of natural materials.

2. Radio. Emergency locator transceiver (AN/PRC-112) and personnel locator system (AN/ARS-6)

a. AN/PRC-112 is frequency-synthesized, providing selection of 3,000 channels with 25 kHz spacing.

(1) Burst transmission (608 milliseconds) capability.

(2) Five selectable frequencies for operations. VHF and UHF guard (121.5 and 243.0 MHz), 282.8 MHz, and two other programmable frequencies in the 225 to 299.975 MHz range.

(3) Powered by a lithium battery that works well in the cold and lasts for seven hours.

(4) It is silent until triggered by a pseudorandom, noise-coded message from the personnel locator system (PLS) in an aircraft. The PRC-112 then responds with its pseudorandom, noise-coded message.

(5) The PRC-112 response includes a six-digit identification code unique to each radio. The PLS transmission will include the code of the intended recipient. The corresponding radio will decode the signal and respond with its own message. Another survival radio receiving the transmission will remain in standby until queried with its own code, which conserves battery life. Captured radios cannot divulge codes or covert operation frequencies.

b. AN/ARS-6 is a PLS designed to rapidly locate personnel equipped with the AN/PRC-112 survival radio.

(1) The system computes the range and bearing to the isolated person's site up to a range of 100 nautical miles. Each radio uses a unique code to resist jamming.

(2) The PLS sends a pseudorandom, noise-coded transmission burst specifically coded for the isolated person. 100,000 possible combinations and nine individual codes can be stored in the PLS.

(3) At ranges of more than one nautical mile, the PLS has an accuracy of one percent of the distance. At less than one nautical mile, the range accuracy is 60 feet.

(4) The PLS provides two-way, UHF voice communication, and provides a direction-finding steer to any CW beacon signal.

(5) The system's resistance to jamming is provided by using spread-spectrum modulation using pseudorandom noise and phase shift keying. Each interrogation response cycle provides steering and range to the survivor in 608 milliseconds.

3. Important Factors. When using a signaling device or technique, remember the following:

a. Know how the signals are used

b. Have the signals ready for use on short notice

c. Use the signals in a manner that will not compromise your position.

5.4.2.5 Water Supply. Water is needed to sustain life, so gather and store as much as possible. Water is very abundant in the northern latitudes and a person can need between 3 to 10 quarts per day, depending on activity. Mountain water should never be assumed safe for consumption. Training in water discipline should be emphasized to ensure personnel drink water only from approved sources. Water is usually available from streams, lakes, or by melting ice or snow. Water obtained from lakes and streams needs to be boiled and/or sanitized. Whenever possible, water should be obtained from running streams or from a lake. If a hole is cut in the ice to obtain water, the hole should be covered by a snow block, board, or poncho to keep it from re-freezing. The hole should be clearly marked. If no free water is available, ice or snow must be melted. Ice produces more water in less time than snow. When melting snow, begin with a small amount of water and snow in the bottom of the pot and add snow as the snow in the pot melts. Boiling or using water-sterilizing tablets should purify the resulting water. Contaminates to be aware of are:

1. Giardia Lamdia, a protozoan parasite, is found in water throughout the world. Giardia is commonly referred to as "beaver fever" and is a debilitating dysentery condition resulting from drinking water that contains animal feces.

2. Glacier worms (Mesenchytreaus Solifugus) are found throughout the world in areas of permanent snow and ice. Normally found in the top few inches of snow, they appear at night and feed on Red pollen, Watermelon algae, and organic debris. Mesenchytracus Solifugus are members of the phylum Annelid (same as earthworms). They are not parasitic, and are digested by the strong acids in the stomach. They are about 1 inch long, very thin, and obtain their reddish color from the pollen and algae they eat. It is recommended to gather snow for drinking during daylight hours and bring all water to the boiling point. It should be noted that the worms cannot be seen in the snow. To test an area for worms, a small amount of snow can be melted in your palm; the movement of the worms can then be seen.

5.4.2.5.1 Potability. All water that is to be consumed must be potable. Water must be assumed to need treatment, unless positive proof is available indicating potability, to avoid disease or illness from polluted water. Non-potable water must not be mistaken for drinking water. Water that is unfit to drink, but otherwise not dangerous may be used for other purposes such as bathing. External cooling (pouring water over the head and chest) is a poor means of rehydration. Drinking water is the only way to maintain a cool and functioning body.

5.4.2.5.2 Water Management. Water is scarce above the timberline. After setting up a perimeter (patrol base, lay-up, defense), a watering party should conduct a search. After sundown, high mountain areas freeze, and snow and ice may be available for melting to provide water. In areas where water trickles off rocks, a shallow reservoir may be dug to collect water (after the sediment settles). Water should be treated with purification tablets (iodine tablets or calcium hypochlorite), or by boiling for ten minutes (longer at higher elevations). Water should be protected from freezing by storing it next to the body or by placing it in a sleeping bag at night. Fruits, juices, and powdered beverages may supplement and encourage water intake. Operators cannot adjust permanently to a decreased water intake. If the water supply is insufficient, physical activity must be reduced. Any temporary deficiency should be resupplied to maintain maximum performance.

5.4.2.6 Food/Survival Diet. Life can be sustained with a diet consisting only of plants. Few plants are completely edible. Select those plants that animals eat. Mushrooms, plants with a milky sap, and grains with a black spore growth should be avoided. Avoid any plant with yellow berries and be aware that 50 percent of all plants with red berries are poisonous. All seaweed is edible except for one that changes color from brown to green when handled or taken out of the water because it contains sulfuric acid. Negative symptoms from plant consumption include extreme bitterness, stinging, and burning.

1. Meat. Meat provides a great source of calories. Mammals and birds are edible. Meat sources should be prepared as follows:

   a. Fish. Cook before eating.

   b. Birds. Cook well before eating.

   c. Insects. May be roasted or boiled. Avoid hairy caterpillars.

   d. Large Game. Skin and gut, do not rupture the gut or bladder. The heart and liver are edible. Do not eat polar bear or wolf liver for it contains a high concentration of vitamin A that, if consumed, could prove fatal.

   e. Sea Life. Urchins, china hats, limpets, chilton or gumboot, blennies (eel-like fish), small crabs, sea cucumbers, and small to medium octopus are all edible and can be found in the tide pools. Caution should be exercised with paralytic shellfish poisoning (most commonly called "red tide") which affects all bivalve shellfish. Symptoms might include dryness of the mouth, nausea, vomiting, shortness of breath, lack of coordination, a choking sensation in the throat, and confused or slurred speech. While it seldom occurs, death can result from respiratory paralysis, usually within 12 hours. Bivalves to avoid are all clams, mussels, scallops, cockles, oysters, and geoduck. Some of these bivalves, such as clams, stay toxic for almost two years. Other sea life to avoid are starfish, coral, sea anemone, sponge, jellyfish, and sand dollars because all have slight toxins and little or no nutritional value.

5.4.2.6.1 Preservation of Food. Preservation of food can be accomplished by three basic methods.

1. Freezing is easy if outside temperatures cooperate.

2. Drying as follows:

   a. Cut into thin strips.

   b. Heavily salt and pepper.

   c. Quickly dry surface of the meat.

   d. Cover with loose cloth bag or screen box to prevent fly or maggot infestation.

   e. Hang and let air-dry. (Must be relatively low humidity or molding will occur.)

3. Smoking (dehydrating) can preserve meat.

   a. Hang meat loosely in strips over a dry, hardwood, smoky fire.

   b. Do not cook.

5.4.2.6.2 Hunting Techniques. If you are hunting game remember the following rules:

1. Use a suppressed/silenced weapon with night-light.

2. Keep the wind in your face (stay downwind).

3. Keep the sun at your back.

4. Hunt mammals in the early morning or late evening.

5. Stay above an expected game area.

6. Avoid crisp snow.

7. Avoid sky lining.

8. Stay off game trails.

9. Approach while game is feeding.

10. Shoot for vital areas between eyes and ears, behind shoulder, or spine.

11. Use the game you catch. Use what you can't eat for bait.

5.4.2.6.3 Trapping Techniques. The seven general tips for trapping are:

1. Know your game. Knowing the habits of the type of animal you want will make it much easier to lure it into the trap. Knowing when and where they move, and where they feed and water, will help to determine where the trap can be most effectively placed.

2. Keep things simple. A survival situation does not allow time to construct elaborate traps and snares. Always build the simplest, fastest, type of trap that will do the job.

3. Set traps in the right place. Look around for an area the animal frequents.

4. Use the right size trap. Adjust the size of the trap to the size of the animal. Too large can be as useless as too small.

5. Use the right type of trap. Some traps work better than others (a simple drag snare works better on rabbits than a baited deadfall). Decide which type of trap will be most effective for the type of animal you want.

6. Check traps. Check traps twice a day; once in the morning and again in the evening. Checking traps more than twice a day may cause the animals to stay away from them.

7. Cover up your scent. Animals will avoid an area that smells strange to them. Smoke from your fire is the best cover to use.

5.4.2.6.4 Traps. Traps require more time and effort to construct than snares. When trapping large game it may be worth the trouble. Have some type of weapon with you in the event that live game crosses your path within killing range. Also, the game in your trap or snare may still be alive. Never handle live game. You may suffer an injury or be subjected to a disease. The following is a list of traps.

1. Deadfall. The deadfall uses a falling weight to kill or pin the animal.

2. Bird Trap. This trap catches the feet of the bird. Use the right size trap. A large bird may fly off with a small trap.

3. Fish Trap and Expedient Hooks. This trap is a barricade of rocks or sticks across a stream. Another barricade located upstream provides a funnel-type entrance into which fish can be driven. Once fish are trapped between these two barricades they may be speared, clubbed, or grabbed. This can be very effective when fish are moving in large groups to spawn.

4. Gill Net. This type of trap entangles the fish as it attempts to swim through the holes in the netting. A gill net could also be used to catch birds by stringing it up between two trees.

5.4.2.6.5 Snares. A snare is a length of wire, rope, or cord with a loop at one end that can be tightened down around an animal's neck (suffocating it), or around its leg or feet (preventing it from escaping). All snares are constructed in the same manner, differing only in the method used for tightening the noose and triggering. Snares are much easier and less time consuming to construct than traps and easier to catch game with. Snare wire should always be carried. It is simple, efficient, and lightweight.

1. Drag or Float Snare. This is a noose that is strung across an animal trail in such a way that the animal will stick its head through it as it moves along the trail. The free end of the snare is attached to a heavy object such as a rock or log that provides resistance when it is dragged by the animal. The resistance tightens the noose and strangles the animal.

2. Trail Set Snare. This is the same as the drag snare except the end away from the loop is attached to an immobile object that will prevent the animal from moving more than the length of the snare line. This may strangle the animal or merely hold it in place until the trap line is checked.

3. Multi-directional Game Snare. This is a series of fixed snares constructed in a circle with a baited center. Regardless of the direction in which the animal approaches to get the bait, it must pass through one of the snares.

4. Squirrel Pole. This is a series of fixed snares on a pole that is leaned against a tree. The squirrel will run along the pole as a route onto or off of the tree and become caught in one of the nooses.

5. Spring Snare. This type of snare uses a bent branch or sapling to provide tension that strangles or immobilizes the animal. A trigger is used to release the branch or sapling. A spring snare may even be used to lift the animal off the ground and out of the reach of predators. This snare is also called the "twitch up."

5.4.2.7 Survival Fires. Fire is needed for warmth, drying, signaling, and purifying water. If matches are available, starting a fire is usually not a problem. But in a survival situation when matches are not available, starting a fire using expedient means can be difficult. Survival time is increased or decreased according to your ability to start a fire when needed. The key to starting a survival fire is to use good tinder.

1. Tinder. Good dry tinder is the basis on which all fires are built. Without it, time and effort will be wasted trying to start a fire. Tinder is nothing more than some type of material that will light readily with minimal time and effort. Tinder is used to ignite the main fuel in the fire. Examples of tinder are:

   a. Dry Punk. This can be obtained from the inside of a rotten tree.

   b. Lint

   c. Rope or twine

   d. Finely shredded bark

e. Bird nests

f. Magnesium shavings from the snowshoe.

2. Kindling. This is the material that is ignited by the tinder. It will burn long enough to ignite the fuel. Examples of kindling are:

   a. Small Sticks/Twigs. Ensure that they are dry.

   b. Pieces of Fuel. Smaller portions of fuel may suffice as kindling.

3. Fuel. Anything that will burn may be used as fuel for a fire.

   a. Wood. If using wood, insure that it is as dry as possible. Green wood gives off excessive smoke that may be seen or smelled over a long distance. Dry wood can be found even in wet weather. Standing dead trees are usually very dry even in the rain. Dead, rotten logs can often be broken or split, and dry wood found in the center. Dry wood may be found in protected areas such as overhanging rocks or underneath thick trees.

   b. Petroleum products. A mixture of gas and oil may be used. Use caution when igniting this mixture.

   c. Coal.

   d. Paper.

   e. Dried animal waste.

4. Site Selection. Select a site where the fire will not spread. Use a platform that will prevent the fire from sinking into the snow and be extinguished.

   a. Windy Weather. Find an area out of the wind or build a windbreak of rocks or logs.

   b. Rainy conditions. Build the fire under an overhanging rock or tree. Insure that the fire won't spread to the tree.

   c. Snow. Don't build a fire under a tree where melting snow may fall and put the fire out.

5. Fire Making. The following steps will assist fire making.

   a. Select a dry area.

   b. Shield a match with a cupped hand, hat, or other object when lighting. Whenever possible, use a candle instead of several matches.

   c. Have dry tinder prearranged before attempting to light and a good supply of kindling.

   d. Start with a tiny fire, adding fuel as it ignites. Don't build too large a fire; several small fires are more effective than one large one.

e. Add fuel in a crisscross pattern to prevent smothering. Blow or fan the flames gently if they don't seem to be spreading.

f. Add fuel above the flames.

g. Learn how to select good dry firewood. Softwood tends to burn quickly with a hot flame, while hard wood burns slower, and produces lasting coals.

6. Field Expedient Techniques of Starting Fires. There are two field expedient techniques:

a. Glass. Some types of glass will magnify sunlight to ignite a fire. Binoculars, riflescope, or eyeglass lenses all work well. Aim the piece of glass at the sun focusing a spot of light on the tinder. Move the glass to or away from the sun until the smallest spot of light appears. Hold the glass in this position until ignition takes place.

b. Flint and Steel. Although flint is preferred to rock for this method, any type of rock that will produce sparks when struck with a piece of steel or rock will suffice. Hold the rock so that when it is struck the sparks land in the center of the tinder. If one rock will not produce sufficient sparks, try another. Using a metal match or magnesium flint bar is much easier and faster than trying to find a rock with flint. With a pocketknife, strike the flint on a 90 degree angle directly into the middle of the tinder.

5.4.2.8 Positive Mental Attitude. The "will to survive" or proper attitude is essential to survival. Immediate action is necessary, especially in a cold environment. Evaluate the situation and start planning based on a survival pattern of: first aid, shelter, fire, food and water, and signals. Care for the injured and organize the platoon and equipment. Remember the enemies of survival and ways to combat them. They are boredom/loneliness, pain, temperature, thirst, fatigue, hunger, and fear. The acronym "SURVIVAL" may be used as follows:

S Size up the situation. Determine all aspects of the situation to evaluate and plan the actions to be taken.

U Undue haste makes waste. Use common sense; don't allow anxiety to cause more duress.

R Remember where you are. Use the four cardinal directions (north, south, east, west,) to your advantage, and to assist you in your efforts in finding safety.

V Vanquish fear and panic. It is essential that emotions be controlled. Giving into emotions may cause added duress or injury.

I Improvise/improve. Continuously improve on the situation. Use imagination to help your situation.

V Value living. Don't give up in any situation. No matter how bad things may look, keep at it. Don't place yourself in the hands of the enemy by undue noise or signaling.

A Act like the natives. Adopt indigenous habits, appearance, and routine.

L Learn basic skills. Learn and apply survival techniques during training.

5.4.3 Navigation. Map reading and navigation in winter follow the same principles as in summer, but are more difficult during cold weather operations. Snow makes the terrain look different, making orientation more difficult. Weather conditions may severely restrict visibility. Route selection is complicated by snow consistency and avalanche danger in mountainous terrain. Despite the fact that normally the PT or platoon navigator and the

PL work out the primary and secondary routes for the patrol, every patrol member should become familiar with the lay of the land by conducting a thorough map study.

1. Planning. Plan your navigation and navigate your plan. If you start modifying your plan you will be courting disaster. When selecting the route to an objective, consider the following:

    a. It is often faster to go around a terrain feature than over it. Check the contours and select a route that requires the least amount of climbing and descending.

    b. Snowshoes or skis should be considered for the terrain that will be encountered.

    c. Heavy packs and sleds will affect movement.

    d. Tracks should be camouflaged. This will be difficult during movement above the tree line in mountainous areas.

    e. Weather. The route should be negotiable under conditions of limited visibility.

    f. Avalanche hazards along the route should be noted.

    g. The route should offer concealment from direct observation by the enemy.

    h. Obstacles to movement will affect navigation and time windows.

2. Detailed route selection should include consideration of the following:

    a. Open Space Terrain. Break only one track, follow a tree line or other natural terrain features.

    b. Dense Forest. Avoid if possible, stay at the edge of wooded areas, or transit less dense portions of forest.

    c. Mountainous Terrain. Use gentle traverses to ascend or descend. Follow slope contours once altitude is gained. Avoid avalanche prone slopes.

    d. Water Routes. These areas can provide excellent routes. Check ice thickness before proceeding and stay close to the shore or bank.

    e. Obstacles. By-pass if possible.

    f. Night Movement. Route must follow the easiest possible terrain and should be well marked.

    g. Enemy Proximity. When approaching the enemy, emphasis shifts from ease of movement to concealment.

5.4.3.1 Navigation Problems. Prior to movement the following factors should be considered when planning the transit to the objective:

1. Long nights, fog, snowfall, blizzards, and drifting snow will drastically limit visibility. An overcast sky and snow-covered ground will create a whiteout condition.

2. Deep snow may completely cover tracks, trails, streams, and roads.

3. Lakes, ponds, marshes, and rivers may be frozen and covered with snow. These terrain features may be difficult to identify or even find, especially on flat or gently rolling terrain.

4. Snowdrifts may hide small depressions in the ground or may appear as small hills.

5. Aerial photography may be difficult to read due to the monotony of detail, absence of relief, and the absence of man-made structures for reference points.

5.4.3.2 Navigational Techniques. The magnetic compass depends on the horizontal intensity of the earth's magnetic field for its directive force. As the poles are approached, this force becomes progressively weaker until at some point the compass becomes useless in determining directions. The charted variation/declination in polar-regions is not of the same order of accuracy as elsewhere due to the constant movement of magnetic poles. Measurements indicate that the north magnetic pole moves up to 100 miles in a generally north-south direction and somewhat less in an east-west direction every 12 hours. This continual motion results in changes as great as 10 degrees from the charted variation /declination of the higher latitudes. The navigation planner, as well as the operator, must be aware of these situations although there is no special technique to counter the effects. Methods of land navigation under winter conditions are the same as under temperate conditions. Methods include:

1. Map reading

2. Map reading and compass combined

3. Dead reckoning

4. Combining the use of map reading, compass work, and altimeter readings

5. Pace counting.

5.4.3.2.1 Dead Reckoning. Dead reckoning in high latitudes (greater than 70 degrees) is complicated by the fact that the elements of dead reckoning (direction and distance) are usually known with less certainty than in lower latitudes. This emphasizes a need for greater accuracy on dead reckoned tracks through more frequent position fixes whether navigating by land or sea. Dead reckoning is the determination of location (land)/position at sea using the elements of speed, time, distance, and direction. When navigating by dead reckoning the following actions must be accomplished:

1. Select the route or track (direction).

2. Plot the intended track on a map or chart.

3. Make out the route card.

4. Maintain a log on the march. Memory will not suffice when fatigued.

5. Adjust the plot as required for obstacles.

6. Trust the compass.

5.4.3.2.2 Distance Measuring. Distance can be measure by pacing or using a cord. When using the pace method, the pace should be checked against a measured distance over terrain similar to that which will be encountered during the movement. Pace will be affected by slopes, depth of snow, wind, weight, and stamina.

1. The Cord Method. The most accurate method of measuring distance is the cord method, particularly when snow makes pacing inconsistent. A 100-meter cord and nine markers may be used as follows:

   a. The lead man moves off dragging the cord.

   b. The rear man jerks the cord when the lead man is at the end. This signals the lead man to drop his first marker.

   c. Both men move out dragging the extended cord between them.

   d. When the rear man reaches the first marker he stops, jerks the cord, and picks up the first marker. When the lead man feels the jerk he looks back to ensure the core is not snagged and drops the second marker.

   e. Both men move out and the procedure is repeated until the rear man has all nine markers.

   f. The lead man stops and the rear man moves to the lead man's position. At this point 1,000 meters have been covered.

   g. Pacers should travel in the rear so that progress is steady and no confusion results. Pacers should not be changed until the end of a leg. Breaks on the trail should be taken when both men are together, i.e., at the end of a 1,000-meter leg.

5.4.3.2.3 Distance Estimates. Visual perception will be affected during operations conducted in mountainous areas. Familiarization with terrain features, size, and distance will increase accuracy when determining routes and times.

1. Distance Underestimation. The following situations will result in objects appearing nearer than they actually are:

   a. When most of the object is visible and offers a clear outline

   b. When looking across a partially concealed depression

   c. When looking down a straight, open road or track

   d. When looking over a smooth, uniform surface, such as snow, water, or desert

   e. When the light is bright and the sun is shining from behind the observer

   f. When the object is in sharp contrast to the background

   g. When seen in the clear atmosphere of high altitude

   h. When looking down from high to low ground.

2. Distance Overestimation. The following situations will result in objects appearing farther away than they actually are:

   a. When only part of the object is seen or it is small in relation to its surroundings

b. When looking across an exposed depression

c. When looking up from low to high ground

d. When vision is narrowly confined

e. When the light is poor, such as dawn, dusk, or low visibility weather; or when the sun is in your eyes, but not behind the object being viewed

f. When the object blends into the background.

5.4.3.2.4 Direction. The standard military, dry card compass functions well in the cold. A liquid filled compass should be kept warm or it will become sluggish. Use aiming marks (terrain features, etc.) when navigating in a cold environment in the same manner as used in temperate environments.

5.4.3.2.5 Altimeter. The altimeter is a very useful tool for navigating in rolling mountainous terrain. It can be used as a second azimuth to obtain a resection. When moving parallel to the contour interval, the unit can maintain an elevation to within 5 to 10 meters. This technique can be used to maintain a certain elevation while traversing a slope. During periods of reduced visibility, identifiable landmarks may not be available. In cold weather, frozen lakes covered with snow are indistinguishable from open fields. Use of a map, combined with an altimeter, can help distinguish one from the other.

5.4.3.3 Survival Navigation. The following are improvised methods of navigation that can assist the unit or individual during extreme conditions.

1. Shadow Stick Method

   a. Obtain a three-foot stick and five pegs.

   b. Place the stick upright in the ground and mark the end of the shadow with a peg.

   c. Wait fifteen minutes and mark the shadow again with a stick.

   d. Repeat this until all pegs are used.

   e. The pegs will form a west-east line.

   f. Bisect this line with another stick to form a cross, and you will be able to determine north and south.

2. Improvised Pocket Navigator. A small piece of paper or any other flat-surfaced material can be used to trace shadow tips; and a pin, nail, twig, matchstick or other similar device can serve as a shadow-casting rod. Secure the material so it will not move and follow these steps:

   a. Set a tiny rod (one or two inches long) upright on the flat piece of material so the sun will cause it to cast a shadow. Mark the position where the base of the rod sits. Use the same spot for later readings. Mark the tip of the rod's shadow.

   b. As the sun moves, the shadow tip moves. Make repeated shadow tip markings. The more marks made, the more accurate the navigator will be.

c. At the end of the day, connect the shadow tip markings. The result will normally be a curved line. If it is not convenient to make a full day's shadow tip markings, observations can be continued on subsequent days.

d. The marking made at exactly noon (sun time) is on a north-south line. The direction of north should be indicated with an arrow on the navigator as soon as it is determined. This north-south line is drawn from the base of the rod to the mark made at solar noon. This line is the shortest line that can be drawn from the base of the pin to the shadow tip curve.

e. To use the improvised navigator, hold it level with the "rod" upright in the same position it was in to make the curve. Slowly rotate the navigator until the shadow tip just touches the curve. The arrow then points true north, from which the operator can orient himself to any desired direction.

f. This improvised navigator will work all day and will not be out of date for several weeks.

3. "On the Run" Navigator. An "on the run" navigator may be used if speed is essential.

a. Make a shadow tip direction reading and mark a true north-south line on the ground.

b. Draw a straight line at any place on the material being used for the navigator. This straight line must connect with the base of the shadow-casting rod. Designate this line as a north line.

c. Lay the material on the ground with its north line coinciding with the north line made on the ground. Mark the rod's shadow tip on the material.

d. After traveling for a half-hour, repeat the procedure.

e. Draw a straight line through these marks and extend them in both directions. This line will serve as the shadow tip "curve" for the day, and can be used to take quick direction readings in the same way as using an improvised navigator.

f. The direction obtained will be identical to the direction achieved when using the shadow stick method without waiting ten minutes for a reading.

4. North and South Using a Watch. This may be one of the best methods but should be considered a rough guide only. In the Northern Hemisphere the watch is held horizontally with the hour hand pointed at the sun. An imaginary line is drawn from the center of the watch through the 12. True South is midway between the hour hand and the 12.

5.4.3.3.1 Nighttime Navigation. The stars and moon can be used in the following manner as aids to navigation.

1. Locating the North Star. There are two methods used in locating the North Star without a navigator.

a. The Big Dipper (Ursa Major). Draw a straight line through the two stars (pointer stars) that form the front of the cup (away from the handle). Extend this line above the dipper five times its original length. The North Star will be located there.

b. Cassiopeia (Big M or W). Draw a line straight out from the center star, approximately half the distance to the Big Dipper. The North Star will be located there.

c. Position in the sky. Because the Big Dipper and Cassiopeia rotate around the North Star, they will not always appear in the same position in the sky.

2. Following the Stars. As the earth rotates, stars seem to move from east to west. This can provide a useful navigation guide in the dark. Place two fixed objects (sticks, weapon, etc.) into the snow over which to watch the stars. If a star is watched for several minutes it will seem to rise, to move to one side or the other, or to sink. If the star:

   a. Appears to be rising, you are looking east

   b. Appears to be falling, you are looking west

   c. Appears to be moving to the right, you are facing south

   d. Appears to move to the left, you are facing north.

3. Moon Navigator. Like the sun, the moon rises in the east and sets in the west, but rarely "due east" or "due west". Therefore, the moon's shadow tip always travels from west to east, everywhere on the earth. Also, like the sun, the moon crosses the observer's meridian at its maximum altitude, and the shadow (which is then the shortest) lies along his meridian.

   Plot the shadow tip's cast by the moon in the same manner as the sun. A moon navigator is made using the same procedure used for a pocket navigator.

5.4.4 Methods of Travel. NSW operators will encounter a variety of terrain features and surface conditions that must be negotiated in a mountain and arctic environment. Movement over rock, snow, and ice requires proficiency with specialized equipment and advanced training. Mountaineering techniques and equipment development are in a state of constant change. It may be unrealistic to expect all NSW cold weather personnel to attain the same level of technical proficiency, but a training cadre should be well versed in all applicable skills. Lead climbers and technical experts may be developed within the platoon structure. These individuals will be responsible for establishing routes, selecting the appropriate equipment, and leading less proficient personnel over difficult terrain. In a mountainous environment, the most direct route to an objective may require the greatest degree of technical expertise. Lead climbers may be employed to negotiate these areas and establish climbing and hauling systems that will expedite the movement of personnel and equipment. Rates of movement should be carefully analyzed to include consideration for terrain, equipment (weight), operational proficiency, and weather. This section will outline the basic types of movement and equipment presently used by mountain and arctic personnel. The techniques that are used when employing this equipment follow general guidelines. Individuals should develop personal and unit proficiency to enhance climbing and casualty rescue techniques. These guidelines and other required skills are currently being disseminated to platoon personnel by the respective NSWG training cadres.

5.4.4.1 Movement Over Rock. The equipment used for this type of movement has been previously discussed. The application of various techniques of movement that climbers should be familiar with will be derived through training. The easiest and most direct route to an objective is not always possible due to terrain, enemy presence, etc., therefore cliff negotiation may be the only way to reach the destination. The evolution may be time consuming, but with knowledge of techniques, safety rules, and commands, the obstacle can be negotiated. Balance climbing is the term applied to the skills required to negotiate movement over rock. Balance climbing requires the application of specific skills and procedures to prevent injury. The use of a belay system and climbing commands is necessary to effect coordination between the climbing and spotting elements. This section will cover rules, techniques, and commands.

1. Safety Rules

   a. Never deviate from the basic safety rules.

   b. Rope up on all exposed areas.

   c. Obey the leader.

   d. Do not climb beyond your ability or skill.

   e. Wear a helmet with chinstrap fastened.

   f. Keep boot soles clean.

   g. Avoid lunging or jumping to reach a hold.

   h. Check hand and foot holds before putting weight on them.

   i. Avoid dislodging rocks and if you do, sound off with "rock!".

   j. Do not look up if a climber above yells "rock;" seek shelter or get flat against the face.

   k. Anchor all belay points and double-check them. Two anchors minimum per belay point.

   l. Remove all watches, rings, and jewelry before climbing.

   m. Avoid wet rocks when possible.

   n. Only wear gloves when needed to keep hands warm.

   o. Stay alert and attentive when on belay.

   p. Try to keep three points of contact on the rock face.

   q. Clip into belay anchors before untying from the rope.

2. Types of Holds. The five basic holds used in balance climbing are:

   a. Push holds

   b. Pull holds

   c. Foot holds

   d. Friction holds

   e. Jam holds.

3. Combination Holds. The application of the five basic holds may be used in a series of combination movements. These include:

a. Chimney Climbing. Use of the entire body when ascending a crack. The entire body is inserted into a crack and a combination of the five holds is used.

b. Lie-back. A combination of pull holds with the hands and friction holds with the feet.

c. Push-pull. Combination of opposing forces.

d. Mantelling. Combination of pulling and pushing with the hands to raise the body.

e. Cross Pressure. Inserting hands in a crack and pulling outward to hold or raise the body.

f. Inverted. Push or pull movement.

g. Pinch. Hold applied to nubbins.

4. General Use of Holds. Individual holds are developed from experience. General guidelines include:

a. Most hand holds can be used as foot holds.

b. Use all hand holds available to conserve energy.

c. Small projections can be used as holds.

d. Use whatever hold is necessary to achieve the objective. (Skill level will determine the appropriate hold.)

5. Body Position. The climber should climb with his body in balance by keeping his weight centered over and between his feet. Don't hug the rock. Don't over extend and become "spread-eagled." Refrain from crossing arms and legs and using knees or elbows except with an extension (forearm, shin). While climbing, keep in mind the acronym "CASHWORTH" for proper body position and movement.

C Conserve energy.

A Always test holds.

S Stand upright on flexed joints.

H Hands kept low; handholds should be waist to shoulder high.

W Watch your feet.

O On three points of contact; avoid using knees because of awkward out-of-balance position.

R Rhythmic movement.

T Think ahead.

H Heels kept lower than toes and inboard.

6. Belay Positions. The belay man uses belay positions to protect the climber from falling during a rapid climb. The belay man should stand to the side of the intended climbing route to observe the climber and avoid

falling rock. The brake hand should never leave the rope. When holding a fall, the brake hand bends the rope away from the belaying device to lock the rope. The belay man must remain alert to anticipate the needs of the climber and be anchored by artificial (chocks, pitons, etc.) or the preferred natural anchor. A static or dynamic belay may be used. The two belay positions are:

a. The sitting belay

b. The standing belay.

7. Route Selection. A detailed route reconnaissance is necessary prior to climbing. The lead climber should pre-select the probable route. The proper equipment should be stowed on an appropriate gear rack, and proper protection and anchor points must be established for the climbers that follow. The following considerations should be addressed:

a. Best approach.

b. Degree of difficulty.

c. Distance between belay positions.

d. Concealment along route.

e. Amount and type of equipment required.

f. Number of personnel required and level of ability.

g. Use two vantage points to obtain a three dimensional picture of the climb.

h. Wet or icy rock can make a route impassable.

i. Moss, lichens, or grass are slippery, especially when wet.

j. Small tufts or bushes do not provide reliable holds.

k. Gullies are subject to rockfalls, stay to the sides.

l. Do not overestimate own ability.

8. Climbing Commands. Climbing commands facilitate coordination and safety between climbers. Voice commands, silent signals, or rope tugs may be used as required by the tactical situation.

a. Belay man starts – "on belay."

b. Climber starts – "climbing."

c. Belay man's response to "climbing" is "climb."

d. Climber needs more rope – "slack."

e. Climber wants a tighter rope – "rope."

f. Climber wants to be held by rope – "tension."

g. Climber thinks he might fall – "watch me."

h. Climber is falling – "falling."

i. Climber is clipped into belay anchor, secure, and won't fall – "off."

j. Belay man's response to "off" is – "off belay."

9. Piton Protection. The following guidelines apply when placing pitons.

a. When selecting a crack, tap on both sides of the crack with the piton hammer to ensure that the rock is solid.

b. Choose a piton that will slide into the crack one third to one half of its length, but which will not bottom out when driven.

c. Attach the piton to the figure-eight loop at the end of the line on the gear rack. Insert it in the crack and drive it in with moderate strokes of the piton hammer.

d. Good piton placement will:

(1) Insert without splitting the rock.

(2) Insert with a rising pitch.

e. Testing a Piton. The two procedures for testing a piton are:

(1) Moderately strike the piton in both directions along crack.

(2) Restrike the piton to check the pitch.

f. Piton Instability:

(1) If the piton moves or the pitch changes, redrive the piton and retest it.

(2) If the pitch is regained and the piton does not move, the placement is good.

(3) If the piton moves or the pitch is not regained, drive another piton in another location.

g. Removing a Piton.

(1) Using the gear rack, attach a carabiner to the piton in the same manner as when driving it.

(2) Tap the piton as far as it will go in one direction then as far as it will go in the other direction along the axis of the crack. Keep repeating this process until the piton is loose.

(3) Obtain a well-braced position.

(4) Gently pull the piton out. Do not try to jerk it out as this may cause loss of balance. Do not attempt to pry the piton out with the piton hammer.

h. Stacking Pitons. At some time it may be necessary to put two pitons back to back and drive them into a crack.

(1) If this is done in a crack where the piton eyes are side to side on the rock face, the carabiner should be attached to the piton that is deepest in the crack.

(2) If this were done in a crack where the piton eyes are from top to bottom on the rock face, the carabiner would be placed in the bottom eye.

(3) Stacking pitons does not affect the tensile strength of the pitons.

i. Hero's Loop (girth hitch). In cases where the crack does not allow the piton to be driven in the correct distance, a "hero's loop" may be used as follows:

(1) Drive the piton into the crack as far as it will go.

(2) Take a runner and form a girth hitch.

(3) Place the girth hitch over the eye of the piton and rest it on the blade of the piton as close to the rock face as possible. This applies strain to the part of the piton that is strongest.

(4) Use the piton as you would any other, but place another piece of protection as soon as possible.

10. Chock Placement. The following guidelines apply when placing chocks:

a. Study the configuration of the crack and select the proper size and shape chock.

b. The chocks should be placed so that there is maximum contact between the chock faces and the rock faces in the crack. The chock must also be steady and not pivot easily.

c. If the chock is well placed, the weakest link will be the wire or cord that the chock is strung with.

d. When placing a chock, ingenuity is of the utmost importance, as there are many ways each chock can be used effectively.

e. Removal of Chocks. The correct steps in removing a chock are as follows:

(1) Pull the chock in the opposite direction from which it was placed.

(2) A long piton may be useful in removing a chock that is in too tight. Care must be exercised however, to prevent damaging the chock.

f. Disadvantage of Chocks. A disadvantage to using chocks is that they are directional. They will provide support in one direction, but if they are inadvertently pulled in the opposite direction, they can be dislodged.

g. Opposed or Offset Chocks. In certain circumstances it may be necessary to oppose or offset chocks.

(1) Opposed chocks are two chocks that are being pulled on in different directions. One chock prevents the other chock from being pulled up and out of the crack as the rope drags against it.

(2) Offset chocks are two chocks placed in the same crack in a different manner.

11. Direct Aid Equipment. Direct aid climbing is the use of anything other than the natural features on a rock face. This technique will increase the rate of ascent for less experienced climbers within a unit. Equipment includes:

a. Etriers. This is a short webbing ladder.

b. Mechanical ascenders/Jumars.

c. Realized Ultimate Reality Piton. Used for thin shallow cracks.

d. Cliffhanger/skyhooks. Hooked over nubbins, flakes, etc.

e. Bashies/copperheads. Soft aluminum or copper devices used in shallow grooves.

f. Daisy Chain. One-inch tubular nylon tied in loops.

g. Fifi-hooks. Small hook attached by a cord to the climbing harness.

h. Stoppers. Excellent, but require reinforcement.

i. Tie-off Loops. Short loops of small diameter tubular nylon.

12. Climbing Sequence

a. The lead climber needs to don sit harness and make ready any climbing protection equipment.

b. Two ropes need to be stacked with running ends on top.

c. The lead climber ties the lead rope onto his harness with a double figure eight knot. This rope will be tended by a belay man on a belay device on locking carabiners.

d. The belay man should be in a standing or sitting braced position. If there is no good position to be braced then the belay man should be clipped into anchors.

e. The second rope (trail rope) will have a double figure eight knot tied into the running end of the line. This bite should be clipped into the back of the lead climber's harness by locking carabiners. This rope is not on belay but paid out as the lead climber advances.

f. The lead climber should have a radio to tactically pass information to his belay man and the OIC.

g. The lead climber performs double safety checks on ropes and harnesses and conducts climbing verbals with his belay man.

h. As the lead climber ascends, the belay man and the trail rope man pay out line.

i. The lead climber places pieces of protection and clips the lead rope into the protection as he ascends.

j. When the lead climber ascends to the end of his rope or to the top of the cliff, he sets belay anchors and ties himself into the belay anchors with the lead rope. He ties a clove hitch to the anchor to allow adjustment without untying.

k. He unclips the trail rope from the back of his harness and attaches it to the belay anchors. He ties a double figure eight bight in the lead rope and attaches it to the belay anchors with a locking carabiner.

l. The lead climber radios to the belay man that he is "off climb" and that both lines are ready for jumar ascent.

m. Climbers are called in from security positions, they check jumar ascenders, clip into line, and ascend.

n. As each jumar climber ascends, the line is held tight by one of the element. This allows the bottom ascender to slide up the line easily.

o. The first man to jumar will take a gear sling to retrieve the lead climber's protection.

p. The last jumar ascender should be the best in technique because no one will be holding tension for him.

q. When the last man is over the top of the cliff, the lines are retrieved.

13. Rappelling. Rappelling is used for rapid descents. The types of rappels that may be used include:

a. The Body Rappel. Friction is absorbed by the body to slow descent.

b. Hasty Rappel. Used on moderate slopes, the climber locks the rope across his body to stop the descent.

c. Seat-hip Rappel. Friction is absorbed by the snaplink to control the rate of descent.

d. Retrievable Double Line Rappel.

(1) Two lines are readied for deployment, standing ends tied together around the anchor with a double fisherman's knot.

(2) Figure eight knots are tied into the running ends of both lines so the first man down does not rappel off the end of the line and the line pulled for retrieval line is identified.

(3) The double fisherman's knots are positioned on the down hill side of the anchor point. The two double figure eight bights are clipped together with a locking carabiner. This isolates each line so both lines can be used at the same time or independently. Both lines are ready.

(4) The first man to rappel off the cliff will have the rappel device clipped onto one line and an ascender with sling clipped to his harness with a locking carabiner. This ascender is clipped to the rappel line above the rappel device and he takes a second ascender with him. The second ascender will allow him to get back on rappel if the first ascender locks him on the line.

(5) The first man descends with a radio and unclips himself from the rappel line when reaching the bottom. He will conduct a quick security check and belay both lines so that the second and third man can rappel at the same time.

(6) The last man rappels down on both lines. On the bottom, he pulls down the identified line. When at the knots, he retrieves the second line while another unties the knots.

5.4.4.2 Movement Over Snow. Movement over snow can be accomplished in a number of ways. Techniques involving the use of snowshoes, skis, skis and pulks, and skijoring are common methods of travel. Each method requires training and is appropriate under specific conditions. In some cases, rock-climbing equipment may be used during movement over snow. NSW personnel should be aware of the methods of movement and become proficient in these methods during pre-deployment training. This section will outline the methods of foot movement over snow.

5.4.4.2.1 Self-arrest Techniques. The ice axe is the most important equipment item used when traversing snow or ice. The axe can be used for stability, establishing foot or handholds, and for arresting a fall.

1. The position you find yourself in when you fall will determine how you self-arrest. You'll likely be sliding in one of four positions. Head uphill or downhill and in either case, face down or on your back. The immediate goal is to get your body in the only useable self-arrest position, with your head uphill and face down. The first move toward that goal is to grasp the axe in both hands, one hand on the axe head and the other at the base of the shaft.

2. Head Uphill Face Down. You're in the desired self-arrest position. All you have to do is get your body over the axe shaft as described earlier.

3. Head Uphill On Your Back. This isn't more difficult than head uphill face down. Roll toward the head of the axe and jab the pick into the snow at the side as you roll over onto the stomach. If the axe head is on your right, roll to the right. If it's on your left, roll left. Beware of rolling the other way, toward the spike that could jam the spike in the snow and rip into your body.

4. Head Downhill Face Down. Self-arrest from head first falls is more difficult because the feet have to be swung downhill. In this face down predicament, reach downhill and off to the axe head side and get the pick into the snow to serve as a pivot to swing the body around. Work to help swing the legs around so they are pointing downhill. Never jab the spike into the snow and pivot on the end of the axe. That would present the adzes in a collision course with your face resulting in an injury.

5. Head Downhill On Your Back. Hold the axe across your torso and slide the pick into the snow, then twist and roll toward it. The pick serves as the pivot point, but merely planting it will not bring your body around into the final self-arrest position. You must work at rolling your chest toward the axe head at the same time you are helping your legs to swing around so they are pointing downhill.

6. Variations. In loose snow the pick may not be able to grab, thus making the usual self-arrest useless. The best brakes in this case are the feet, knees, and elbows widely spaced and deeply pressed into the snow.

5.4.4.2.2 Glissading. Glissading is the intentional, controlled; rapid descent, or slide of a mountaineer down a steep slope that is covered with snow. Glissading is similar to skiing. The same balance and control is needed. Instead of skis, the soles of the feet or buttocks are used. The standard ice axe is the only piece of equipment required and acts as the rudder, brake, and guide for the glissade. There are three methods of glissading:

1. Sitting Glissade. Simply sit down and go while holding the ice axe in a self-arrest position. The spike of the axe can be used as a rudder by placing it off to one side. Applying downward pressure to the shaft will force the spike to dig in and reduce forward speed.

2. Squatting Glissade. Hold axe in self-arrest position with legs bent and balanced. Dragging the spike in the snow gives three points of contact.

3. Standing Glissade. Saves clothing from getting wet and is similar to downhill skiing. Feet can be together or spread, with one foot slightly forward of the other. If you hit snow of different consistency you must adjust your footing. It is most effective on a firm base with a softer layer on top.

5.4.4.2.3 Low Angle Slopes. There are two methods for walking on low angle slopes:

1. Rest Step. The rest step is used to allow the leg muscles to relax. After each step, the rear leg supports the full body weight by extending and locking of the knee. The pace should be slow and steady with methodical breathing.

2. Step-kicking. Step kicking is done on moderate angle slopes when crampons are not worn. Hold the ice axe like a cane with the palm over the adz and the thumb under the pick. Plant the ice axe, with the spike down, into the snow. With one foot, step up and kick a foothold into the snow. With the other foot, step up and kick a second foothold. With the body in balance, withdraw the axe and replant it an arm's length in front of the body. The ice axe can also be held in the stake position, holding the head in one hand with the other grasping the shaft. Center the ice axe in front of the body. As the angle becomes steeper, advance diagonally. As the slope becomes steeper, the axe is carried diagonally across the body and the pick may be used instead of the spike.

5.4.4.2.4 High Angle Slopes. As the slope angle increases, the individual may choose one of the following methods:

1. Diagonal Ascent. Move across the face of the slope. Start from a position of balance. The most balanced position is with uphill leg slightly bent and downhill leg fully extended to make use of the leg bones and skeleton bearing most of the weight. The uphill foot is in front of and above the downhill foot. Bring your downhill foot forward. You are now momentarily in an out of balance position. To switchback, you now move the uphill foot in the new direction. When performing a diagonal ascent with the ice axe, remember to keep the axe in the uphill hand.

2. Step-cutting. Use the ice axe to establish footholds.

3. Plunge Stepping. A downhill aggressive and confident walking technique used by stepping assertively away from the slope and landing solidly on the heel. Knees should be slightly bent. Toes are facing out in a herringbone fashion.

5.4.4.2.5 Snow Anchors. These anchors are used for ascending and descending steep snow-covered slopes.

1. Pickets. Stakes driven into the snow in pairs to provide support.

2. Flukes. Metal plates with a wire sling.

3. Snow Bollards. Constructed of snow and used as an anchor.

4. Ice Axe Anchor. May be used as a belay point.

5.4.4.2.6 Roped Climbing. Unroped travel is the fastest and best under favorable conditions, (daylight, gentle slope, and sturdy snow) but roping-in may be required. Roping-in provides a secure means of belayed travel when the experience of the climbing unit is questionable, light and weather conditions are poor, and the route is steep and slippery. When climbing on ice or snow, the members of a unit tie into a climbing rope for safety. When crevasses may be encountered, a three-man rope team is best. The others arrest the fall of one member.

5.4.4.2.7 Belaying. Belay systems used during rock movement may be applied to movement over snow. Three types of belays that may be used in snow include:

1. Boot-axe Belay. The rope is routed around the ice axe and boot. This is the least secure method.

2. Standing Snaplink/Ice Axe Belay. The rope is routed through a nylon runner attached to the ice axe with a snaplink. The rope is routed around the hips of the belay man.

3. Sitting Hip Belay. This method is the same as that used for rock climbing.

5.4.4.3 Movement Over Ice. An ice area may be beneath a snowfield, or lightly covered with snow. Although the ice can provide an easy and safe route on glaciers, ice climbing is required wherever water freezes on slopes. This may include snowmelt and subsequent freezing, frozen waterfalls, glaciers, and ice fields. Ice climbing may include portions of ice, rock, snow, or combinations of these. The composition of ice may change as will the slope and weather. Different techniques may be used depending on the angle of the slope, ice composition, weather, light conditions, and equipment available. Walking on level ice slopes is the same as on snow (balance climbing) except that crampons are usually worn. Slopes are classified as the following:

Gentle	0 to 30 degrees	Vertical	80 to 90 degrees
Moderate	30 to 45 degrees	Overhangs	over 90 degrees
Steep	45 to 80 degrees		

5.4.4.3.1 Step-cutting. Step-cutting on ice gives climbers a quick and easy avenue of ascent. Even with crampons, steps make foot positions more secure. Hard ice requires that the pick of the ice axe be used. The adz may be used on soft ice. The axe should do the work. The climber should swing the axe with the entire arm. He assumes a stable position and begins by directing blows at right angles to the slope that makes a fracture line along the base of the intended step. The following are step-cutting techniques applied to ice.

1. Diagonal Traverse. Ascent step-cutting. Used to ascend gentle slopes.

2. Alternating Parallel Steps. A zigzag pattern used to ascend steep slopes.

5.4.4.3.2 Crampons. When walking on crampons, the same principles are used as in mountain walking. When a leg is advanced, it is swung in an arc around the fixed foot to avoid locking the crampons or catching them on clothing. The two methods of using crampons on slopes are flat footing and front pointing. Both methods may involve traversing and vertical climbing. They can be used in combinations to allow the muscles to rest and to adapt to changes in ice composition and slope angle. Crampons used with the ice axe provide another point of contact that adds to individual security. Balance and rhythm coupled with the proper use of ice tools is key to successful negotiation of any slope. Careful study of the ice while walking is needed to identify positions for foot and axe placement. Ice is usually more compact in depressions and offers better hooking action on the lip of the depression. When using the ice axe, the climber swings from the shoulder and lets the weight of the axe head do the work. Clear ice is usually harder than opaque ice and is more brittle. Opaque ice has air inside and is softer, less brittle, and holds well. The ice should be tapped with the ice axe before emplacement to determine an adequate placement.

1. Flat-footing. Flat-footing is good for moving on slopes up to 50 degrees and on hard ice. The crampon points must be carefully and deliberately placed. All ten sole crampon points should be placed flat against the surface. The ankles may be rolled so that the crampons are edged flat against steeper surfaces. On snow the points penetrate easily. On ice the foot must be stamped firmly to obtain maximum penetration. At the turning points of a traverse, direction is changed with the uphill foot as in mountain walking.

   a. Ascending. Feet should be kept shoulder-width apart with all crampon points flat against the ice. The ice axe is held in the cane position (self-belay grip). As the route becomes steeper, the feet are in a herringbone position with the ice axe in the cane position. Following a diagonal line of ascent provides an easier, less tiring, and safer (less chance of falling backwards) means of ascending steep slopes.

   b. Descending. On gentle slopes, face directly away from the slope with the feet pointing downhill and crampon points flat against the surface. Hold the ice axe in the cane position to aid in balance, control, and movement. As the slope becomes steeper, bend the knees slightly and keep the body weight centered over the feet. Position the feet in a herringbone position and hold the ice axe in the cane position or at the side with two hands. Drop the buttocks lower with the ice axe firmly planted in ice or snow before moving the feet. Crampon points maintain contact flat against the surface. Hold the ice axe in one hand low on the shaft. Thrust the pick below the feet into the ice surface and walk past the ice axe, maintaining control of it.

2. Front-pointing. On steep pitches, the technique of front-pointing can be used. It is more efficient than step cutting. The two front points of the crampons are used as the primary foot-to-ice contact. Since the angle of the ice is steep, this technique requires at least one ice axe. Rhythm and balance are required to maintain control for a smooth ascent. A careful route reconnaissance is required before the start of an ascent or descent. The condition of the ice is crucial since this provides handholds and footholds for the crampons and ice axes. While climbing, place the crampons and ice axes carefully, choosing flat or in-cut holds. Three points of contact are preferred while moving on vertical and overhanging ice.

   a. Ascending. Twelve-point crampons are needed to front-point (ten on the sole and two that serve as fangs). For vertical climbs over 1,000 feet, rigid crampons are recommended since the rigidity allows the calf muscles to rest. A stiff mountain ice-climbing boot is recommended. The boot should have a rigid shank to provide needed support. Keep the feet about shoulder-width apart. Point the toe slightly downward and kick into the ice. One kick should suffice. Push the heel down to a horizontal level to firmly plant the front points. Keep toes square against the ice. To remove points, lift the heel and pull the points outward. With two ice axes securely planted and one foot in a front point, rest the other leg flat-footed against the ice wall. Use front-pointing and flat-footing together if the ice condition dictates. The ice axe is used to give support and balance to the upper body. If the pitch is steep or long, two ice axes should be used. Traversing may be required when changing routes. Instead of crossing the feet, a side-to-side step should be used.

   b. Down climbing is much more difficult than ascending. When possible, face away from the slope and walk straight down, maintaining full sole crampon point contact with the surface. If the pitch is vertical or overhanging, it is best to rappel down the slope. A front-point descent is a dangerous technique. For short vertical distances of less than 10 feet, assume the X position with the crampons and ice axes shoulder-width apart. Position the ice axes at head level. Step down, one foot at a time, and position the crampon. Do not step so low as to cause the heel to raise and front points to shear out. A diagonal descent is preferable to a direct vertical descent. Always use a belay to conduct the descent.

5.4.4.3.3 Maritime Ice Movement. Movement near shore, across lakes, and between islands will be affected by ice. Visibility, sound, cover available, and ice thickness will determine techniques used to move across ice surfaces. A pike, 20-meter safety line, and ice drill should be used during movement. Skis may be used if noise is not a factor. Wood skis are preferred since wax on the base of the ski will be rapidly scraped off when moving over ice. Thin ice will be encountered near points of land, centers of channels, piers, currents, and tidal areas. The unit should move to land for rest breaks. Recommendations are:

1. Prior to commencing the patrol, obtain an ice core with the drill. Two inches of ice are required to support a fully loaded platoon.

2. Ice cores are taken periodically throughout the patrol. Samples obtained with the pike can be taken frequently by the PT to determine ice quality. On solid ice the pike should chip without sticking and have little penetration. On ice of lesser quality, the pike will penetrate further and stick into the ice. Poor ice will sound hollow when struck with the pike.

3. The PT will carry the pike and be attached to the second man (PL) with the safety line. The interval between personnel should be roughly 15 yards.

4. When visibility is good, the unit must stay close to shore, jump tracks frequently, and use available cover. Movement next to shore should be slow with maximum distance between personnel to minimize the sound created by the skis.

5. When visibility is poor, noise becomes the major factor that may compromise the unit. When moving between islands, stay away from land by splitting the distance between shores. When establishing a LUP on an island, ski past it and hook back. Remove skis and walk in to minimize noise.

6. If an individual (most likely the PT) breaks through the ice, he may be rescued by using his ice pick or line attached to follow-on patrol members. Once out of the water, he should roll away from the hole to a place where the ice is firm. All wet clothing should be removed and dry clothing donned. The individual should begin skiing vigorously to warm himself. Platoon members should erect temporary shelter and heat water. After fluids have been administered, the unit should move out rapidly to assist the warming process. The platoon must know first aid procedures for cold-water immersion.

7. Reconnaissance team or flank personnel may be used to protect the main unit. Before the main body moves to land the reconnaissance team should check for an ambush. Each flank or reconnaissance team should carry a pike.

5.4.4.4 Movement Over Water. Craft TTP are presented in NSW Combatant Craft TACMEMO XL-2080-2-89. The northern maritime environmental conditions make marine transits and OTB operations the most demanding. Knowledge, experience, proper planning, and preparation are needed for survival. Meteorological factors that affect operations in this environment are presented in detail in section 5.2.3. Those with the most effect on transits over water are as follows.

1. Temperature. The ability of a platoon to successfully complete a mission in temperature extremes encountered in the northern region will vary greatly depending on experience, discipline, and to a lesser extent, equipment. How many hours a CRRC can transit safely in these temperature extremes is impossible to predict. It is difficult, if not impossible, to treat hypothermia or freezing injuries in transit over water.

2. Wind. The two biggest dangers in transit are:

a. Wind chill significantly increases the chilling effect of the ambient temperatures.

b. Surface winds generate waves and swells. Experience and preparation will dictate how strong the wind can be for a transit. Before sea state builds up, surface winds to 35 knots can be handled safely.

3. Sea Condition. Waves greater than 20 feet can occur anywhere in the North Pacific. Seas exceed 20 feet as much as 10 percent of the time in the Gulf of Alaska. Similarly, swells greater than 12 feet occur 35 percent of the time in the Southern Gulf of Alaska. Sea conditions depend primarily on three factors.

a. Wind Speed. The higher the speed of the wind, the greater is the sea disturbance.

b. Duration of the Wind. The sea condition will increase the longer the wind blows until a maximum sea state is reached.

c. Fetch. This is the length of the sea surface that the wind acts on from the same direction. If the fetch is small, the effect will be relatively small no matter how great the wind speed or duration.

d. Effects of high sea conditions on your transit.

(1) Increased fuel consumption (as much as 18 to 20 gallons per nautical mile).

(2) Extremely hard communications.

(3) Possible seasickness, injuries, or death of personnel.

(4) CRRCs become harder to detect for completing rendezvous at sea.

e. Sea condition guidelines for CRRC transits.

(1) Guidelines are set by platoon experience.

(2) If you haven't experienced heavy seas, do not try to deal with them without safety back-ups.

(3) Wind and sea conditions of 35-41 knots with waves of 10-20 feet have been transited by platoons with little experience, but this is not recommended without experienced safety back-up.

(4) Foul weather transit plans should include tactical BLSs if needed until sea conditions calm.

4. Precipitation.

a. Rain will affect water levels at rivers and estuaries, activity at BLS, and performance of optics.

b. Snow affects visibility, detectability, camouflage requirements, and may increase or decrease signature across BLS.

5. Fog. Heavy fog makes flying impossible and rendezvous at sea difficult, but concealment is great.

6. Ice. Many areas are subject to months of being ice-bound.

a. The freezing temperature of salt water is usually 28.6 degrees Fahrenheit.

b. CRRCs will not pass through a quarter inch of ice with a 55 hp engine.

c. Sea ice can form as fast as 3.5 inches (10 cm) in 24 hours.

d. Two inches of freshwater ice will support the weight of a heavy man. The same thickness of newly formed sea ice will not support more than 10 percent of this weight.

7. Tides. Mean tide ranges in these regions can vary from less than one to more than thirty feet. Large tidal ranges must be considered when selecting BLSs. A beach with a shallow gradient could leave you with two miles of mud to cross, or 50 yards of rocky beach to carry a boatload of gear across. Rivers and tidal estuaries may be navigated miles up stream or into areas not accessible to people at lower tides.

   Tidal currents. Horizontal water movement in areas of large tidal ranges can cause extreme tidal currents, especially in channeled areas. The currents will affect dead reckoning navigation tracks. Current vector calculations are difficult, if not impossible, in some areas due to the many directions the tidal currents will be moving around the landmass.

8. Daylight. The seasonal tilt of the earth causes the northern latitudes to have long summer days and long winter nights. Charts are available to calculate these times.

9. Bioluminescence is caused by luminescent euphausids and copepoda that give off very noticeable amounts of light in the water.

10. Magnetic variation should not interfere with the use of the boat compass below the Arctic Circle. Local charts should be studied, as there are bays with local disturbances with as much as 180 degrees error.

5.4.4.4.1 Local Population's Maritime Activities. Choice of navigational tracks and BLSs should not be made until a study of the local maritime activities have been made. Considerations include:

1. Fishing activities

   a. Are they on-or offshore?

   b. Are they at day or night? Huge lights can light up large areas of bays.

   c. Are they foreign or domestic?

   d. Is small craft fishing off beaches or streams?

   e. Fishing nets and aquaculture farms can be located close to shore or as far out as five miles.

2. Marine transportation routes include ferry and small craft avenues of movement to villages, towns, and fishing sites.

3. Planning requirements

   a. Contacts. Encounters with military, security forces, and civilian vessels in the AO can be a problem enroute to the objective. Detailed plans must be prepared to deal with the possible contact or sighting and briefed to all hands involved with the mission.

b. Communications. How the transit element will communicate with headquarters and other elements involved must be identified. Radios, batteries, antennas, and environmental factors are presented in detail in Section 3.4.

c. Insertion/extraction. Back-up locations for all sites, contingencies for foul weather, equipment failures, loss of communications, and missed rendezvouses must be planned, with courses of action understood by all personnel.

d. Reference information for planning

(1) Chart Number One. An explanation of all abbreviations and symbols used on nautical charts and interpretation of principal foreign abbreviations.

(2) American Practical Navigator. A large, two-volume encyclopedia of navigational information and useful tables.

(3) Coast Pilot. Nine-volume publication for the waters of the U.S. and its possessions, containing information on navigation aids and hazards, tides and currents, and prominent landmarks.

(4) Sailing Directions. Consists of 48 publications for planning open ocean transits and piloting in coastal waters.

(5) Light List. Contains detailed information on lights and sound signals in the waters of the U.S. and its possessions. Published annually in seven volumes.

(6) List of Lights. Contains detailed information on lights and sound signals in foreign waters, similar to the Light List.

(7) Tide Tables lists heights and times of high and low waters in selected areas around the world. It is published annually in four volumes. It also lists times for sunrise/set and moonrise/set in those areas.

(8) Tidal Current Tables list times, velocities, and directions of currents in selected areas around the world. Published annually in two volumes.

(9) Nautical Almanac. Contains information on all celestial phenomena.

(10) World Port Index. Gives information about specific ports in the world, such as tides, currents, fuel positions, mooring positions, etc.

5.4.5 Patrolling Techniques/Considerations. Many of the patrolling techniques that are used in temperate climates may be applied to operations conducted in the arctic. Routines in a mountain and arctic environment will be affected by temperature, hours of light/darkness, and terrain conditions. Movement over snow, whether on foot or skis, will be more easily observed. Camouflage, concealment, and deception are tools that NSW units must employ during movement to the objective. In many instances, the situation will dictate the appropriate actions, but a detailed route reconnaissance and knowledge of the enemy threat will enhance preparedness. SOPs are essential to ensure survivability, cohesiveness, and morale within the unit. Individual responsibilities must be well defined to overcome the effects of the weather and to accomplish the mission.

5.4.5.1 Patrol Routine. During the transit to the objective, a routine should be established to ensure that individual responsibilities are coordinated, the effects of the cold are minimized, and a tactical profile is maintained. Events to consider include:

1. Dusk minus one and half-hours
    a. Shake wake
    b. Stand to
    c. Prepare breakfast
    d. Change clothing
    e. Change sentry
    f. Pack equipment
    g. Fill stoves
    h. Pull poles
    i. Pack tents
    j. Level site
    k. Fill and camouflage latrines
    l. Wax skis
    m. Check area
    n. Stand to five minutes before departure.
2. After ten minutes, stop and vent if necessary.
3. During normal movement, stop every hour for five to ten minutes.
4. Stop if required for the communication schedule.
5. Change trailbreakers as conditions dictate (e.g., when personnel in the back of the patrol become cold).
6. When stops are required, loop on track and ambush (Fish hook).
7. Dawn minus one and half-hours.
    a. Detour into lay-up area.
    b. Carry out routine for lay-up.

5.4.5.2 Track Discipline. Tracks in the snow are visible from a great distance, particularly during soft snow conditions. This drives the requirement for night movement. Rules for and adherence to good track discipline include:

1. The line of march should be concealed from air and ground observation. The route should follow the natural contours of terrain features and take advantage of shadows. Never make straight tracks through open areas.

2. Existing tracks, providing the risk of ambush is considered, can be used to expedite movement.

3. A deception plan may be implemented by establishing other tracks.

4. The use of available overhead cover and prevailing shadows will help keep the track camouflaged. Towing brushwood behind the last man will disrupt the trail pattern and prevent proper analysis by the enemy.

5. Ski tracks are very obvious from the air, particularly when the sun is low causing a shadow. Hard snow, ice, and frozen ground should be used to keep tracks to a minimum.

6. Track discipline, deception, and concealment must be rigidly enforced. The platoon can easily mine or ambush their tracks.

7. Common mistakes doing track camouflage are:

    a. Making un-natural edges in clean snow sheet.

    b. Not wiping out the edge enough to remove shadows from the track.

    c. Tracking up free snow spots with boot prints.

    d. Disruption of snow on vegetation in the area.

    e. Leaving a path of broken vegetation.

5.4.5.3 Track Intelligence. NSW units may encounter tracks that must be immediately analyzed for their intelligence value. Several methods of interpreting ski and snowshoe tracks can be used to determine the number of enemy personnel and their direction of travel. The enemy may also employ track deception procedures. Two men on skis can lay a deception track(s) that will appear as if a larger force had skied through by planting ski poles continuously and stomping on skis.

1. Ski Poles

    a. Direction. As the ski pole is planted and the skier moves forward, the basket will also angle forward causing the basket to dig into the snow and leave an indentation on the forward edge indicating the direction. The point of the ski pole will also contact the snow before the pole is planted making a line pointing away from the direction of movement.

    b. Number of Personnel. Count the number of ski pole plant marks. A reasonable estimation can be made regarding the number of personnel on the track. On one side of a track, a skier on level ground will plant his pole approximately 3-4 times every ten yards.

    c. Terrain. Personnel will plant poles less on downhill runs and more frequently moving uphill.

2. Skis

   a. Direction of Travel. Direction of travel will be indicated in the ski track by small ridges caused by walking on skis.

   b. Kick Phase of Diagonal Stride. The ski being depressed into the snow on the kick phase of a diagonal stride causes ridges. The tail of the ski will create a small step or ridge. The lower portion is in the direction of travel.

   c. Unit Size. The size of the unit that has used the track is determined by the basket marks, the width of the track, and hardness of the snow. Up to a squad size, the basket marks can be counted and the snow at the base of the ski tracks can be penetrated with one finger. When a unit of greater than 20 personnel uses a track, the basket marks will be difficult to distinguish and, if the surface is not icy, the base of the ski track can be penetrated with two fingers. If a unit larger than 30 personnel has used the track, the basket marks will resemble a ditch and the base of the ski track can be penetrated with the thumb. Larger units will create a broader ski track and the edges of the track will be obscured at bends and curves.

3. Snowshoes Direction. The tail of the snowshoe being in the rear indicates direction.

5.4.5.4 Camouflage and Concealment. Effective concealment depends on a unit's ability to blend in with the surrounding terrain. During the winter, the white background will highlight objects that differ in color, texture, and tone. Terrain features most likely encountered in arctic and sub-arctic regions are barren, wooded, and rocky. Coastal areas include beaches and sea ice. Camouflage in a snow-covered environment is achieved by a unit's awareness of its shape, shine, silhouette, shadow, and spacing.

1. Shape. Attempt to blend bivy and tent areas into surrounding terrain. Nature rarely produces large symmetrical or right angle objects. Be aware of shapes; rounding out rough edges will minimize leading lines and shadows.

2. Shine. Use care when preparing equipment for the field. The human eye will see a glimmer of light before it will detect movement.

3. Silhouette. Patrol and bivouac in tree covered areas and use the military crests of mountains when patrolling. Consider the use of netting and natural materials. Camouflage nets can be erected to obscure the silhouette of the item being camouflaged. If the snow depth is sufficient, "dig in."

4. Shadow. When possible, patrol in the shadows cast from natural terrain features. Tracks create shadows easily spotted from the air.

5. Spacing. Proper spacing depends on the terrain, weather, and threat. A good operator always anticipates what the spacing will be as he observes terrain he is approaching.

6. Sound. Sound carries very long distances in cold, dry air. The sound of skis and ski poles, snowshoes, and ice crust breaking can warn the enemy of your approach. Keep noise to a minimum. Noise discipline is everyone's responsibility.

7. During planning, the predominant terrain feature will dictate the appropriate camouflage techniques.

   a. Above the tree line, wear an all white uniform.

b. At the top of the tree line, wear dark trousers and a white top.

c. At the bottom of the tree line, wear white trousers and a dark top.

d. Below the tree line, the terrain will dictate uniform.

5.4.5.5 Weather Conditions. Changing weather conditions can enhance camouflage and concealment. Drifting snow changes the appearance of terrain. Weather can restrict visibility to a point where movement becomes unnoticed. It may erase tracks left in the snow. Certain conditions can muffle the sound of weapons firing. Conversely, the voice can be heard at extreme distances in conditions of extreme-cold when the air is clear and still. Weather is basically a combination of sky and ground conditions. An object will appear much closer on a clear, cold day and farther away on a dull, cloudy day. In various combinations, these conditions can aid in concealment and movement or greatly increase the risk of detection.

1. Very Clear Conditions. The sky is without cloud cover and the surface is calm and cold. In these conditions, the air is extremely transparent and visual detection of objects can be made from a greater distance. Sounds travel much further.

2. Broken-clear Conditions. The sky is partially clear and the ground conditions are completely clear. This occurs in areas of open water and in the proximity of large areas of ice and snow covered terrain. These conditions are better than very clear for movement, but still have increased visual and hearing detection ranges.

3. Overcast Conditions. The combination of overcast conditions and low sun on the horizon can create flat, dull days without shadows. Objects seem further away and movement is more difficult to detect.

4. Drifting Snow. Drifting snow is blown by the wind to a height of less than six feet above the ground surface. It does not impair vision, but a man in a prone or near-prone position is more difficult to detect.

5. Blowing Snow. In conditions of blowing snow, it is lifted by the wind to heights great enough to restrict horizontal visibility. During this condition, movement and sound are almost impossible to detect.

6. Falling Snow. Heavy falling snow will erase your movement trail and can cause navigation landmarks to disappear. Caution is required as cliffs, depressions, and avalanche areas will be difficult to locate.

5.4.5.5.1 Cold Weather Problems. The cold can impact mobility of the patrol. Plan for the worst. Use temperature updates while planning, and prepare all weapons, optics, and communications equipment for the intense cold.

1. Ice Fog. Ice fog is caused by the condensation of super-cooled droplets of water. Weapons firing, vehicle exhaust, and a man's breath can cause ice fog. The formation of ice fog can reveal the position of weapons and bivouac sites.

2. Wind. A strong wind can blow off overwhite pack covers; gear and clothing can be carried off and disappear into the snow. Keep gear secure.

3. Light Discipline. The long arctic nights accentuate the need for light from flashlights, candles, etc. Light discipline is required during movement and when in bivouac. Face tent entrance away from probable enemy approach. Chemlights have proven satisfactory for use inside tents with a fly, allowing no visible light outside the tent.

4. Thermal Signature. Heat radiated by stoves and human bodies will contrast with the cold background temperatures in the arctic. The probability of detection by IR devices is increased. Obtain intelligence on the enemy's thermal device capabilities and specifications.

5. Smoke Signature. Fires must be carefully controlled to minimize the smoke signature. Wood fires can be smelled from a greater distance than liquid fuel fires. Open fires should only be used in an emergency situation.

6. Radiation fog. This condition occurs on cold, clear, calm, winter nights. These conditions allow for maximum radiation from the ground. Air is cooled by conduction through ground contact. When the surface air is cooled below its dew point, fog will occur. This fog is thickest right after sunrise, but can burn off quickly when the sun penetrates it.

7. Advection fog. This occurs by the movement of warm air over a colder surface, either land or water. It is very prominent in northern maritime regions and does not have a cycle like radiation fog. Advection fog may burn off in the morning, return in the early afternoon, or may last for days. This is the fog encountered on boat transits that offers great concealment, but can present serious navigation problems.

8. Storms. While storms offer opportunities to surprise the enemy and excellent concealment, they present special difficulties. Ability to move as a platoon can be effected by wind chill and frigid temperatures. Planned routes to and from an area may become unusable. Know the history of storms in the area and how they may affect camouflage, concealment, and the mission. Don't get caught in a winter storm.

5.4.5.6 Breaks. A break is usually a brief halt of only a few minutes in duration. It is an enroute control measure that can be used to vent, change clothing and camouflage, wax skies, mount climbing skins, and rest. It must be used to keep the patrol fresh or overall effectiveness will deteriorate. Breaks should be short at the start of the patrol and longer as the patrol gets fatigued. Actions are as follows:

1. Set Security. The first vent break is usually ten to fifteen minutes after moving out from the IRP or a sterilized LUP. The PL should pre-brief the perimeter formation the patrol will use during the halt. The emphasis will be on maintaining 360-degree coverage of the patrol's position. Normally the patrol will drop the RS and APL to cover the rear. The main body will ski ahead one to three hundred meters and form a perimeter. The main body should maintain visual contact if possible or work out a squelch code on the inter squad radio. If the patrol must remain together, the patrol must fish hook or jump off into the perimeter position and set up a circular or cigar shape perimeter.

2. Vent clothing. The patrol will utilize the buddy system to remove or vent clothing.

3. Rest. Normal rest breaks are ten minutes for every sixty minutes of patrolling.

4. Pass the Word. This is an opportune time to disseminate new or updated information and instructions to all the patrol members. It is also important to ensure that the patrol members have an opportunity to give any information that they have on what they may have seen or heard during the move from the last halt.

5. Map Check. Every time that the patrol stops for a brief period, the PL should conduct a map check to pinpoint the patrol's exact location. The position should be confirmed by using the portable GPS, resection of prominent terrain features, and time of travel. All of the above data should correlate to provide a very accurate position plot. The PT, RS, APL, navigator, and assistant navigator should all be a part of the process.

6. Communications. Radio communications can be performed more easily from a halt than while on the move, especially if satellite and burst communications are being used. Make communications the final event before the patrol moves out to reduce the enemy's ability to DF your location. Never make your last communication of the day from your LUP.

5.4.5.7 Movement. During movement, unit personnel will be required to perform specific tasks. If necessary, flank protection may be positioned on high ground while the main unit transits open terrain. Spacing will increase in open terrain and decrease in close terrain. The number of personnel will determine the appropriate formation. The basic formation and organization is represented in Figure 5-1. Duties include:

1. Trail Breaker(s). This individual establishes the trail and acts as the navigator for the patrol. In some cases an assigned navigator who continuously checks the navigation plan may follow him. Snow conditions will determine how often this position needs to be changed. A well-established routine will preclude the necessity for words during the change. When it is snowing or snow is particularly deep, it may be necessary to use a four-man trail breaking team.

2. Number Two Man/PL. Depending on the size of the unit, the next man in the formation will straighten curves and improve the trail. In a larger unit, the PL will follow the number two man. The PL will select the main route, establish security routines, and initiate the changing of tasks. If a pulk is being used, the PL will establish the "third" ski trail.

3. Main Party. The unit or pulk team will follow in this position. Pulk teams need room to increase speed for up hill travel.

4. RS. This individual is responsible for security and track camouflage.

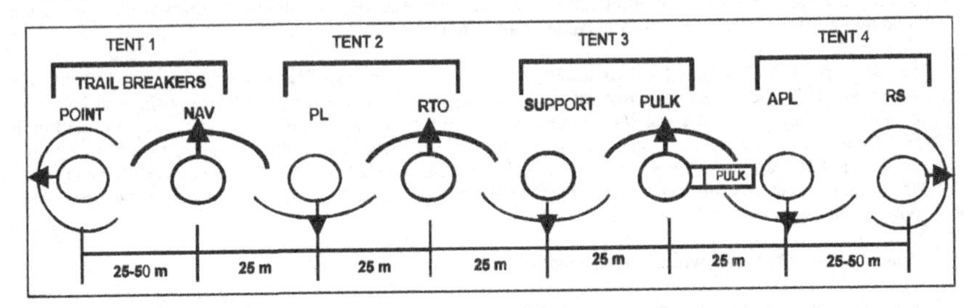

Figure 5-1. Patrol Formation and Organization

5.4.5.7.1 Day Movement. During day movement a high degree of caution and good route selection is required. Lead personnel should use binoculars frequently. Conduct SLLS upon entering new danger zones or changes in terrain. If remaining in the same zone, conduct SLLS at least every fifteen minutes. The terrain will determine the tactical method(s) of movement. The distance between men will depend on the size, speed, and control of the unit. Day movement is feasible when:

1. The terrain provides good cover.

2. The weather is inclement.

3. The route is isolated.

4. The tactical situation allows.

5.4.5.7.2 Night Movement. Movement at night requires greater control. All rendezvous points should be briefed. During platoon-size movements, the PL should maintain the third position in the file. The following points should be considered:

1. Select a route that is within the ability of all personnel.

2. Actions will take longer to accomplish.

3. Extra care is required when checking an area prior to departure.

4. If flashlights are required, use filters.

5. Movement is more difficult and much slower. Spacing needs to be increased for falls.

6. More time is required for listening when moving over new ground.

7. Small obstacles can become difficult to negotiate.

8. Little dips and bumps become invisible.

9. Downhill skiing is more difficult.

5.4.5.7.3 Speed. As the patrol moves out of the IRP as well as subsequent patrol break perimeters, the patrol should move out slowly and gradually pick up speed until a happy medium is reached. The golden rule is not to go faster than the slowest man. The speed at which a team travel depends on the following:

1. Tactical situation.

2. Size and experience of the unit.

3. Degree of difficulty of the terrain. A patrol on skies averages three kilometers per hour on flat terrain plus one hour for every three hundred meters of ascent and eight hundred meters of descent.

4. Weight and size of the loads.

5. Overall duration and distance to be covered.

6. Snow and weather conditions.

5.4.5.7.4 Linear Danger Areas. When a patrol encounters a linear danger area and has determined it secure enough to cross, the following procedures apply:

1. The PT will halt the patrol, establish a cigar shaped perimeter, and conduct SLLS. The PL will move to the PT's position and assess the situation.

2. The PL will determine the type of crossing that will be conducted and pass the appropriate hand and arm signal.

3. The PL and the man behind him will assume left and right security as the PT moves across the danger area.

4. Once the PT has reached the far side, he will conduct SLLS.

5. The PT will signal the PL across and the PT moves forward and drops down within visual range.

6. The PL then signals the next man across which bumps the PT and PL forward. As each man crosses he bumps the man in front forward to the next position until all are across. All men crossing maintain visual contact with the men to their front and rear position across the danger area.

7. The last two men in the patrol will cross together, initiate the head count, and continue movement.

5.4.5.7.5 Open Danger Areas.

1. Small open danger areas may be skirted around, staying within the tree line for cover.

2. Large open areas, especially those dominated by high ground or by terrain that affords the enemy cover and concealment, are likely kill zones and should be avoided. If these areas have to be crossed, use the bounding over watch technique.

5.4.5.8 River and Stream Crossing. Methods for river and stream crossing can be either wet or dry. Wet methods are discussed in detail in section 2.4.7.5. Equipment and terrain unique to operating in the northern maritime environment drives the need for special techniques and considerations. Dry methods are essential so personnel will not have to recover from being wet (and, therefore, physically vulnerable) and have to stabilize core temperatures with dry insulation before proceeding with the mission. Crossing methods and considerations are as follows:

1. Natural bridges of dirt may be in place and rocks and trees can be maneuvered to build an improvised bridge. This will depend on water depth, velocity, distance, and location.

2. Snow and Ice Crossing Precautions

   a. Use the PT to probe with an ice axe or ski pole when crossing snow or ice.

   b. Always have someone check the thickness of ice before attempting to cross. Use good intervals between people and unbuckle all gear no matter what type water.

   c. Ice is always thinner where rivers or streams enter or exit lakes and ponds, outside bends of rivers or streams, at the outer edge of lakes or large rivers, and on rivers with steep gradients. Have someone standing by with a rope for emergencies.

   d. Belay PT when possible.

   e. Spread weight out by using snowshoes, skis, ski poles, and branches.

   f. Know the weather history. This will give a better idea of the conditions.

g. If you do fall through the ice, DON'T PANIC. Carry out the following procedures quickly:

(1) Retain ski poles.

(2) Throw rucksack, weapon, and skies onto the ice.

(3) Use spikes of ski poles to drag yourself onto the ice and crawl to safety.

(4) Get out of wet clothing quickly. Dry off, put on dry clothes, and rewarm with hot drink.

3. One Rope Tyrolean Traverse with Z-Pulley System. An excellent way to cross with an injured person or transport heavy gear across a river.

a. Individual gear:

(1) Rucksack

(2) Six feet of one-half-inch tubular nylon for pack

(3) Two carabiners with one being a locking type

(4) Ascender with a three-foot runner of one-half-inch tubular nylon

(5) Weapon with sling and bungee cord

(6) Chest harness with extra carabiner

(7) Three feet runner of one-half-inch tubular nylon for slinging pack on rope.

b. Squad gear

(1) The PT will need a dry suit.

(2) One 165 feet length of 11-millimeter rope.

(3) Four pulleys.

(4) Five carabiners (one pear shaped).

(5) Three ascenders.

(6) 25 feet of one-inch tubular nylon for system brake.

(7) Two each 8-, 10-, and 15-foot lengths of one-inch tubular nylon runners for anchoring system.

(8) One-inch tubular nylon adjustable length for riggers of the system.

(9) Two-foot loop of 7 or 9 millimeter accessory cord for heart knot.

(10) 50 feet of one-half-inch tubular nylon for extra line (as back-up to other one-half-inch tubular line requirements).

c. Steps and considerations for choosing anchor points

(1) Points must be suitable for both loading and unloading.

(2) When possible, load point should be higher than unloading point.

(3) Pick a strong anchor, the smaller it is the lower you must attach the runner.

(4) Rope length example: 165 feet divided by 2 = 82.5 feet. Subtract ten feet for knots and you have 72.5 feet. Add line if the rope isn't long enough by using a double becket bend.

d. A minimum of three people is needed for set-up. Use tactics for danger crossing and maintain security. Break into elements of security, point and rear crossing teams, and rigging team.

e. Procedure

(1) Security element sets up security at the crossing point.

(2) Send PT, dressed in drysuit and fins, to swim the rope across.

(3) PT will put runner around anchor and secure with a carabiner, tie a figure eight on the rope, secure to the carabiner, and signal the element to tighten rope.

(4) While the PT is setting up the far side anchor point, the riggers are setting the near side anchor point. Use a runner on the near anchor with a carabiner and pulley. As far up the rope as possible, put an ascender on the line with a carabiner and a pulley on it. Once the other side is done, set up the pulley system. Get the rope as tight as possible without over stressing the rope.

(5) While the rope is being tightened one person needs to tie a heart knot with the two-foot runner of accessory cord between the anchor and the ascender. Attach a carabiner to the heart knot loop and wrap it around the anchor taking up all the slack. Start to slacken up on the rope and clean up the pulley system.

(6) Take the rope around the anchor and to the carabiner that is on the heart knot loop. Taking all the slack off the rope, tie a slipknot onto the carabiner and secure the loop of the slipknot to the side of the line the ascender is on with a carabiner. Make sure the bight of the slipknot is not over six inches. The system is ready.

(7) As the system is being finished, the first man going across should don his harness and his and the point man's pack. Have someone help people on and off the rope. Using the six feet of one-half-inch tubular nylon, secure each man’s pack to him before he crosses so he can tow it. If the incline is less than 30 degrees, a man should lead his pack. If the incline is greater than 30 degrees, send the pack first and the man should go feet first after it.

(8) As the first man goes, the next man should be ready at the load point. Only one man on the rope at a time. The next to the last man will take RS pack. Last man will sterilize the area.

(9) RS will "S" fold excess rope onto the loading rope. He will tie a figure eight knot at the end and attach it to himself. Before he gets on the rope he must take the safety carabiner off, make sure the slipknot is not slipping, and go. When on the other side he will untie the figure eight knot and pull. When the knot breaks, he can pull the line across.

5.4.5.9 Fish Hook into a Break. See Figure 5-2. The fish hook maneuver involves a 180 degrees turn offset and parallel to the direction of travel. This allows the patrol to observe their trail for enemy trackers, puts them at an advantage, and negates or minimizes the enemy's capabilities. Procedures for the fish hook maneuver are:

1. Halt the Patrol. If not in a file formation, the PL gives the signal for a file formation .
2. Leave the direction of travel at a 45 degree angle and proceed into a concealed area.
3. The formation circles around until moving parallel and 180 degrees from the original direction of travel.
4. The formation will halt in an area where they can observe the trail for enemy trackers.
5. The PL will signal for the patrol to move into a staggered file formation, set security, and conduct SLLS.
6. To resume movement, the patrol will return to the track in file formation and move out.

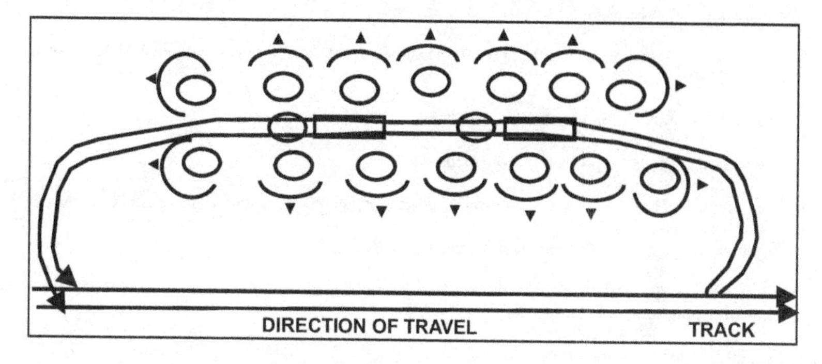

Figure 5-2. Fish Hook into a Break

5.4.5.10 Lay-up Points and Observation Posts. The selection of LUPs and OPs is critical if the unit is to maintain a clandestine profile. Planning should include map, air photography, and enemy capability studies to select potential LUP/OP areas. The location of the LUP or OP should take advantage of terrain features to allow for movement during daylight hours if necessary. Consider cover on opposite sides of terrain features like tree lines. Routes to each position must be carefully selected as unnecessary tracks create an opportunity for detection. Details of LUPs and OPs are presented in section 2.4.10. The following sections outline techniques that apply when establishing LUPs or OPs in cold weather environments. Figure 5-3 is a typical LUP site.

1. Insertion and exfiltration routes. The approach and withdrawal routes from the LUP and OP site should follow the natural cover provided by terrain features and vegetation.

2. Avoid avalanche slopes. Lee slopes have more snow accumulation and are excellent for digging in and camouflage, but are possible avalanche areas.

3. Hillsides. Low areas should be avoided, cold air descends and enemy units are likely to follow lower routes. Do not locate a LUP or OP on top of a ridge to avoid being silhouetted.

4. Move into position during hours of low light.

5. Vegetation. Dense trees can provide overhead and oblique cover.

6. When moving off high features, use level ground.

7. Use local guide and/or information where acceptable or possible.

8. When constructing tents, consider the following:

   a. Duration of stay.

   b. Number of tents.

   c. Establish defensive positions.

   d. Establish cover from view and fire. Overwhite suits, pack covers, and camouflage nets help conceal dark colored tents.

   e. Establish communication trenches between tents.

   f. Use natural features for protection from elements.

   g. Dig in tents and build a snow wall to provide protection against wind and improve concealment.

   h. Ensure there are exits in the wall at either end of the tent.

   i. Allow enough space between the wall and tent for movement inside the wall. This will prevent a heavy snowfall from pressing in on the sides.

   j. Prepare fire positions around the wall.

   k. Create space between tents to prevent a heavy snowfall from crushing them.

   l. Use overhead cover for movement underneath and around the perimeter of the tent.

   m. Dig tents into the sides of large drifts.

   n. Use smooth lines to camouflage.

   o. Do not allow unnecessary movement outside the perimeter.

   p. Create a large enough gap between the tent and camouflage net. Heat will melt the natural snow cover.

9. Open areas. If LUP or OP must be in an open area, look for natural depressions or small knobs and dig in. The site should blend in with surroundings. Areas needing the least modification and disruption of natural features are the best locations. Tracks to open area LUPs should zigzag back and forth to delay the enemy tracker and facilitate observation of approaching enemy by the patrol.

10. Natural defense. The best form of defense is proper camouflage and concealment. Early warning is important to give an opportunity to escape or prepare hasty ambushes.

11. Lines of drift. Bivouac up and as far away as possible.

12. Water. A water source is vital while in bivouac. Cooking and eating in a winter environment require a substantial amount of water. Snow can be melted, but it is easier to obtain water from a lake or stream even if it means breaking through the ice. All water purification standards still apply in winter.

5.4.5.10.1 Lay-up Point Routine. Before reaching the final position, loop back on the track, and prepare to ambush it. Watch and listen for approximately ten minutes, then reconnoiter the final approach and position. During the move into the LUP, a sentry is left to cover the approach route. The rest of the unit will establish the position and check the immediate area. The following LUP drills are recommended:

1. Lay deception trails and camouflage the trail. Fish hook into the area. Set out Claymores if carried.

2. Indicate the bolt route. This should be downhill into cover. Plan rendezvous points.

3. Tents should be dug in and camouflaged with 360 degree security and designated fields of fire.

4. Keep the majority of equipment packed.

5. Set up the radio antenna in a concealed position.

6. Set out skis and weapons for immediate use.

7. Dig out the latrine.

8. Store fuel outside tent.

9. Establish a sentry routine.

10. Rig communication lines.

11. Pack all equipment before sleeping.

12. Eat, and schedule sentries for warming and eating periods.

13. Check boots, socks, and feet.

14. Rest and conduct observation.

15. At dawn, stand to.

16. Check camouflage.

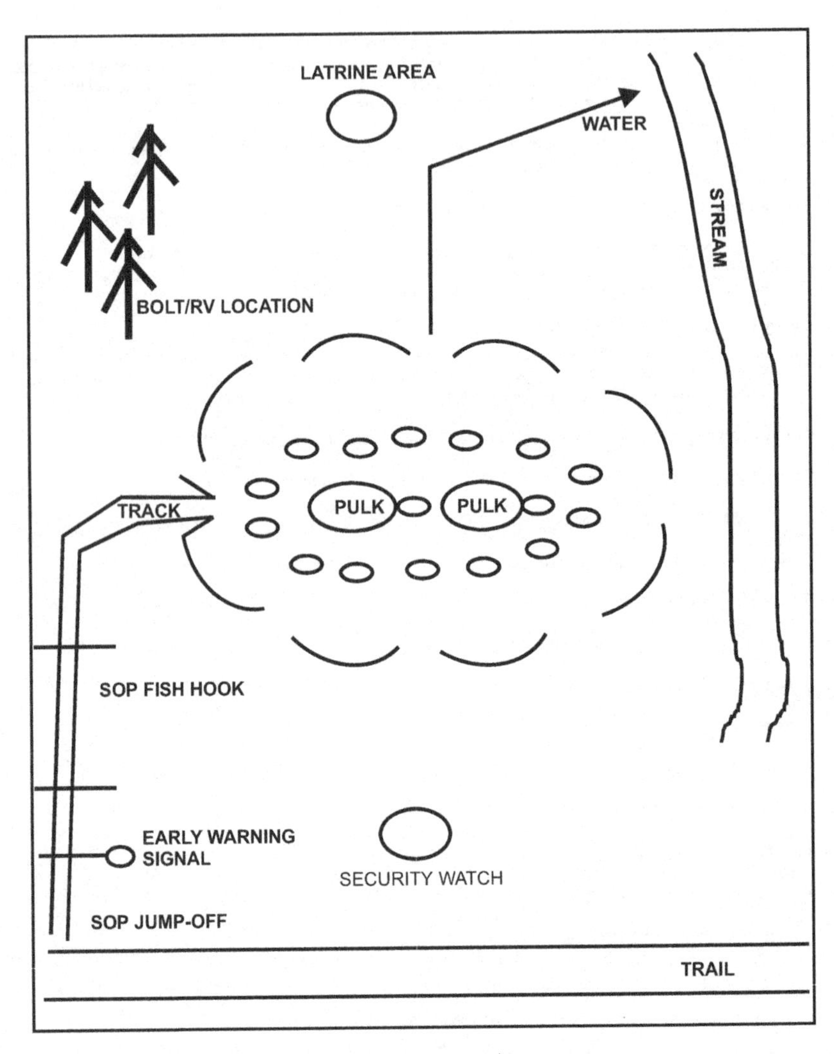

Figure 5-3. Lay-up Point

17. Conduct communications as required.

18. During the day rest, eat, and prepare hot drinks for those on watch.

19. Brief plans, progress, routines, track deception, etc.

20. Sanitize area before leaving.

5.4.5.10.2 Observation Post Routine. The following points should be considered when establishing an OP:

1. Prepare the OP position.

2. Conduct observation.

3. Commence OP routine.

4. First and second line equipment is worn while in OP.

5. Keep weapons and equipment packed and close at hand at all times.

6. Maintain a watch log and know watch rotation times.

7. Know the route to and from OP to LUP.

8. Follow radio schedules.

9. Use correct camouflage at all times; stay at OP until relieved.

10. Boots must be worn at night. Unlace them to allow ventilation and circulation.

11. No cooking or sleeping.

12. Eyes on target and behind.

13. The enemy will gain intelligence from the amount and type of waste left behind. Everything that goes into the field must come out. In some tactical situations, this may include fecal matter.

5.4.5.11 Jump-off Standard Operating Procedures. A jump-off is a location on the trail where a patrol can change direction of travel and not be noticed. See Figure 5-4. These are tentatively identified during route planning. Jump-offs are best done at a diversion point like a downhill slope or other distinctive terrain feature where it will be easy to conceal the change of direction. Track discipline is crucial. Jump-off considerations are as follows:

1. If skins are to be used they should be put on at a rest stop prior to entering the diversion area.

2. Diversion area considerations:

   a. Down hill slope. Speed on a downhill slope may cause an enemy tracker to miss the jump-off location.

   b. Snow free spots. These areas won't show tracks if the ground is frozen or rocky.

c. Natural lines like fences, walls, forest lines, or wild life trails conceal tracks.

d. Wind drift areas. Wind blown snow can wipe out a ski trail.

3. Common mistakes doing jump-offs

a. Breaking the track discipline. Increased snow disturbance and pole plants at the jump-off point will draw the attention of enemy trackers and make it harder for the jump-off team to cover the tracks.

b. Breaking vegetation and disruption of snow. Good trackers can determine direction of travel from disturbed snow and vegetation.

c. The new track direction is in plain view of the old track. Use the opposite side of a tree line, fence, or snowdrift to obscure the view of the new track from the old.

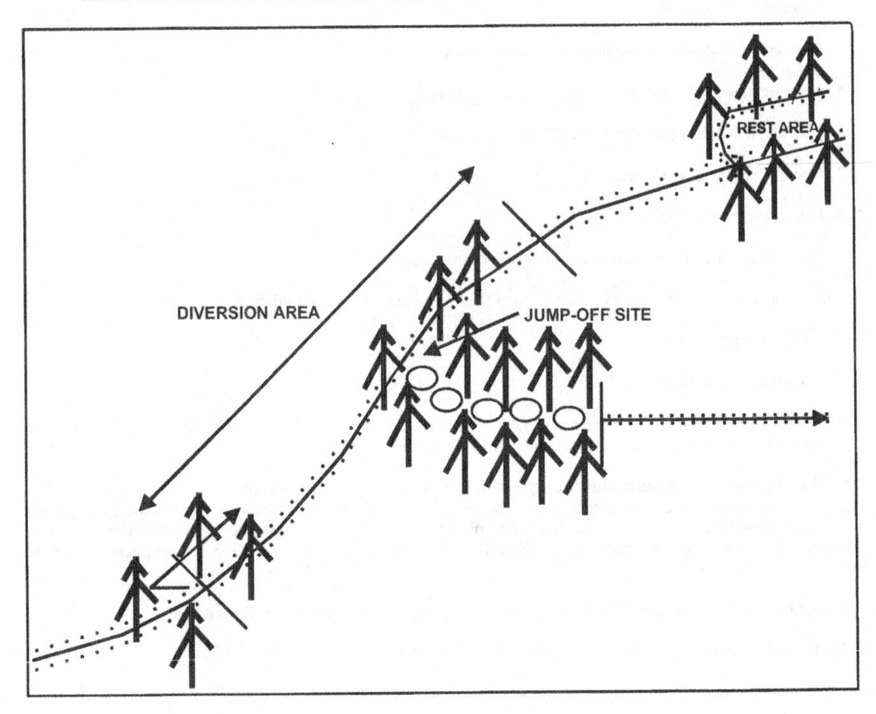

Figure 5-4. Jump-off Standard Operating Procedures

5.4.5.12 Deception Trail Standard Operating Procedures. A deception trail is used to buy time and deceive the enemy regarding target and patrol direction. See Figure 5-5. These are tentatively identified in the route planning for the mission. Again, track and trail discipline are crucial. Considerations are as follows:

1. Route selection. Select a route in a treeline for overhead cover.

2. The deception trail should continue for 20 to 45 minutes beyond the jump-off point.

3. The deception trail team needs be practice track deception by increasing pole plants and making false trails to further delay the enemy.

4. False camp, rest stop. Creating a false camp can further confuse and delay the enemy.

5. Direction changes and trail jump-on/off points must be sharp and clean.

6. Be careful not to disturb snow or vegetation on return along deception trail.

7. Common mistakes making deception trails

    a. The track is too small and not long enough (20-45 minutes). The deception trail needs to be approximately 1,000 meters long and the turn around point out of view to enemy trackers.

    b. The track breaks with earlier track discipline, and tactical route choices

    c. Fewer pole marks and shallower trail depths.

    d. Twice as many pole marks at the jump-off point and end of deception trail.

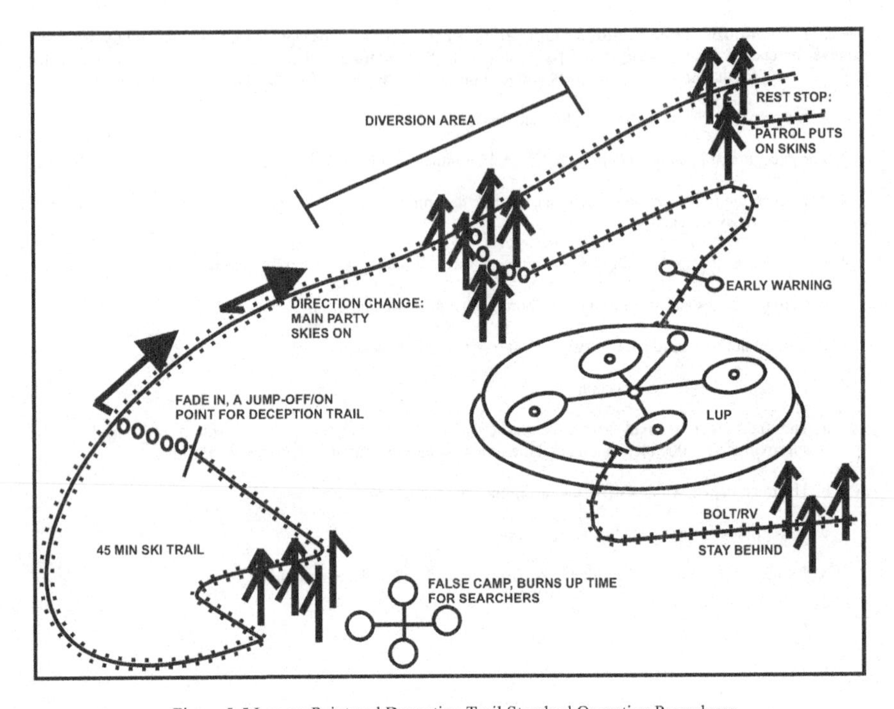

Figure 5-5 Lay-up Point and Deception Trail Standard Operating Procedures

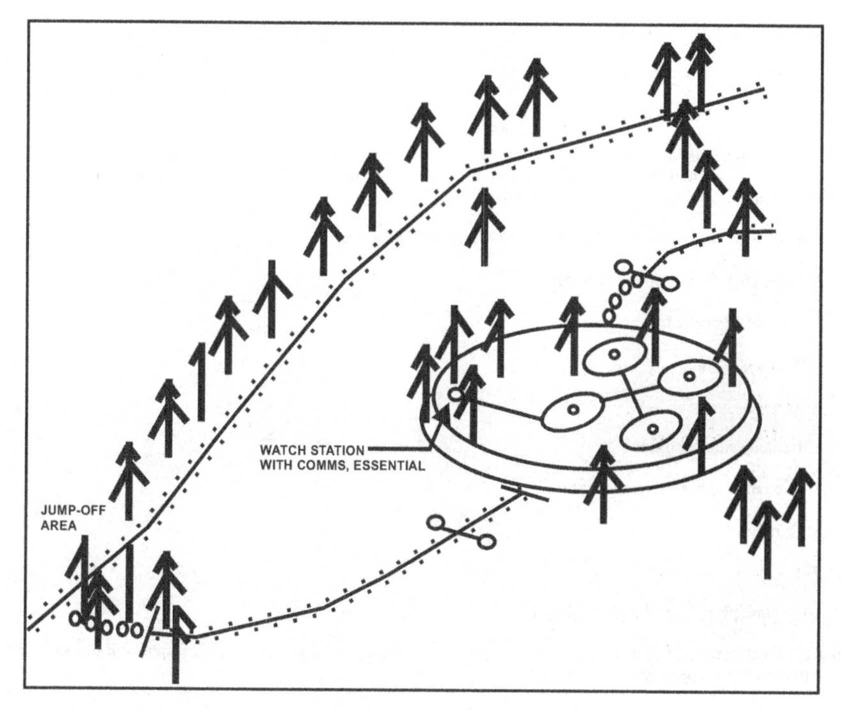

Figure 5-6. Fish Hook into Lay-up Point

5.4.5.13 Fish Hook into a Lay-up Point. See Figure 5-6. Considerations are as follows:

1. Plan LUP site. Potential sites will be identified in the mission planning.
2. Ski in plain view of site. This will enable the patrol to get an overall view of the site, and once set up, observe the trail for enemy trackers.
3. Plan on a 20 to 45 minute fish hook route into the LUP.
4. Practice good jump-off/on point, deception trail, and rest stop SOPs
5. Identify bolt route and rendezvous points.
6. Common mistakes doing fish hooks

a. The fish hook is too short.

b. The track is to obvious leading to the LUP.

c. The track doesn't lead into a clear area past the LUP site

5.4.6 Diving Operations. The environment and water temperatures will play a major role during diving operations conducted in the arctic. Cold-water training and rehearsals will improve efficiency and establish proper SOPs. Other factors that will affect the mission include:

1. Time of year (Summer will be a time of 24-hour daylight.)
2. Snow conditions (depth, dry/wet)
3. Equipment needed
4. Air temperature
5. Terrain
6. Ice and water conditions
7. Type of insertion (staging area)
8. Duration
9. Recovery procedures
10. Time (things take longer in the cold).

5.4.6.1 Unit organization. During diving operations, the unit must establish specific responsibilities to ensure that the mission is conducted efficiently.

1. Main dive team. This team will conduct the dive phase of the mission.
2. Back-up team will provide the following:

   a. Equipment carrying

   b. Route finding

   c. Security for all stages of the mission

   d. Preparation of swimmers

   e. De-servicing of swimmers' equipment

   f. Caching of equipment

   g. Signals, if any.

5.4.6.2 Stages of Operation. The following tasks will facilitate a coordinated effort by mountain and arctic personnel during the conduct of diving operations.

1. Perform all SOPs enroute to the LUP/OP.
2. Arrive at the final LUP/OP at least 24 hours prior to the dive to allow for rest if tactical situation permits.
3. If possible, observe target the night and day prior to the dive.
4. If possible, service and check all equipment during daylight.
5. Establish a forward base area within 1,500 yards of the target. This might be a tent to allow for final preparation and minimal transit distance to the water entry point.
6. Once the swim teams are rigged, the back-up team will provide security at all times, and break the trail enroute to the diver insertion point.
7. Back-up personnel will assist the swim team into the water.
8. If the water exit rendezvous point is different from the entry point, a back-up team must move to it, secure it, and receive the swim team.
9. Back-up personnel will assist the swim team from the water, undress them, and assist them in donning extra clothing. Weather conditions and area security will determine whether teams undress at the water exit point or move back to the staging area.
10. After de-servicing the dive teams, move back to the LUP/OP area. Final packing is done in this area before moving out.
11. Cache equipment.
12. All stores and equipment are to be packed and ready to move at all times. Divers will quickly become cold and incapacitated.

5.4.7 Arctic Parachute Operations. Wind and precipitation will be the primary limiting factors that affect parachute operations in the arctic environment. The following points should be considered.

1. Frozen lakes and rivers may be the best DZs.
2. Parachuting with skis or snowshoes requires special rigging and practice.
3. SOPs will apply during water jumps. Proper clothing and equipment will be necessary to protect the individual from freezing temperatures and wind chill during surface transits.
4. Land jumps will require the following considerations:

    a. Backward landings in deep snow will decrease the chance of a broken leg.

    b. DZ procedures should be carried out quickly to avoid compromise.

c. Equipment must be accessible and easily removed from the DZ.

d. The base of a load may be designed similar to a sled. This will facilitate the rapid movement of equipment from the DZ.

e. Parachutes must be cached.

5.4.8 Demolition. NSW personnel use demolitions in support of numerous missions. Demolition charges may be used during arctic operations to deny the enemy avenues of approach (lakes and rivers), protect defensive positions, and ambush troops and vehicles. Ice breaching techniques may require special equipment to assist charge emplacement. The following points should be considered when using demolitions in an arctic environment:

1. At low temperatures explosives are difficult to handle. Plastic explosives become brittle and flaky. Prepare in warm surroundings if molding is required.

2. Detonating cord becomes brittle and may not explode along its length if kinks are not removed. Tails at joints and ends should be increased by one-half.

3. Certain detonators (blasting caps, clocks) may crack if subjected to sharp temperature changes. Check prior to use.

4. Batteries will lose their charge if not kept warm.

5. Snow lessens the effect of shrapnel.

6. Flying ice is effective against personnel.

7. Equipment will be lost if put down in snow and not marked.

8. Waterproof everything left in the snow.

9. Chemicals work more slowly in the cold. Test chemical time delays (time pencils), allowing extra time, as the temperature will constantly be changing.

10. Remember all rules and regulations governing explosives. These will be even more critical under extreme conditions.

11. Always allow plenty of time for preparation because gloves will reduce dexterity.

12. Layers of snow tend to move, so light charges may slide unless properly secured.

5.4.8.1 Ice Breaching. This capability will deny the enemy an ideal crossing point for troops and vehicles. Additionally, equipment can be cached or disposed of through holes created by appropriate demolition placement. Windblown snow will help to conceal the hole until it freezes over. The location of the cache must be remembered if equipment is to be retrieved.

1. If the water depth is over eight feet, the charges should be suspended four feet under the ice.

2. If the water is less than eight feet deep, the charges are suspended halfway between the ice and the bottom.

3. Charges can be suspended from a stake that bridges the hole.

4. Where currents are a factor, the charges should be weighted to keep them stationary at the proper depth.

5.4.8.2 Ice Ambushes. There may be a need to blow ice without marking the snow on top of it. Employing the following techniques can do this:

1. Near the bank, remove a section of ice 1.5 x 1-meter in size.

2. Cut a number of saplings up to 10 meters in length and trim the branches.

3. Attach explosives to the saplings and connect the detonating cord.

4. Slide saplings through the hole in the ice with the explosives attached. Lash the end of one sapling to the front of the next until the length of ice that is to be blown is covered. The charges are kept afloat and flush against the ice by the buoyancy of the saplings.

5. Blow the charge from the bank as required.

APPENDIX A

Basic Field Craft

A.1 INTRODUCTION

There are various field skills that, when properly employed, complement tactics. This appendix provides some of the more basic aspects of SEAL field craft that will be useful in conducting operations. These include:

1. Camouflage
2. Movement
3. Observation techniques
4. Tracking and counter-tracking techniques
5. Caching
6. Munitions placement.

A.2 CAMOUFLAGE

Camouflage is the measures that you use to conceal yourself, your team, weapons, equipment, and positions. Camouflage uses the environment and other natural and manmade materials. No matter how well covered a position is, if it is not well concealed, it can be seen and either avoided or attacked.

1. Take advantage of all available natural concealment. Leaves and bushes can hide your position.
2. Camouflage against ground and air observation. Check or have your position checked from various angles or views to ensure you are well hidden.

A.2.1 Lay-up Points and Ambush Positions. Camouflage your position as you prepare it. Do the following:

1. Study the terrain and vegetation in the area. Arrange grass, leaves, brush, and other natural materials you use to conform to the area. For example, do not expect tree branches stuck into the ground in an open field to provide effective camouflage.
2. Use only as much material as needed. Excessive use of material (natural or artificial) can reveal your position.
3. Obtain your natural material over a wide area. Do not strip an adjacent area of all foliage because such bared areas could attract enemy attention.

4. Dispose of excess soil by covering it with leaves and grass or by dumping it under bushes, into streams, or into ravines. Piles of fresh dirt indicate an area is occupied and reduce effectiveness of camouflage.

5. After your camouflage is completed, inspect it from the enemy's viewpoint. Check often to ensure it remains natural looking and actually conceals your position. If it does not look natural, rearrange or replace it.

6. Practice camouflage discipline. The best camouflage fails to provide concealment if tracks lead to your position or if MRE bags and trash are scattered about. Every man in your unit shares this responsibility.

A.2.2 Equipment. Equipment must blend with the natural background. Remember, vegetation changes color with the season.

1. Your LBE, belt, canteen cover, etc. blends well with most terrain unless it is badly faded. If faded, color it to blend with the surrounding terrain. If no paint is available, use mud, charcoal, or crushed grass. Remember to paint in bold, irregular patterns.

2. Headgear should alter the natural contour of the head. A floppy hat is best for this purpose. Scarves are generally inappropriate as head wear because they do not offer an irregular contour for the head. (Scarves are also inappropriate as headgear because the tied ends may flap in the breeze - such movement is easily detected.) If you must wear a helmet, alter the distinctive silhouette of your helmet with a cover made of cloth or burlap colored to blend with the terrain. Let foliage stick over the edges to alter the outline, but do not use too much. If material for covers is not available, disguise and dull the surface of your helmet with irregular patterns of paint or mud. Use a camouflage band, string, burlap strips, or rubber bands to hold the foliage in place.

3. Camouflage nets are useful as covers for packs, equipment, and personnel in lay-up, reconnaissance, or ambush positions.

4. Use mud, dirt, or cloth wrappings to dull the shiny surface of your weapon. Be careful, however, to keep the working parts clean and free so your weapon will function properly.

5. Wear overwhites and color your equipment white when operating against a background of unbroken, snow-covered terrain.

6. Color your equipment black or forest green when operating in the jungle. Flat black, rust, and gray are also a good color combination.

A.2.3 Uniforms. Fieldwork uniforms also blend well with most terrain unless they are badly faded. If your uniform is faded, camouflage it with mud, soot, paint, or dye.

A.2.4 Exposed Skin. Exposed skin reflects light and attracts the enemy's attention. Even very dark skin will reflect light because of its natural oil. Cover it or use camouflage paint.

1. Camouflage face paint sticks are issued in three standard two-tone sticks as follows:

   a. Loam and light green for light-skinned troops, in all but snow regions.

   b. Sand and light green for dark-skinned troops.

   c. Loam and white for all troops in snow-covered terrain.

2. Paint the shiny areas (forehead, cheekbones, nose, and chin) with a dark color. Paint the shadow area (around the eyes, under the nose, and under the chin) with light color. Paint the exposed skin on the back of your neck and hands. The issue-type face paint camouflage stick can be used to apply a two-color combination in irregular pattern. A hunter's three-tone camouflage pack or individual tubes of face paint are easier to use than the stick-type camouflage.

3. Burnt cork, charcoal, or lampblack can be used to tone down exposed areas of skin when issue-type face paint sticks are not available.

4. Mud may be used in an emergency. Remember that mud changes color as it dries and may peel off, leaving the skin exposed. Remember that mud may also contain harmful bacteria.

5. When applying camouflage, use the buddy system; work with another man and check each other.

A.3 MOVEMENT

A.3.1 Individual Stealthy Movement. Follow these general rules to move without being seen or heard by the enemy:

1. Prepare yourself and your equipment.

   a. Camouflage yourself and your equipment.

   b. If wearing ID tags, tape them together and to the chain so they cannot slide or rattle.

   c. Tape or pad any parts of your weapon or equipment that rattles or is so loose that it may snag. Be sure that the tape or padding does not interfere with operation of the weapon or equipment.

   d. Wear soft well-fitting clothes. Starched clothing swishes as you move and loose, baggy clothing may snag on brush. Use rigger tape on your thighs on field trousers to take up slack and the ankle ties when you do not blouse the trousers. Do not tie them too tightly. No other tie-downs should be used at any time. They may interfere with circulation in the legs and feet. In cold weather, poor circulation may lead to frostbite or other cold injury.

   e. Wear a soft cover with a less distinctive outline. (The helmet has a distinctive silhouette and also may muffle or distort sounds, especially if there is a slight breeze.)

   f. Do not carry unnecessary equipment. You cannot move rapidly when weighted down.

2. Move by bounds; that is, short distances at a time. Halt. Listen. Observe. Then move again.

3. Look for the next spot where you will stop before leaving the concealment of one position. Observe that area carefully for enemy activity. Select the best available covered and concealed route to the new location; take advantage of darkness, fog, smoke, or haze to assist in concealing your movement.

4. Change direction slightly from time to time when moving through tall grass. Moving in a straight line causes the grass to wave with an unnatural motion that attracts attention. The best time to move is when wind is blowing the grass.

5. If you alarm birds or animals, remain in position and observe briefly. Their flight or movement may attract the enemy's attention.

6. Take advantage of the distraction provided by noises.

7. Cross roads and trails where there is the most cover and concealment. Look for a large culvert, a low spot, or a curve. Cross quickly and quietly; do not run.

8. When moving over a plowed field with deep furrows, follow the furrows as much as possible. When you must cross the furrows, look for a low section in the field; crawl down a furrow to that section and make your cross-furrow movement.

9. Avoid steep slopes and areas with loose stones.

10. Avoid cleared areas to prevent being silhouetted.

A.3.2 The Crawl. There are times when you must move with your body close to the ground to avoid being seen. There are two ways to do this, the low crawl and the high crawl. Use the method best suited to the conditions of visibility, cover and concealment available, and speed required.

A.3.2.1 The Low Crawl. Use the low crawl when cover and concealment are scarce, when visibility permits good enemy observation, and when speed is not essential.

1. Keep your body as flat as possible against the ground. Grasp the rifle sling at the upper sling swivel. Let the balance of the rifle rest on your forearm and let the butt of the rifle drag on the ground. Keep the muzzle off the ground.

2. To start forward, push your arms forward and pull your right leg forward.

3. To move forward, pull with your arm and push with your right leg.

4. Change your pushing leg frequently to avoid fatigue.

A.3.2.2 The High Crawl. Use the high crawl when cover and concealment are available, when poor visibility reduces enemy observation, and when you require more speed.

1. Keep your body free of the ground and rest your weight on your forearms and lower legs. Cradle the rifle in your arms, keeping its muzzle off the ground. Keep your knees well behind your buttocks so it stays low.

2. Move forward by alternately advancing your right elbow and left knee, left elbow and right knee.

A.3.3 Stealthy Movement Tips. Use the methods explained below when extremely quiet movement is necessary, such as when you are very near the enemy. These movements must be made slowly, are very tiring, and require great patience to perform properly.

A.3.3.1 Walking. Hold your weapon at the ready (port arms). Make your footing sure and solid by keeping your weight on one foot as you step. Raise your other leg high enough to clear brush or grass. With your weight on the rear leg, gently let your foot down toe first. Feel with your toe to pick a good spot. Lower your heel after finding a solid place. Shift your weight and balance to your forward foot and continue. Take short steps to avoid losing your balance. At night, and when moving through dense vegetation, avoid making unnecessary noise by holding your weapon with one hand and extending the other hand forward, feeling for any obstruction as you move.

A.3.3.2Going into the Prone Position. Hold your weapon under one arm and crouch slowly. Feel for the ground with your free hand and make sure it is clear. Lower your knees, one at a time, until your weight is on your knee and free hand. Shift your weight to your free hand and opposite knee. Raise your free leg up and back and lower it gently to the ground, feeling with your toe for a clear spot. Roll gently to that side and move your other leg into position the same way. Roll quietly into the prone position.

A.3.3.3Crawling. The low crawl and high crawl are often not suitable when you are very near the enemy. They make a shuffling noise that is easily heard. Crawl on your hands and knees. Lay your weapon on the ground by your side. With your right hand, feel for or make a clear spot for your knee. Keep your hand on the spot and bring your right knee forward until it meets your hand. Clear a spot with your left hand and move your left knee up in the same way. Be sure your weapon is always within reach! To move your weapon, feel for a place, clear it, and lift the weapon into position. Crawl very slowly and keep your movements absolutely silent. A low profile can be maintained for short periods by extending the arms past the shoulders, but this is very tiring.

A.3.4 Actions under Flares. There are two general types of flares; ground flares that burst and burn in place, and overhead flares that burst and burn high in the air.

A.3.4.1General Considerations

1. If caught by a flare while crossing an obstacle such as barbed wire, crouch low and stay still until the flare burns out.

2. Continue the mission as soon as the flare burns out and night vision is regained.

3. If caught by a flare during an assault, continue the assault.

A.3.4.2Ground Flares. Ground flares are placed in the same manner as antipersonnel mines, so you will set them off or the enemy can set them off by pulling wires. The areas in which they are placed are often under enemy observation and fire. A ground flare set off nearby usually means the enemy has seen you or suspects your presence.

1. If caught in the light of a ground flare, move quickly out of the lighted area.

2. Keep moving until you are well away from the area, reorient yourself, and continue the mission.

3. If caught by a flare during an assault, continue the assault.

A.3.4.3Overhead Flares. Overhead flares may be placed and set off in the same manner as ground flares. They may be fired from rifles, mortars, artillery, hand projectors, or may be dropped from aircraft. All are placed high in the air before bursting and burning.

1. If you set off an overhead flare or hear one fired, get down while it is rising and conceal yourself before it bursts.

2. If caught in the light of the burst where you blend well with your background, freeze in place until the flare burns out.

3. If among trees, step quickly behind one.

4. If caught in the open, crouch low or hit the ground. The burst of light is temporarily blinding and may prevent your being seen.

A.4 OBSERVATION TECHNIQUES

Observe with all your senses. On patrol, notice every sign of movement, marks on paths, and footprints or disturbed areas in the sand. Every few minutes plan to stop, look, listen, and smell. All of your senses' capabilities are enhanced at night.

A.4.1 Smell. Smells can help or endanger you. Odors from gasoline, cooking food, or burning tobacco can warn you of the enemy's presence or alert him to your presence. Patrol members must be free of all competing odors like cigarettes, gum, after shave, perfumed soaps, deodorant, perfumed insect repellant, and such. Odors from these may reveal your presence to the enemy. Patrol members should be able to distinguish indigenous smells like tobacco, cooking food, or smoke. This can help the patrol to find the enemy or avoid contact as appropriate.

A.4.1.1 Diet. Different diets produce different smells that can also affect the patrols detectability. Some recommendations are:

1. Eat foods indigenous to AO if considered safe.

2. Do not smoke or dip smokeless tobacco on patrol.

A.4.1.2 Smelling Techniques. A simple method of helping to detect smells is to:

1. Face into the wind at a 45-degree angle.

2. Relax and breath normally, taking quick sniffs.

3. Concentrate on specific odors. For example, cigarette smoke can be smelled up to 150 meters away.

A.4.2 Vision. You can see much more in the dark than you may realize. However, to take maximum advantage of this ability you must understand how to see best under conditions of poor visibility.

A.4.2.1 Searching Terrain. Visually search terrain in two steps.

1. Make a quick overall search for obvious targets and unnatural colors, outlines, or movements.

    a. Look from ground level for a better view in order to silhouette personnel or equipment.

    b. Look straight down the center of the area you are observing, starting just in front of your position.

    c. Raise your eyes quickly to the maximum distance you wish to observe.

    d. If the area is wide, further subdivide it.

2. Next, cover all areas as follows, first searching the ground nearest you:

a. Search a strip 50 meters deep, looking from right to left parallel to your front.

b. Search from left to right over a second strip farther out, but overlapping the first strip.

c. Continue in this manner until you have studied the entire area.

d. When you see a suspicious spot, stop and search it thoroughly.

A.4.2.2 Night Vision. Night vision is different than day vision because the eye uses the rods to differentiate between black and white. The cones, located in the middle of the eye, recognize colors and are used during the day. Rods cannot differentiate color and are easily blinded when exposed to light. This creates a central blind spot causing larger and larger objects to be missed as distances increase.

1. Protect night vision by doing the following:

   a. Avoid exposure to bright direct or reflected sunlight.

   b. Wear medically approved sunglasses with quality lenses.

   c. Close one eye prior to exposure to white light.

   d. Cover one eye with your hand or an eye patch.

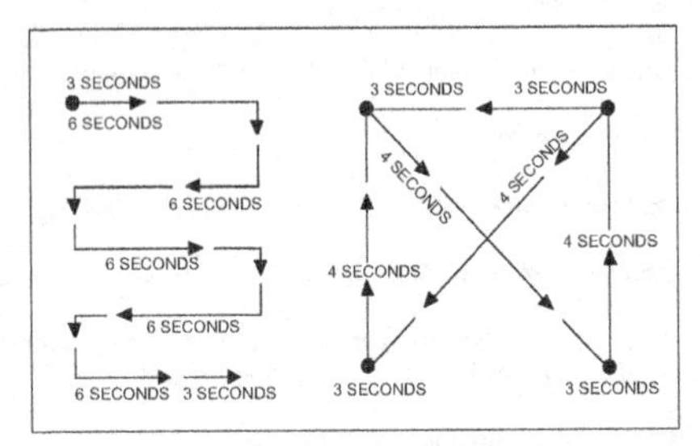

Figure A-1. Scanning

2. Scanning. Dark adaptation or "night vision" is only the first step toward maximizing the ability to see at night. Night vision scanning can enable you to overcome many of the physiological limitations of your eyes and can reduce confusing visual illusions. The technique involves scanning from right to left or left to right using a slow, regular scanning movement. Although both day and night searches employ scanning movements, at night it is essential to avoid looking directly at a faintly visible object when trying to confirm its presence. See Figure A-1.

3. Off-center vision. Viewing an object using central vision during daylight poses no limitation, but this technique is ineffective at night. This is due to the night blind spot existing during periods of low illumination. To compensate for this limitation, use off-center vision. This technique requires that an object

be viewed by looking 10 degrees above, below, or to either side of it, rather than directly at the object. This allows the peripheral vision to maintain contact with an object. See Figure A-2.

4. Bleach-out effect. Even when off-center viewing is practiced, the image of an object viewed longer than two to three seconds tends to bleach out and become one solid tone. As a result, the object is no longer visible and can produce a potentially unsafe operating condition. To overcome this limitation, you must be aware of the phenomenon and avoid looking at an object longer than two to three seconds. By shifting your eyes from one off-center point to another, you can continue to pick up the object in your peripheral field of vision.

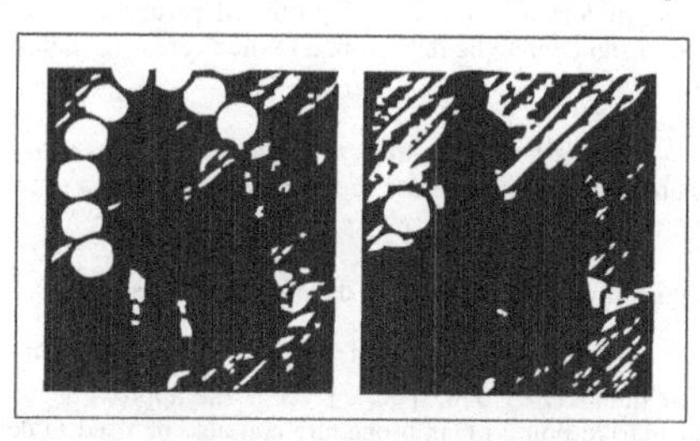

Figure A-2. Off-Center Vision

5. Shape or silhouette. Visual sharpness is significantly reduced at night; consequently, objects must be identified by their shape or silhouette. Familiarity with architectural design of structures common to the AO will determine one's success using this technique. For example, the silhouette of a building with a high roof and a steeple can be recognized in the U.S. as a church, while churches in other parts of the world may have entirely different architecture.

6. Light Sources. Some average distances that typical light sources can be seen at night with the naked eye are listed below:

a. Vehicle headlights	4 to 8 km
b. Muzzle flashes from small arms	1.5 to 2 km
c. Bonfire	6 to 8 km
d. Flashlight	up to 1.5 or 2 km
e. Lighted match	up to 1.5 km
f. Lighted cigarette	0.5 to 0.8 km.

A.4.3 Sound. Learn to identify, evaluate, and react to common battlefield noises and the common sounds in your AO. All sounds, the snap of a twig, the click of a bolt, the bark of a dog, the calls of wild and domestic animals and fowls, are information that may be valuable to you. You can learn a lot by listening.

1. Train yourself to be patient; it may be necessary to listen in complete silence for long periods.

2. If you are wearing a hood or ear flaps, remove them temporarily to listen.

3. If any unusual sound is noted the patrol must "freeze", then slowly ease down out of sight. Sudden movements catch the eye and may give away the position.

A.4.4 Estimating Distances. There are several methods that an observer can use to determine distance while on patrol.

A.4.4.1 Laser. Laser range finders are one of the preferred means of determining the observer-to-target distance. When a laser is used, distance may be determined to the nearest 10 meters. Be aware that laser detection devices may be employed on target.

A.4.4.2 Flash-to-bang. When it is necessary to verify distance, the flash-to-bang technique is helpful when the impact of ordnance is visible or lightning can be seen. Sound travels at a speed of approximately 350 meters per second. Use the following equation:

(Elapsed time between impact and sound) X (350) = distance in meters.

Multiply the number of seconds between round impact or lightning strike (flash) and when the sound reaches the observer (bang or thunder) by 350. The answer is the approximate number of meters between the observer and the round impact/strike point. (This procedure can also be used to determine the distance to enemy weapon muzzle flashes.)

A.4.4.3 Estimation. In the absence of a more accurate method of determining distance to a target, the observer must estimate distance. The degree of accuracy in this method depends on several factors, such as terrain relief, time available, and the experience of the observer. Generally, the longer the observer remains stationary, the better he can use this technique. Some methods of estimating distance are discussed below.

1. Mental estimation can be made by using a known unit of measure. Distance is estimated to the nearest 100 meters by determining the number of known number units of measure, such as a football field (100 yards), between the observer's position and a target. For longer distances, the observer may have to progressively estimate distance. To do this, he determines the number of units of measure (for example, 100 yards) to an intermediate point and doubles the value.

2. The observer should consider the following effects of estimating distance:

   a. Object appears nearer when:

      (1) In bright light.

      (2) In clear air at higher altitude

      (3) The observer is looking down from a height

      (4) The observer is looking over a depression, most of which is hidden.

   b. Object appears more distant when:

      (1) In poor light or in fog.

(2) Only a part or portion of the object can be seen.

(3) The observer is looking over a depression, most of which is visible.

(4) The background is similar in color to that of the object.

(5) Observing from a kneeling or sitting position on hot days, especially when the ground is moist.

3. When visibility is good, distances can be estimated by using the appearance of tree trunks, their branches, and foliage (using the naked eye) in comparison with map data. See the table in Figure A-3.

DISTANCE	TREE DESCRIPTION
1,000 meters	Trunk and main branches are visible. Foliage appears in cluster-like shape Daylight may be seen through foliage.
2,000 meters	Trunk is visible, main branches are distinguishable, foliage appears as smooth surface. Outline of foliage of separate trees is distinguishable.
3,000 meters	Lower half of trunk is visible. Branches blend with foliage. Foliage blends with adjoining trees.
4,000 meters	Trunk and branches blend with foliage. Foliage appears as a continuous cluster. Motion caused by wind cannot be detected.
5,000 meters and beyond	Whole area covered by trees; appears smooth and dark.

Figure A-3. Estimation by Appearance of Trees

4. Distance can be estimated by using known dimensions of vehicles and the mil relation formula (W=R x mils). By applying the width of a vehicle appearing perpendicular to an observer as the lateral distance (W) and measuring the width in mils, the distance can be determined by solving the formula for range (R) in thousands, or R = W/mils. This data, when compared with map data, will help an observer estimate distance. The dimensions of selected equipment are shown in Figure A-4.

A.4.5 Determining Direction. Determining direction is an essential skill for SEAL operators. Direction is an integral part of terrain-map association, adjustment of fire, and target location. There are four methods that can be used to determine direction.

A.4.5.1 Estimating. With a thorough terrain-map analysis of his zone of operation, the patrol member can estimate direction on the ground. As a minimum, he should be able to visualize the eight cardinal directions (N, NE, E, SE, S, SW, W, and NW). Because of the inaccuracy of this method, it is the least preferred method of determining direction.

A.4.5.2 Using a Compass. Using a lensatic compass, the observer can measure direction to an accuracy of 10 mils. Care must be taken using a compass around radios or large metal objects.

EQUIPMENT	DIMENSIONS IN METERS	
	SIDE VIEW	FRONT VIEW
Tank T-62 T-72	 6.6 6.9	 3.3 3.6
Reconnaissance Vehicle BRDM-2 BTR-60	 5.7 7.2	 2.4 2.8
Armored Personnel Carrier BMP	 6.8	 2.9
Air Defense Weapon ZSU 23-4	 6.5	 3.0
EXAMPLE An observer sees an armored personnel carrier (BMP). He measures its width (as seen from the side view) as mils. Using the formula, he determines that the distance is as follows: R = (W/m)(1000); m = 2 mils; W = 6.8 meters; R = (6.8/2)(1000) = 3,400 meters.		

Figure A-4. Equipment Dimensions

A.4.5.3 Measuring from a Reference Point. Using a reference point with a known direction, you can measure horizontal angular deviations and apply them to the reference direction. Angular deviations may be measured with binoculars or with the hand. In measuring with binoculars, angular deviation is determined to the nearest 1 mil.

1. The horizontal scale of the binocular reticle pattern is divided into increments of mils.

2. The vertical scale on the right of the lens is not used to determine data for target location.

3. The vertical scales on the left and in the center of the lens are divided into increments of 5 mils and are used in height-of-burst adjustments.

4. The vertical scale in the center of the lens is divided into increments of 10 mils and is used for height of burst adjustments.

5. The horizontal and vertical scales on the laser ranger finder reticle are divided into increments of mils.

A.4.5.4Hand Measurement. When it is necessary to measure angular deviations, the observer may use hand and fingers as a measuring device. See Figure A-5 for the general hand measurements of angular deviation.

1. Since finger width and hand size varies for each person, each observer should calibrate his hand and fingers to determine the values of the angles for the various combinations of fingers and hand positions shown.

2. When using your hand or fingers in measuring angular deviation, fully extend your arm (locking the elbow) so that your hand and fingers are always the same distance from the eyes. The palm of the hand is always pointed toward the target area.

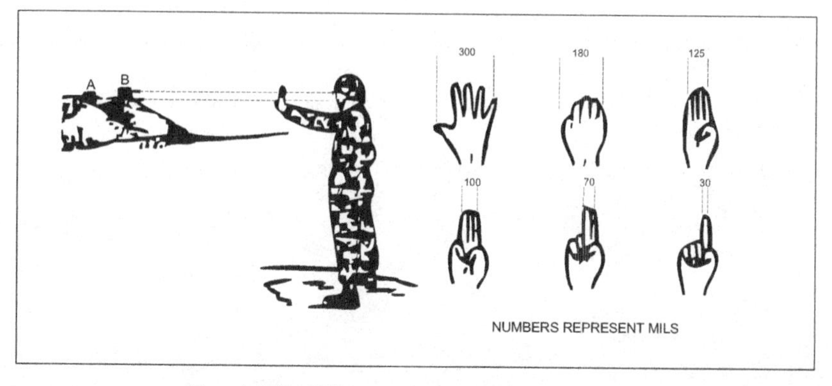

Figure A-5. Hand Measurement of Angular Deviation

A.5 TRACKING AND COUNTER-TRACKING TECHNIQUES

There are good reasons for the SEAL patrol to learn tracking and stealth:

1. Finding signs (tracking). The success of a mission may depend on the ability of the SEAL patrol to recognize and interpret trail signs left by the enemy and thus gain an advantage without being detected.

2. Not leaving signs. A SEAL patrol must constantly be alert to avoid leaving signs of their own activities that may forfeit surprise and give an advantage to the enemy.

A.5.1 Training to Track. Tracking requires careful study and practice. There will likely be one or more patrol members who can develop their natural talents and become good trackers if enough time is invested. Some pointers are:

1. An abundance of sign is essential for the beginning tracker.

2. Beginning tracking is best taught immediately after a fresh fall of snow or after a hard rain when fresh tracks show plainly and old trail signs have been covered or washed away.

3. Vary the time between making the trail and following it; first by a few hours, later by a day or two or longer, so the trackers learn something of the time element involved in weathering.

4. Note which men become most adept at reading trail sign so they can serve as the patrol's PT during training and combat operations.

A.5.2 Concepts of Tracking. Even a rudimentary knowledge of these basic concepts will be helpful to any SEAL patrol. These will not make anyone an expert tracker, but they may help to pick up a trail or locate the enemy. Perhaps just as important, these concepts will help you to move with more stealth and apply counter-tracking principles throughout the patrol.

A.5.2.1 Displacement. Displacement takes place when anything is moved from its original position. These are some of the more common types of displacement:

1. Footprints displace the soil by compression. Some of the more basic information that you may find in footprints includes:

    a. The type of load they are carrying: heavy, moderate, light, none

    b. Gender

    c. Direction of travel

    d. Number of personnel

    e. Condition of footgear

    f. Rate of movement

    g. Knowledge of being followed.

2. Sticks or stones moist underside, attached soil, or darker underside.

3. Grass or other vegetation - bent in direction of movement - tree leaves are lighter color underneath.

4. Movement of brush or treetops.

5. Wild animals or birds flushed.

6. Cleared trails.

7. Covered ant holes.

8. Spider web (broken).

A.5.2.2 Staining. Staining occurs when any substance from one organism or article is smeared or deposited on something else.

1. Mud staining the water in a stream indicates movement.

2. Muddy water in footprints may indicate recent movement. Normally the mud will clear in approximately 1 hour (will vary with terrain).

3. Roots, stones, and vines may be stained where leaves or berries are crushed on them by moving feet.

A.5.2.3 Weathering. The effect of weather on a particular indicator will indicate the age of that indicator.

1. Fresh blood is bright red, old blood has dark brown crust.

2. Sap from trees will harden after a period of time.

3. Footprints - initially, the moisture in the ground holds the edges of the print together. As it gets older and dries the edges will crumble into the print.

4. Light rain will round edges of footprints. Knowledge of the time of the last rainfall and condition of prints will provide vital age information.

5. Wind:

   a. Litter in tracks - age is determined by time of windstorm.

   b. Carries sound and odors.

6. Broken limbs/twigs - drying time indicates the age of the break.

A.5.2.4 Litter. Litter indicates movement; weathering of litter can help determine age (e.g., rust on ration cans, paper or pulp litter). Litter includes:

1. Gum or candy wrappers

2. Ration cans or wrappers

3. Cigarette butts

4. Remains of fire

5. Human feces, etc.

A.5.3 Stealth. Stealth implies the use of artful actions in order to pass through an enemy's territory without detection. While patrolling there is often the need to avoid any behavior likely to be sensed by the enemy. To be seen, heard, or smelled usually means giving up the element of surprise against the enemy and being surprised instead. Stealthy movement is usually most difficult during darkness, but unlimited visibility increases the probability of visual detection by the enemy. The aims must always be to:

1. Move as silently as possible commensurate with the speed required.

2. Use cover, concealment, and camouflage to avoid visual detection.

3. Avoid smoking, fire building, cooking, and the use of scents on the individual, which might provide the enemy with an odor with which to detect your activity.

4. Leave as few trail signs as possible.

A.5.3.1 Stealth Training. To master stealthy movement:

1. Practice initially during unlimited visibility. This will permit men to identify the objects that cause noise during movement and to develop techniques to minimize such noises.

2. Short patrols in relatively safe areas of a new environment will permit the necessary learning if the leader focuses the attention of his men and emphasizes the need to learn quickly. Men in a new environment are at a disadvantage against an enemy who is native to the AO.

A.5.4 Counter-tracking Tips. Counter-tracking is using the knowledge and techniques of tracking to avoid being tracked or lessen the effectiveness of the enemy's ability to track your movements. Techniques are included below. The overriding principle of counter-tracking is to never set a pattern. If one technique does not work, change to another.

A.5.4.1 Skip Method. This is a very effective method of counter-tracking that requires practice to employ. The patrol will stop in place and on command will move left or right of present route. Each member moves to the designated flank for a distance of 20 or 30 meters, then resumes the patrol's original route of march. It is very important that everyone move as carefully as possible so as not to make a trail or leave any telltale signs of the move. The PL should send the PT ahead to make a false trail for 30 to 50 meters before executing the skip method. This method takes practice to perfect and patrol members must be careful not to leave any signs as they move. See Figure A-6

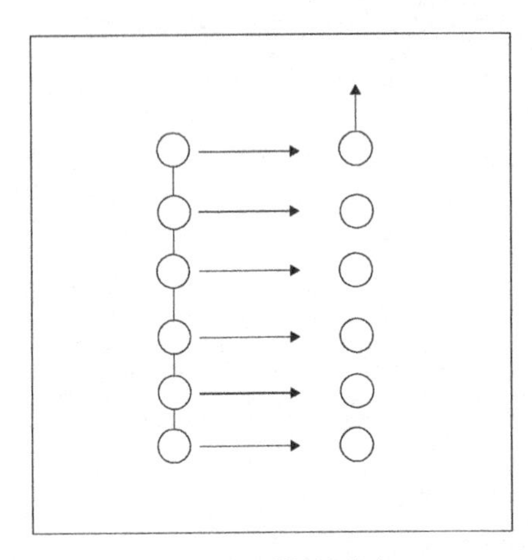

Figure A-6. Skip Method

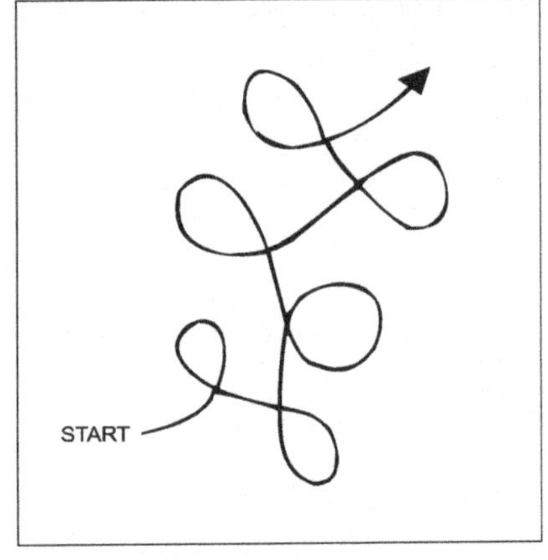

Figure A-7. Figure Eight Method

A.5.4.2 Figure Eight. The figure eight method is very similar to the box technique. The basic concept is the same, in that the patrol circles back on or across its own trail while maintaining the overall direction of movement. This concept is shown in Figure A-7 below. This technique is more difficult to execute accurately than boxing because you must maintain a constant pace while also maintaining a constant incremental course change until you return to your base course.

A.5.4.3 Angle Technique. This is another effective method for evading effective enemy tracking attempts. The patrol will change the direction of movement from the present patrol route of march in a series of angular course changes. For example, the patrol will make an angular course change of 30, 45, or 70 degrees, for a fixed distance of a 100 meters or so and then make a course correction back to their original line of march. A variation of this technique is to pre-plan the route with multiple angular course changes between two known points on the route. See Figure A-8

A.5.4.4 Step Technique. This is another fairly simple method to execute on the march. The patrol makes right angle (90 degrees) course changes, always stepping forward on the intended patrol course. See Figure A-9.

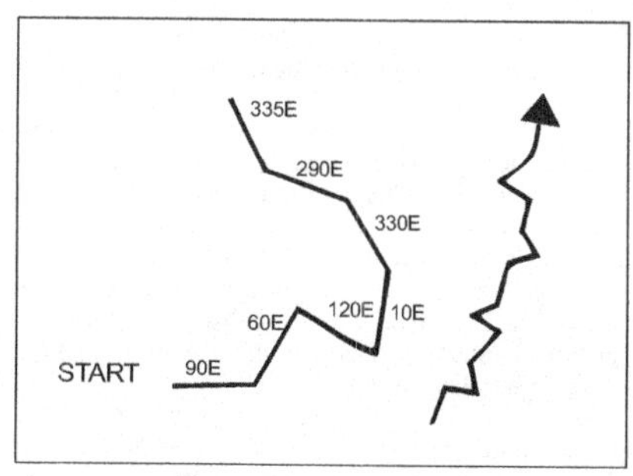

Figure A-8. Angle Method

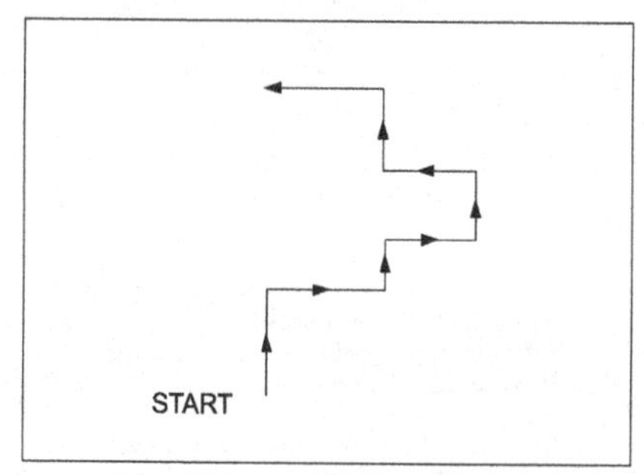

Figure A-9. Step Method

A.6 CACHING

Caching is the process of hiding equipment or materials in a secure storage place with the view to future recovery for operational use. SEAL missions are seldom long enough to justify caching, but the procedure may be required to:

1. Extend mission length
2. Lighten loads in extreme weather or terrain conditions (mountain, jungle, and desert environments)
3. Provide for E&E contingencies
4. Cache materials for the advance of larger forces.

A.6.1 Planning for a Caching Operation. Caching involves selecting the items to be cached, procuring them, and selecting a cache site. Selection of the items to be cached requires a close estimate of what will be needed by particular units for particular operations.

1. Purpose and Contents of the Cache. Planners must determine the purpose and contents of each cache because these will influence both the location and method of hiding the cache. For instance, small barter items can be cached at any accessible and secure site because they can be concealed easily on the person once recovered. However, it would be difficult to conceal rifles for a guerrilla band once recovered. Therefore, this site must be in an isolated area where the band can establish at least temporary control. Certain items, such as medical stock, have limited shelf life and require periodic rotation or special storage

considerations, necessitating easy access to service these items. Sometimes it is impossible to locate a cache in the most convenient place for an intended user. Planners must compromise between logistical objectives and actual possibilities when selecting a cache site. Security is always the overriding consideration.

2. Anticipated Enemy Action. In planning the caching operation, planners must consider the capabilities of any intelligence or security services not participating in the operation. They should also consider the potential hazards the enemy and its witting or unwitting accomplices present. If caching is done for wartime operational purposes, its ultimate success will depend largely on whether the planners anticipate the various obstacles to recovery, which the enemy and its accomplices will create if the enemy occupies the area. What are the possibilities that the enemy will preempt an ideal site for one reason or another and deny access to it? A vacant field surrounded by brush may seem ideal for a particular cache because it is near several highways, but such a location may also invite the enemy to locate an ordnance depot where the cache is buried.

3. Activities of the Local Population. Probably more dangerous than deliberate enemy action are all of the chance circumstances that may result in the discovery of the cache. Normal activity, such as construction of a new building, may uncover the cache site or impede access to it. Bad luck cannot be anticipated, but it can probably be avoided by careful and imaginative observation of the prospective cache site and of the people who live near the site. If the cache is intended for wartime use, the planners must project how the residents will react to the pressures of war and conquest. For example, one very likely reaction is that many residents may report the cache to avoid having their personal funds and valuables seized by the enemy. If caching becomes popular, any likely cache site will receive more than normal attention.

4. Intended Actions by Allied Forces. Using one cache site for several clandestine operations involves a risk of mutual compromise. Therefore, planners should probably rule out otherwise suitable caching sites if they have been selected for other clandestine purposes, such as drops or safe houses. A site should not be located where it may be destroyed or rendered inaccessible by bombing or other allied military action, should the area be occupied by the enemy. For example, installations likely to be objects of special protective efforts by the occupying enemy are certain to be inaccessible to the ordinary citizen. Therefore, if the cache is intended for wartime use, the caching party should avoid areas such as those near key bridges, railroad intersections, power plants, and munitions factories.

5. Packaging and Transportation Assets. Planners should assess the security needs and all of the potential obstacles and hazards that a prospective cache site can present. They should also consider whether the operational assets of the organization are sufficient to overcome those obstacles and hazards securely. Planners must consider the assets that could be used for packaging and transporting the package to the site. Best results are obtained when experts at a packaging center do the packaging. The first question, therefore, is to decide whether the package can be transported from the headquarters or the field packaging center to the cache site securely and soon enough to meet the operational schedules. If not, the packaging must be done locally, perhaps in a safe house located within a few miles of the cache site. If such an arrangement is necessary, the choice of cache sites may be restricted by limited safe house possibilities.

6. Personnel Assets. All who participate directly in emplacement will know where the cache is located. Therefore, only the fewest possible and the most reliable persons should be used. Planners must consider the distance from the person's residence to the prospective cache site and what action cover is required for the trip. Sometimes transportation and cover difficulties require the cache sites to be within a limited distance of the person's residence. The above considerations also apply to the recovery personnel.

7. Caching Methods. Which cache method to use depends on the situation. It is unsound to lay down any general rules, with one exception. Planners should always think in terms of suitability. For example, use the method most suitable for each cache, considering its specific purpose.

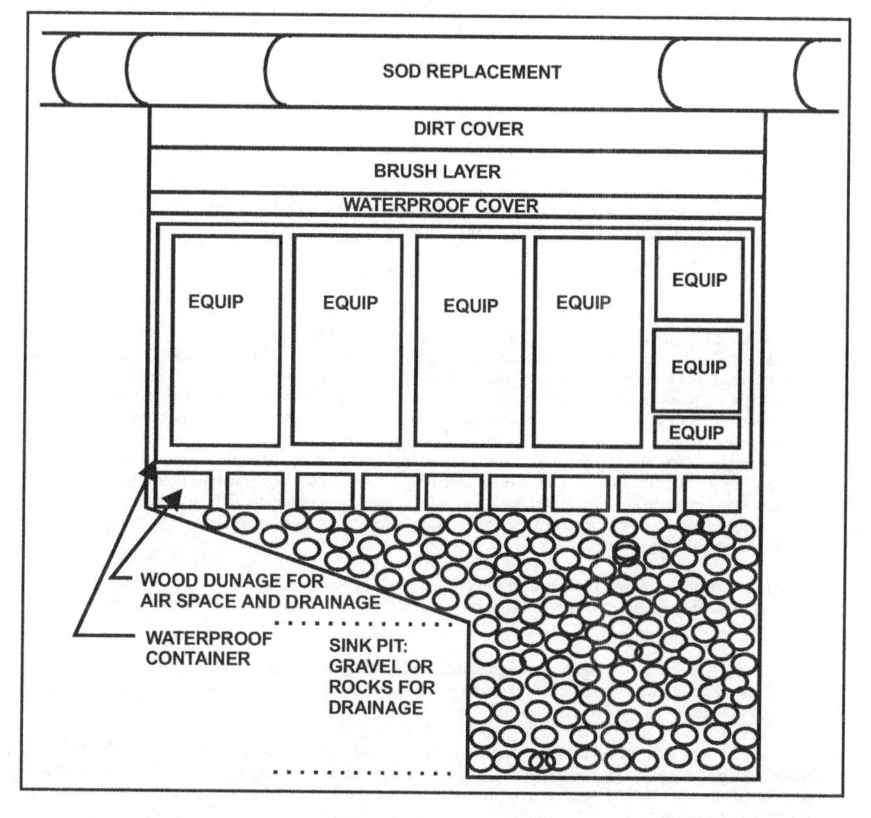

Figure A-10. Burial Cache

a. Concealment. Concealment requires the use of permanent manmade or natural features to hide or disguise the cache. It has several advantages. Both emplacement and recovery usually can be done with minimum time and labor, and cached items concealed inside a building or dry caves are protected from the elements. Thus, they require less elaborate packaging. Also, in some cases a concealed cache can be readily inspected from time to time to ensure that it is still usable. However, there is always the chance of accidental discovery in addition to all the hazards of wartime that may result in discovery or destruction of a concealed cache or denial of access to the site. The concealment method, therefore, is most suitable in cases where an exceptionally secure site is available or where a need for quick access to the cache justifies a calculated sacrifice in security. Concealment may range from securing small gold coins under a tile in the floor to walling up artillery in caves.

b. Burial. See Figure A-10. Adequate burial sites can be found almost anywhere. Once in place, a properly buried cache is generally the most secure. In contrast to concealment, however, burial in the ground is a laborious and time-consuming method of caching. The disadvantages of burial are that:

(1) Burial almost always requires a high-quality container or special wrapping to protect the cache from moisture, chemicals, and bacteria in the soil.

(2) Emplacement or recovery of a buried cache usually takes so long that the operation must be done after dark unless the site is exceptionally secluded.

(3) It is especially difficult to identify and locate a buried cache.

c. Submersion. Sites that are suitable for secure concealment of a submerged cache are few and far between. Also, the container of a submerged cache must meet such high technical standards for waterproofing and resistance to external pressure that the use of field expedients is seldom workable. To ensure that a submerged cache remains dry and in place, planners must determine not only the depth of the water, but the type of bottom, the currents, and other factors that are relatively difficult for non-specialists to ascertain. Emplacement likewise requires a high degree of skill. At least two persons are needed for both emplacement and recovery. Especially when a heavy package is involved, recovery is often more difficult than emplacement and requires additional equipment. In view of the difficulties - especially the difficulty of recovery - the submersion method is suitable only on rare occasions. The most noteworthy usage is the relatively rare maritime resupply operation where it is impossible to deliver supplies directly to a reception committee. Caching supplies offshore by submersion is often preferable to sending a landing party ashore to bury a cache.

A.6.2 Selection of the Site. The most careful estimates of future operational conditions cannot ensure that a cache will be accessible when it is needed. However, criteria for a site selection can be met when three questions are answered.

1. Can the site be located by simple instructions that are unmistakably clear to someone who has never visited the location? A site may be ideal in every respect, but if it has no distinct, permanent landmarks within a readily measurable distance it must be ruled out.

2. Are there at least two secure routes to and from the site? Both primary and alternate routes should provide natural concealment so that the emplacement party and the recovery party can visit the site without being seen by anyone normally in the vicinity. An alternate escape route offers hope of avoiding detection and capture in an emergency.

3. Can the cache be emplaced and recovered at the chosen site in all seasons? Snow and frozen ground create special problems. Snow on the ground is a hazard because it is impossible to erase a trail in the snow. Planners must consider whether seasonal changes in the foliage will leave the site and the routes dangerously exposed.

A.6.2.1 The Map Survey. Finding a cache site is often difficult. Usually, a thorough systematic survey of the general area designated for the cache is required. The survey is best done with as large-scale a map of the area as is available. By scrutinizing the map, the planners can determine whether a particular sector must be ruled out because of its nearness to factories, homes, busy thoroughfares, or probable military targets in wartime. A good military-type map will show the positive features in the topography: proximity to adequate roads or trails, natural concealment (for example, surrounding woods or groves), and adequate drainage. A map also will show the natural and manmade features in the landscape. It will provide the indispensable reference points for locating a cache site: confluence of streams, dams and waterfalls, road junctures and distance markers, villages, bridges, churches, and cemeteries. When possible, it is a good idea to pass the actual map used to place the cache (with appropriate markings) to the patrol that will recover it.

A.6.2.2 The Personal Reconnaissance. A map survey normally should show the location of several promising sites within the general area designated for the cache. To select and pinpoint the best site, however, a well-qualified observer must examine each site firsthand. If possible, whoever examines the site should carry adequate maps, a compass, a drawing pad or board for making sketch maps or tracings, and metallic measuring line. (A wire knotted at regular intervals is adequate for measuring. Twine or cloth measuring tapes should not be used because stretching or shrinking will make them inaccurate if they get wet.) The observer should also carry a probe rod for probing prospective burial sites, if the rod can be carried securely. Since an observer seldom completes a field survey without being noticed by local residents, his action cover is of great importance. His cover must offer a natural explanation for his activity in the area. Ordinarily, this means that an observer who is not a known resident of the area can pose as a tourist or a newcomer with some reason for visiting the area. However, his action cover must be developed over an extended period before he undertakes the actual reconnaissance. If the observer is a known resident of the area, he cannot suddenly take up hunting, fishing, or wildlife photography without arousing interest and perhaps suspicion. He must build up a reputation for being a devotee of his sport or hobby.

A.6.2.3 Reference Points. When the observer finds a suitable cache site, he prepares simple and unmistakable instructions for locating the reference points. These instructions must identify the general area (the names of generally recognizable places, from the country down to the nearest village) and the immediate reference point. Any durable landmark that is identified by its title or simple description can be the immediate reference point (for example, the only Roman Catholic church in a certain village or the only bridge on a named road between two villages). The instructions must also include a final reference point (FRP), which must meet four requirements:

1. It must be identifiable, including at least one feature that can be used as a precise reference point.
2. It must be an object that will remain fixed as long as the cache may be used.
3. It must be near enough to the cache to pinpoint the exact location of the cache by precise linear measurements from the FRP to the cache.
4. It should be related to the immediate reference point by a simple route description, which proceeds from the immediate reference point to the FRP.

Since the route description should be reduced to the minimum essential information, the ideal solution for locating the cache is to combine the immediate reference point and the FRP into one landmark readily identifiable, but sufficiently secluded. The following objects, when available, are sometimes ideal reference points: small, unfrequented bridges and dams; boundary markers, kilometer markers and culverts along unfrequented roads; a geodetic survey marker; battle monuments; and wayside shrines. When such reference points are not available at an otherwise suitable cache site, natural or manmade objects may serve as FRPs:

distinct rocks, posts for power or telephone lines, intersections in stone fences or hedge rows, and gravestones in isolated cemeteries.

A.6.2.4 Pinpointing Techniques. Recovery instructions must identify the exact location of the cache. These instructions must describe the point where the cache is placed in terms relative to the FRP. When the concealment method is used, the cache ordinarily is placed inside the FRP, so it is pinpointed by a precise description of the FRP. A submerged cache usually is pinpointed by describing exactly how the moorings are attached to the FRP. With a buried cache, any of the following techniques may be used.

1. Placing the cache directly beside the FRP. This is the simplest method since pinpointing is reduced to specifying the precise location of the FRP

2. Sighting the cache by projection. This method may be used if the FRP has one flat side long enough to permit precise sighting by projecting a line along the side of the object. The burial party places the cache a measured distance along the sighted line. This method may also be used if two precise FRPs are available, by projecting a line sighted between the two objects. In either case, the instructions for finding the cache must state the approximate direction of the cache from the FRP. Since small errors in sighting are magnified as the sighted line is extended, the cache should be placed as close to the FRP as other factors permit. Ordinarily this method becomes unreliable if the sighted line is extended beyond 50 meters.

3. Placing the cache at the intersection of measured lines. If two FRPs are available within several paces, the cache can be on single lines projected from each of the FRPs. If this method is used, state the approximate direction of the cache from each FRP. To ensure accuracy, neither of the projected lines (from the FRPs to the point of emplacement) should be more than twice as long as the base line (between the two FRPs). If this proportion is maintained, the only limitation on the length of the projected lines is the length of the measuring line that the recovery party is expected to carry. The recovery party should carry two measuring lines when this method is used.

4. Sighting the cache by compass azimuth. If the above methods of sighting are not feasible, one measured line may be projected by taking a compass azimuth from the FRP to the point where the cache is placed. To avoid confusion, use an azimuth to a cardinal point of the compass (north, east, south, or west). Since compass sightings are likely to be inaccurate, a cache that is pinpointed by this method should not be placed more than 10 meters from the FRP.

A.6.2.5 Measuring Distances. The observer should express all measured distances in a linear system that the recovery party is sure to understand - ordinarily the standard system for the country where the cache is located. He should use whole numbers (6 meters, not 6.3 or 6.5) to keep his instructions as simple as possible. To get an exact location for the cache in whole numbers, take sightings and measurements first. If the surface of the ground between the points to be measured is uneven, the linear distance should be measured on a direct line from point to point, rather than by following the contour of the ground. This method requires a measuring line long enough to reach the full distance from point to point and strong enough to be pulled taut without breaking.

A.6.2.6 Marking Techniques. The emplacement operation can be simplified and critical time saved if the burial point for the cache is marked during the reconnaissance.

1. If a night burial is planned, the point of emplacement may be marked during a daylight reconnaissance. This method should be used whenever operational conditions permit.

2. The marker must be an object that is easily recognizable but that is meaningless to an unwitting observer. For example, use a small rock or a branch with its broken end placed at the selected emplacement point.

A.6.2.7 Additional Data Required for Emplacement. During a personal reconnaissance, the observer must not only pinpoint the cache site, but also gather all the incidental information required for planning the emplacement. It is especially important to determine the best route to the site and at least one alternate route, the security hazards along these routes, and any information that can be used to overcome the hazards. Since this information is also essential to the recovery operation, it must be compiled after emplacement and included in the final cache report.

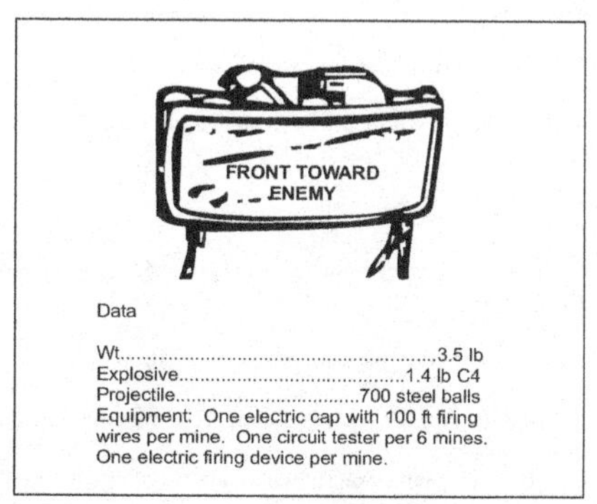

Figure A-11. Claymore Mine Data

A.7 CLAYMORE MINES

Claymore mines are ideal ambush initiators. They are also effective when used in series, with the second set of explosions taking place several seconds after the first to kill enemy who may have survived initially and have begun to flee. Claymores can effectively increase the size of the kill zone. (See Figure A-11.)

Claymores should be emplaced first on the flanks of the ambush site, with priority given to the flank most likely to be used by the enemy. A Claymore may be emplaced to the rear of the ambush site to cover rear security. If the patrol withdraws from the ambush site in this direction, the Claymore can be used to cover the withdrawal of the ambush force. If the purpose of the ambush is to obtain prisoners for intelligence purposes, a gap should be left in the effective fire zones of the Claymores.

A.7.1 Mines for Security. If using Claymores for security or to increase the kill zone, consider the following:

A.7.1.1 Placement Security. While one man places the Claymores, another must stand security over the area.

A.7.1.2 Sequence of Placement

1. Set only one flank at a time, starting with the flank of the enemy's most likely approach. If time is critical, both flanks can be set simultaneously.

2. Place the rear security Claymore last.

3. Add more Claymores to increase the size of the kill zone.

4. If placing them in trees, ensure they are well concealed.

5. Don't forget to allow 16 meters for back blast.

   a. This could cause a tree to fall on personnel.

   b. Back blast will cause dust and confusion.

6. Interlock the Claymore fans of fire unless prisoners are being taken.

7. For prisoners: Leave a gap, fire the Claymores. This may wound potential prisoners, but should not kill all personnel.

A.7.1.3 System Wiring

1. Any rewiring required should be performed prior to insertion. Additional changes required should be done while initially setting up the Claymores.

2. Never bring Claymore-firing wires directly back to the firing position.

   a. This prevents the enemy from turning Claymores 180 degrees and creating self-ambush.

   b. It also prevents the enemy from knowing exactly where ambush personnel are.

3. Tug lines. Pass out the tug lines when the Claymores are set up. These will be used to signal platoon members.

A.7.2 Installing, Firing, or Recovering an M18A1 Claymore Mine

A.7.2.1 Positioning the Claymore

1. Place the shorting plug on the firing wire connector. Then tie the shorting plug end of the firing wire to a fixed object such as a stake, tree, or root at the firing position. Select a site close to the desired kill zone. During the installation, the firing device must be kept by the SEAL installing the mine to preclude accidental firing by another person. Ensure friendly personnel are not within 16 meters (55 feet) of the rear of the mine unless cover such as an embankment or sandbagging is available and used. Friendly personnel must be a minimum of 250 meters away from the front of the mine.

2. Place the wire reel on the shoulder and unroll the firing wire to the selected site for emplacing the mine.

3. Remove the mine from the reel. Open both pairs of legs on its base to a 45-degree angle with two legs facing the front and two legs facing the rear of the mine.

4. Position the mine on the ground with the words "FRONT TOWARDS ENEMY" towards the kill zone (or tie the mine to a tree or supporting structure).

5. Push the legs about one-third of the way into the ground. (It may be necessary to spread the legs to the maximum width under adverse conditions such as heavy wind or when the legs cannot be pressed into the ground). See Figure A-12.

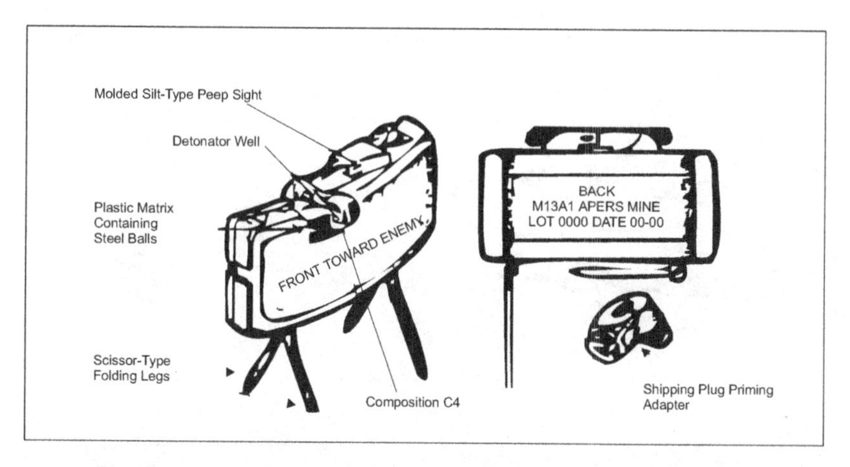

Figure A-12. M18A1 Claymore Mine

A.7.2.2 Aiming the Claymore

1. Use the slit-type peep sight or knife-edge sights to select an aiming point about 150 feet to the front of the mine and about eight feet above the ground.

2. Position the eye about six inches to the rear of the sight and aim the mine toward the center of the target area. The groove of the sight should be in line with the aiming point.

A.7.2.3 Circuit Test to Ensure it is Functioning. Directions are outlined on the cover of each Claymore bag.

1 Remove the dust covers from the M57 firing device and the M40 test set.

2. Plug the M40 test set into the M57 firing device and swing the safety bail to the FIRE position.

3. Depress the firing handle, and watch the window of the test set for a flash of light that indicates the circuit is good. Remember this flashing light only indicates the M57 and M40 are functioning properly.

4. Turn the safety bail to the SAFE position.

5. Remove the dust cover from the test set and the shorting plug connector of the firing wire.

6. At this time place the blasting cap behind a tree, in a hole, or under a large rock to avoid injury in the event the blasting cap detonates during a circuit check.

7. Plug the connector of the firing wire into the test set. Now swing the safety bail to the FIRE position.

8. Depress the firing handle and observe the window of the M40 test set for a flash of light, which indicates a good circuit. See Figure A-13.

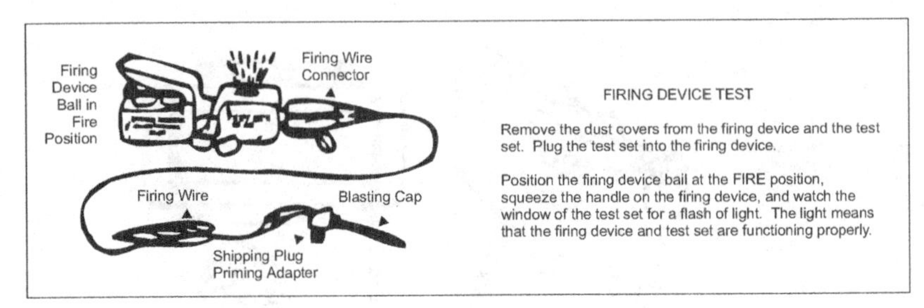

Figure A-13. Claymore Firing Test

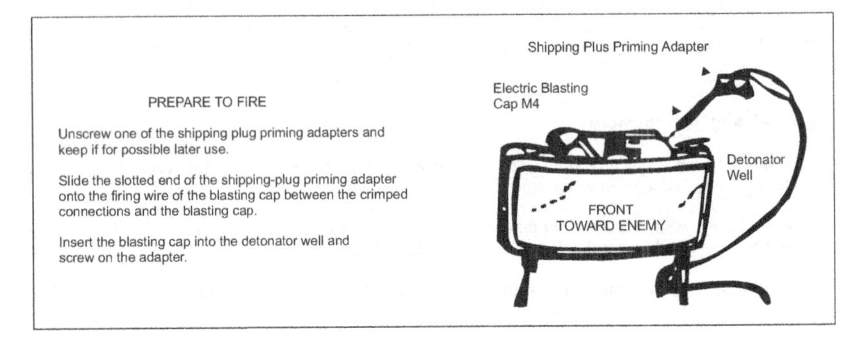

Figure A-14. Preparation to Fire

A.7.2.4 Arming the Claymore

1. Making certain the firing device is in the hands of the person designated to fire the Claymore, unscrew one of the shipping-plug priming adapters from the mine. Keep the adapter for later use.

2. Carefully slide the slotted end of the priming adapter onto the firing wire of the blasting cap between the crimped connection and the blasting cap.

3. Insert the blasting cap into the detonator well and screw on the adapter. See Figure A-14.

4. Secure the firing wire about three feet to the rear of the mine so the mine will not become misaligned if the wire is disturbed.

A.7.2.5 Camouflage of Claymore and the Firing Wire

1. Camouflage the mine with leaves, sticks, and other natural debris from the area.

2. Retest the circuit with the M40 test set after the firing wire is laid out and after the cap is placed inside the mine.

3. Remove the M40 test set and reconnect the firing device, keeping it in a safe position. Ensure that all SEALs (to include the designated firer) within 250 yards to the front and 16 yards to the rear are under cover before testing the circuit. (Greater distances should be used whenever possible.)

A.7.2.6 Firing the Claymore

1. Ensure the safety bail is on the firing device in the SAFE position until it is time to detonate the Claymore.

2. Position the firing device safety bail to the FIRE position when the lead element of the target approaches within 10-30 feet of the Claymore. See Figure A-15.

3. Squeeze the firing device handle with a firm, quick squeeze when the enemy is within the desired range. See Figure A-16.

A.7.2.7 Recovering an Unfired Claymore

1. Place the firing device safety bail in the SAFE position. Ensure the firing device is kept with the SEAL recovering the mine throughout the recovery evolution.

2. Disconnect the firing wire from the firing device, and replace the dust covers.

3. Unscrew and remove the priming adapter from the mine.

4. Remove the blasting cap from the priming adapter.

5. Screw the adapter back into the detonating well.

6. Remove the firing wire from the stake; reroll it; replace the wire and cap inside the cardboard container.

7. Recover the mine. Repack the mine and accessories into the bandoleer.

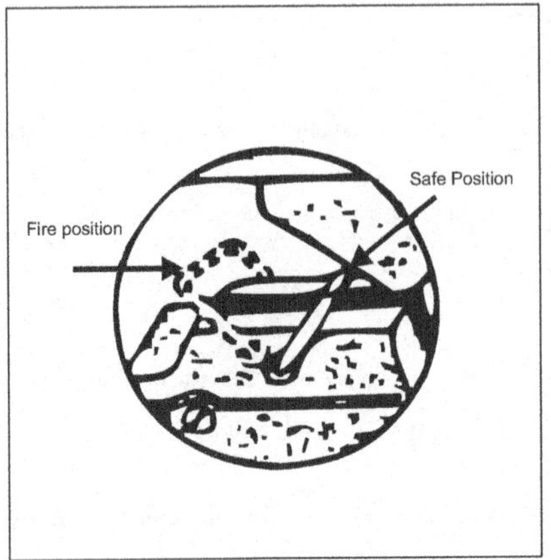

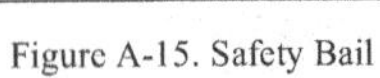
Figure A-15. Safety Bail

Figure A-16. Firing

APPENDIX B

Manpack Radios

RADIO	FREQ	RANGE (NM)	MODE	WEIGHT (LB)	POWER (WATT)	CRYPTO
PRC-104	2-30 MHZ	25-400*	USB, LSB, AM, CW	16	20	KY-65, 57, 99, CSZ-1
PRC-138	1.6-60 MHZ	25-400*	USB, LSB, FM, CW	10	20	(SAME AS 104)
PRC-117	30-90 MHZ	15	FM	15	10	EMBEDDED KY-57
PRC-119	30-88 MHZ	5	FM	20	4	(SAME AS 77)
PRC-113	116-155 VHF, 225-400 UHF	LOS	AM	14.2	10	(SAME AS 77)
URC-110	225-400 MHZ (25KHZ CHNL ONLY)	SATCOM	AM/FM	22	20	KY-57
PSC-3	225-400 MHZ (5 OR 25KHZ)	SATCOM	AM/FM	22.5	35	KY-57, 99
VSC-7 IS SAME AS PSC-3 BUT IS HARD MOUNTED FOR BASE STATION						
AN/PSC-5	30-400 MHZ	VHF-15 UHF-5-7	AM/FM/FSK/PSK/ SBPSK/SOQPSK/ DEQPSK	11.7	AM-5 FM-18	EMBEDDED KY-58,KG-84, ANDVT
LST-5	225-400 MHZ (5 OR 25KHZ)	SATCOM	AM/FM	11.8	18	KY-57, 99, CSZ-1
LST-5D	225-400 MHZ		AM/FM	11.5	AM-5 FM-18	EMBEDDED KY-58, KY84, ANDVT
(PSC-10 WAS THE PROTOTYPE FOR LST-5 AND IS USED FOR BASE STATION)						
MST-20	225-400 MHZ (5 OR 25KHZ)	SATCOM	AM/FM	10	20	KY-57, 99, CSZ-1
(PSC-7 IS THE SAME AS THE MST-20, HST-4 WAS FIRST GENERATION MST-20)						
* RANGE DEPENDS ON TYPE OF ANTENNA USED.						
HANDHELDS						
PRC-112	*121.5, *243, *282.8, 225-300	LOS	VOICE/BEACON	1.6	1	NONE
AN/PRC-137	2-60 MHZ	0-2500	USB/LSB/SSB/AME/ CW/FM	5	10	EMBEDDED KG-84C
MX-300	136-174	LOS		20.5	6	DES CODER
MX-300	136-14, 406-420, 450-512	LOS		2.5	2	DES CODER
MX-300R	136-174	LOS		2.5	6	DES CODER
MBITR	30-512 MHZ	5-7	AM/FM/PM	31 OZ	5	EMBEDDED INDICATOR
MOTSR	136-174, 406-420	LOS	VHF/UHF	2.5	6	FASCINATOR
SABER I	12 SLOTS	LOS	VHF/UHF	1.6		DES CODER
SABER II	48 SLOTS	LOS	VHF	1.6		FASCINATOR
SABER III	72 SLOTS	LOS	VHF	1.6		FASCINATOR
*PRESET						

APPENDIX C

Hand and Arm Signals

There are a wide variety of hand and arm signals in use throughout NSW. Those listed below provide a starting point. Individual platoons should train extensively to ensure all members understand each signal. Confusing signals that are not clear at night should be avoided.

1. ENEMY SEEN or SUSPECTED ("I See")

Two fingers pointing at eyes.

2. NO ENEMY IN SIGHT or IT IS ALL CLEAR ("OK/Good")

Thumb pointed upwards from a closed fist.

3. ME

Point down at own chest.

4. YOU

Point at man concerned.

5. HALT

Closed fist.

6. FREEZE

Point man freezes. All freeze (no signal required); emulate man in front.

7. LISTEN or I HEAR

Cupped hand held to ear.

8. DEPLOY/MOVE OFF TRACK

Hand held low and flat, moved side to side. Then indicate to which flank or side of the track the patrol is to move.

9. DANGER AREA

Hand slashing at throat.

10. STREAM

Hand moved out at waist level in a "wavy" motion.

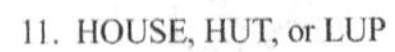

11. HOUSE, HUT, or LUP

Thumb pointing down from an open hand with fingers extended and open so as to form an inverted V between.

12. RECONNAISSANCE

Hand held up to eye as though using a monocle.

13. COME AND LISTEN TO ME

Hand moved in a "yakity-yak" manner.

14. HASTY AMBUSH

Clenched fist pumping vigorously in direction ambush is required.

15. REQUEST HEAD COUNT

Hand tapping top of head.

16. HURRY UP

Clenched fist pumped up and down.

17. SPREAD OUT

Hand held against rifle and moved outwards several times.

18. SLOW DOWN

As in driving, hand moved slowly.

19. GET DOWN

Hand moves fast and urgently (or man in front gets down - no signal required).

20. REMOVE/REPLACE PACKS

A pack removing gesture.

21. BREAK FOR 5 MINUTES

    Imitate breaking a stick, then display 5 fingers.

22. STAGGERED FILE

    Move arms up and down slowly.

23. CLOSE ON ME

    Beckon with hand. Used to pass information, especially when very dark. Man following goes forward, receives word, then waits for next man.

24. PARAMETER

Circle index finger while pointing up.

25. RALLY POINT

Circle index finger while pointing down.

26. BOOBYTRAP

Cupped hand facing up.

APPENDIX D

Weapon and Accessory List

ITEM	WEIGHT (LB)
M14	
M14 W/FULL MAGAZINE	13.1
M14 W/FULL MAGAZINE	10.1
M14 MAGAZINE	0.5
M14 MAGAZINE W/ 20 7.62 ROUNDS	1.5
M2 BIPOD	1.7
M4A1	
M4A1 Mk 4	6.5
M4A1 W/FULL MAGAZINE	7.5
M4A1 MAGAZINE W/ 20 5.56 ROUNDS	0.7
M4A1 MAGAZINE W/ 30 5.56 ROUNDS	1.0
M203	
M203 UNLOADED	3.0
M203 LOADED	3.5
M4A1 /203 W/ FULL MAGAZINE LOADED	11.0
Mk 43 Mod 0	
Mk 43	18.5
M112 TRIPOD W/ T&E MECHANISM	15.0
100 LINKED 7.62 ROUNDS	7.0

ITEM	WEIGHT (LB)
H&K MP5N	
MP5N W/O MAGAZINE	6.3
MP5N MAGAZINE OF 30 9 MM ROUNDS	1.3
SIG/Sauer P226	
P226 W/ EMPTY MAGAZINE	1.7
P226 W/ MAGAZINE OF 15 9 MM ROUNDS	2.1
H&K Mk 23	
Mk 23 W/ EMPTY MAGAZINE	1.9
Mk 23 W/ MAGAZINE OF 12 .45 CAL ROUNDS	6.3
MINES	
CLAYMORE	3.5
GRENADES	
FRAGMENTATION	1.0
CONCUSSION	1.0
SMOKE	1.2
WP	1.7
M224 60 MM MORTOR	
M224 ASSEMBLED CONVENTIONAL	46.5
M224 ASSEMBLED HANDHELD	18.0
M225 CANNON	114.4
M170 BIPOD	15.2
M7/M8 BASE PLATE	14.4/3.6
M64 SIGHT UNIT	2.5
ROCKETS	
AT4 ANTI-TANK WEAPON	13.0
LIGHT ANTI-ARMOR WEAPON (LAW)	5.5
BARREL	24.0
M3 TRIPOD W/T&E MECHANISM	44.0

APPENDIX E

Equipment List

FIRST LINE (ON THE PERSON)	
CLOTHING	
Floppy brimmed hat	Camouflage uniform
Net/scarf	Belt/ suspenders
Gloves	Boots and socks
EQUIPMENT	
Weapon and rounds (e.g. M4 + one 30-round magazine)	
Notebook and pencil in waterproof container (e.g., plastic bag or use waterproof paper, wet notes, etc.)	
Two heat tablets	Survival saw
Booby trap wire	Money: U.S. and local
Swiss seat: 12 ft of 1-in tubular nylon	Unlubricated condom (for water)
Watch	Compass: wrist and /or neck
Map and protractor	Pace counter/pace line
Waterproof matches/lighter (in tube with an O ring)	Flashlight: small, penlight size
Signal mirror	Pencil flares
Camouflage stick or compact (cammie paint)	20 feet parachute line (550 cord)
Pocket knife/survival knife (with can opener)	Whistle
Marker panel (sewn inside hat and cammie shirt: hats get lost)	
Rations: one meal (for example, two power bars or one cut-down MRE)	
Emergency medical kit: battle dressing, ace bandage, drugs: codeine, morphine, iceflex	
SECOND LINE (ON THE LOAD BEARING EQUIPMENT)	
Magazines and ammunition	Grenades
Weapons cleaning equipment	Insect repellent
Lightweight handcuffs/tie-ties	Field dressing
Two 2-quart canteen/cup and purification tabs	Strobe light and IR cover
Secondary weapon	Rations: 24 hours
Battle dressing (in same location for all personnel)	Snap links
Flotation jacket (UDT life jacket or equivalent)	Rigger's tape
20 ft of .5-in nylon	Chemlites (red, green, infrared)
Medical kit (see next page for list of items)	Knife

THIRD LINE (IN THE PACK)	
Radio equipment and spare batteries	Demolition materials and mines
Plastic trash bags (dark green or black)	Poncho and poncho liner
Individual cammie net (approx. 6 ft x 6 ft)	Change of uniform and socks
Water bladders and/or spare water bottles	Ground sheet/pad
Swiss seat, toilet articles and towel strip	
OPTIONAL ITEMS (NOT NECESSARILY REQUIRED BY ALL PERSONNEL)	
Heavy duty wire cutters	Heavy duty pruning shears
Crimpers	Binoculars
Night vision	Machete
Pistol and ammunition	PRC-112 survival radio/batteries
Squad radios	Katadyn water filter pump
Entrenching tool	Hammock
Stove and heat tabs	Tool kit
Additional ammunition and grenades	Insect net
Two 16-guage quick cath intravenous (IV)	Bungee cord
Macro connecting tube	1-inch Penrose drain (constricting band)
500 cc bag lactated ringers (carry at least one bag for every two men)	
PERSONAL MEDICAL KIT	
One of everything, except as noted. Keep items together and easily accessible	
Knuckle band-aids	10-inch surgical tubing
1-inch tape roll, cotton, white	Large battle dressing
Medium battle dressing	Small battle dressing
Cravat	4-inch kerflex
4-inch ace bandage	3-inch petroleum gauze
CORPSMAN MEDICAL KIT	
4 500-cc lactated ringers	8 sixteen gauge quick cath IV
6 macro connecting tubes	4 1-inch penrose drains
4 1-inch tape rolls	4 large battle dressings
4 medium and 4 small battle dressings	8 4-inch kerflex
4 4-inch ace bandage	4 3-inch adhesive tape rolls
2 adult oralpharygeal airways	1 penlight
1 CRIC (cricothyroid) kit (plastic)	4 3-inch petroleum gauze

APPENDIX F

Tactical Lessons Learned

F.1 PATROL LEADER

1. Lead the way you want to be led.

2. No individual or platoon can practice or train too much or too often.

3. Teamwork is the key to success and will only come through constant training and rehearsal.

4. While on a mission, attempt to minimize fatigue. Fresh men go further.

5. Show confidence in your decision making, and your platoon will be more likely to have confidence in you.

6. Don't be afraid to take advice from your platoon members, encourage it.

7. Don't lose your temper; it will affect your judgement. Keep cool.

8. Always have an alternate plan. Think ahead.

9. Fight aggressively, but think intelligently.

10. Inject realism into all phases of training.

11. Conduct at least half of your training at night.

12. Implement a good physical training program and your platoon will have fewer health problems.

13. Develop and use a pre-mission and post-mission checklist to ensure nothing is neglected.

14. Don't set patterns in your operations.

15. Never do the obvious.

16. Stay alert at all times when on patrol. You are never 100 percent safe until you are back at the base.

17. Don't arbitrarily make all tactical lessons learned your platoon SOPs. Always consider METT-T.

18. Don't forget to factor in the effect of cold body temperatures on platoon members' energy levels; keep personnel warm with the correct protective clothing.

19. Danger areas are not one-dimensional. They are 360 degrees and may have terrain features and trees/vegetation to hide the enemy.

20. If involved in unintended enemy contact, use fire superiority to make the enemy believe they have stumbled upon something "bad".

21. Use the KISS standard. Go back to basics and do them right. Any time you split the patrol, you have increased the potential for compromise and problems.

22. Be aggressive, bold, and creative when planning.

23. Do thorough, detailed mission planning.

24. Intelligence is perishable and often incorrect. Never trust intelligence 100 percent.

25. Be aware of possible "set-ups" when agents are involved.

26. Plan for contingencies.

27. Don't underestimate the enemy.

28. Understand the construction of structures you attack. Bullet penetration of flimsy structures may require changing the fire support location to prevent self-inflicted injury.

29. Know which direction doors and windows open: right, left, up, or down.

30. Prepare psychological leaflets, if deemed appropriate.

31. Ensure OPSEC is paramount in all aspects of operational planning.

F.2 MEDICAL

1. Take and use insect repellent. If mites or other insects become a problem, try to find appropriate repellent. Mud can be used as a last resort. Carry repellant in zip-lock bags and isolate it from other items in your rucksack. Squeeze air from the container and screw the cap on firmly.

2. Carry tweezers to remove anything that sticks. In areas with large thorns (e.g., Panama) carry needle nose pliers.

3. Devise or obtain a field litter and practice carrying wounded personnel during training. Even light personnel become very heavy when being carried in a litter. Practice carrying for at least 2 km.

4. Become knowledgeable on the how, why, and ways for MEDEVAC.

5. Ensure the personal medical kit is in a standard location and has standard contents for all platoon personnel.

6. Carry throat lozenges to stop coughing on patrol/ambush. (Be careful of odors caused by some lozenges.)

7. Use a wounded individual's medical field dressing before using your own.

8. Take care of your feet. If a platoon member starts to get foot problems while on a mission, stop in a secure position, remove boots, dry and treat the feet with powder and medicated ointment. Tape and proceed as able.

9. Narcotic painkillers, antibiotics, and anti-diuretics should be taken on patrol and carried in non-collapsible waterproof containers (key float containers available at marine stores work well).

F.3 DIET

1. Drink large amounts of water to hydrate before going on a mission and ensure enough water is taken and drunk throughout the mission. Stay ahead of water needs.

2. Carry water purification tablets and use them.

3. Do NOT smoke or use other tobacco products before or during operations, as it ruins night vision and has a distinctive odor.

4. Use vitamin supplements only if advised to do so by medical officer or HM. Most SEALs have a sufficient vitamin supply stored up in their bodies already.

5. Adjust to type of rations used; don't eat them for the first time on an operation.

6. If concerned with smelling "American", eat local area foods at least 72 hours before a mission; this allows both the sweat and the system to adjust from regular diet.

F.4 SURVIVAL

1. Carry survival gear as first line equipment. Include a mirror, aircraft panel, compass, flares, penlight, E&E kit, map of area, water purification, diarrhea medication, two power bars, or one cut-down food ration.

2. Ensure a sharp, compact knife is readily available.

3. Prior to deployment, learn and become familiar with the types of plants and animal existing in AO. These are possible survival foods.

4. Do not neglect to take food. It is unrealistic to expect to hunt or fish successfully.

F.5 UNIFORM

1. Wear a floppy hat to break up the outline of the head. A headband or bandanna by its self is ineffective for camouflage; however, a bandanna provides good neck protection and has other uses.

2. An "Australian net" is effective when worn around the neck as a utility item. An alternative to this is an OD towel.

3. Wearing gloves with padded palms and fingers cut off at mid-finger is effective for protecting hands from hot barrels and thorny plants. Full finger gloves are another option, especially for the AW man.

4. To maximize hearing, don't cover ears.

5. Ensure any non-standard clothing has grommet holes to allow for water drainage.

6. Do not wear different types of uniforms than expected by U.S. or allied forces (e.g., tiger stripes when everyone else is wearing jungle cammies). This may cause friendly forces to confuse you with the enemy.

7. Tie trouser blousing strings to boot laces or wrap cloth around top of boot (e.g., carrying handles from Claymore mine bag) to keep legs free of leeches.

8. The "blue and gold" tee shirt is not a good substitute for the OD T-shirt.

9. Any diving watch with a fluorescent face must be covered to prevent the glow from being seen in the dark. Sweatbands or a cut-up sock will work well for covering watches. Beware of velcro covers; they are noisy. Fluorescent watch faces can be used to assist SEALs following each other in very dark conditions, if other identification is not in use.

10. Suspenders for the trousers are very handy, they can be pre-made or improvised. In desert ops these allow clothing to be worn more loosely, which will assist with cooling.

11. Full-length trousers and shirts are a must to keep insect bites to a minimum.

12. Break in a set of boots and then set them aside for operations.

13. Wear good boots with laces. Running shoes have a tendency to pull off in the mud. Going barefoot can be hazardous to you and your platoon.

14. Enemy trackers can be successfully confused as to the size of the patrol if the entire SEAL platoon has the same sole pattern on their boots.

15. "Velcro" closures make too much noise. Zippers are another item to avoid. Buttons are much quieter than velcro or zippers. Pull-the-dot fasteners work well but require lubrication if used in and around salt water.

16. Two-piece camouflage uniforms are better than those modified to one piece (jumpsuit fashion).

17. Green socks that blend with equipment are better than white ones, as you can tie them to your pack to let them dry out. State-of-the-art dark backpacking/athletic socks are very effective. Wool or polypropylene is better than cotton, which absorbs too much water.

18. Don't sew luminous glint tape to floppy hats or uniforms, unless required for a specific mission.

19. Tuck your jacket into your pants. You can't use the lower pockets because of your LBE but you can temporarily stuff expended magazines inside your shirt during contact.

20. Tie down all survival equipment to the uniform or LBE to prevent loss if pockets should tear.

F.6 WEAPONS

1. Carry spare bolt groups for the M60; know how to change out rapidly. The M60 is not a good 20-pound club.

2. Ensure buoyancy is correct by dip testing personnel with all of their equipment on; adjust buoyancy as weight changes. Use LIVE AMMUNITION, or correct for live ammo weight.

3. Carry linked ammo in a carrying case (e.g., canteen case or other improvised case) to keep it from getting out of line. Ensure the case has a grommet to drain water.

4. Bring at least one weapons cleaning kit per patrol.

5. When modifying the issued sling, improvise but be aware nylon can melt on a hot barrel. Slings can be used as "ropes" to secure splints and as tourniquets. Ensure that tape on the weapon does not hinder its operation in any way.

6. Change the round in the chamber after a night in the field.

7. Clean and lube all magazines and rounds daily and when returning from operations; never assume they are clean enough.

8. Using something to separate 30-round magazines, such as a small piece of wood, and then taping them together allows a quick magazine changeover. The magazines should be in the same directions with enough space to allow them both to seat properly.

9. Don't fire the M4 on full automatic, unless you are the PT responding to an immediate ambush.

10. Fire your weapon shooting low, particularly at night. Ricochets will kill too, and most people will hit the ground when the shooting starts.

11. M203 GNs must practice and fully understand arming distances, types of 40mm rounds, and shooting in the open or in confined vegetation.

12. Always carry the weapon with the selector switch on SAFE.

13. Use a plastic muzzle cap or tape to keep water and dirt out of the barrel.

14. Place magazines upside down in their pouches, primers inboard, to keep out dirt.

15. Do not retrieve your first expended magazine during contact; it will consume valuable time.

16. The mission will dictate proper weapons mix.

17. Shotguns and 9mm sub-machine guns are not as effective in a jungle environment as the M60 and M14 or M4. Ensure personnel are cross-trained in all weapons.

18. Point all weapons in a safe direction while on board the insertion vehicle:

    a. Upward in a boat

    b. Downward or outboard in a helicopter

    c. Outward or upward in a vehicle.

19. In the desert, each individual should carry cleaning equipment for his own weapon on every patrol, regardless of the length, type or nature of mission. Cleaning must be strictly controlled and staggered so that at least 75 percent of the weapons in the patrol are ready for use at any time. Check with the PL before cleaning a weapon. Use non-oil base lubricants, such as graphite to prohibit accumulations of dust and grit in the weapon. You must have your weapon with you at all times. Whenever possible, keep two hands on it. When sleeping or relaxing, keep it within arm's length.

20. Test fire all weapons prior to departing a friendly area on patrol to ensure proper functioning. The tests should be done under similar situations that will be encountered on the patrol.

21. The variety of weapons should be limited to make ammunition redistribution easier and assures familiarity with each other's weapons. Also, identification of your own personnel from the enemy by the weapon sounds is made more possible.

F.7 ORDNANCE AND DEMOLITIONS

1. A Claymore mine with a 30 or 60-second fuse attached is a handy, quick ambush or "get away" technique. The Claymore mine can be one of the most useful weapons; learn all there is to know about them. Attaching a CS grenade to the Claymore is also very effective.

2. When employing Claymores, use two men. One places the mine and the other stands guard.

3. Never place the Claymore where you cannot observe it. Consider cutting the firing wire in half when operating in dense vegetation. You won't use more than 50 feet of wire. This makes emplacement and recovery easier and cuts weight.

4. Place the Claymore so the blast parallels your personnel and the firing wire does not lead straight to your position.

5. Bring two cap crimpers per patrol when using demolitions.

6. Ensure no shiny rounds of linked ammo are worn outside the uniform, as this will destroy camouflage effectiveness.

7. Ensure hand-thrown grenades are thrown clear of vegetation, so they do not bounce back and explode near the thrower or his platoon mates.

8. Make continuous daily checks on all grenades when on patrol to ensure the primers are not coming unscrewed.

9. Don't bend the pins on the grenades flat. The rings are too hard to pull when needed.

10. Paper tape through the rings of grenades and tape the ring to the body of the grenade. The paper tape will tear for fast use, while plastic or cloth tape will not. This also keeps the ring open for your finger, stops noise, and prevents snagging.

11. Fragmentation grenades are good for inflicting casualties.

12. CS grenades are ideal for stopping or slowing down enemy troops and dogs pursuing you and are effective in damp and wet weather, whereas CS powder will dissipate.

13. WP grenades have a great psychological effect against enemy troops and can be used for the same purpose as CS grenades. The use of both CS and WP at the same time will more than double the effectiveness of each.

14. Thermite grenades will destroy either friendly or enemy equipment. Recommend one per patrol.

15. Violet and red are the smoke colors most visible from the air, but in dense vegetation or wet weather use white HC smoke to signal aircraft. Notify the aircraft prior to signaling with WP, as they may mistake it for a marking rocket indicating an enemy position and attack you.

16. The AT-4 is very effective against mud bunkers.

F.8 LOAD BEARING EQUIPMENT/RUCKSACK

1. Be sure all snaps and buckles are taped. Don't use paper tape.

2. Wear your LBE buckled when not sleeping. If you're wounded, your teammates can drag you by your LBE shoulder straps.

3. Use a waterproof bag to protect equipment. Ensure at least one man carries an extra plastic bag; it weighs very little and can be used for many things.

4. Test the shoulder straps on the rucksack before packing it for patrol. Always carry 550 cord to repair straps.

5. Sew a Claymore bag onto the top flap of your rucksack to carry Claymores, binoculars, a camera, or handset.

F.9 COMMUNICATIONS

1. Communications is the responsibility of everyone in the patrol, not just the RTO.

2. Pre-set frequencies so a quick turn of the dials will put you on the desired frequency. This is especially helpful at night when trying to avoid using light.

3. Perform pre-mission radio checks with the radio and crypto device (e.g., KY-57) hooked up and packed in the rucksack, just the way it would be in the field. Test the radio after the crypto device has been loaded, with and without the crypto device hooked up, with the NSWTG/TU and various support elements.

4. Always use fresh batteries in all communications gear. Carry spare batteries, but do not remove the plastic wrapping prior to using them.

5. Claymore firing wire can be used as a field expedient antenna for HF communications.

6. When using intra-squad radios (e.g., Saber) use a headset on both ears with the center cut out of the non-functioning earpiece. Use a push-to-talk microphone instead of a sound activated one. Put the push-to-talk button on the right side of "H" gear for a right-handed shooter so it can be operated with the left hand.

7. Always remain calm and professional, no matter what happens.

8. Signals to be used on patrol should be prearranged and understood by all patrol members. They should be based on standard hand and arm signals and kept simple.

9. Always keep talk to a minimum. Use the standard hand and arm signals whenever practical.

10. While patrolling, pass simple instead of complex instructions. Allow time for the word to be passed before executing.

11. When giving the PLO, visual aids are of great value. Examples include a sand table, chalkboard, or a sketch with a stick on cleared ground.

12. Consider the use of emergency recall signals, such as an aircraft circling a certain feature for five minutes at a set time. This eliminates the need for coming up on a radio frequency, which is a potentially compromising evolution.

13. Pass word when on patrol. Ensure the platoon knows what is happening.

14. Whenever identifying your position to support elements with smoke or flares, always have support elements authenticate to make certain that the enemy is not using deception.

15. Whisper into the handset while in the field. Exhale first, then speak, or your transmission will sound like a tire leaking air.

16. The operational base must avoid making unnecessary, unscheduled radio checks just because they haven't heard from a team.

17. Repeat grid coordinates sent to you to ensure accurate copy.

18. Do not send "same" or "no change" when reporting team location. Always send your coordinates.

19. Constantly check your CEOI to ensure your authentication tables are folded open to the page showing the most current set. This will prevent dangerous delays when an aircraft requests authentication.

20. Waterproof your CEOI and authentication tables by laminating them with acetate or putting them in a plastic zip-lock bag.

21. Clean all contacts daily with the eraser end of a pencil.

F.10 SPECIAL EQUIPMENT

1. A good pair of garden branch cutting shears can be used to cut vegetation.

2. NVEO can be very helpful; however, ensure someone not wearing NVEO accompanies the NVEO user. This is especially true if the SEAL wearing the NVEO is patrolling as a point element.

3. When in an observation position, scan with NVEO every five minutes or so. If you scan continuously, you increase the chance of being spotted. Two persons using NVEO in the passive mode looking directly at each other will see glowing "cat eyes."

4. NVEO and thermal viewers complement each other and should be used in combination; for example, the PT could use the M845 riflescope and the PL could use a thermal imaging rifle sight.

5. Carry the Litton 911 in a 5.56 mm magazine pouch or on a lanyard around your neck.

6. Noise suppressors ("silencing" devices for weapons) have application in night ambush, but they require well-understood SOPs.

7. A Swiss seat of tubular nylon, worn or taped to web gear, is a good addition to operational equipment.

8. Everyone should carry a 20-foot length of 550 cord.

9. Wear equipment above the waist to prevent it from chafing the legs, and for safety.

10. Carry a gas mask in a 2-quart canteen pouch when planning to use CS gas or if the threat of enemy gas is high.

F.11 OPERATIONAL TECHNIQUES

F.11.1 Insertion

1. Use OPDEC by never using the same routes and using false inserts and extracts.

2. Ensure multiple AOs have been cleared with blanket times to allow OPSEC flexibility.

3. Use false insertions and extractions.

F.11.2 Extraction

1. Do not set patterns.

2. Use correct landing signals to assist helo pilots when landing.

3. Ensure all signals are pre-briefed and understood.

F.11.3 Navigation

1. Use compass heading and pace count. Trust the compass, but be aware of areas with magnetic anomalies.

2. Use a pre-measured line (e.g., 550 cord; 10-25 meters) to work critical navigation situations.

3. Learn to use the stars for navigation. While doing so, confirm your location periodically with a compass.

4. Every SEAL should know the methods of finding the North Star and know how to use the sun's shadow to determine the direction of north.

5. Do not mark any of your field maps/charts with friendly information; you can, however mark them with enemy information.

6. Preset your compasses prior to departure on patrol if personnel are not proficient at setting the compass in the dark.

7. When planning a patrol route, remember to use offsets whenever applicable. An offset is planned magnetic deviation to the right or left of the straight-line azimuth to an objective. It is used to verify your exact location relative to the objective.

8. Use the PT as a point, not as a compass man. He is primarily concerned with the security of the patrol. Have the second or third man responsible for navigation.

9. In an LUP note the direction on the compass of how you are lying so that if you need to move off quickly, you will be somewhat oriented where you are in relation to the camp and patrol intentions.

F.11.4 Patrolling

1. Know personnel limits. Take time of day, weight, terrain type, thickness of vegetation, swamps and so forth into consideration when planning distance to be covered. Underestimate distance and overestimate time.

2. Memorize routes.

3. Develop SOPs that include stealth.

4. Listen for animal noises, or the lack of them, as indications of possible enemy activity. Know how local animals react, e.g., monkeys scream initially, then quiet down.

5. Watch for small, domesticated animal fences around buildings.

6. Do not uncover luminescent watch dials (or compass dials) to follow platoon members at night.

7. Pack out any item you pack in; minimize amount carried.

8. Don't allow tobacco in the field.

9. Go as quietly as you can as long as you can. Don't make noise, don't make contact, don't go hot until in place and ready to attack. Exception: if compromised, go immediately into VIOLENT ACTION.

10. Whether on a planned assault or an IAD, use extreme and overwhelming violence of action. The initial 10-15 seconds of engagement should be so violent that those who are not killed immediately will be in a state of shock. The subsequent selective fire clean-up will take care of the shock victims.

11. Apply vigorous ACCURACY at all times. Don't just put holes in every leaf without drawing blood.

12. Don't try to do battle with a superior force in a firefight. Lay down a solid base of sustained suppressive fire (15-30 seconds maximum). If still receiving significant return fire, break contact. Once contact is broken stop shooting immediately; it only continues to give away your position.

13. Eliminate pocket litter; check all pockets prior to departing. Wear only taped dog tags.

14. Carry maps and notebooks in waterproof containers; use wetnotes or waterproof paper.

15. Use pencils, not pens, to make notes on patrol. Ink smears when it gets wet.

16. Never take pictures of platoon members while on patrol. Valuable intelligence can be gained if photos are lost or captured.

17. Reapply camouflage at least three times per day or as needed; absolutely necessary at an ambush or raid site and after crossing a beach.

18. Each patrol member should keep a list of his own lessons learned for after-action reports and personal use.

19. Use deception techniques on patrol to deceive the enemy and/or catch him in a hasty ambush. The following techniques can be useful:

a. Double back (See Figure F-1).

b. Leave a false trail.

c. Execute a false extraction.

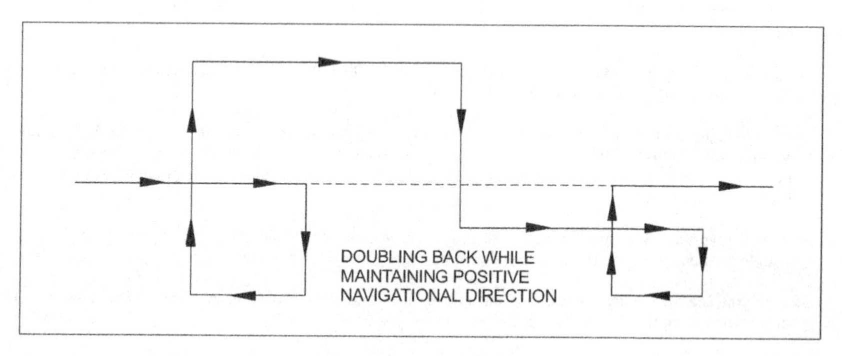

Figure F-1. Doubling Back

20. On desert patrols, take turns carrying the heavy equipment so as not to unduly wear out any one patrol member.

21. It is best to patrol on nights that are dark, rainy, and windy.

22. When conducting a recon of an enemy position, maintain a covering force within supporting distance of the recon element.

23. Whenever breaking the patrol at night in terrain that is difficult for control, break the patrol in place. Each man faces in the direction of responsibility, and kneels.

24. While moving at night, take advantage of ambient noise such as wind, vehicles, planes, shelling, battle sound, and sounds of insects.

25. Whenever possible, allow all patrol members sufficient time for their eyes to adjust to darkness. It takes over 30 minutes before your eyes fully adjust.

26. When using any light on night patrol or when passing through a lighted area, close one eye to prevent total loss of night vision and warn other patrol members whenever light is to be used.

27. When infiltrating enemy lines, an alternate rendezvous point should be selected in the event the first point is occupied by the enemy.

28. When a patrol is moving in a file, each man follows the man's footsteps in front to minimize detection of footprints and knowledge of how many personnel there are in the patrol if they are detected. This is particularly important in the desert.

30. In mountainous terrain, plan to use the upper part of the ridges for movement if the situation permits but do not get on the skyline. This allows for good visibility downward and the chance to evade over the other side of the ridge if required.

31. On small patrols RS should automatically send the count forward after each extended break or passage of any obstacle.

32. Whenever stopping, establish an "all around defense". Obey this rule for short breaks, long breaks, in the LUP, when setting up an ambush - in fact, at all times.

33. Depending on the mission tasked, the aim is often to avoid contact with the enemy. SEALs operate in small numbers and their job is often to reconnoiter or conduct clandestine attacks. Nevertheless, every patrol must be prepared for this eventuality and thoroughly understand and rehearse the procedures. Practice your break-contact drills.

34. When working with friendly agents such as partisans, never take along the entire patrol to make contact with them. Send only one man to make the contact and cover that one man.

35. If you see an enemy, freeze like an animal does and then slowly ease down out of sight. Sudden movements catch the eye and are giveaways to your position.

36. Be conscious of the natural habits, rhythms, and sounds of animals, birds and insects and learn to recognize them when they have been disturbed.

F.12 MISCELLANEOUS

1. Use unit SOPs for sterilizing equipment, team numbers, letters, or whatever. Avoid names and ranks on equipment.

2. As experience increases, the weight of carried equipment will decrease. More is not always better; "fight light" for greatest effectiveness.

3. Learn a few key language phrases of the enemy tongue, so in an unexpected meeting with an enemy or local, they will hesitate a second or two.

4. Take two or more ponchos in any element; they can be used as litters, for construction of rafts, concealing lights, and shelters.

F.13 TRAINING

1. Practice day and night; emphasize night training.

2. Rehearse for missions.

3. Acclimate personnel before missions.

4. Learn about enemy cultures, superstitions, and habits to get the "feel" for planning operations.

5. Ensure all equipment and any personal additions are sound proofed by testing them. Wrap items with tape or pad them to prevent noise. This is especially true of magazines.

6. Practice getting under low obstacles and over higher obstacles, remembering not to catch parts of rucksack and other equipment on the obstacle or on overhanging vegetation.

F.14 SPECIAL OPERATIONS MISSION CRITERIA

1. Is this an appropriate SOF mission?

2. Does it support the CINC's campaign plan?

3. Is it operationally feasible?

4. Are required resources available to execute?

5. Does expected outcome justify the risk?

F.15 SPECIAL OPERATIONS FORCES TRUTHS

1. **HUMANS ARE MORE IMPORTANT THAN HARDWARE**

2. **QUALITY IS BETTER THAN QUANTITY**

3. **SPECIAL OPERATIONS FORCES CANNOT BE MASS PRODUCED**

4. **COMPETENT SPECIAL OPERATIONS FORCES CANNOT BE CREATED AFTER EMERGENCIES OCCUR.**

APPENDIX G

Standard Loadout Plan

This is a worksheet that can be used to design a Standard Loadout for a Platoon of 16 SEALs. Planners should be prepared to discuss the specific techniques/procedures underlying loadout choices.

Position (e.g. Point, AW, RTO)	Weapons (e.g. M4, M203)	Rounds (Amount, Magazines)	Other/Equipment (Water, Claymores, Illumination)	Total Weight
1				
2				
3				
4				
5				
6				
7				
8				

Position (e.g. Point, AW, RTO)	Weapons (e.g. M4, M203)	Rounds (Amount, Magazines)	Other/Equipment (Water, Claymores, Illumination)	Total Weight
9				
10				
11				
12				
13				
14				
15				
16				

APPENDIX H

FIRE CONTROL MATRIX

COC	NAME	WEAPONS	PYROTECHNICS DEMOLITIONS	OPENING CONTACT	SUSTAINED RATE OF FIRE

APPENDIX I

Wind-chill Chart

Wind Speed		Cooling Power of Wind Expressed as "Equivalent Chill Temperature"																				
Knot	MPH	Temperature (°F)																				
Calm	Calm	40	35	30	25	20	15	10	5	0	-5	-10	-15	-20	-25	-30	-35	-40	-45	-50	-55	-60
		Equivalent Chill Temperature																				
3-6	5	35	30	25	20	15	10	5	0	-5	-10	-15	-20	-25	-30	-35	-40	-45	-50	-55	-60	-70
7-10	10	30	20	15	10	5	0	-10	-15	-20	-25	-35	-40	-45	-50	-60	-65	-70	-75	-80	-90	-95
11-15	15	25	15	10	0	-5	-10	-20	-25	-30	-40	-45	-50	-60	-65	-70	-80	-85	-90	-100	-105	-110
16-19	20	20	10	5	0	-10	-15	-25	-30	-35	-45	-50	-60	-65	-75	-80	-85	-95	-100	-110	-115	-120
20-23	25	15	10	0	-5	-15	-20	-30	-35	-45	-50	-60	-65	-75	-80	-90	-95	-105	-110	-120	-125	-135
24-28	30	10	5	0	-10	-20	-25	-30	-40	-50	-55	-65	-70	-80	-85	-95	-100	-110	-115	-125	-130	-140
29-32	35	10	5	-5	-10	-20	-30	-35	-40	-50	-60	-65	-75	-80	-90	-100	-105	-115	-120	-130	-135	-145
33-36	40	10	0	-5	-15	-20	-30	-35	-45	-55	-60	-70	-75	-85	-95	-100	-110	-115	-125	-130	-140	-150
Winds above 40 have little additional effect		LITTLE DANGER					INCREASING DANGER (Flesh may freeze within 1 minute)						GREAT DANGER (Flesh may freeze within 30 seconds)									

APPENDIX J

Rates of Movement Table

TABLE J.1 RATES OF MOVEMENT OVER FLAT TO GENTLY ROLLING TERRAIN

Expected rates of march carrying rucksacks over flat to gently rolling terrain.

Day Movement

	Condition	
Movement Mode	**Unbroken Trail**	**Broken Trail**
On foot, less than 1 ft (30cm) snow	1.5-3 kph	2-3 kph
On foot, more than 1 ft snow	0.5-1 kph	2-3 kph
Snowshoeing	1.5-3 kph	3-4 kph
Skiing	1.5-3 kph	3-6 kph
Pulk	0.5-2.0 kph	1-4 kph
Skijoring	N/A	8-24 kph

Night Movement

	Condition	
Movement Mode	**Unbroken Trail**	**Broken Trail**
On foot, less than 1 ft (30cm) snow	0.5-3 kph	2-3 kph
On foot, more than 1 ft snow	0.25-1 kph	2-3 kph
Snowshoeing	0.5-3 kph	3-4 kph
Skiing	1.5-4 kph	3-6 kph
Pulk	0.5-2 kph	1-4 kph
Skijoring	N/A	8-24 kph

TABLE J.2 RATES OF MOVEMENT OVER MODERATE TO STEEP TERRAIN

Expected rates of march carrying rucksacks over moderate to steep terrain.

Day Movement

Movement Mode	Condition: Unbroken Trail	Condition: Broken Trail
On foot, less than 1 ft (30cm) snow	1.5-3 kph	2-3 kph
On foot, more than 1 ft (30cm) snow	0.5-1 kph	2-3 kph
Snowshoeing	1.5-2.5 kph	3-4 kph
Skiing	1.4-4 kph	3-6 kph
Pulk	0.25-2 kph	1-4 kph
Skijoring	N/A	4-12 kph

Night Movement

Movement Mode	Condition: Unbroken Trail	Condition: Broken Trail
On foot, less than 1 ft (30cm) snow	0.5-3 kph	2-3 kph
On foot, more than 1 ft snow	0.25 kph	2-3 kph
Snowshoeing	0.5-3 kph	3-4 kph
Skiing	0.5-4 kph	3-6 kph
Pulk	0.5-2 kph	1-4 kph
Skijoring	N/A	4-12 kph

TABLE J.3 RATES OF MOVEMENT FOR VERTICAL ASCENT ROCK AND ICE CLIMBING

Expected rates of march carrying rucksacks over terrain requiring vertical ascent rock and ice climbing.

Day Movement

	Condition	
Movement Mode	**Snow Covered Rock**	**Ice Covered Rock**
On foot	0.025-0.1 kph	0.025-0.1 kph

Night Movement

	Condition	
Movement Mode	**Snow Covered Rock**	**Ice Covered Rock**
On foot	0-0.1 kph	0-0.1 kph

APPENDIX K

Centigrade/Fahrenheit Conversion Table

Centigrade	Fahrenheit
100	**212**
90	194
80	**176**
70	158
60	**140**
50	122
40	**104**
30	86
20	**68**
10	50
0	**32**
-10	14
-20	**-4**
-30	-22
-40	**-40**
-50	-58

To convert centigrade degrees to Fahrenheit degrees:

F=(9C/5)+32

To convert Fahrenheit degrees to centigrade degrees.

C=5(F-32)/9

Example: Convert 75° C into Fahrenheit.

F=(9C/5)+32
F=(675/5)+32
F=(135)+32
F=167°

Example: Convert 50°F into centigrade.

C=5(F-32)/9
C=5(50-32)/9
C=5(18)/9
C=90/9
C=10°

APPENDIX L

Mountain and Arctic Warfare Hand and Arm Signals

Hand and arm signals are used to communicate quickly, quietly, and clearly within the patrol without using voice commands. This is especially useful in winter in snow covered terrain, given the often-optimum light conditions. Correctly passing the word is everyone's responsibility. The following are hand and arm signals for mountain and arctic.

1. ADVANCE – Hand waves forward.

2. HALT/MOVE INTO FIELD OF FIRE – The operator holds up a clenched fist as he moves into his field of fire.

3. STOP, LOOK, LISTEN, AND SMELL – Remove your hat.

4. CHANGE DIRECTION OF PATROL – Extend hand and arm into the new direction.

5. DANGER AREA – Extended hand-slicing motion across throat. For roads or land type linear danger areas, the hand and arm signal for danger area is followed by both arms extended from your sides.

6. RIVER/STREAM DANGER AREA – For rivers and streams, the hand signal for danger area is followed by one hand and arm horizontal making a waving motion.

7. OPEN DANGER AREA – For large and small open danger areas, both arms crossing the chest follow the hand and arm signal for danger area.

8. HOUSE/STRUCTURE – The hand and arm signal for danger area is followed by one hand cupped with fingers facing downward.

9. BOOBY TRAP/MINE – The hand and arm signal for danger area is followed by one hand cupped with fingers facing upward.

10. AVALANCHE DANGER – The hand and arm signal for danger area is followed by extended hand bent at the elbow slamming down, then pointing to the danger area.

11. PATROL RALLY POINT – Circular movement above the head with finger and arm followed by pointing to the location of the patrol rallying point.

12. PERIMETER – Same as patrol rallying point hand and arm signal, followed by hand and arm straight up and bent at the elbow for a cigar perimeter.

13. TWO MINUTES – Either two fingers or the thumb extended out from a mitten.

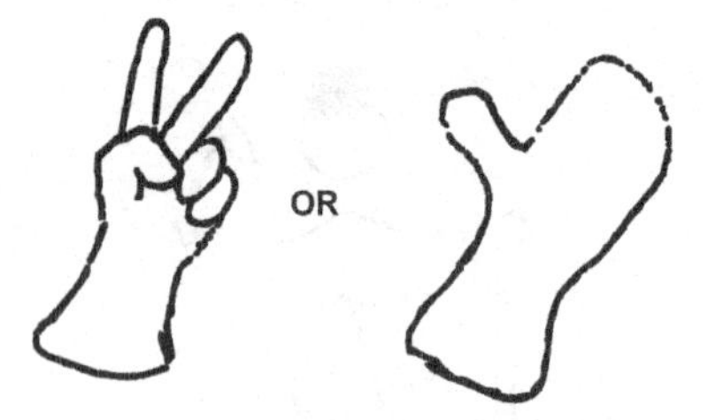

14. HEAD COUNT – Pat top of head.

15. I SEE – Use your gloved or mittened hand as a visor over your eyes.

16. ENEMY – Usually following the "I SEE" hand and arm signal. Hand over the face followed by pointing at the threat.

17. SPEED UP – Pump arm and fist up and down above head.

18. SLOW DOWN – Extended hand pressing down.

19. OPEN/CLOSE PATROL INTERVAL – Open and close arms.

20. MOVE TO STAGGERED FILE – Ski poles held over-head, parallel to the ground.

21. HASTY AMBUSH – Pump arm and fist into the direction the patrol is to set up.

APPENDIX M

Mountain and Arctic Equipment List

FIRST LINE (ON THE PERSON)	
Snow/glacier goggles	Eye and ear protection
Compass/protractor/map	Altimeter
Pace counter	Flashlight: small, penlight size
Medical blow out kit	Pocket knife/survival knife (with can opener)
Balaclava or face masque	Waterproof matches/lighter (in tube with an O ring)
Camouflage stick or compact (cammie paint)	Pencil flares
Signal mirror	12 hour food ration
Whistle	30 feet of 550 cord
Avalanche beacon	Emergency space blanket
Notebook and pencil in waterproof container (e.g., plastic bag or use waterproof paper, wet notes, etc.)	
	20 feet parachute line (550 cord)
	Whistle
Marker panel (sewn inside hat and camouflage shirt: hats get lost)	
Rations: one meal (for example, two power bars or one cut-down MRE)	
Emergency medical kit: battle dressing, ace bandage, drugs: codeine, morphine, iceflex	
SECOND LINE (ON THE LOAD BEARING EQUIPMENT)	
Magazines and ammunition	Grenades
Weapons cleaning equipment	Weapon with muzzle cover
Patrol ski wax, skins, and glue	Heat tab stove with heat tabs
Two 2-quart canteen/cup and purification tabs	Strobe light and IR cover
Bivy bag, fly or poncho	Rations: 24 hours
Fixed blade knife	2 Snap links
Medical kit	

THIRD LINE (IN THE PACK)	
Radio equipment and spare batteries	Demolition materials and mines
Whisperlite stove with fuel and utensils	Sleeping bag inside bivy bag
Individual camouflage net (approx. 6 ft x 6 ft)	Change of uniform, socks, gloves
Snow shovel	Ski repair kit
Overwhites	Snow shoes
OPTIONAL ITEMS (NOT NECESSARILY REQUIRED BY ALL PERSONNEL)	
Heavy duty wire cutters	Heavy duty pruning shears
Crimpers	Binoculars
Night vision	Machete
Pistol and ammunition	PRC-112 survival radio/batteries
Squad radios	Katadyn water filter pump
Entrenching tool	Hammock
Stove and heat tabs	Tool kit
Additional ammunition and grenades	Insect net
Two 16-guage quick cath intravenous (IV)	Bungee cord
Macro connecting tube	1-inch Penrose drain (constricting band)
500 cc bag lactated ringers (carry at least one bag for every two men)	
PERSONAL MEDICAL KIT	
One of everything, except as noted. Keep items together and easily accessible	
Knuckle band-aids	10-inch surgical tubing
1-inch tape roll, cotton, white	Large battle dressing
Medium battle dressing	Small battle dressing
Cravat	4-inch kerflex
4-inch ace bandage	3-inch petroleum gauze
CORPSMAN MEDICAL KIT	
4 500 cc lactated ringers	8 sixteen gauge quick cath IV
6 macro connecting tubes	4 1-inch penrose drains
4 1-inch tape rolls	4 large battle dressings
4 medium and 4 small battle dressings	8 4-inch kerflex
4 4-inch ace bandage	4 3-inch adhesive tape rolls
2 adult oralpharygeal airways	1 penlight
1 CRIC (cricothyroid) kit (plastic)	4 3-inch petroleum gauze

APPENDIX N

Sources of Medical Intelligence

The following organizations are recipients of medical intelligence from AFMIC and may also be the sources of other data that can be used with medical intelligence for development of support plans or medical threat risk assessments.

1. Office of The Surgeon General, U.S. Army

   a. Category of information: Not applicable

   b. Tasking requirement: Direct contact

   c. Maintains limited data. Specific areas of interest may be addressed.

2. Walter Reed Army Institute of Research (Preventive Medicine)

   a. Category of information: For Official Use Only/Not Releasable to Foreign Nationals (FOUO/NOFORN)

   b. Tasking requirement: Direct contact

   c. Maintains area specific information to brief epidemiological survey teams. Area specific information reports may be addressed in two to three days.

3. Natick Research Laboratory

   a. Category of information: FOUO/NOFORN

   b. Tasking requirement: Direct contact

   c. Maintains documents on weather survey teams and effects of weather on soldiers. Also maintains climatic and geodetic information. Does not maintain specific medical intelligence.

4. Foreign Service and Technology Center

   a. Category of information: FOUO/NOFORN

   b. Tasking requirement: Direct contact

   c. Maintains foreign military equipment data and some medical information in the preventive medicine field such as the water purification apparatus.

5. Defense Pest Management Information Analysis Center

a. Category of information: Unclassified

b. Tasking requirement: Direct contact

c. Maintains arthropod, vector, and pest biology, ecology, and geographical distribution data; arthropod-borne disease data; and rodent, venomous vertebrate and invertebrate, hazardous marine organisms, and toxic flora data.

6. United States Army Medical Research Institute of Infectious Diseases. Operational Medicine Branch

a. Category of information: FOUO/NOFORN

b. Tasking requirement: Direct contact

c. Maintains information on biological warfare agents and medical countermeasures to such agents. Also maintains a deployable aeromedical isolation team to evacuate a patient with a highly infectious disease under biocontainment conditions.

7. United States Army Medical Research Detachment-Brooks

a. Category of information: FOUO/NOFORN

b. Tasking requirement: Direct contact

c. Maintains information on directed energy threat and countermeasures. Also maintains capability for management of laser eye injuries.

The following are civilian sources for procuring various types of medical intelligence products.

1. Department of State

a. Category of information: Secret

b. Tasking requirement: Direct contact

c. Maintains annual updates of endemic diseases and prophylaxes. Lists embassy medical personnel and medical capabilities at each location.

2. World Health Organization

a. Category of information: Unclassified

b. Tasking requirement: Direct contact

c. Primarily maintains statistical data on endemic diseases (information is subject to skepticism). Data compiled from each country's self-prepared reports. Country may report incorrect data because of national pride, impact on tourism, or lack of surveillance.

The following are source locations at Fort Bragg, North Carolina, for procuring various types of medical intelligence products:

1. XVIII Airborne Corps

   a. Category of information: Secret

   b. Tasking requirement: Assistant Chief of Staff, Intelligence

   c. Maintains National Security Agency material with general medical and epidemiological information.

2. XVIII Airborne Corps Surgeon's Office.

   a. Category of information: Secret

   b. Tasking requirement: None

   c. Maintains AFMIC capability studies and monthly scientific intelligence review produced by the National Foreign Assessment Center.

3. United States Army John F. Kennedy Special Warfare Center and School (Assistant Chief of Staff, Intelligence)

   a. Category of information: Secret

   b. Tasking requirement: None

   c. Maintains Defense Intelligence Agency and National Security Agency reports with limited medical information (normally several years old).

4. United States Army John F. Kennedy Special Warfare Center and School Surgeon's Office

   a. Category of information: Secret

   b. Tasking requirement: None

   c. Maintains AFMIC weekly wire that contains specific and current update of medical information, area studies produced by the Defense Intelligence Agency on specific countries, State Department reports, World Health Organization weekly reports, and reference library on most endemic disease groups, files, slides, and studies.

5. Fourth Psychological Operations Group

   a. Category of information: FOUO/NOFORN

   b. Tasking requirement: None

   c. Maintains basic Defense Intelligence Agency and National Security Agency documents. Army experts on psychological operations are very knowledgeable on specific cultures, psychological back-ground data, nutritional basics, taboos, and folk medicine.

6. Regional Studies Course

a. Category of information: FOUO

b. Tasking requirement: None

c. Maintains no source documents; the Regional Studies Course is attended by officers with extensive experience in the regions they study. Generally, those officers have lived in those regions. The point of contact is Commander, Co D, 3d Bn, 1st SPWAR TN Group (A), Fort Bragg, North Carolina 28307-5000.

7. General Intelligence Production Detachment

a. Category of information: Top Secret

b. Tasking requirement: None

c. Maintains order of battle material with continuous update. Intelligence production request must be channeled through G-2. Medical information is limited.

8. United States Army John F. Kennedy Center and School Library

a. Category of information: FOUO/NOFORN

b. Tasking requirement: None

c. Maintains general area information: Department of State reports, DA Form 550 series, and area handbooks.

9. Threat Manager, U.S. Army Medical Department Center and School

a. Category of information: Unclassified through top secret

b. Tasking requirement: Requests for information to Commander, U.S. Army Medical Department Center and School, ATTN: Threat Manager, Fort Sam Houston, TX 78234-6100.

c. Maintains finished intelligence and United States Army Training and Doctrine Command threat databases. Produces medical threat risk assessment projections (MEDTRAP). The MEDTRAP methodology is a tool used to assess and project risk from assorted medical threats to U.S. forces operating in different geographical areas and during the execution of a variety of missions across the operational continuum. These assessments are based on historical data, open source, published medical information, insights from subject matter experts, and current, finished medical intelligence. The MEDTRAP methodology allows for timely and accurate vulnerability assessment for development and execution of effective passive and active protective measures prior to exposure of U.S. forces to high-risk medical threat agents.

The following publications are also a source of medical information:

1. NWP 4-02, Operational Health Service Support.

2. NWP 4-02.2 (Part A), Naval Expeditionary Forces Medical Regulations.

3. NWP 4-02.3 (Part B), Planning, Operations, and Medical Intelligence.

4. NWP 4-02.4 (Part A & C), Deployable Health Service Support Platforms.

5. NWP 4-02.5 (Part C), Marine Corps Health Service Support Operations.

The following military commands have medical information available on the world wide web:

1. U.S. Navy Environmental Health Center http://www-nehc.med.navy.mil/index.htm

2. Naval Medical Logistics Command (AMAL listing) http://www-nmlc.med.navy.mil/

3. Virtual Navy Hospital (linked to BUMED) http://www.vnh.org/

4. U.S. Army NBC Medical http://www.nbc-med.org/ie40/

5. U.S. Navy Plans, Operations and Medical Intelligence Homepage http://navymedicine.med.navy.mil/pomi/

Made in the USA
Monee, IL
02 May 2025